Biology and Biotechnology of Patagonian Microorganisms

Nelda Lila Olivera • Diego Libkind
Edgardo Donati
Editors

Biology and Biotechnology of Patagonian Microorganisms

Springer

Editors
Nelda Lila Olivera
Patagonian Institute for the Study of Continental Ecosystems (IPEEC-CENPAT)
National Scientific and Technical Research Council (CONICET)
Argentina

Diego Libkind
Andean-Patagonian Institute for Biological and Geoenvironmental Technologies (IPATEC) CONICET
National University of Comahue
Argentina

Edgardo Donati
Industrial Fermentation Research and Development Center (CINDEFI)
CONICET-UNLP, School of Science
National University of La Plata
La Plata, Argentina

ISBN 978-3-319-42799-7 ISBN 978-3-319-42801-7 (eBook)
DOI 10.1007/978-3-319-42801-7

Library of Congress Control Number: 2016954633

Printed on acid-free paper

This Springer imprint is published by Springer Nature
The registered company is Springer International Publishing AG
The registered company address is: Gewerbestrasse 11, 6330 Cham, Switzerland

Preface

Patagonia is a vast and heterogeneous territory that extends along the southernmost part of the Americas (approximately from 40°S to 56°S). This region occupies part of Chilean and Argentinean territories. Argentinean Patagonia, in which this book is focused, extends east of the Andes and south of the Colorado River, having an area of about 790,000 km^2. This region includes large extents of arid plains, forests, mountains and glaciers, fertile valleys, and wide seashores. These ecosystems offer a great diversity of selective environments scarcely explored that are suitable for the bioprospection of biotechnologically relevant microorganisms.

The aim of *Biology and Biotechnology of Patagonian Microorganisms* is to provide readers from the academic and biotechnological communities with a concise and clearly illustrated treatment of outstanding topics of Patagonian microbiology and biotechnology. The investigations included in this book represent interesting examples of the microbial world as a highly significant source of valuable technology that can modify and boost regional economy and progress.

The book is subdivided into three parts, preceded by a geo-historical introduction to Patagonia. The first part introduces the reader to the field of environmental microbiology, including insights into microbial communities from hydrocarbon- and heavy metal-polluted sites and lands degraded by natural events and human economic activities. The second part gives an overview of the biotechnological potential of Patagonian microorganisms. Bioprospection of antimicrobials, enzymes, and other bioactive compounds produced by extremophilic microorganisms is discussed. Important aspects of Patagonian microorganism application in biomining, winery, brewing, and aquaculture are also addressed. Moreover, included is a discussion on antibiotic-resistant isolates. The third part of this book describes Patagonian yeasts with biotechnological potential, particularly their application in industrial fermentations (e.g., wine and beer industry) and food biopreservation.

The editors wish to thank all the authors for their willingness to participate in this book and for their excellent contributions. We also acknowledge the Associate Editor João Victor Pildervasser (Life Sciences, Springer-SBM), who invited us to be editors of this book, and the Springer International Publishing AG staff for their outstanding assistance and guidance. Finally, we extend our gratitude to the funding sources Ministerio de Educación y Deportes, Ministerio de Ciencia, Tecnología e Innovación Productiva and CONICET, Argentina.

Puerto Madryn, Argentina — Nelda Lila Olivera
Bariloche, Argentina — Diego Libkind
La Plata, Argentina — Edgardo Donati

Contents

Contributors

Federico Acosta Centro de Investigaciones y Trasferencia de Catamarca CITCA-CONICET, San Fernando del Valle de Catamarca, Catamarca, Argentina

Facultad de Ciencias Exactas y Naturales, Universidad Nacional de Catamarca FACEN-UNCa, San Fernando del Valle de Catamarca, Catamarca, Argentina

Verónica Elizabeth Álvarez Laboratorio de Investigaciones de los Mecanismos de Resistencia a Antibióticos, Instituto de Microbiología y Parasitología Médica (IMPaM, UBA-CONICET), Facultad de Medicina, Universidad de Buenos Aires, Ciudad Autónoma de Buenos Aires, Argentina

Gonzalo V. Arnau Planta Piloto de Procesos Industriales Microbiologicos PROIMI-CONICET, San Miguel de Tucumán, Tucumán, Argentina

Jaime D. Babot Planta Piloto de Procesos Industriales Microbiologicos PROIMI-CONICET, San Miguel de Tucumán, Tucumán, Argentina

Nicolas Bellora Laboratorio de Microbiología Aplicada, Biotecnología y Bioinformática de Levaduras, Instituto Andino-Patagónico en Tecnologías Biológicas y Geoambientales (IPATEC), CONICET y Universidad Nacional del Comahue (CRUB), San Carlos de Bariloche, Argentina

Cecilia Bernardelli CINDEFI (CCT La Plata-CONICET, UNLP), Facultad de Ciencias Exactas, Universidad Nacional de La Plata, La Plata, Argentina

Natalia Bourguignon Planta Piloto de Procesos Industriales Microbiológicos (PROIMI-CONICET), San Miguel de Tucumán, Tucumán, Argentina

Sebastián Mario Ezequiel Bravo Instituto de Investigación y Desarrollo en Ingeniería de Procesos, Biotecnología y Energías Alternativas (PROBIEN), Consejo Nacional de Investigaciones Científicas y Técnicas de la República Argentina (CONICET)- Universidad Nacional del Comahue, Neuquén, Provincia del Neuquén, Argentina

Facultad de Ciencias y Tecnología de los Alimentos, Universidad Nacional del Comahue, Villa Regina, Provincia de Río Negro, Argentina

Bárbara M. Bravo-Ferrada Laboratorio de Microbiología Molecular, Instituto de Microbiología Básica y Aplicada, Departamento de Ciencia y Tecnología, Universidad Nacional de Quilmes, Bernal, Buenos Aires, Argentina

Federico del Brio Centro para el Estudio de Sistemas Marinos (CESIMAR-CONICET), Puerto Madryn, Chubut, Argentina

Natalia S. Brizuela Laboratorio de Microbiología Molecular, Instituto de Microbiología Básica y Aplicada, Departamento de Ciencia y Tecnología, Universidad Nacional de Quilmes, Bernal, Buenos Aires, Argentina

Adriana Catalina Caballero Instituto de Investigación y Desarrollo en Ingeniería de Procesos, Biotecnología y Energías Alternativas (PROBIEN), Consejo Nacional de Investigaciones Científicas y Técnicas de la República Argentina (CONICET), Universidad Nacional del Comahue, Neuquén, Provincia del Neuquén, Argentina

Facultad de Ciencias y Tecnología de los Alimentos, Universidad Nacional del Comahue, Villa Regina, Provincia de Río Negro, Argentina

Facultad de Ingeniería, Universidad Nacional del Comahue, Neuquén, Provincia del Neuquén, Argentina

Viviana Andrea Carreño Facultad de Ciencias y Tecnología de los Alimentos, Universidad Nacional del Comahue, Villa Regina, Provincia de Río Negro, Argentina

Marcelo Hernán Cassini Grupo GEMA, Departamento de Ciencias Básicas, Universidad Nacional de Luján, Luján, Buenos Aires, Argentina

Laboratorio de Biología del Comportamiento, IBYME, Ciudad Autónoma de Buenos Aires, Argentina

Gabriel Alejandro Castro Hospital Regional de Ushuaia, Ushuaia, Argentina

Daniela Centrón Laboratorio de Investigaciones de los Mecanismos de Resistencia a Antibióticos, Instituto de Microbiología y Parasitología Médica (IMPaM, UBA-CONICET), Facultad de Medicina, Universidad de Buenos Aires, Ciudad Autónoma de Buenos Aires, Argentina

Patricia Chiacchiarini PROBIEN (CONICET-UNCo), Facultad de Ingeniería, Universidad Nacional del Comahue, Buenos Aires, Neuquén, Argentina

Fernando Coronato Patagonian Institute for the Study of Continental Ecosystems (IPEEC-CENPAT), National Scientific and Technical Research Council (CONICET), Argentina

Yolanda Leticia Curilén Facultad de Ciencias y Tecnología de los Alimentos, Universidad Nacional del Comahue, Villa Regina, Provincia de Río Negro, Argentina

Mariana E. Danilovich Planta Piloto de Procesos Industriales Microbiologicos PROIMI-CONICET, San Miguel de Tucumán, Tucumán, Argentina

Lucrecia Delfederico Laboratorio de Microbiología Molecular, Instituto de Microbiología Básica y Aplicada, Departamento de Ciencia y Tecnología, Universidad Nacional de Quilmes, Bernal, Buenos Aires, Argentina

Osvaldo Delgado Centro de Investigaciones y Transferencia de Catamarca CITCA-CONICET, San Fernando del Valle de Catamarca, Catamarca, Argentina

Facultad de Ciencias Exactas y Naturales, Universidad Nacional de Catamarca FACEN-UNCa, San Fernando del Valle de Catamarca, Catamarca, Argentina

Sergio F. Martínez Díaz Microbiology and Molecular Biology Laboratory, Interdisciplinary Center of Marine Science, CICIMAR-IPN, La Paz, Mexico

Hebe M. Dionisi Laboratorio de Microbiologia Ambiental, Centro para el Estudio de Sistemas Marinos (CESIMAR, CONICET), CCT Centro Nacional Patagonico, Puerto Madryn, Chubut, Argentina

Edgardo Donati Industrial Fermentation Research and Development Center (CINDEFI), CONICET-UNLP, School of Science, National University of La Plata, La Plata, Argentina

Melania Fernández Centro para el Estudio de Sistemas Marinos (CESIMAR-CCT CENPAT-CONICET), Puerto Madryn, Chubut, Argentina

Secretaría de Ciencia, Tecnología e Innovación Productiva de la Provincia de Chubut, Rawson, Chubut, Argentina

Marcela A. Ferrero Planta Piloto de Procesos Industriales Microbiológicos (PROIMI-CONICET), San Miguel de Tucumán, Tucumán, Argentina

Facultad de Bioquímica, Química y Farmacia, Universidad Nacional de Tucumán (UNT), San Miguel de Tucumán, Tucumán, Argentina

Melisa González Flores Instituto de Investigación y Desarrollo en Ingeniería de Procesos, Biotecnología y Energías Alternativas (PROBIEN, Consejo Nacional de Investigaciones Científicas y Técnicas de la República Argentina–Universidad Nacional del Comahue), Neuquén, Argentina

Marisa E. Garcés Patagonian Institute for the Study of Continental Ecosystems (IPEEC-CENPAT), National Scientific and Technical Research Council (CONICET), Argentina

Virginia de Garcia Laboratorio de Microbiología Aplicada, Biotecnología y Bioinformática de Levaduras, Instituto Andino-Patagónico en Tecnologías Biológicas y Geoambientales (IPATEC), CONICET y Universidad Nacional del Comahue (CRUB), San Carlos de Bariloche, Argentina

Alejandra Giaveno PROBIEN (CONICET, UNCo), Facultad de Ingeniería, Universidad Nacional del Comahue, Buenos Aires, Neuquén, Argentina

Fiorella F. Gomez Planta Piloto de Procesos Industriales Microbiológicos PROIMI-CONICET, San Miguel de Tucumán, Tucumán, Argentina

Facultad de Ciencias Exactas y Naturales, Universidad Nacional de Catamarca FACEN-UNCa, San Fernando del Valle de Catamarca, Catamarca, Argentina

María Soledad Vela Gurovic CERZOS UNS-CONICET CCT-Bahía Blanca, Bahía Blanca, Argentina

Danay Valdés La Hens Laboratorio de Microbiología Molecular, Instituto de Microbiología Básica y Aplicada, Departamento de Ciencia y Tecnología, Universidad Nacional de Quilmes, Bernal, Buenos Aires, Argentina

Axel Hollmann Laboratorio de Microbiología Molecular, Instituto de Microbiología Básica y Aplicada, Departamento de Ciencia y Tecnología, Universidad Nacional de Quilmes, Bernal, Buenos Aires, Argentina

Laboratorio de Biointerfaces y Sistemas Biomiméticos-Universidad Nacional de Santiago, del Estero (UNSE) y Centro de Investigaciones y Transferencia de Santiago del Estero (CITSE)-CONICET-UNSE, Santiago del Estero, Argentina

Martín S. Iglesias Patagonian Institute for the Study of Continental Ecosystems (IPEEC-CENPAT), National Scientific and Technical Research Council (CONICET), Argentina

Paula Isaac Centro de Investigación y Transferencia – Villa María (CIT-UNVM) CONICET, Córdoba, Argentina

Laura Lavalle PROBIEN (CONICET-UNCo), Facultad de Ingeniería, Universidad Nacional del Comahue, Buenos Aires, Neuquén, Argentina

Diego Libkind Andean-Patagonian Institute for Biological and Geoenvironmental Technologies (IPATEC), CONICET National University of Comahue, San Carlos de Bariloche, Argentina

María Alejandra Lima CINDEFI (CCT La Plata-CONICET, UNLP), Facultad de Ciencias Exactas, Universidad Nacional de La Plata, La Plata, Argentina

Christian Ariel Lopes Facultad de Ciencias Agrarias, Universidad Nacional del Comahue, Argentina

Instituto de Investigación y Desarrollo en Ingeniería de Procesos, Biotecnología y Energías Alternativas (PROBIEN, Consejo Nacional de Investigaciones Científicas y Técnicas de la República Argentina–Universidad Nacional del Comahue), Neuquén, Argentina

Sofía Lopez Centro de Investigación y Extensión Forestal Andino Patagónico (CIEFAP), Chubut, Argentina

Gustavo A. Lovrich Centro Austral de Investigaciones Científicas CADIC-CONICET, Ushuaia, Tierra del Fuego, Argentina

Mariana Lozada Laboratorio de Microbiologia Ambiental, Centro para el Estudio de Sistemas Marinos (CESIMAR, CONICET), CCT Centro Nacional Patagonico, Puerto Madryn, Chubut, Argentina

María C. Lutz Facultad de Ciencias Agrarias, Universidad Nacional del Comahue, Neuquén, Argentina

Daniela Maizel Planta Piloto de Procesos Industriales Microbiológicos (PROIMI-CONICET), San Miguel de Tucumán, Tucumán, Argentina

Magalí S. Marcos Patagonian Institute for the Study of Continental Ecosystems (IPEEC-CENPAT), National Scientific and Technical Research Council (CONICET), Argentina

Ramona del Carmen Maturano Instituto de Investigación y Desarrollo en Ingeniería de Procesos, Biotecnología y Energías Alternativas (PROBIEN), Consejo Nacional de Investigaciones Científicas y Técnicas de la República Argentina (CONICET)- Universidad Nacional del Comahue, Neuquén, Provincia del Neuquén, Argentina

Martin Moliné Laboratorio de Microbiología Aplicada, Biotecnología y Bioinformática de Levaduras, Instituto Andino-Patagónico en Tecnologías Biológicas y Geoambientales (IPATEC), CONICET y Universidad Nacional del Comahue (CRUB), San Carlos de Bariloche, Argentina

Silvana María del Mónaco Instituto de Investigación y Desarrollo en Ingeniería de Procesos, Biotecnología y Energías Alternativas (PROBIEN), Consejo Nacional de Investigaciones Científicas y Técnicas de la República Argentina (CONICET)-Universidad Nacional del Comahue, Neuquén, Provincia del Neuquén, Argentina

Marina L. Nievas Centro para el Estudio de Sistemas Marinos (CESIMAR-CONICET), Puerto Madryn, Chubut, Argentina

Facultad Regional Chubut, Universidad Tecnológica Nacional, Puerto Madryn, Argentina

Nelda Lila Olivera Patagonian Institute for the Study of Continental Ecosystems (IPEEC-CENPAT), National Scientific and Technical Research Council (CONICET), Argentina

Andrea Cecilia Origone Instituto de Investigación y Desarrollo en Ingeniería de Procesos, Biotecnología y Energías Alternativas (PROBIEN, Consejo Nacional de Investigaciones Científicas y Técnicas de la República Argentina–Universidad Nacional del Comahue), Neuquén, Argentina

María B. Pildain Centro de Investigación y Extensión Forestal Andino Patagónico (CIEFAP), Chubut, Argentina

Rosana Polifroni Centro para el Estudio de Sistemas Marinos (CESIMAR-CONICET), Puerto Madryn, Chubut, Argentina

Graciana Willis Poratti CINDEFI (CCT La Plata-CONICET, UNLP), Facultad de Ciencias Exactas, Universidad Nacional de La Plata, La Plata, Argentina

María Paula Quiroga Laboratorio de Investigaciones de los Mecanismos de Resistencia a Antibióticos, Instituto de Microbiología y Parasitología Médica (IMPaM, UBA-CONICET), Facultad de Medicina, Universidad de Buenos Aires, Ciudad Autónoma de Buenos Aires, Argentina

María Eugenia Rodríguez Instituto de Investigación y Desarrollo en Ingeniería de Procesos, Biotecnología y Energías Alternativas (PROBIEN, Consejo Nacional de Investigaciones Científicas y Técnicas de la República Argentina–Universidad Nacional del Comahue), Buenos Aires, Neuquén, Argentina

Facultad de Ciencias Médicas, Universidad Nacional del Comahue, Neuquén, Argentina

Leandro A. Sánchez Planta Piloto de Procesos Industriales Microbiologicos PROIMI-CONICET, San Miguel de Tucumán, Tucumán, Argentina

Marcela P. Sangorrín Instituto de Investigación y Desarrollo en Ingeniería de Procesos, Biotecnología y Energías Alternativas (PROBIEN, Consejo Nacional de Investigaciones Científicas y Técnicas de la República Argentina-Universidad Nacional del Comahue), Neuquén, Argentina

Ana Belén Segretin CINDEFI (CCT La Plata-CONICET, UNLP), Facultad de Ciencias Exactas, Universidad Nacional de La Plata, La Plata, Argentina

Liliana Semorile Laboratorio de Microbiología Molecular, Instituto de Microbiología Básica y Aplicada, Departamento de Ciencia y Tecnología, Universidad Nacional de Quilmes, Bernal, Buenos Aires, Argentina

Marcela A. Sepúlveda Centro para el Estudio de Sistemas Marinos (CESIMAR-CONICET), Puerto Madryn, Chubut, Argentina

Cynthia Sequeiros Centro para el Estudio de Sistemas Marinos (CESIMAR-CCT CENPAT-CONICET), Puerto Madryn, Chubut, Argentina

Adriana Beatriz Simes Facultad de Ciencias y Tecnología de los Alimentos, Universidad Nacional del Comahue, Villa Regina, Provincia de Río Negro, Argentina

Faustino Siñeriz Planta Piloto de Procesos Industriales Microbiologicos PROIMI-CONICET, San Miguel de Tucumán, Tucumán, Argentina

Andrea Trochine Laboratorio de Microbiología Aplicada, Biotecnología y Bioinformática de Levaduras, Instituto Andino-Patagónico en Tecnologías Biológicas y Geoambientales (IPATEC), CONICET y Universidad Nacional del Comahue (CRUB), San Carlos de Bariloche, Argentina

Elizabeth E. Tymczyszyn Laboratorio de Microbiología Molecular, Instituto de Microbiología Básica y Aplicada, Departamento de Ciencia y Tecnología, Universidad Nacional de Quilmes, Bernal, Buenos Aires, Argentina

Ricardo Ulloa PROBIEN (CONICET-UNCo), Facultad de Ingeniería, Universidad Nacional del Comahue, Buenos Aires, Neuquén, Argentina

María Sofía Urbieta CINDEFI (CCT La Plata-CONICET, UNLP), Facultad de Ciencias Exactas, Universidad Nacional de La Plata, La Plata, Argentina

María L. Villalba Instituto de Investigación y Desarrollo en Ingeniería de Procesos, Biotecnología y Energías Alternativas (PROBIEN, Consejo Nacional de Investigaciones Científicas y Técnicas de la República Argentina-Universidad Nacional del Comahue), Neuquén, Argentina

Chapter 1
Introducing Patagonia: An Approach from Geo-History

Fernando Coronato

Abstract The uniqueness of Patagonia stems not only from its remote location on the world map but from several other geographic factors briefly discussed on this chapter, starting from some climatic features and sharp contrasts not found anywhere else. The last glaciation caused drainage deviations that complicated international boundary delimitation, which, in turn, hastened expansionism based on sheep farming of not always suitable lands. Widespread desertification is the major ecological consequence of this unwise process, developed at the expense of Native Patagonian peoples, some of whom have completely disappeared. From a terrestrial point of view, Patagonia is a geographic cul-de-sac, so aboriginal cultures were materially rudimentary; on the other hand, to seafarers Patagonia was on the interoceanic route and was a dangerous coast to be left behind as soon as possible. The unavoidable contact was dramatically contrasting and not always peaceful. With very low demographic density, and despite isolated cases of ecological mismanagement, the region still keeps an aura of pristine nature, enhanced by distance and legend. This chapter may seem incongruent in this volume, but it aims to place the readers in the wider scope of Patagonian geo-history, the human environment in which the research presented here was carried out.

1.1 Climate and Physical Environment

Patagonia, the southernmost region of the Americas, extends approximately from 40°S, where the width of the continent is about 1000 km, and gradually narrows southward until disappearing in Cape Horn at 56°S (Fig. 1.1). Less than 900 km separate this point from the northernmost tip of the Antarctic Peninsula, but the Drake Passage is wide enough to sensibly warm the Antarctic air masses that may reach South America, and in such a way, Patagonian winters are much milder than those at equivalent latitudes in North America. These maritime polar air masses can

F. Coronato, Ph.D. (✉)
Patagonian Institute for the Study of Continental Ecosystems (IPEEC-CENPAT), National Scientific and Technical Research Council (CONICET), Argentina
e-mail: coronato@cenpat-conicet.gob.ar

N.L. Olivera et al. (eds.), *Biology and Biotechnology of Patagonian Microorganisms*, DOI 10.1007/978-3-319-42801-7_1

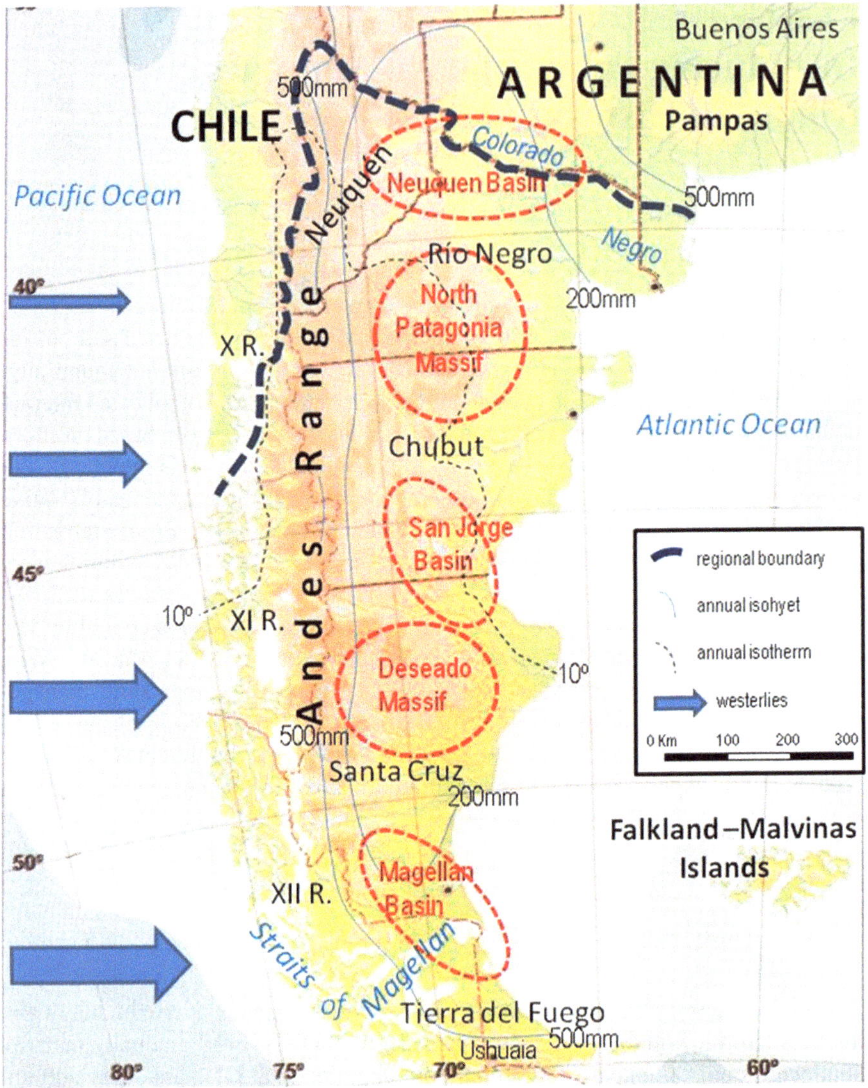

Fig. 1.1 Main physical features and political boundaries of bi-national Patagonia

reach Patagonia during the whole year, so they are one main cause for the relatively low thermal level in southern Patagonia (Weischet 1985).

Because of its latitude Patagonia is fully located in the belt of the west winds, which here are put in motion by the interplay between the permanent high-pressure cells on the South Atlantic and Southeast Pacific (centered about on 30°S) and the subpolar trough at 60°S. The seasonal N–S shift of the high cells is quite noticeable although the sub-Antarctic low belt hardly changes its position. Thus, the pressure gradient increases in summer, when the anticyclones move southward, and consequently wind speed is higher during that season. Moreover, as no other continental landmass stands in the way of the westerlies at these latitudes, wind speed may be particularly high in most of Patagonia. At the time of sailing ships, seamen referred to this latitudinal belt as "the roaring forties" and "the screaming fifties." Assertions such as "In few parts of the world is the climate of the region and its life so determined by a single meteorological element, as is the climate of Patagonia by the constancy and strength of the wind" (Prohaska 1976) seem well justified and, beyond any statement, wind strength indications are numerous in Patagonia, such as wind erosion, leeward-sided trees, and architectonic details.

One of the main consequences of such winds is the cooling effect and the perceived systematic reduction of the temperature. Because (as already stated), wind speed is higher in the warm season, the wind cooling effect is greater in this season too, and thus, there is an apparent shrinking in the annual range of temperature and so the climate is perceived as being cooler (and more maritime) than it actually is. Considering Patagonia as a whole, Coronato (1993) calculated that average wind cooling effect is 4.2 °C in annual figures. Yet, the major consequence of the overwhelming prevalent west winds is the abrupt contrast between windward, well-watered, and forested western Patagonia and the leeward, rain-shadowed, dry steppe eastward in the Andes that covers plateaus and peneplains until the Atlantic shores. Although south of 40°S the Andes rarely exceed 3000 m height, the mountain range intersects the westerly flux perpendicularly, blocking the disturbances embedded in it and producing orographic precipitations over the Pacific flank. On the western side of the Andes of Patagonia rainfall may be as much as 20 times greater than on the eastern flank, where moreover subsidence leads to arid and highly evaporative conditions (Garreaud et al. 2013). This effect creates a marked climatic contrast that entails one of the sharper vegetation gradients in the world (Endlichter and Santana 1988; Warren and Sugden 1993). Any W–E transect crossing over the Andes from the Pacific shore to the eastern foothills (no more than 200–300 km) starts in a wet temperate forest, grades into Alpine forests and grasslands, changing again into moderate continental forests, to merge finally into an arid environment that spans further hundreds of kilometers to the Atlantic (Bailey 1996). The vegetation gradient is clearly visible in the glacial lakes of the eastern side, whose fjord-like western shores deeply penetrate between snow-capped mountains, covered by a humid and cool forest, whereas the distal eastern end, in pebbled wide beaches on morainic hills, is surrounded by xerophytes and active sand dunes.

If precipitation amounts change dramatically at both sides of the Andes (10:1 on average), rainfall seasonality and cloudiness and temperature regime do not, or change much less markedly. As the poleward movement of the subtropical high-pressure cells is clearly manifested in South America until about 40°S, summers are usually sunnier and drier in the northern half of Patagonia (i.e., 45°–46°S). This Mediterranean-like climate is fully expressed in northwestern Patagonia, as a marginal area of the truly Mediterranean climate that reigns over Central Chile. On the other side, in northeastern Patagonia, although keeping the same annual distribution, the scarcity of rainfall allows only considering this climate as an arid-degraded Mediterranean (Le Houérou 2004). Further east, on the Atlantic Coast north of 45°S, summer drought may be less marked because of sporadic irruptions of oceanic air masses and consequent rainfall, in the (few) occasions when South Atlantic anticyclonic circulation prevails over the Pacific air masses that have dried crossing the Andes.

In contrast, in the southern half of Patagonia, the westerly flux is not counteracted at any season and thus rainfall is more evenly distributed throughout the year. Eventually, cloudiness and precipitation values are greater in summer south of 50°S, which is indicative of the subpolar pattern of this regime (Prohaska 1976). The Patagonian case of the unfettered westerlies impacting transversally on the N–S--oriented mountain range is a textbook example of the mechanic effect of deviation of the original flux into a roughly anticyclonic circulation as described by Flohn (1969). Such deviation entails a much higher frequency of SW winds leeward, that is, in Argentine stations, compared to windward (Chilean) stations, where NW winds prevail (Carrasco et al. 1998).

As already mentioned briefly, the Andes are far from affecting the cloudiness and temperature regimes in the same degree they affect wind and precipitation. For instance, the annual isonephs of sky cover show a roughly latitudinal pattern whose values range from 50 % in northern Patagonia to 70 % in the south, with a noticeable decrease toward inland (Prohaska 1976). No marked differences in sky cover are recorded at both sides of the mountains, but instead, differences in the type of cloudiness are indeed observable. Mostly, the middle and high cloudiness recorded in eastern Patagonia as far as the Atlantic Coast is only residual cloudiness generated by the orographic uplift of the wet Pacific air and the consequent precipitation.

Regarding the temperature pattern, although the Andean cordillera restricts the inland extent of oceanic influences, it is evident that some differences exist between western and eastern Patagonia beyond the rainfall amount. However, it cannot be said that leeward Patagonia has a definite continental climate because temperature range between summer and winter remains rather restricted (12 °C on average in eastern Patagonia). In fact, there is disagreement among authors about the continentality-oceanity of the Patagonian climate: although for some it is "definitely maritime" (Walter and Box 1983), for others "it has distinct continental features" (Mensching and Akhtar 1995). According to Miller (1976) the attenuation of its continental characteristics stems on the narrowness of the continent, which is obviously a hindrance to the formation of continental air masses (Taljaard 1969). Besides, it is clear that although the Atlantic and Pacific Oceans are not too far away from each other in Patagonia, the influence of the former is almost negligible because of the prevailing westerly winds.

As already stated, the longitudinal extent of Patagonia exceeds 15° in latitude, that is, more than 1600 km in the N–S direction. In Europe, it would be equivalent to the distance between Madrid and Edinburgh (an equivalence not merely geometric but also environmental), which, most obviously, entails noticeable differences in the incoming solar radiation. It ranges from an annual average of slightly above 180 W m^{-2} (annual average) in the north (Neuquén, 39°S) to only 100 W m^{-2} in Tierra del Fuego (Ushuaia, 55°S) (Paruelo et al. 1998a, b) or even less in the outer islands on the Pacific shore. According to the increasing latitude, the ratio between summer and winter incoming solar radiation augments from 4:1 at Neuquén, to 13:1

at Ushuaia, yet the temperature follows an opposite pattern because of the narrowing of the continental mass. The mean annual temperature range varies from 16 °C in the north to 8 °C in the south, and it can be as low as 4 °C in the most exposed islands of the west coast, a "hyper-oceanic" area (Tuhkanen 1992). The extreme temperatures also reflect the widening of the continent northward. Maxima of 38 °C are reached almost every summer in north Patagonia, and peaks of 35 °C are far from uncommon inland and were recorded as south as 46°S. In Tierra del Fuego the maximum temperature does not exceed 30 °C, and it does not even reach 20 °C in the hyper-oceanic islands. On the other hand, minimum temperatures of −30 °C were recorded in the central plateaus as north as 41°S whereas in the extreme south they seldom drop below −20 °C (SMN 1960). Along the Pacific Coast at sea level, absolute minima never are lower than −10 °C (Zamora and Santana 1979).

Year-to-year variation of the temperature is not the same in all Patagonia, but two main areas of isofluctuation were detected: north and south (Coronato and Bisigato 1998). Thus, the main factor of divergence in this matter is not the Andean Cordillera, which acts as a secondary-level factor, but latitude. The two meteorological stations that best correlate with the isofluctuative areas are Trelew (43°S) and Rio Gallegos (51°S), both on the Atlantic Coast.

Regarding long-term changes, a clear warming has been reported in most of Patagonia since 1950 (Villalba et al. 2003; Vincent et al. 2005). On average, the annual mean temperature increased 0.4 °C in the whole region, although the rising of the maxima was between 0.5 ° and 1 °C and that of the minima 0.4–0.8 °C, depending on the area. This trend is also evidenced by a decline in the number of cold days and an augmentation of warm days since 1960 (Rusticucci and Barrucand 2004).

Whether a definite trend is observable in temperature, that is not the case with precipitation. No significant changes were observed in the past 50 years in most of Patagonia; also, year-to-year variability overrides long-term changes throughout the region. Notwithstanding, different models project up to 10% less rainfall within the near future scenarios, especially in northern Patagonian Andes (Masiokas et al. 2008). This possibility is important in terms of hydrological ecosystem services because the major river basins of Eastern Patagonia have their headwaters in that mountain range.

From everything explained up to this point, the point emerges why eastern Patagonia fits badly in a global climatic classification. Elsewhere, the eastern side of a continent in equivalent latitudes would present a cold temperate climate, rather continental and with a moderate annual rainfall amount (a Cfb or Dfb climate in Köppen's classification). Instead, eastern Patagonia exhibits a temperate dry climate with moderate annual temperature range (a BSk climate or even a BWk in drier areas). The uniqueness of the eastern Patagonian climatic pattern is reinforced by the (also very rare) sharp contrast with western Patagonia, which displays a strongly oceanic climate (no doubt a Cfc climate), a little colder than its counterparts elsewhere in the world, especially because of the lack of real summer heat (Weischet 1985).

All in all, it can be said that the Andean Range is the main geographic feature of Patagonia even if the eastern tablelands have a totally opposite landscape and are quite far away.

1.2 A Geographic Approach

The influence of the Andes in the climate of Patagonia has been repeatedly pointed out already. However, its importance goes far beyond the climate, because the mountain range is both the continental watershed and, in many areas, the international boundary between Argentina and Chile. In the turn of the nineteenth century, when the border between the two nations was being demarcated, a problem arose because the line of high peaks does not always match the *divortium aquarum*, that is, the continental divide. These anomalies result from drainage diversions during postglacial times, which, in turn, stem from the intensive modeling of glacier action during the Pleistocene. Several hydrographic basins of the Pacific slope have their headwaters in the morainic systems frankly east of the main range, and thus this line was claimed by Chile as being the boundary. In contrast, the Argentinean claim followed the line of summits of the main range, very close to the Pacific shores. The controversial boundary was eventually settled in 1902 by the arbitration of Edward VII, the King of England, and it resulted in a mixture of partial (mostly intermediate) solutions, amidst which a plebiscite among pioneer settlers was also taken into account. In such a way, Andean lakes that empty into the Atlantic belong exclusively to Argentina while lakes that drain into the Pacific are split in two. This rule is not satisfied in the refereed area, where some Pacific draining lakes, such as Futalaufquen, were entirely given to Argentina.

Regardless of the international boundary, the total surface area of bi-national Patagonia is around 1,000,000 km^2. The figure is approximate because of the poorly defined north border on the Chilean side, where only a part of the administrative Region X (Los Lagos) is considered as Patagonia without question. Chilean Patagonia also includes Regions XI and XII (Aysén and Magallanes, respectively), a total of 260,000 km^2. On the other side, east of the Andes, the northern frontier of Patagonia is clearly demarcated by the Colorado River, which crosses Argentinean territory from the Andes to the Atlantic. The portion of Argentina south of the Colorado River, that is, Argentinean Patagonia, has an area of 790,000 km^2 and includes the provinces of Tierra del Fuego, Santa Cruz, Chubut, Río Negro, and Neuquén, as well as a single county of Buenos Aires province (Fig. 1.1). The total population for the whole binational region is a little less than 2.5 million inhabitants, of which 2.1 million are in Argentina and 0.3 million in Chile (censuses of 2010 and 2012, respectively). Although demographic density remains very low, the region exhibits at present a rapid demographic growth, mainly because of migrations from the core of the respective countries. Suffice it to say that demographic density reached 1 inhabitant/km^2 as late as the late 1970s. Although the Patagonian population is overwhelmingly urban, it is clear that human impact on most of the territory is really yet very small and the whole region has an aura of pristine and well-preserved nature (even if this is not true in some spots). National parks and national reserves and protected areas currently occupy almost 90,000 km^2 of the whole of Patagonia, or 8.7 % of the regional territory. When provincial reserves or lesser categories are considered, almost 10 % the proportion of the entire region has some protection status.

Despite quite a good level of protection, natural ecosystems in vast areas of Patagonia face a major threat, desertification, which seems to be the direct consequence of the unwise application of a sedentary pastoral system conceived in wetter and more productive rangelands. Indeed, owing to the scarcity of precipitation, the Patagonian steppes have a modest aboveground net primary productivity, roughly ranging from 350 kg ha^{-1} $year^{-1}$ in the driest areas to 900 kg ha^{-1} $year^{-1}$ in the best rangelands in the foothills of the Andes (Paruelo et al. 1998b). Patagonian rangelands were among the last such lands in the world to be devoted to animal husbandry. Commercial sheep farming in the area started only 100–120 years ago and boomed until the 1930 world crash, triggering the hasty colonization of grasslands of diverse productivity, corseted in closed paddocks. Thus, less than a century after the beginning of pastoral colonization, much of the steppes of Patagonia were transformed into desert-like areas, especially in its central and eastern parts (i.e., the regions less favored by rainfall), to the point that sheep grazing became unsustainable and has ceased on many ranches that are now abandoned (between one third and two thirds of them, depending on the area). However, some farms manage to survive with a carrying capacity as low as one sheep for each 8–12 ha.

Along with widespread desertification, natural hazards such as droughts or heavy snowfall coupled to the vagaries of economic policies have always made sheep farming in Patagonia an uncertain activity. Consequently, sheep gradually lost ground as the leading socioeconomic activity as compared with oil and gas production, fishing, and lately, tourism. The situation worsened during the last quarter of the twentieth century to the point that, at present, the farm contribution to regional GDP is less than 4 %. However, no other activity can ensure the (sparse) occupation of the whole region, and owing to its pioneer character, sheep farming still plays an important part in the Patagonian identity.

Widespread sheep farming and the omnipresent scarcity of rainfall have created a quite monotonous landscape everywhere in East Patagonia. However, this uniformity conceals a wide variety of geologic features because the Patagonian substratum is far from being uniform. Besides the Andean cordillera, already mentioned, four fifths of the territory is mainly molded on sedimentary and volcanic rocks of Mesozoic and Cenozoic age, forming a mosaic of plateaus, hills, and closed basins. In some areas Paleozoic rocks or even the crystalline basement also appear. Despite the heterogeneity of the bedrock substratum, the geology of Patagonia can be summarized as two main outcrops of the Patagonian Craton, namely, the North Patagonian Massif and the Deseado Massif further south (Coronato et al. 2008) (Fig. 1.1). Both ancient massifs are flanked by sedimentary basins, the Neuquén Basin in the far north of Patagonia, the San Jorge Basin, separating the two massifs, and the Magellan Basin in the far south (Fig. 1.1): this gives a distribution pattern of B-M-B-M-B where B are the basins and M the massifs. All the five large structures are backed by the Andean Range, indeed the spine of Patagonia. As elsewhere in the world, massifs are rich in metal ores whereas the basins are good oil and gas producers. In fact, once sheep farming waned, hydrocarbons have become the main source of wealth in Patagonia during the past 60 years, whereas metal mining is currently thriving and it is often considered as a threat to natural conservancy, one of the regional paramount values.

In Patagonia, natural conservancy also concerns the marine coasts and adjacent waters, especially because of the richness and diversity of coastal wildlife and marine mammals, such as seals, fur seals, and sea lions (*Otaria* sp.) or phocids (*Mirounga* sp.), as well as right whales (*Eubalaena australis*) and killer whales (*Orcinus orca*). Both the Pacific and the Atlantic waters adjacent to Patagonia are rather cold owing to the flow of sub-Antarctic waters carried either by the early stages of the Humboldt Current on the Chilean shores or by the Malvinas (Falklands) Current on the Atlantic, beyond the very wide Patagonian continental shelf. Even the shallow waters on the shelf are quite cold because the west wind blowing from the mainland generates upwelling situations most of the time. Thus, the Patagonian Atlantic is a productive fishery, more exploited by overseas fleets rather than regional ones. There are few fishing ports in Patagonia, perhaps because there is little natural anchorage in these straight and wide open coasts.

Nor has agriculture has ever been very important in Patagonia because of the want of irrigable lands (and population as well). Just in the foothills of the Andes crops may grow without watering, but woodlands and mountains restrict the extent of arable land; however, Andean wheat was successful during the first decades of the twentieth century. In eastern Patagonia, farmlands are restricted to riparian oases in the main rivers: the Colorado, Negro, and Chubut. Further south, the shortness of the growing season becomes a limiting factor, so commercial agriculture is not really developed.

1.3 A Glance at History

Europeans first saw Patagonia from the Atlantic coastline: solitary, arid, and windy. From AD 1520, when discoverer Magellan gave those inhospitable shores the name of "Patagonia," the term identified the vast plains of the American southern end, whose cliffy and waterless coasts were inaccessible. Moreover, fierce and tall nomad natives, the "giant" Patagonians, deterred any attempt for Europeans to settle in. It was only in the nineteenth century that the concept of Patagonia included the Southern Andes and the Pacific Coast. Also, it was only in the second half of the nineteenth century when alien people first succeeded in permanently settling in the region south of 40°S: this happened on the east coast, a Chilean penitentiary settlement in the Straits of Magellan (53°S) in 1843, and a Welsh Colony, under the Argentine flag, in Chubut (43°S) in 1865. Thus, in 1832–1833, by the time Charles Darwin explored Patagonia on his initial trip, no nonnative population existed anywhere south of the Negro River.

Inner Patagonia remained unexplored until as late as 1870, when the English explorer George Musters accompanied an Indian crew in their annual south–north migration: his book, "At home with the Patagonians" (London, 1871), is a masterpiece in regional ethnology. Shortly after, once Argentina and Chile achieved internal organization, Patagonia came under dispute by the two young republics. Sheep farming was the tool that both governments employed to occupy huge extents of land (not always suitable for such goals) in a sort of expansionist race. The fast growth of sheep flocks was done at the expense of several thousands of Native American victims, who were killed, reduced to servitude, or scattered on the margins of the national "new order." In this way Patagonia was "freed," ready for its occupation by settlers of European ancestry, pioneers coming from the Falklands, Buenos Aires, or central Chile, for the greater benefit of British, Flemish, or German wool companies (Coronato 2010).

In agreement with the austere and only slightly productive natural environment already described, Native Patagonians never have been numerous. The most accepted figures are around 10,000 people for the entire region south of 40°S at the time of the beginning of colonization by the mid-nineteenth century. The aboriginal populations of the steppes were all hunters-gatherers, nomadic tribes that followed the guanaco (*Lama guanicoe*) herds, which were then their livelihood. It was these people whom the first Spaniards seamen named "Patagones," and the name of their country followed. The Patagones (also called Tehuelche) themselves distinguished two groups among them: the *Aoniken*, in the south, and the *Guenaken*, in the north. Once they got horses, after the contact with Europeans, Aoniken and Guenaken moved freely in the entire region, as far north as Buenos Aires. An isolated group of the same culture inhabited in the plains of the island of Tierra del Fuego, the *Selknam*, but they never had horses and the rich plains where they dwelled were fiercely and quickly occupied by sheep farming companies in the late decade of the nineteenth century. Thus, the fate of the Selknam was especially tragic, and constitutes one of the most shameful chapters in Patagonian history (Coronato and Tourrand 2010).

In the jagged coastline of the Pacific, where overland displacements are practically impossible, a few thousands of coastal nomadic canoeists dwelled in the fjords and archipelagoes. From north to south they were the *Chonos*, *Alakaluf*, and *Yahgan*, although with few differences among them. These peoples were almost isolated from each other because of the difficulty of crossing some stretches of open water imposed by the geography. However, because of their location on the inter-oceanic route, they were openly exposed to contact with Europeans, always in extremely unequal conditions. Darwin was horrified to see their primitivism, and at the same time, aboard the same ship, the *HMS Beagle*, three Yahgan having spent 3 years (1830–1833) in England were repatriated to their homeland after an amazing anthropological experiment of cultural transplant performed by Captain Robert Fitz-Roy.

In another kind of cultural transplant, evangelization, some Christian missions were established during the second half of the nineteenth century, first by Anglicans (Patagonian Missionary Society) based on the Falklands and later by Catholics (Salesian priests). Despite the good intentions of the clergymen, compulsory sedentary life added to crowded dormitories were lethal to these people used to outdoor and free displacements (Casali et al. 2006).

Needless to say, all these Native Patagonian peoples hardly survived to "civilization," either by an agonic cultural decline, introduced epidemics, or open military campaigns against them, as occurred in the so-called "Conquest of the Desert" in Argentina (1879–1884), or the cynically named "Pacification of Araucania" in Chile at the same period. Even if Araucania is not in Patagonia, its people

(Araucanians or *Mapuche*) was a strong "First Nation" that deeply influenced Native Patagonians, and their rebellious spirit is still active in Northwest Patagonia.

Even nowadays Patagonia remains a peripheral region, scarcely populated and lately incorporated to the respective national identities, and for which full integration to their national spaces is still in progress. It can be said that Patagonia is defined as a geographic space by the deep feeling of regional belonging of its inhabitants, linked by a common history and sharing a feeling of aloofness regarding the national cores. Perhaps, because of the uttermost position of Patagonia in the world map, the region bears a strong mythic charge that largely supports its identity. It is easily understood that this converse *Ultima Thule* plays the same psychological meaning, and—as in Darwin's times—is still chosen as the goal of initiatory trips. If the book "In Patagonia" by Bruce Chatwin (1977) is not a mere book of travel literature but anything deeper, the present volume, concerned with Patagonian microorganisms, might be something more than biotechnology.

References

Bailey R (1996) Ecosystem geography. Springer, New York

Carrasco J, Casassa G, Rivera A (1998) Climatología actual del campo de hielo sur y posibles cambios por el incremento del efecto invernadero. Anales del Instituto de la Patagonia 26: 119–128. Punta Arenas, Chile

Casali R, Fugassa M, Guichón R (2006) Epidemiological approach to European–Native contact in northern Tierra del Fuego. Magallania (Chile) 34:87–101

Chatwin B (1977) In Patagonia. J. Cape, London

Coronato F (1993) Wind chill factor applied to Patagonian climatology. Int J Bioclimatol 37:1–6

Coronato F (2010) Moutons et colons en Patagonie. Editions Universitaires Européennes, Sarrebruck, Germany

Coronato F, Bisigato A (1998) A temperature pattern classification in Patagonia. Int J Climatol 18:765–773

Coronato F, Tourrand J (2010) Le chasseur chassé ou le drame de l'identité au bout du monde. In: Michaud M-C, Delhom J (eds) Guerres et identités dans les Amériques, vol 35, Mondes Hispanophones. Presses Universitaires de Rennes, France, pp 163–176

Coronato A, Coronato F, Mazzoni E, Vázquez M (2008) Physical geography of Patagonia. In: Rabassa J (ed) The late Cenozoic of Patagonia and Tierra del Fuego, vol 11, Development in quaternary sciences. Elsevier, Amsterdam, pp 13–55

Endlichter W, Santana A (1988) El clima del sur de la Patagonia y sus aspectos ecológicos. Anales del Instituto de la Patagonia 18:57–86

Flohn H (1969) Local wind systems. In: Flohn H (ed) General climatology, vol 2, World survey of climatology. Elsevier, Amsterdam, pp 139–171

Garreaud R, Lopez P, Minvielle M, Rojas M (2013) Large-scale control on the Patagonian climate. J Clim 26:215–230

Le Houérou H (2004) An agro-bioclimatic classification of arid and semiarid lands in the isoclimatic Mediterranean Zones. Arid Land Res Manag 18:301–346

Masiokas M, Villalba M, Luckman B, Lascana M, Delgado S, Stepanek P (2008) 20th-century glacial recession and regional hydroclimatic changes in the northwestern Patagonia. Global Planet Change 60:85–100

Mensching H, Akhtar M (1995) Desertification and changes in the geomorphic processes. Ann Arid Zone 34:79–85

Miller A (1976) The climate of Chile. In: Schwertfeger W (ed) Climates of Central and South America, vol 12, World survey of climatology. Elsevier, Amsterdam, pp 113–145
Paruelo J, Beltran A, Jobbágy E (1998a) The climate of Patagonia: general patterns and controls on biotic processes. Ecol Austral 8:85–101
Paruelo J, Jobbágy E, Sala O (1998b) Biozones of Patagonia. Ecol Austral 8:145–153
Prohaska F (1976) The climate of Argentina, Paraguay and Uruguay. In: Schwertfeger W (ed) Climates of Central and South America, vol 12, World survey of climatology. Elsevier, Amsterdam, pp 13–112
Rusticucci M, Barrucand M (2004) Observed trends and changes in temperature extremes in Argentina. J Clim 17:4099–4107
SMN (1960) Atlas climático de la República Argentina. Servicio Meteorológico Nacional, Buenos Aires
Taljaard J (1969) Air masses of the Southern Hemisphere. Notos 18:79–104
Tuhkanen S (1992) The climate of Tierra del Fuego from a vegetation geographical point of view and its ecoclimatic counterparts elsewhere. Acta Bot Fenn 145:1–64
Villalba R, Lara A, Boninsegna J, Masiokas M, Delgado S, Aravena J, Roig F, Schmelter A, Wolodarsky A, Ripalta A (2003) Large-scale temperature changes across the southern Andes: 20th-century variations in the context of the past 400 years. Clim Change 59:177–232
Vincent L, Peterson T, Barros V (2005) Observed trends in indices of daily temperature extremes in South America 1960–2000. J Clim 18:5011–5023
Walter H, Box E (1983) Climate of Patagonia. In: West N (ed) Ecosystems of the World, vol 5, Deserts and semideserts of Patagonia. Elsevier, Amsterdam, pp 440–454
Warren C, Sugden D (1993) The Patagonian icefields: a glaciological review. Arct Alp Res 25:316–331
Weischet W (1985) Climatic constraints for the development of the Far South of Latin America. GeoJournal 11:79–87
Zamora E, Santana A (1979) Características climáticas de la costa occidental de la Patagonia. Anales Instituto Patagonia 10:109–144

Part I
Environmental Microbial Biotechnology

Chapter 2
Molecular Biological Tools for the Assessment of Hydrocarbon-Degrading Potential in Coastal Environments

Mariana Lozada and Hebe M. Dionisi

Abstract This chapter includes the advances achieved so far in the design and implementation of molecular biological tools (MBTs) for the assessment of hydrocarbon-degrading potential in microbial communities from coastal environments of Patagonia. A brief introduction on the role of hydrocarbon-degrading bacteria in marine environments follows the basic concepts of MBTs, methods, and applications. The review then focuses on studies performed on the Patagonian coast to identify functional biomarker genes associated with hydrocarbon biodegradation, with emphasis on polycyclic aromatic hydrocarbons (PAHs): (a) advances on determining the identity, abundance, and biogeographic distribution of dioxygenase gene variants from known obligate PAH-degrading marine bacteria as well as yet uncultured microorganisms; (b) testing of selected variants in experimental systems; and (c) results of recent metagenomic analyses revealing the genetic context and PAH-degrading capabilities of uncultured microorganisms from Patagonia carrying an ecologically relevant biomarker gene. Alkane biodegradation biomarker genes are also covered, as well as analyses based on phylogenetic biomarker genes. Obligate and specialized hydrocarbon degraders are identified in microbial communities from Patagonia by culture-independent approaches based on the 16S rRNA gene. Last, the design and testing of a community-level ecological indicator based on high-throughput sequencing of the 16S rRNA gene and perspectives on the use of MBTs in coastal regions of Argentinean Patagonia are discussed.

M. Lozada (✉) • H.M. Dionisi
Laboratorio de Microbiología Ambiental, Centro para el Estudio de Sistemas Marinos (CESIMAR, CONICET), CCT Centro Nacional Patagónico, Bvd. Brown 2915, Puerto Madryn, U9120ACD, Chubut, Argentina
e-mail: lozada@cenpat-conicet.gob.ar

N.L. Olivera et al. (eds.), *Biology and Biotechnology of Patagonian Microorganisms*, DOI 10.1007/978-3-319-42801-7_2

2.1 Molecular Biological Tools

Microorganisms have a key role in the fate of environmental pollutants, mainly participating in their removal through the use of these compounds as carbon and energy sources (Jeon and Madsen 2013). The assimilation of these compounds leads to a rapid increase in the abundance of these populations and higher biodegradation rates. Therefore, information of the presence, abundance, or activity of pollutant-degrading microbial populations in a polluted site could aid in decision making at each stage of the remediation process, as a complement of contaminant concentrations and geochemistry of the site (Lebron et al. 2011). By increasing the efficiency and predictability of the biodegradation process, this information could reduce the time and costs of bioremediation and accelerate the adoption of these technologies. Because it is impossible or difficult to culture most of the environmental microorganisms, traditional culture-dependent methods are not able to provide this information accurately and rapidly. Molecular biological tools (MBTs) targeting pollutant-degrading microorganisms, often used to increase our understanding of pollutant biodegradation processes, can also be adapted to develop novel tools for environmental site management (Interstate Technology and Regulatory Council Environmental Molecular Diagnostics 2011).

The MBTs target biomolecules of microbial populations that participate in pollutant biodegradation processes, such as nucleic acids, proteins, or lipids (Table 2.1). Each of these biomolecules will provide a different type of information, such as revealing the presence of microbial populations with degrading potential [e.g., polymerase chain reaction (PCR), fingerprinting methods such as denaturing gradient gel electrophoresis (DGGE), terminal restriction fragment length polymorphism (tRFLP), clone library construction and sequencing], assessing the abundance of key microorganisms, and monitoring population growth [quantitative PCR (qPCR), next-generation sequencing-based fingerprinting such as pyrotags or I-tags, microarrays], providing evidence of pollutant biodegradation (SIP), or estimating the activity of specific microbial populations (RT-qPCR). As the microorganisms involved in the

Table 2.1 Molecular biological tools, the biological molecules on which they are based, and the type of information they provide

	Molecule			
Information	DNA	RNA	Protein	Substrate
Presence	PCR, clone libraries, fingerprinting	FISH		
Abundance	qPCR, NGS[a]-fingerprinting, microarrays	FISH		
Activity		RT-qPCR, CARD-FISH	Enzymatic activity	
Degradation process				SIP

[a]Next-generation sequencing
[b]Stable isotope probing

biodegradation process could vary among sites, even for the same type of environment (Lozada et al. 2014b), an in-depth knowledge of the system is required for the development of molecular diagnostic tools that can provide accurate information in bioremediation applications. In spite of the challenges for the development and implementation of MBTs, these tools are increasingly being used in multiple cleanup sites in various countries (Interstate Technology and Regulatory Council Environmental Molecular Diagnostics 2013), particularly when a limited number of pollutants and microorganism types are involved, such as the anaerobic biodegradation of chlorinated solvents (Lebron et al. 2011). The biodegradation of hydrocarbon pollutants in coastal environments represents one of the most challenging scenarios. It involves the presence of multiple chemical structures, varying with the contamination source, as well as dynamic environmental conditions that influence the structure and function of the involved microbial communities.

2.2 Biodegradation of Hydrocarbons in Marine Environments

Although hydrocarbons are energy-rich compounds, their use requires the activation of a highly stable molecule. The most energetically efficient way to capture this energy is by its oxidation with molecular oxygen (Austin and Callaghan 2013). Aerobic hydrocarbon degradation initiates via enzymatic complexes, which include a terminal oxygenase and accompanying enzymes forming an electron transport chain (Peng et al. 2008; Rojo 2009). Marine microorganisms have been exposed to hydrocarbons during millions of years, such as in natural oil seeps. As a consequence, the capability of hydrocarbon biodegradation has been acquired rather frequently through evolution, and hydrocarbon-degrading microorganisms are widespread in nature (Prince et al. 2010). In the marine environment, important bacterial groups participating in hydrocarbon biodegradation are members of the Gammaproteobacteria (e.g., *Pseudomonas*, *Alteromonas*, *Neptunomonas*, *Marinobacter*, *Alcanivorax*, *Cycloclasticus*) (Vila et al. 2015), the Alphaproteobacteria (the sphingomonads) (Kertesz and Kawasaki 2010, the Roseobacter clade (Kim and Kwon 2010), and the Actinobacteria (*Rhodococcus*, *Mycobacterium*, *Nocardioides*) (Vila et al. 2015). Of special interest is a group of marine microorganisms that have specialized exclusively in the utilization of hydrocarbons, or almost exclusively, named obligate hydrocarbonoclastic bacteria (OHCB): these are represented mainly by members of the Gammaproteobacteria class (Yakimov et al. 2007). Examples are *Cycloclasticus*, which degrades polycyclic aromatic hydrocarbons, and *Alcanivorax* and *Oleispira*, which feed on alkanes. Oil-degrading microorganisms can be a small proportion of the bacterial community. However, when these compounds are available in the environment, these populations grow at their expense increasing their number, which allows their monitoring through space and time, if appropriate tools are used (Head et al. 2006). Moreover, these populations can be stimulated to increase biodegradation rates in the natural environment, which are often too slow to prevent damage to vulnerable ecosystems and human health (Lozada et al. 2014a; Ron and Rosenberg 2014).

When released to the marine environment, oil suffers a number of abiotic and biotic processes that are collectively called *weathering*, and include spreading, evaporation, photooxidation, emulsification, dissolution, sedimentation, adsorption, and biodegradation (McGenity et al. 2012). Among these processes, biodegradation is a major mechanism of hydrocarbon removal from the marine environment. Hydrocarbons vary in their biodegradability, as a result of differences in their physicochemical properties such as their solubility, hydrophobicity, and capacity to adsorb to matrices, which affect their bioavailability for microorganisms (McGenity et al. 2012). Bacteria have evolved strategies for efficient uptake of hydrocarbons, such as the modification of cell membranes and the release of biosurfactants (Harms et al. 2010). In addition, the presence of nitrogen and phosphorus nutrients in sufficient amounts is a key factor affecting the rate of biodegradation in natural conditions (Ron and Rosenberg 2014). Oxygen is another key factor driving this process. In contrast to seawater, which is mainly aerobic, sediments are aerobic only on their surface and become rapidly anoxic with depth, giving rise to steep redox gradients. Therefore, aerobic biodegradation processes only occur in the sediment surface, while in deeper layers anaerobic processes are predominant, coupled to nitrate, iron, and/or sulfate reduction (Acosta-González et al. 2013). The latter are less understood at the molecular level than aerobic processes, specially for PAHs (Estelmann et al. 2015). In fact, sediments constitute a complex matrix in which diverse and spatially structured microbial communities take part in biotic and abiotic interactions that ultimately affect the biodegradation process (Cravo-Laureau and Duran 2014).

Although individual hydrocarbon-degrading strains typically exhibit the ability to degrade only a limited number of hydrocarbons, a natural microbial community can display an important biodegradation potential through synthrophy (cross-feeding), making these communities a suitable target for the depuration of the complex mixture present in oil (McGenity et al. 2012). Moreover, in coastal sediments cells are more concentrated than in open waters, increasing the probability and complexity of their interactions (McGenity et al. 2012). The use of culture-independent approaches targeting phylogenetic (e.g., 16S rRNA gene) and/or functional (e.g., hydrocarbon-activating oxygenases) marker genes, has allowed the characterization of hydrocarbon-degrading communities in various marine environments. The catastrophic spill resulting from the *Deepwater Horizon* blowout in 2010 provided a unsought opportunity for the scientific community to develop multidisciplinary analysis tools, including chemical and environmental analyses coupled to targeted-gene surveys, metagenomics, and meta transcriptomics. This development has resulted in unprecedented knowledge regarding the response of marine microbial communities to oil input and their potential for remediation (Kimes et al. 2014). Future perspectives involve systems biology, which includes multiple-level assessment of microbial communities and the environmental factors controlling mass fluxes across its members (Roling and van Bodegom 2014). This knowledge will aid in gaining predictability on these systems and advance into the application of knowledge-based bioremediation tools (de Lorenzo 2008).

2.3 Hydrocarbon-Degrading Bacteria on the Patagonian Coast

A series of studies have shown the presence of hydrocarbon pollution in several sites along the Patagonian coast, as a result of oil extraction and transportation activities, as well as port and vessel operations (Commendatore et al. 2000, 2012). Increasing our understanding of hydrocarbon biodegradation processes in the coastal environments of Patagonia, and in particular the identification of the microbial populations that are key for these processes, is essential for the design of MBTs specific for this region. A series of culture-independent approaches were used to characterize hydrocarbon-degrading bacteria in sediments from coastal environments of Patagonia (Lozada et al. 2008, 2014b; Marcos et al. 2009, 2012; Dionisi et al. 2011; Guibert et al. 2012, 2016; Loviso et al. 2015). These studies considered both spatial and temporal variations that could occur in these populations. PCR-based approaches targeted both functional and phylogenetic biomarker genes, which revealed the identity of hydrocarbon-degrading bacterial populations, as well as their abundance and biogeographic distribution. In addition, metagenomic approaches allowed the analysis of genome fragments from some of these microorganisms, exposing their particular adaptations to pollutant biodegradation (Loviso et al. 2015; Guibert et al. 2016).

2.3.1 Polycyclic Aromatic Hydrocarbons (PAHs)

PAHs constitute the compounds of most concern among the components of crude oil or petroleum refined products due to their toxic and carcinogenic properties as well as their accumulation and persistence in the environment (Rojo-Nieto and Perales 2015). Various PAHs were identified in intertidal sediments of Patagonia, often exceeding the levels recommended for this matrix (Marcos et al. 2012). The biomarker gene most commonly used for the detection of bacterial populations with the potential to degrade PAHs encodes the large subunit of the terminal component of class A ring-hydroxylating oxygenases (RHOs) (Chakraborty et al. 2012). Using primer sets with different specificities, we amplified fragments of these genes from sediment DNA and cloned these amplification products to construct PCR clone libraries. This analysis revealed a high PAH-degrading potential in the sediments of polluted sites, as 25 distinct RHO α-subunit gene variants (sharing $<80\,\%$ identity at the amino acid level) were detected in the PCR clone libraries (Lozada et al. 2008; Marcos et al. 2009; Dionisi et al. 2011; Loviso et al. 2015). Remarkably, 22 of the gene variants shared low or moderate identity values with previously identified genes. Most of the novel gene variants were only identified in sediments of South Patagonia, where they were found to be very abundant, outnumbering archetypical genes such as *nahAc* and *phnAc* (Marcos et al. 2012). These results suggest a biogeographic distribution of these populations restricted to cold environments, which was confirmed for some of these variants by the use of qPCR analyses (Fig. 2.1). In contrast, one of the gene

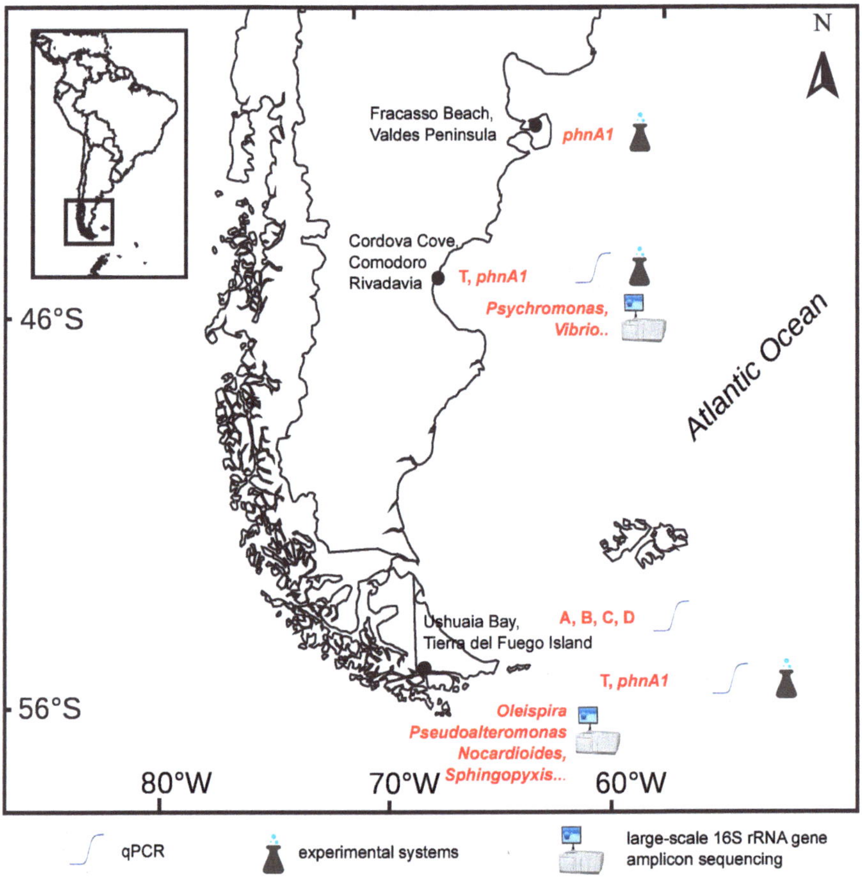

Fig. 2.1 Map of the Patagonian coast showing the distribution of hydrocarbon-degrading microorganisms and related gene variants, as identified using different approaches. *A*, *B*, *C*, *D*, *T* novel RHO α-subunit gene variants quantified by qPCR in the sediments, *phnA1* gene variant described in the obligate PAH-degrader *Cycloclasticus* spp.

variants identified in the sediments, named gene variant T, was found to have a distribution extending at least to North and South Patagonia (Loviso et al. 2015). The relative abundance of this gene was high in chronically polluted sediments, and further increased in experimental systems after PAH exposure in sediments of both regions. Archetypical marine PAH-degrading bacteria belonging to the genus *Cycloclasticus* were also identified in coastal sediments of North and South Patagonia, and their abundance in polluted sediments correlated with low molecular weight PAH concentrations (Marcos et al. 2012). These populations also increased their abundance after PAH exposure in experimental systems (Loviso et al. 2015). Overall, these results suggest that both *Cycloclasticus* and the uncultured populations carrying the gene variant T could be among the key players for PAH biodegradation in the coastal environments of Patagonia.

PCR-based approaches are useful for the recovery of biomarker genes from microbial populations associated to specific metabolic processes. However, as this approach can only recover fragments of the targeted genes, the complete sequence of these genes and their genomic context are lost. Furthermore, it is very difficult to infer the potential host of the identified gene fragments, which precludes linking the structure with the function of the microbial community. Metagenomic libraries, the cloning of fragments of environmental DNA into appropriate vectors, can provide this type of information. We constructed a metagenomic library from chronically polluted sediments retrieved near an oil jetty in Ushuaia Bay, in South Patagonia (Loviso et al. 2015). The screening of the library with a broad specificity primer set previously used to construct PCR clone libraries allowed the identification of a fosmid clone (M117) containing a 37-kb fragment with a gene almost identical to gene variant T. This fragment could only be affiliated to the Proteobacteria phylum, as it is highly divergent from the genome sequences of currently described strains. Besides this gene, this metagenomic fragment contained five additional RHO α-subunit sequences, which indicates the high aromatic hydrocarbon-degrading potential of this microorganism (Loviso et al. 2015). These sequences shared low identity values at the protein level (19.4–42.5 %), and were most closely related to sequences identified in *Cycloclasticus* strains. Five of these genes had a codirectional β-subunit sequence, characteristic of RHOs with a $\alpha_n\beta_n$ hetero-multimeric structure (Chakraborty et al. 2012). Several highly specialized PAH-degrading microorganisms, such as *Cycloclasticus* strains, carry multiple genes coding for α- and β-subunits of RHOs (Lai et al. 2012; Cui et al. 2013). The analysis of the phylogenetic relationship of the identified metagenomic sequences with α-subunit sequences identified in complete genomes from three *Cycloclasticus* strains (*Cycloclasticus* sp. P1, *Cycloclasticus pugetii* PS-1, and *Cycloclasticus zancles* 7-ME) showed that the metagenomic sequences were clearly divergent from dioxygenases from these highly specialized organisms, also identified in coastal sediments of Patagonia (Fig. 2.2). Three of the oxygenases contained in the metagenomic fragment were classified as class A RHOs, which include enzymes that catalyze the dioxygenation of PAHs (Loviso et al. 2015). The modeling of the three-dimensional structures of these enzymes suggested that this microorganism could degrade high molecular weight PAHs, as two of these modeled proteins presented a catalytic pocket able to accommodate large PAH molecules. In particular, the catalytic cavity dimensions of one of the proteins (modeled from sequence M117-38; Fig. 2.2) were comparable with the ones from NidAB from *Mycobacterium vanbaalenii* PYR-1, which presents pyrene as the preferential substrate (Kweon et al. 2010).

Overall, these analyses showed the presence of a large diversity of microorganisms with PAH-degrading potential in chronically polluted sediments of Patagonia, and allowed the identification of genes from key bacterial populations that could be used as targets for field assessment. In particular, two of the designed qPCR assays, targeting gene variants T (from an uncultured proteobacterium: Loviso et al. 2015) and *phnA1* gene (from *Cycloclasticus* spp.: Marcos et al. 2012) are promising MBTs that could next be validated in a large-scale field study.

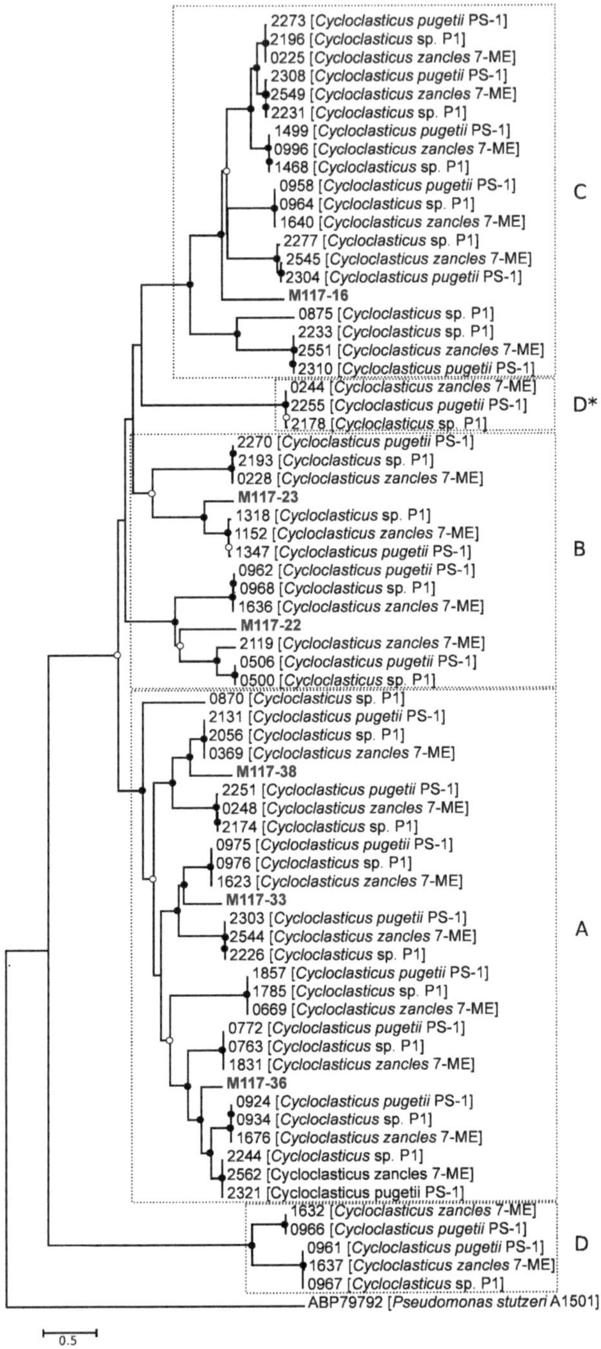
2273 [Cycloclasticus pugetii PS-1]
2196 [Cycloclasticus sp. P1]
0225 [Cycloclasticus zancles 7-ME]
2308 [Cycloclasticus pugetii PS-1]
2549 [Cycloclasticus zancles 7-ME]
2231 [Cycloclasticus sp. P1]
1499 [Cycloclasticus pugetii PS-1]
0996 [Cycloclasticus zancles 7-ME]
1468 [Cycloclasticus sp. P1]
0958 [Cycloclasticus pugetii PS-1]
0964 [Cycloclasticus sp. P1]
1640 [Cycloclasticus zancles 7-ME]
2277 [Cycloclasticus sp. P1]
2545 [Cycloclasticus zancles 7-ME]
2304 [Cycloclasticus pugetii PS-1]
M117-16
0875 [Cycloclasticus sp. P1]
2233 [Cycloclasticus sp. P1]
2551 [Cycloclasticus zancles 7-ME]
2310 [Cycloclasticus pugetii PS-1]
C
0244 [Cycloclasticus zancles 7-ME]
2255 [Cycloclasticus pugetii PS-1]
2178 [Cycloclasticus sp. P1]
D*
2270 [Cycloclasticus pugetii PS-1]
2193 [Cycloclasticus sp. P1]
0228 [Cycloclasticus zancles 7-ME]
M117-23
1318 [Cycloclasticus sp. P1]
1152 [Cycloclasticus zancles 7-ME]
1347 [Cycloclasticus pugetii PS-1]
0962 [Cycloclasticus pugetii PS-1]
0968 [Cycloclasticus sp. P1]
1636 [Cycloclasticus zancles 7-ME]
M117-22
2119 [Cycloclasticus zancles 7-ME]
0506 [Cycloclasticus pugetii PS-1]
0500 [Cycloclasticus sp. P1]
B
0870 [Cycloclasticus sp. P1]
2131 [Cycloclasticus pugetii PS-1]
2056 [Cycloclasticus sp. P1]
0369 [Cycloclasticus zancles 7-ME]
M117-38
2251 [Cycloclasticus pugetii PS-1]
0248 [Cycloclasticus zancles 7-ME]
2174 [Cycloclasticus sp. P1]
0975 [Cycloclasticus pugetii PS-1]
0976 [Cycloclasticus sp. P1]
1623 [Cycloclasticus zancles 7-ME]
M117-33
2303 [Cycloclasticus pugetii PS-1]
2544 [Cycloclasticus zancles 7-ME]
2226 [Cycloclasticus sp. P1]
1857 [Cycloclasticus pugetii PS-1]
1785 [Cycloclasticus sp. P1]
0669 [Cycloclasticus zancles 7-ME]
0772 [Cycloclasticus pugetii PS-1]
0763 [Cycloclasticus sp. P1]
1831 [Cycloclasticus zancles 7-ME]
M117-36
0924 [Cycloclasticus pugetii PS-1]
0934 [Cycloclasticus sp. P1]
1676 [Cycloclasticus zancles 7-ME]
2244 [Cycloclasticus sp. P1]
2562 [Cycloclasticus zancles 7-ME]
2321 [Cycloclasticus pugetii PS-1]
A
1632 [Cycloclasticus zancles 7-ME]
0966 [Cycloclasticus pugetii PS-1]
0961 [Cycloclasticus pugetii PS-1]
1637 [Cycloclasticus zancles 7-ME]
0967 [Cycloclasticus sp. P1]
D
ABP79792 [Pseudomonas stutzeri A1501]
0.5

2.3.2 Aliphatic Hydrocarbons

Although not as toxic as aromatic hydrocarbons, alkanes are a major component of crude oil (Head et al. 2006). Their low water solubility and poor reactivity, and their tendency to accumulate on sediments, constitute a challenge for their effective removal (Rojo 2009). Various microbial strategies, however, have evolved to access these compounds, even the longer molecules that are solid at room temperature (paraffins) (Wentzel et al. 2007). In addition to genes coding PAH-degrading enzymes, we have analyzed the functional biomarker gene *alkB*, which codes for the initial oxygenase of aliphatic hydrocarbons (alkane-1-monooxygenase). AlkB is an integral-membrane non-heme di-iron monooxygenase, which hydroxylates the alkane molecule in the terminal position. It requires two electron transport proteins, a rubredoxin and a rubredoxin reductase. AlkB family enzymes are highly extended among oil-degrading bacteria (van Beilen and Funhoff 2007). Using a PCR-based method with a broad-specificity primer set targeting conserved regions of the gene, we could uncover a remarkable diversity of AlkB sequences in South Patagonian sediments (Guibert et al. 2012). The majority of the sequences were found to be related to uncultured microorganisms from cold marine sediments or soils from high-latitude regions, suggesting a major role of temperature in the selection of bacterial populations with this capability. The analysis of experimental systems constructed from these sediments and amended with crude oil showed the specific enrichment of various *alkB* variants, related to genes described in members of the Gammaproteobacteria and Actinobacteria. The majority of these variants were also detected in the corresponding environmental samples, highlighting their ecological relevance (Guibert et al. 2012). A complementary approach involving large-scale sequencing of 16S rRNA gene amplicons also allowed the identification of various genera that could not be targeted by the functional gene approach. For example, the obligate oil-degrading genus *Oleispira*, probably psychrophilic *Oleispira antarctica*, was detected in South Patagonian sediments by this method (Guibert et al. 2012). More recently, the analysis of a metagenomic shotgun sequencing dataset involving more than 6000 putative AlkB sequences from subtidal sediments from this site evidenced that AlkB diversity could be more than one order of magnitude higher than estimated by PCR-based methods (Guibert et al. 2016). The AlkB sequences identified by metagenomics exhibited high phylogenetic diversity,

Fig. 2.2 Phylogenetic tree of RHO α-subunit sequences identified in fosmid M117 and in *Cycloclasticus* spp. genomes. The neighbor-joining tree includes metagenomic (in *gray*) and genomic deduced amino acid sequences. Sequence number within each genome (NCBI numbering) and strain name (in *brackets*) is indicated in each case. RHO classification according to the scheme proposed by Chakraborty et al. (2012) is indicated on the *right*. The phylogenetic tree was built with Mega 6 (Tamura et al. 2013) using the Jones-Taylor-Thornton (JTT) substitution model. Bootstrap values were calculated as percentage of 1000 repetitions, with only values ≥50 % indicated in the figure (*black circles*, >75 %; *white circles*, 50–75 %). *Bar* represents inferred amino acid changes per position

spanning the whole AlkB phylogenetic tree known up to date. Furthermore, completely novel sequences only moderately related to putative AlkBs from genomes of Bacteroidetes and *Alphaproteobacteria* were found in high abundance. Not all these enzymes have been characterized, and to date, the role of these bacterial groups in aliphatic hydrocarbon biodegradation has been largely underestimated. In the same work, the previously mentioned metagenomic library was also analyzed to search for *alkB* genes using a molecular approach. This analysis rendered two fosmid clones containing genomic fragments from uncultured bacteria belonging to the phylum Planctomycetes, allowing the description for the first time of alkane-degrading potential in members of this group (Guibert et al. 2016). These results highlight the power of metagenomic approaches for uncovering new potential features of microbial communities, as they are not as dependent on previous knowledge as the PCR-based methods. However, it must be noted that this approach still presents challenges related to functional annotation of metagenomic sequences as a result of the still-remaining knowledge gaps (Temperton and Giovannoni 2012).

2.3.3 Ecological Index of Hydrocarbon Exposure

When we identify microorganisms based on a phylogenetically informative gene (e.g., 16S rRNA gene), we normally infer their metabolic capabilities by searching the characteristics reported for the genus or species in the scientific literature. This type of inference is performed either advertently or inadvertently. Moreover, when we analyze the taxonomic composition of a microbial community, we draw conclusions about its metabolic capabilities and the potential processes occurring in the analyzed environment. For example, the presence of genera known to harbor sulfate-reducing strains is evidence that this process is probably occurring. An example of this inference carried out in a systematic manner is the *picrust* program, a software that predicts the metagenome of a microbial community based on the information obtained with the phylogenetic marker gene 16S rRNA (Langille et al. 2013). As it relies upon the existing information of microbial genomes for the prediction, the method is specially effective when the microbial community contains genera for which many sequenced genomes are available (e.g., the human microbiome) (Langille et al. 2013).

In the marine environment, hydrocarbonoclastic bacteria such as *Alcanivorax* were detected through their phylogenetic marker genes, allowing the inference of hydrocarbon-degrading processes following a pollution event (Kostka et al. 2011). This detection was possible because of the narrow substrate preferences of these microorganisms, which allowed the rapid linking of phylogenetic information to function. Interestingly, there are a number of well-described genera, which can be easily detected with community-level approaches, known to carry out biodegradation of hydrocarbons in the marine environment, allowing the monitoring of the whole process at the microbiological level (Dubinsky et al. 2013). Based on the fact that, in theory, these genera would normally be present at low abundances but that will increase with pollution if optimal conditions are met (Head et al. 2006), they can constitute good indicators of the environmental state of a certain site. We therefore developed an ecological indicator, the Ecological Index of Hydrocarbon Exposure (EIHE), which is defined as the sum of the relative abundances (proportion with respect to total community members) of potential hydrocarbon-degrading genera in a certain sample (Lozada et al. 2014b). This index involves an arbitrary list of genera harboring hydrocarbon-degrading strains normally encountered associated with biodegradation in the marine environment (Lozada et al. 2014b). A high EIHE value would be evidence of a previous exposure to pollutants, whereas a low value would be typical of a non-impacted site. Moreover, the variation of the index could be followed in polluted sites to monitor the growth of the populations in the presence of environmental constraints, for example, nutrient limitations. The estimation of the abundances should be ideally performed by high coverage techniques such as large-scale sequencing of 16S rRNA gene amplicons, which allow the rapid identification of a number of genera in relatively low abundance in the context of a diverse community, a condition that is found, for example, in sediments (Lozada et al. 2014b). The taxonomic information derived from the bioinformatic analysis of sequences is extracted automatically and processed to calculate the EIHE (Lozada et al. 2014b) (Fig. 2.3). A strength of this approach is that different populations of degrading bacteria arising in different environments or conditions can be detected, thus contributing to the index value (Fig. 2.3). This indicator has the potential to be utilized as a MBT, as it relies on standardized procedures during the whole process and has the possibility of high-throughput analysis of various samples.

The preliminary evaluation of this tool was performed in samples from the Patagonian coast, as well as in public sequence datasets from other oil-exposed marine samples, including sediments and seawater from experimental systems and field studies. In all cases, the EIHE was significantly higher in oiled than in unpolluted samples (Lozada et al. 2014b). Moreover, in sediment samples from acute pollution events such as the *Deepwater Horizon* spill, the EIHE index reached a value of approximately 30, which means that 30 % of the bacterial community was composed of potential hydrocarbon degraders; for a nearly non-impacted site, values remained as low as 2 (Lozada et al. 2014b). Our results suggest that this ecological indicator could be a promising tool for environmental diagnostics in marine systems. The next steps involve its assessment at field scale, in samples from the Patagonian coast subjected to chronic oil pollution.

2.4 Perspectives on the Use of MBTs in the Patagonian Coastal Region

The identified functional biomarker genes probably do not represent the complete diversity of hydrocarbon-degrading bacteria present in this region, as methodological challenges still limit the access to all the microorganisms present in an environmental sample. Through the use of multiple approaches, we identified potential key

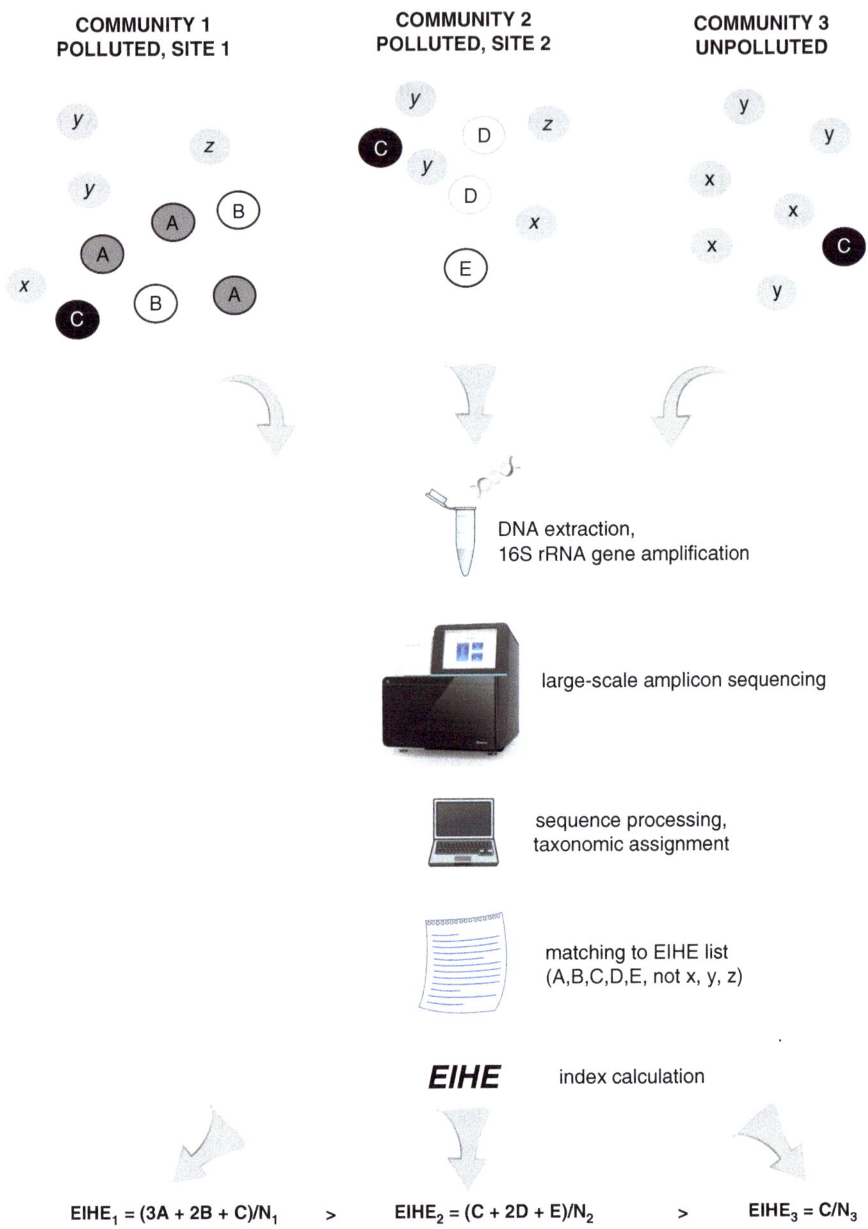

Fig. 2.3 Schematic representation of the Ecological Index of Hydrocarbon Exposure (EIHE) index concept and method. Three different samples are analyzed: two from different polluted sites, and an unpolluted sample. In the samples, the microbial communities are composed of hydrocarbon degraders (*A* to *E*) and non-degraders (*x* to *z*), of different types and abundances according to the history and environmental state of the site. The DNA is extracted, a hypervariable region of the phylogenetic marker gene (16S rRNA) is amplified with conserved primers and subjected to large-scale sequencing, and sequences are processed through bioinformatic methods, and taxonomically assigned to the lowest possible level relying on public databases for this gene. The taxonomic assignment of each sample is matched against the "EIHE" list to extract the abundances of potential hydrocarbon-degrading genera. The abundances are summed and expressed as proportion of the total bacterial community (total sequences, *N*), which is the EIHE value. Note that the individual genera contributing to EIHE may vary in different polluted samples

members of hydrocarbon biodegradation processes, which allowed the design of MBTs able to detect these microbial populations, estimate their abundance, and evaluate changes as a result of pollution events or the use of bioremediation technologies. The next steps are the validation of these tools and the development of standardized guidelines for their application in polluted sites. We envision the use of these tools to generate baseline information on hydrocarbon exposure in marine environments before offshore oil exploitation, as well as to evaluate hydrocarbon-degrading potential and the dynamics of key bacterial populations during natural attenuation of chronically polluted sites or after an accidental spill. There is also a need to increase the awareness in stakeholders such as government agencies and companies operating ports and oil terminals on the Patagonian coast on the availability of these tools. Many highly vulnerable environments can be threatened with an oil spill similar to that which occurred in Cordova Cove in December 2008, which affected 4 km of the coastline of an inlet used for recreation and for artisanal fisheries. Although enhanced bioremediation is not an alternative currently considered in these accidents, a more active evaluation of natural biodegradation processes and eventually the implementation of these technologies are necessary to limit the damage to human populations and environmental health in coastal environments of Patagonia.

Acknowledgments M.L. and H.D. are staff members of the Argentinean National Research Council (CONICET). The work at the Environmental Microbiology Laboratory (Laboratorio de Microbiología Ambiental, CESIMAR, CONICET) has been supported by grants from CONICET, the National Agency for Science and Technology Promotion of Argentina (ANPCYT), The Secretary of Science, Technology and Innovation from the Chubut Province, Argentina, and the Community Sequencing Program, Joint Genome Institute (JGI-DOE, USA).

References

Acosta-González A, Rosselló-Móra R, Marqués S (2013) Characterization of the anaerobic microbial community in oil-polluted subtidal sediments: aromatic biodegradation potential after the Prestige oil spill. Environ Microbiol 15:77–92

Austin RN, Callaghan AV (2013) Microbial enzymes that oxidize hydrocarbons. Front Microbiol 4:338

Chakraborty J, Ghosal D, Dutta A, Dutta TK (2012) An insight into the origin and functional evolution of bacterial aromatic ring-hydroxylating oxygenases. J Biomol Struct Dyn 30:419–436

Commendatore MG, Esteves JL, Colombo JC (2000) Hydrocarbons in coastal sediments of Patagonia, Argentina: levels and probable sources. Mar Pollut Bull 40:989–998

Commendatore MG, Nievas ML, Amin O, Esteves JL (2012) Sources and distribution of aliphatic and polyaromatic hydrocarbons in coastal sediments from the Ushuaia Bay (Tierra del Fuego, Patagonia, Argentina). Mar Environ Res 74:20–31

Cravo-Laureau C, Duran R (2014) Marine coastal sediments microbial hydrocarbon degradation processes: contribution of experimental ecology in the ‘omics’ era. Front Microbiol 5:39

Cui Z, Xu G, Li Q, Gao W, Zheng L (2013) Genome sequence of the pyrene-and fluoranthene-degrading bacterium *Cycloclasticus* sp. strain PY97M. Genome Announc 1:e00536-13

de Lorenzo V (2008) Systems biology approaches to bioremediation. Curr Opin Biotechnol 19:579–589

Dionisi HM, Lozada M, Marcos MS, Di Marzio WD, Loviso CL (2011) Aromatic hydrocarbon degradation genes from chronically polluted subantarctic marine sediments. In: de Bruijn FJ (ed) Handbook of molecular microbial ecology. II: Metagenomics in different habitats. Wiley, Hoboken, NJ, pp 461–473

Dubinsky EA, Conrad ME, Chakraborty R, Bill M, Borglin SE, Hollibaugh JT, Mason OU, Piceno YM, Reid FC, Stringfellow WT, Tom LM, Hazen TC, Andersen GL (2013) Succession of hydrocarbon-degrading bacteria in the aftermath of the *Deepwater Horizon* oil spill in the Gulf of Mexico. Environ Sci Technol 47:10860–10867

Estelmann S, Blank I, Feldmann A, Boll M (2015) Two distinct old yellow enzymes are involved in naphthyl ring reduction during anaerobic naphthalene degradation. Mol Microbiol 95:162–172

Guibert LM, Loviso CL, Marcos MS, Commendatore MG, Dionisi HM, Lozada M (2012) Alkane biodegradation genes from chronically polluted subantarctic coastal sediments and their shifts in response to oil exposure. Microb Ecol 64:605–616

Guibert L, Loviso C, Borglin S, Jansson J, Dionisi H, Lozada M (2016) Diverse bacterial groups contribute to the alkane degradation potential of chronically polluted subantarctic coastal sediments. Microb Ecol 71:100–112

Harms H, Smith KEC, Wick LY (2010) Microorganism–hydrophobic compound interactions. In: Timmis KN (ed) Handbook of hydrocarbon and lipid microbiology. Springer, Berlin, pp 1479–1490

Head IM, Jones DM, Röling WFM (2006) Marine microorganisms make a meal of oil. Nat Rev Microbiol 4:173–182

Interstate Technology and Regulatory Council (ITRC) Environmental Molecular Diagnostics (EMD) (2011) Environmental Molecular Diagnostics Fact Sheets

Interstate Technology and Regulatory Council (ITRC) Environmental Molecular Diagnostics (EMD) (2013) Technical and Regulatory Guidance

Jeon CO, Madsen EL (2013) *In situ* microbial metabolism of aromatic-hydrocarbon environmental pollutants. Curr Opin Biotechnol 24:474–481

Kertesz M, Kawasaki A (2010) Hydrocarbon-degrading sphingomonads: *Sphingomonas*, *Sphingobium*, *Novosphingobium*, and *Sphingopyxis*. In: Timmis K (ed) Handbook of hydrocarbon and lipid microbiology. Springer, Berlin, pp 1694–1705

Kim SJ, Kwon KK (2010) Marine, hydrocarbon-degrading Alphaproteobacteria. In: Timmis KN (ed) Handbook of hydrocarbon and lipid microbiology. Springer, Berlin, pp 1707–1714

Kimes NE, Callaghan AV, Suflita JM, Morris PJ (2014) Microbial transformation of the Deepwater Horizon oil spill: past, present, and future perspectives. Front Microbiol 5:603

Kostka JE, Prakash O, Overholt WA, Green SJ, Freyer G, Canion A, Delgardio J, Norton N, Hazen TC, Huettel M (2011) Hydrocarbon-degrading bacteria and the bacterial community response in Gulf of Mexico beach sands impacted by the Deepwater Horizon oil spill. Appl Environ Microbiol 77:7962–7974

Kweon O, Kim S-J, Freeman JP, Song J, Baek S, Cerniglia CE (2010) Substrate specificity and structural characteristics of the novel rieske nonheme iron aromatic ring-hydroxylating oxygenases NidAB and NidA3B3 from *Mycobacterium vanbaalenii* PYR-1. Mol Microbiol 1:e00135-10

Lai Q, Li W, Wang B, Yu Z, Shao Z (2012) Complete genome sequence of the pyrene-degrading bacterium *Cycloclasticus* sp. strain P1. J Bacteriol 194:6677

Langille MGI, Zaneveld J, Caporaso JG, McDonald D, Knights D, Reyes JA, Clemente JC, Burkepile DE, Vega Thurber RL, Knight R, Beiko RG, Huttenhower C (2013) Predictive functional profiling of microbial communities using 16S rRNA marker gene sequences. Nat Biotechnol 31:814–821

Lebron C, Petrovskis E, Loffler F, Henn K (2011) Application of nucleic acid-based tools for monitoring monitored natural attenuation (MNA), biostimulation and bioaugmentation at chlorinated solvent sites. DTIC Document, Port Hueneme, California

Loviso CL, Lozada M, Guibert LM, Musumeci MA, Sarango Cardenas S, Kuin RV, Marcos MS, Dionisi HM (2015) Metagenomics reveals the high polycyclic aromatic hydrocarbon-degradation potential of abundant uncultured bacteria from chronically polluted subantarctic and temperate coastal marine environments. J Appl Microbiol 119:411–424

Lozada M, Mercadal JPR, Guerrero LD, Di Marzio WD, Ferrero MA, Dionisi HM (2008) Novel aromatic ring-hydroxylating dioxygenase genes from coastal marine sediments of Patagonia. BMC Microbiol 8:50

Lozada M, Marcos M, Dionisi H (2014a) La Biorremediación de ambientes costeros contaminados con hidrocarburos. Fondo Editorial Provincial, Provinica del Chubut, Argentina

Lozada M, Marcos MS, Commendatore MG, Gil MN, Dionisi HM (2014b) The bacterial community structure of hydrocarbon-polluted marine environments as the basis for the definition of an ecological index of hydrocarbon exposure. Microbes Environ 29:269–276

Marcos MS, Lozada M, Dionisi HM (2009) Aromatic hydrocarbon degradation genes from chronically polluted subantarctic marine sediments. Lett Appl Microbiol 49:602–608

Marcos MS, Lozada M, Di Marzio WD, Dionisi HM (2012) Abundance, dynamics, and biogeographic distribution of seven polycyclic aromatic hydrocarbon dioxygenase gene variants in coastal sediments of Patagonia. Appl Environ Microbiol 78:1589–1592

McGenity TJ, Folwell BD, McKew BA, Sanni GO (2012) Marine crude-oil biodegradation: a central role for interspecies interactions. Aquat Biosyst 8:1–19

Peng R-H, Xiong A-S, Xue Y, Fu X-Y, Gao F, Zhao W, Tian Y-S, Yao Q-H (2008) Microbial biodegradation of polyaromatic hydrocarbons. FEMS Microbiol Rev 32:927–955

Prince RC, Gramain A, McGenity TJ (2010) Prokaryotic hydrocarbon degraders. In: Timmis KN (ed) Handbook of hydrocarbon and lipid microbiology. Springer, Berlin, pp 1671–1692

Rojo F (2009) Degradation of alkanes by bacteria. Environ Microbiol 11:2477–2490

Rojo-Nieto E, Perales JA (2015) Estimating baseline toxicity of PAHs from marine chronically polluted sediments and bioaccumulation in target organs of fish hypothetically exposed to them: a new tool in risk assessment. Environ Sci Process Impacts 17:1331–1339

Roling WFM, van Bodegom PM (2014) Toward quantitative understanding on microbial community structure and functioning: a modeling-centered approach using degradation of marine oil spills as example. Front Microbiol 5:125

Ron EZ, Rosenberg E (2014) Enhanced bioremediation of oil spills in the sea. Curr Opin Biotechnol 27:191–194

Tamura K, Stecher G, Peterson D, Filipski A, Kumar S (2013) MEGA6: molecular evolutionary genetics analysis version 6.0. Mol Biol Evol 30:2725–2729

Temperton B, Giovannoni SJ (2012) Metagenomics: microbial diversity through a scratched lens. Curr Opin Microbiol 15:605–612

van Beilen JB, Funhoff EG (2007) Alkane hydroxylases involved in microbial alkane degradation. Appl Microbiol Biotechnol 74:13–21

Vila J, Tauler M, Grifoll M (2015) Bacterial PAH degradation in marine and terrestrial habitats. Curr Opin Biotechnol 33:95–102

Wentzel A, Ellingsen TE, Kotlar H-K, Zotchev SB, Throne-Holst M (2007) Bacterial metabolism of long-chain n-alkanes. Appl Microbiol Biotechnol 76:1209–1221

Yakimov MM, Timmis KN, Golyshin PN (2007) Obligate oil-degrading marine bacteria. Curr Opin Biotechnol 18:257–266

Chapter 3
Indigenous PAH-Degrading Bacteria in Oil-Polluted Marine Sediments from Patagonia: Diversity and Biotechnological Properties

Paula Isaac, Natalia Bourguignon, Daniela Maizel, and Marcela A. Ferrero

Abstract The diversity of indigenous bacteria recovered from polluted sediment samples from several coasts of Patagonia is described in this chapter. Selective enrichment cultures supplemented with naphthalene, phenanthrene, and pyrene allowed isolation of bacteria with the capability to degrade polycyclic aromatic hydrocarbons (PAHs). Bacterial communities of different composition (analyzed by denaturing gradient gel electrophoresis, DGGE) showed changes along with enrichment culture conditions. The ability of isolates to grow and remove different low and high molecular weight PAHs was demonstrated by detection of the residual substrate by HPLC. The presence and differential expression of naphthalene and catechol dioxygenase genes in several isolates suggest biodegradation potential in these sediments. Successful bacterial isolation with the ability to degrade PAH in pure and mixed cultures allows discussing the possibility to study and to further consider strategies to increase the intrinsic bioremediation opportunities on polluted coasts of Patagonia. Other biotechnological properties are also considered in this chapter, such as biosurfactant production and other biotransformations.

P. Isaac
Centro de Investigación y Transferencia - Villa María (CIT-UNVM) CONICET,
Arturo Jauretche 1555, Villa María 5900, Córdoba, Argentina

N. Bourguignon • D. Maizel
Planta Piloto de Procesos Industriales Microbiológicos (PROIMI-CONICET),
Av. Belgrano y Pje. Caseros, San Miguel de Tucumán T4001MVB, Tucumán, Argentina

M.A. Ferrero (✉)
Planta Piloto de Procesos Industriales Microbiológicos (PROIMI-CONICET),
Av. Belgrano y Pje. Caseros, San Miguel de Tucumán T4001MVB, Tucumán, Argentina

Facultad de Bioquímica, Química y Farmacia, Universidad Nacional de Tucumán (UNT),
Ayacucho 471, San Miguel de Tucumán T4001MVB, Tucumán, Argentina
e-mail: mferrero@proimi.org.ar

N.L. Olivera et al. (eds.), *Biology and Biotechnology of Patagonian Microorganisms*, DOI 10.1007/978-3-319-42801-7_3

3.1 Diversity of Indigenous PAH-Degrading Bacteria and Their Biodegradation Capabilities

Extending south of 40°S latitude, the Patagonian coast holds an exceptional biodiversity, sustaining one of the most productive marine ecosystems (Olson and Dinerstein 2002). However, the occurrence of coastal or marine pollution is a permanent risk because of daily oil transportation. Anthropogenic hydrocarbons have been detected in sediments at several locations along the Patagonian coast (Commendatore et al. 2000; Esteves et al. 2006), and high levels of polycyclic aromatic hydrocarbons (PAHs) were found in marine mammals, mainly as the result of intensive oil exploitation and transportation (Barragán Muñoz et al. 2003). The characterization of indigenous hydrocarbon-degrading microbial populations is therefore necessary for a better understanding of natural biodegradation processes in this vulnerable ecosystem and for the successful application of bioremediation technologies (Lozada et al. 2008).

Surficial intertidal sediment samples were collected at 12 different locations along the coastline of Patagonia, Argentina. Seven different intertidal marine sediments along the Patagonian coast of Argentina were analyzed for their intrinsic capability for degrading polycyclic aromatic hydrocarbons (PAHs). Three of the sampling sites (north to south: MP, RW, and CR) are situated along the eastern coast of Patagonia, at the Chubut Province, next to the Atlantic Ocean. In contrast, the five remaining sites (west to east: CS, EM, OR, and OL) are located on the south coast of the Big Island of Tierra del Fuego, next to the Beagle Channel (Lozada et al. 2008).

PAH concentrations as well as sampling procedures were as previously described (Lozada et al. 2008; Marcos et al. 2012). Enrichment cultures were set up for 15 days in minimal medium supplemented with naphthalene or phenantrene as a sole carbon and energy source using coastal sediment as inoculum (Isaac et al. 2013). The total DNA of each naphthalene- or phenantrene-enriched bacterial population was extracted, and all 14 samples were analyzed by a combination of polymerase chain reaction (PCR) and denaturing gradient gel electrophoresis (DGGE) (Fig. 3.1).

Analyzing the populations based on the rDNA 16S-DGGE band profiles, we noted that they were not clustered according to the seven sites sampled along Patagonia. Instead, two clusters were fixed depending only on the hydrocarbon used as carbon and energy source (phenantrene or naphthalene). Cluster 1 was composed of all the phenantrene-growing populations and only two naphthalene-enriched samples, and cluster 2 included all the other naphthalene-growing populations.

Within cluster 1, we found three informative clades (Fig. 3.1b). Clade I is the most important and contained the phenantrene-growing samples from Chubut (Comodoro Rivadavia, Piedrabuena, Rawson, Caleta Sara). Clade II grouped two Tierra del Fuego samples (Olivia and Orion), and clade III included the two naphthalene-enriched populations of this cluster.

The majority of the bands were reamplified and sequenced to determine the phylogenetic status of the populations. Regarding DGGE sequence analysis, the bacterial community developed in enrichment cultures from different sediments

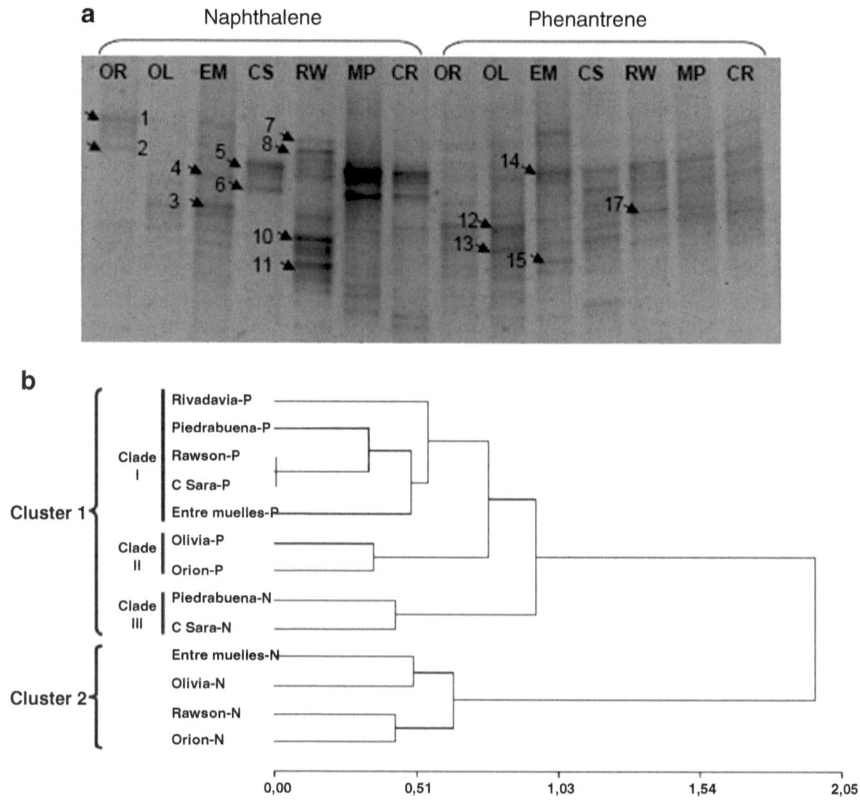

Fig. 3.1 **a** Image of denaturing gradient gel electrophoresis (DGGE) of bacterial community developed in each enrichment culture supplemented with naphthalene or phenanthrene. **b** Ward's dendrogram of rDNA 16S of the populations. *Cophenetic correlation* = 0.856

was mainly composed of *Gammaproteobacteria*. Approximately 25 % of bands recovered resulted in operational taxonomic units (OTUs) with 100 % identity to *Cycloclasticus* sp., and these were found in samples of all the seven sites studied. Even though *Cycloclasticus* is a versatile PAH-degrading bacterium and many sequences were found by culture-independent methods, isolates were not found at these coasts. OTUs related to *Pseudomonas* (bands 1, 2, 7) were found only in populations grouped in cluster 2; sequences related to *Vibrio* and *Shewanella* were found principally in cluster 1. *Photobacterium* (bands 10 and 11) was present in all the naphthalene enrichment cultures, and OTUs related to *Pseudolateromonas* were excised from phenantrene-growing populations (band 17).

Nutrient amounts in the surrounding marine environments, especially those based on nitrogen and phosphorus, are deficient to support some microbial growth requirements, and even more after an oil spill, which is associated with an increase in the hydrocarbon level in water (Harayama et al. 2004). Microbial communities present in PAH-contaminated soils are generally enriched by the microorganisms that are able to use them as a carbon and energy source (Koutny et al. 2003). However, the microbial communities often include marine microorganisms surviving in cell debris or intermediaries in PAH metabolism.

DGGE fingerprinting is useful to analyze the population dynamics in different environments (Kao et al. 2010), because it allows direct determination of changes in band profiles and correlation with changes in the bacteria community (Fromin et al. 2002): these have been reported to be relatively rapid methods of community analysis by comparing fingerprinting profiles (Konopka et al. 1999; Nakatsu et al. 2000). Also, the identification of key organisms in pollutant biodegradation is important to evaluate and develop bioremediation strategies in situ whose effectiveness could be evaluated by DGGE (Harayama et al. 2004; Molina et al. 2009; Ríos et al. 2010; Simarro et al. 2012).

Bacteria populations recovered from selective enrichment cultures supplemented with PAH and those from polluted sediment were monitored by DGGE and their members were identified by sequencing of bands. These bacteria probably were in low proportion in the original sediment and their growth was stimulated by the selective enrichment procedure in the laboratory. This observation represents an important point in this study because indigenous bacteria with catabolic capabilities to degrade PAH could be stimulated in situ to increase the number of cells and thus the degradation activity in the native microbial community living on contaminated Patagonian coasts.

For isolation procedures, *Pseudomonas* and gram-positive Actinobacteria strains were isolated from enrichment cultures from all oil-polluted marine sediments and selected according to their capability to grow with crystals of naphthalene or phenanthrene on solid media. These bacteria can grow on media supplemented with naphthalene, phenanthrene, and pyrene as the sole carbon source under aerobic conditions, indicating that those bacteria have the potential to degrade PAH under those conditions.

Some morphological and physiological features of the isolates allowed their preliminary characterization. Sequence analysis of the 16S rRNA gene of the isolates P26, N3, and N12 allowed determining their relationship with the genus *Pseudomonas* (*P. monteilii* P26, *Pseudomonas* sp. N3, *P. xanthomarina* N12), and P7, P18, H19, F27, HT1A, HT2B, and HT3N showed the closest relationship with the Actinobacteria phylum (*Arthrobacter* sp. P7, *Rhodococcus* sp. P18, *Gordonia* sp. H19, *Rhodococcus* sp. F27, *Arthrobacter* sp. HT1A, *Arthrobacter* sp. HT2B, *Rhodococcus* sp. HT3N) (Isaac et al. 2013, 2015).

Pseudomonas monteilii P26 and *P. xanthomarina* N12 were able to remove 100 % of naphthalene and 65 % and 43 % of phenanthrene, respectively; *Pseudomonas* sp. N3 removed 100 % of both PAHs. Furthermore, although Actinobacteria strains were not able to remove naphthalene and phenanthrene efficiently, some of them were able to remove up to 19 % of pyrene in 21 days. P18, H19, and F27 Actinobacteria strains showed higher pyrene removal from the culture medium, but only if other carbon and energy sources were available. Cometabolic degradation may occur in strains that cannot use PAH as their source of carbon and energy but can effectively degrade them (Zhou et al. 2008) (Fig. 3.2).

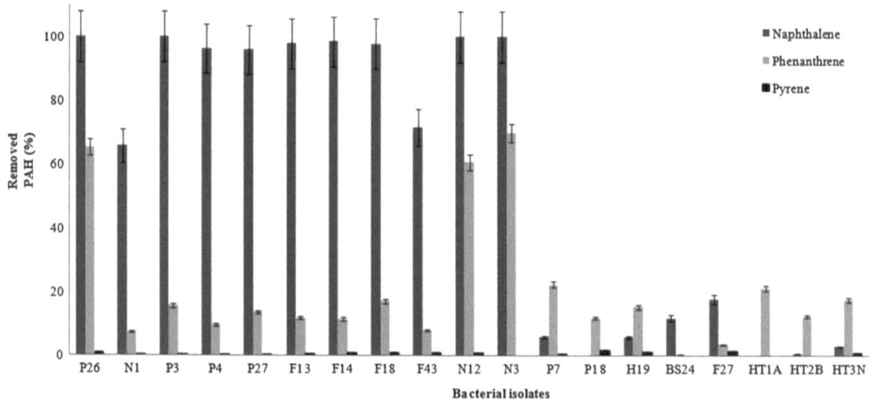

Fig. 3.2 Removal of polycyclic aromatic hydrocarbons (PAHs) of pyrene by strains in pure culture. Pyrene removal was determined after 21 days of incubation. Strains: *P. monteilii* P26, *P. stutzeri* N3, *P. xanthomarina* N12, *Arthrobacter* sp. P7, *Rhodococcus* sp. P18, *Gordonia* sp. H19, *Rhodococcus* sp. F27, *Arthrobacter* sp. HT1A, *Arthrobacter* sp. HT2B, *Rhodococcus* sp. HT3N. Hydrocarbon abiotic loss was considered in all cases. Values are average of triplicate samples

In the environment, however, it is probable that PAH produces different effects on the microbiological properties of contaminated marine sediments, inhibiting their biodegradation in situ. In addition, the ability to degrade these chemicals in the environment may be provided through the action of many organisms acting in concert (Ghazali et al. 2004; Gonzalez et al. 2011), and frequently microorganisms may act synergistically to degrade aromatic compounds in the biodegradation of complex hydrocarbon mixtures (Trzesicka-Mlynarz and Ward 1995; Kanaly and Harayama 2000).

On the other hand, the effectiveness of hydrocarbon degradation usually requires the addition of surface-active compounds (Gonzalez et al. 2011). Biodegradation of PAHs is often limited by their low bioavailability (low aqueous solubility and high sorption to soil particles), which affects the removal rate in a two-liquid-phase system, including substrate uptake and efflux (Cerniglia 1984). To overcome the low aqueous solubility of phenanthrene, the strains might produce extracellular polymeric substances that promote the availability and uptake of the hydrophobic compounds.

In this context, to increase the apparent PAH solubility, different approaches such as addition of synthetic surfactants or biosurfactants help the biodegradation processes. However, these compounds usually are toxic for microorganisms (Sartoros et al. 2005). The relative toxicity, low biodegradability, and low efficiency at low concentrations of synthetic surfactants reduce their application in contaminated sites (Desai and Banat 1997). Bacteria that produce extracellular surfactants are desirable because they enhance droplet formation and "pseudo-solubilization" during hydrocarbon degradation by stabilizing water-in-oil or oil-in-water emulsions and by reducing limitations from substrate mass transport (Ron and Rosenberg 2001). Bioemulsifying activity increases the bioavailability of oils (strongly hydrophobic) by providing access of microorganisms to such compounds for degradation.

Table 3.1 Bioemulsifier production by individual strains

Samples	EI: 24 h	ES
Water	0.00	0.00
JPP	0.00	0.00
Tween 20	66.82	79.72
P26	21.85	50.21
N12	0.00	0.00
N3	27.14	54.41
P7	12.82	100.00
P18	0.00	0.00
F27	0.00	0.00
H19	0.00	0.00
HT1A	0.00	0.00
HT2B	0.00	0.00
HT3N	0.00	0.00

Emulsification Index at 24 h (EI-24 h) and Emulsion Stability (ES) were determined in all strains. Values are presented as percentage of triplicate samples. Tween 20 was used as positive control

Bioemulsifying activity was detected in *Pseudomonas monteilii* P26, and *Pseudomonas* sp. N3 supernatants produced stable and compact emulsions, according to the emulsification index at 24 h (EI-24) (Table 3.1). The emulsifying activity and removal capabilities demonstrated by both microorganisms make these *Pseudomonas* strains promising candidates for use in bioremediation protocols.

3.2 Defined Mixed Cultures Would Greatly Improve the Removal Efficiency of PAHs

PAHs are found in nature as a complex heterogeneous mixture, wherein each compound has the ability to influence another, affecting their bioavailability and increasing difficulties for degradation by microorganisms (Marsili 2000; Chávez et al. 2004). In a real event of oil contamination, usually low molecular weight (LMW) PAHs such as naphthalene and phenanthrene are easily degradable, but high molecular weight (HMW) PAHs are very difficult to degrade (Commendatore et al. 2000).

Even though a large number of bacteria capable of degrading a wide range of PAHs have been isolated from contaminated sites (Mrozik 2003; Janbandhu and Fulekar 2011; Zhong et al. 2011), not many genera have been reported be able to degrade both LMW and HMW polycyclic hydrocarbons (Song et al. 2011). Some recent studies have employed the individual degradation capabilities of different bacterial genera to define mixed cultures with cooperative interactions to improve the efficiency of degradation on a mixture of PAHs used as substrate (Mrozik 2003; Chávez et al. 2004; Janbandhu and Fulekar 2011; Zhong et al. 2011; Mikesková et al. 2012).

Recent work concerning pollutant biodegradation by mixed cultures has been reviewed (Senthilvelan et al. 2014), and their application for degradation of a wide range of contaminants show the mixed culture as the better strategy in all assayed conditions (Barsing et al. 2011; Jin et al. 2012). Regarding PAH biodegradation, a consortium of *Pseudomonas putida*, *Flavobacterium* sp., and *Pseudomonas aeruginosa* showed higher removal of less water soluble PAHs than strains in pure culture (Trzesicka-Mlynarz and Ward 1995), and Janbandhu and Fulekar (2011) tested a microbial consortium with interesting phenanthrene and other PAH degradation capabilities. At the same time, Arun and Eyini (2011), Gonzalez et al. (2011), and Simarro et al. (2012) reported the use of mixed cultures as an alternative strategy for PAH remediation.

To improve the degradation process of a mix of PAHs, 16 combinations of bacterial strains previously evaluated according to LMW and HMW PAHs removal capabilities in liquid media were assayed. The aim of these experiments was to improve the degradation process of a mix of PAHs by enhancing individual biodegradation performances after the formulation of a defined mixed culture in which synergistic activity would be observed. These combinations of *Pseudomonas* and *Actinomycetes* strains allowed formulating a defined mixed culture able to efficiently degrade a mix of LMW and HMW PAHs in JPP medium.

PAH removal performance was evaluated according to the efficiency (*E%*) (Isaac et al. 2015). To compare the PAH-degrading performance of defined mixed cultures versus pure cultures, relative removal activity (observed/expected, *E %*) was determined according to a modification of the data analysis technique described by Fuentes et al. (2011). The observed *E%* values of each mixed culture were compared with the expected removal activity, calculated as the average of the *E%* of each individual strain forming the complex biological system (Table 3.2).

All mixed cultures assayed showed maximum *E%* values for naphthalene degradation, in accordance with expected results. Also, all consortia evaluated reached 100 % efficiency in phenanthrene degradation, increasing in most cases the expected value

Table 3.2 Removal of polycyclic aromatic hydrocarbons (PAHs) by defined mixed cultures

Mixed culture	Isolates	Pyr E% Expected	Pyr E% Observed	Relative activity Observed/expected
C1	P26/P18	19.87±2.02	41.43±10.03	2.08 AB
C2	P26/F27	8.97±1.36	24.27±2.17	2.70 ABCDEFGH
C3	P26/H19	5.87±0.79	25.26±1.40	4.29 ABCDEFG
C4	N3/P18	19.88±2.02	25.87±12.81	1.30 ABCDEFG
C5	N3/F27	8.97±1.36	29.79±4.45	3.32 ABCDEF
C6	N3/H19	5.87±0.79	13.93±1.02	2.37 FGHI
C13	P26/N3/P18/F27	14.42±1.69	22.52±3.59	1.56 CDEFGHI
C14	P26/N3/P18/H19	12.87±1.41	25.61±6.09	1.98 ABCDEFG
C15	P26/N3/H19/F27	7.42±1.08	41.91±1.54	5.64 A

Expected values were calculated as an average of removal capabilities of strains in pure culture. Means that do not share a letter are significantly different. *Pyr* pyrene

calculated from the individual performance. Pyrene removal values by mixed cultures of two strains were higher than those obtained with the corresponding pure cultures. Combinations of *P. monteilii* P26/*Rhodococcus* sp. P18 (C1) showed pyrene removal value of 41 %, whereas *P. monteilii* P26 showed no pyrene removal capability when grown in pure culture. Interestingly, C3 (*P. monteilii* P26/*Gordonia* sp. H19) removed 25 % of pyrene, improving the expected performance more than fourfold. Maximum removal of the mix of hydrocarbons by a mixture of three strains was obtained in a defined mixed culture of *P. monteilii* P26/*Rhodococcus* sp. P18/*Rhodococcus* sp. F27 (C7) capable of degrading nearly 39 % of pyrene from the culture medium.

A mixed culture of *P. monteilii* P26, *Pseudomonas* sp. N3, *Gordonia* sp. H19, and *Rhodococcus* sp. F27 (C15) showed the highest pyrene biodegradation activity with removal values close to 42 %, almost six times higher than removal values obtained with these strains in pure culture, whose removal did not exceed 9 % (Table 3.2).

Given these results, the capacity of removal of C15 mixed culture against a wide range of PAHs was evaluated. A mixture of six polyaromatic hydrocarbons of low and high molecular weight was used, differing in the amount (2–5 rings) and molecular geometry of the aromatic rings (naphthalene, biphenyl, acenaphthene, phenanthrene, fluoranthene, pyrene).

The C15 culture showed an interesting ability to degrade PAHs of two and three rings (naphthalene, biphenyl, acenaphthene, phenanthrene), removing 100 % thereof. It also showed 34 % and 25 % removal of high molecular weight hydrocarbons pyrene and fluoranthene, respectively (Fig. 3.3).

Removal rates of naphthalene, phenantrene, and pyrene by C15 mixed culture were compared with those obtained with pure cultures. It could be observed that the removal rate of naphthalene by the mixed culture C15 (1.047 mg l^{-1} h^{-1}) was significantly lower than the value reached by the pure culture of *Pseudomonas* sp. P26 (2.563 mg l^{-1} h^{-1}). The removal rate of phenanthrene by C15 showed similar values compared to that obtained by the more efficient phenantrene-degrading bacteria in

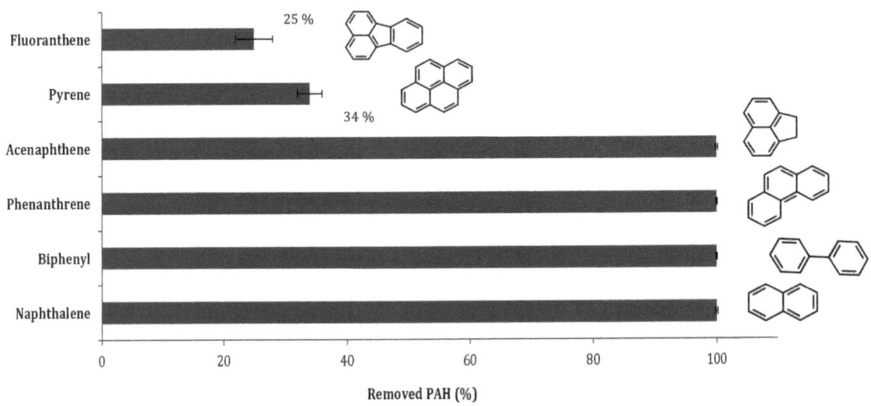

Fig. 3.3 Removal of a complex mixture of PAHs by the mixed culture C15. The study was performed in triplicate

pure culture, in this case *Pseudomonas* sp. N3 (0.495 mg l^{-1} h^{-1}). On the other hand, the calculated pyrene removal rate value was 64 % greater than the maximum value obtained in a pure culture of *Rhodococcus* sp. P18, and effective removal of pyrene was substantially improved: *P. monteilii* P26, *Pseudomonas* sp. N3, *Gordonia* sp. H19, and *Rhodococcus* sp. F27 (C15) showed the highest pyrene biodegradation activity with removal values close to 42 %, almost sixfold higher than removal values obtained with these strains in pure culture, whose removal did not exceed 9 %.

Certainly the C15 defined microbial consortium showed greater removal efficiency of a mixture of naphthalene, phenanthrene, and pyrene than individual strains. Also, it was capable of removing a mix of PAHs consisting of naphthalene, biphenyl, acenaphthene, phenanthrene, fluoranthene, and pyrene at the same proportion.

Many such mixed cultures obtained results in the degradation of different compounds plus hydrocarbons such as dyes, phenols, pesticides, etc. (Khehra et al. 2005; Arun and Eyini 2011; Ghanem et al. 2011; Hamitouche et al. 2011; Pino and Penuela 2011). This improvement in efficiency could be explained as the result of the potential obtained by combining different enzymes and metabolic pathways. However, researchers should pay special attention to the development and optimization of mixed culture, because the culture must be able to survive and also remain biologically active throughout the degradation process (Mikesková et al. 2012).

With respect to metabolic cooperation, a simple combination of microorganisms based on growth inhibition assays is not a sufficient tool to verify that the system power formulated the capacity because of interspecies interactions of the microorganisms (Seneviratne et al. 2008). We could not specifically determine the type of interaction that occurs between individual Patagonian strains in C15 defined mixed culture.

Because intermediary metabolites of pyrene degradation could be substrates for the enzymes of the degradation pathways of simpler aromatic compounds in *Pseudomonas* (Moscoso et al. 2012), the end result is a significant and interesting acceleration of the complete process of degradation. In this manner, we could reasonably argue that even though *Pseudomonas* strains seem to have a metabolic pathway to initiate the degradation of pyrene, their enzymes may accelerate the process when the actinomycetes have reached a certain degree of degradation of pyrene.

In that process, in the first stage, the polyaromatic hydrocarbons of low molecular weight (naphthalene and phenanthrene) are degraded by *Pseudomonas* strains, and active metabolism tends to be elevated. These strains release bioactive compounds (emulsifier/surfactant) in the culture medium and make available all its enzymatic machinery to assist pyrene degradation initiated by actinomycetes strains. Thus, a consortium that effectively improves the individual capacities of degradation and increases the potential of microorganisms for use in biotechnology for bioremediation of sites contaminated with polycyclic aromatic hydrocarbons is obtained. However, other complementary tests are needed to support this hypothesis.

This study contributes to the knowledge of the indigenous microbiota present in marine sediments off the Patagonian coast and its potential for bioremediation of polycyclic aromatic hydrocarbons, the main components of oil and its derivatives. The use of oil-contaminated sediments allowed the isolation of bacteria capable of degrading native PAHs of low and high molecular weight; the selective enrichment process was fundamental to this isolation.

References

Arun A, Eyini M (2011) Comparative studies on lignin and polycyclic aromatic hydrocarbons degradation by basidiomycetes fungi. Bioresour Technol 102:8063–8070

Barragán Muñoz JM, Dadon JR, Matteuchi SD, Morello JH, Baxendale C, Rodríguez A (2003) Preliminary basis for an integrated management program for the coastal zone of Argentina. Coast Manag 31:55–77

Barsing P, Tiwari A, Joshi T, Garg S (2011) Application of a novel bacterial consortium for mineralization of sulphonated aromatic amines. Bioresour Technol 102:765–771

Cerniglia CE (1984) Microbial metabolism of polycyclic aromatic hydrocarbons. Adv Appl Microbiol 30:31–71

Chávez FP, Lunsdorf H, Jerez CA (2004) Growth of polychlorinated-biphenyl-degrading bacteria in the presence of biphenyl and chlorobiphenyls generates oxidative stress and massive accumulation of inorganic polyphosphate. Appl Environ Microbiol 70:3064–3072

Commendatore MG, Esteves JL, Colombo JC (2000) Hydrocarbons in coastal sediments of Patagonia, Argentina: levels and probable sources. Mar Pollut Bull 40:989–998

Desai JD, Banat I (1997) Microbial production of surfactants and their commercial potential. Microbiol Mol Biol Rev 61:47–64

Esteves JL, Commendatore MG, Nievas ML, Paletto VM, Amín O (2006) Hydrocarbon pollution in coastal sediments of Tierra del Fuego Islands, Patagonia, Argentina. Mar Pollut Bull 52:582–590

Fromin N, Hamelin J, Tarnawski S, Roesti D, Jourdain-Miserez K, Forestier N, Teyssier-Cuvelle S, Gillet F, Aragno M, Rossi P (2002) Statistical analysis of denaturing gel electrophoresis (DGGE) fingerprinting patterns. Environ Microbiol 4:634–643

Fuentes MS, Sáez JM, Benimeli CS, Amoroso MJ (2011) Lindane biodegradation by defined consortia of indigenous *Streptomyces* strains. Water Air Soil Pollut 222:217–231

Ghanem KM, Al-Garni SM, Biag AK (2011) Statistical optimization of cultural conditions for decolorization of methylene blue by mono and mixed bacterial culture techniques. Afr J Microbiol Res 5:2187–2197

Ghazali FM, Rahman RNZA, Salleh AB, Basri M (2004) Biodegradation of hydrocarbons in soil by microbial consortium. Int Biodeter Biodegr 54:61–67

Gonzalez N, Simarro R, Molina MC, Bautista LF, Delgado L, Villa JA (2011) Effect of surfactants on PAH biodegradation by a bacterial consortium and on the dynamics of the bacterial community during the process. Bioresour Technol 102:9438–9446

Hamitouche A, Amrane A, Bendjama Z, Kaouah F (2011) Phenol biodegradation by mixed culture in batch reactor: optimization of the mineral medium composition. Desalin Water Treat 25:20–24

Harayama S, Kasai Y, Hara A (2004) Microbial communities in oil-contaminated seawater. Curr Opin Biotechnol 15:205–214

Isaac P, Sanchez L, Bourguignon N, Cabral ME, Ferrero M (2013) Indigenous PAH-degrading bacteria from oil-polluted sediments in Caleta Cordova, Patagonia Argentina. Int Biodeter Biodegr 82:207–214

Isaac P, Martínez FL, Bourguignon N, Sanchez L, Ferrero MA (2015) Improved PAHs removal performance by a defined bacterial consortium of indigenous *Pseudomonas* and Actinobacteria from Patagonia, Argentina. Int Biodeter Biodegr 101:23–31

Janbandhu A, Fulekar MH (2011) Biodegradation of phenanthrene using adapted microbial consortium isolated from petrochemical contaminated environment. J Hazard Mater 187:333–340

Jin DF, Hu H, Liu DF, Ding HT, Xia XM, Zhao YH (2012) Optimization of a bacterial consortium for nitrobenzene degradation. Water Sci Technol 65:795–801

Kanaly RA, Harayama S (2000) Biodegradation of high-molecular weight polycyclic aromatic hydrocarbons by bacteria. J Bacteriol 182:2059–2067

Kao CM, Chen CS, Tsa FY, Yang KH, Chien CC, Liang SH, Yang C, Chen SC (2010) Application of real-time PCR, DGGE fingerprinting, and culture-based method to evaluate the effectiveness of intrinsic bioremediation on the control of petroleum-hydrocarbon plume. J Hazard Mater 178:409–416

Khehra MS, Saini HS, Sharma DK, Chadha BS, Chimni SS (2005) Decolorization of various azo dyes by bacterial consortium. Dyes Pigments 67:55–61
Konopka A, Zakharova T, Bischoff M, Oliver L, Nakatsu C, Turco RF (1999) Microbial biomass and activity in lead-contaminated soil. Appl Environ Microbiol 65:2256–2259
Koutny M, Ruzicka J, Chlachula J (2003) Screening for phenol-degrading bacteria in the pristine soils of south Siberia. Appl Soil Ecol 23:79–83
Lozada M, Riva Mercadal JP, Guerrero LD, Di Marzio WD, Ferrero MA, Dionisi HM (2008) Novel aromatic ring-hydroxylating dioxygenase genes from coastal marine sediments of Patagonia. BMC Microbiol 8:50
Marcos MS, Lozada M, Di Marzio WD, Dionisi HM (2012) Abundance, dynamics, and biogeographic distribution of seven polycyclic aromatic hydrocarbon dioxygenase gene variants in coastal sediments of Patagonia. Appl Environ Microbiol 78:1589–1592
Marsili L (2000) Lipophilic contaminants in marine mammals: review of the results of ten years' work at the Department of Environmental Biology, Siena University (Italy). Int J Environ Pollut 13:416–452
Mikesková H, Novotný C, Svobodová K (2012) Interspecific interactions in mixed microbial cultures in a biodegradation perspective. Appl Microbiol Biotechnol 95:861–870
Molina MC, González N, Bautista LF, Sanz R, Simarro R, Sánchez I, Sanz JL (2009) Isolation and genetic identification of PAH degrading bacteria from a microbial consortium. Biodegradation 20:789–800
Moscoso F, Teijiz I, Deive FJ, Sanromán MA (2012) Efficient PAHs biodegradation by a bacterial consortium at flask and bioreactor scale. Bioresour Technol 119:270–276
Mrozik A (2003) Bacterial degradation and bioremediation of polycyclic aromatic hydrocarbons. Polish J Environ Stud 12:15–25
Nakatsu KH, Torsvik V, Øvreås L (2000) Soil community analysis using denaturing gradient gel electrophoresis (DGGE) profiles of 16S rDNA PCR products. Soil Sci Soc Am J 64:1382–1388
Olson DM, Dinerstein E (2002) The global 2000: priority ecoregions for global conservation. Ann Mo Bot Gard 89:199–224
Pino M, Penuela G (2011) Simultaneous degradation of the pesticides methyl parathion and chlorpyrifos by an isolated bacterial consortium from a contaminated site. Int Biodeter Biodegr 65(6):827–831
Ríos SM, Barquín M, Nudelman N (2010) Hydrocarbons characterization in coastal sediments of the Argentine Patagonia using the nuclear magnetic resonance (NMR) spectroscopy. Environ Chem Lett 8:223–229
Ron EZ, Rosenberg E (2001) Natural role of biosurfactants. Environ Microbiol 3:229–236
Sartoros C, Yerushalmi L, Béron P, Guiot SR (2005) Effects of surfactant and temperature on biotransformation kinetics of anthracene and pyrene. Chemosphere 61:1042–1050
Seneviratne G, Zavahir JS, Bandara WMMS, Weerasekara MLMAW (2008) Fungal-bacterial biofilms: their development for novel biotechnological applications. World J Microbiol Biotechnol 24:739–743
Senthilvelan T, Kanagaraj J, Panda RC, Mandal AB (2014) Biodegradation of phenol by mixed microbial culture: an eco-friendly approach for the pollution reduction. Clean Technol Environ Policy 16:113–126
Simarro R, Gonzalez N, Bautista L, Molina C, Schiavi E (2012) Evaluation of the influence of multiple environmental factors on the biodegradation of dibenzofuran, phenanthrene, and pyrene by a bacterial consortium using a orthogonal experimental design. Water Air Soil Pollut 223:3437–3444
Song X, Xu Y, Li G, Zhang Y, Huang T, Hu Z (2011) Isolation, characterization of *Rhodococcus* sp. P14 capable of degrading high-molecular-weight polycyclic aromatic hydrocarbons and aliphatic hydrocarbons. Mar Pollut Bull 62:2122–2128
Trzesicka-Mlynarz D, Ward OP (1995) Degradation of polycyclic aromatic hydrocarbons (PAHs) by a mixed culture and its component pure cultures, obtained from PAH-contaminated soil. Can J Microbiol 41:470–476

Zhong Y, Luan T, Lin L, Liu H, Tam NFY (2011) Production of metabolites in the biodegradation of phenanthrene, fluoranthene and pyrene by the mixed culture of *Mycobacterium* sp. and *Sphingomonas* sp. Bioresour Technol 102:2965–2972

Zhou HW, Luan TG, Zou F, Yee Tam NF (2008) Different bacterial groups for biodegradation of three- and four-ring PAHs isolated from a Hong Kong mangrove sediment. J Hazard Mater 152:1179–1185

Chapter 4
Hydrocarbon Remediation by Patagonian Microbial Consortia

Marina L. Nievas, Rosana Polifroni, Federico del Brio, and Marcela A. Sepúlveda

Abstract Among the technologies for recovering hydrocarbon-polluted sites, bioremediation is particularly interesting for wastewater and residue treatment. In this chapter, we first introduce hydrocarbon bioremediation focusing on the hydrocarbon biodegradation capabilities of microorganisms of the marine environment. Then, the context of petroleum hydrocarbons in the Argentinean Patagonia coast is depicted, and recent advances in bioprospection of hydrocarbon-degrading microorganisms from polluted sediments are reviewed. Finally, we discuss bilge waste bioremediation by Patagonian autochthonous microbial consortium as a treatment alternative. In particular focusing on the extent of hydrocarbon biodegradation, bioremediation trials at different experimental scales and bilge waste microbial community members are reviewed.

4.1 Introduction

Hydrocarbon pollution is widespread in the marine environment, causing serious environmental problems by catastrophic oil spills, such as those of the *Deepwater Horizon*, *Prestige*, or *Exxon Valdez* accidents (NRC 2003; Head et al. 2006; Díez et al. 2007; Camilli et al. 2010). Further, chronic pollution caused by the massive usage of petroleum-derived products, of which part reaches the environment as a sum of small-sized but frequent spills, is significant in the sea (NRC 2003; Head et al. 2006). Microorganisms have a key role in the biogeochemistry cycle of organic

M.L. Nievas (✉)
Centro para el Estudio de Sistemas Marinos (CESIMAR-CONICET),
Blvd. Brown 2915, Puerto Madryn, Chubut, Argentina

Facultad Regional Chubut, Universidad Tecnológica Nacional,
Av. del Trabajo 1536, Puerto Madryn, Argentina
e-mail: nievas@cenpat-conicet.gob.ar

R. Polifroni • F. del Brio • M.A. Sepúlveda
Centro para el Estudio de Sistemas Marinos (CESIMAR-CONICET),
Blvd. Brown 2915, Puerto Madryn, Chubut, Argentina

N.L. Olivera et al. (eds.), *Biology and Biotechnology of Patagonian Microorganisms*, DOI 10.1007/978-3-319-42801-7_4

matter in the biosphere (Kaiser and Attrill 2011). Biodegradation involves the breakdown of organic compounds carried out by the metabolic activity of organisms, the role of microorganisms being dominant. Despite their toxicity, most organic pollutants undergo biodegradation by specialized microorganisms, either through biotransformation into less complex metabolites or through mineralization into innocuous compounds (McGenity et al. 2012). Hydrocarbon biodegradation occurs naturally within a variety of environments, in aerobic and anaerobic conditions (McGenity et al. 2012; Cravo-Laureau and Duran 2014). Moreover, microbial biodegradation is believed to be the most important hydrocarbon cleanup process in the aquatic environment (NRC 2003; Head et al. 2006).

In addition to the relevance of biodegradation at the environmental level, this natural process is the basis of bioremediation, a technology aiming to increase the extent and the rate-limiting steps of biodegradation by modifying the environmental factors affecting pollutant microbial transformations. Environmental restoration by microbial-based processes has become a global bioremediation industry. Hydrocarbon remediation is actually a competitive technology for recovering impacted environments and also for wastewater and residue treatment (USEPA 2000; Zhu et al. 2001; Khan et al. 2004). Hence, optimizing hydrocarbon bioremediation processes in a variety of environmental conditions is a relevant research topic.

The metabolic capabilities of microorganisms have a central role in bioremediation. Thus, the bioprospection of hydrocarbon-degrading microorganisms in sites with selective pressure such as hydrocarbon pollution and the assessment of its biodegradation potential is one of the main tools not only to broaden our bioremediation knowledge background but also to expand the biotechnological applications. We present here the recent advances of the bioprospection of hydrocarbon-degrading microorganisms from marine hydrocarbon-impacted sites of Argentinean Patagonia. First, we introduce the bioremediation issues of petroleum composition, the metabolic capabilities of hydrocarbon-degrading microorganisms, and biodegradation by microbial consortia. Then, we focus on bioprospection of hydrocarbon-degrading microorganisms in marine sediments, and finally, we discuss bilge waste bioremediation by a Patagonian autochthonous microbial consortium as a treatment alternative.

4.2 Hydrocarbon Bioremediation in the Marine Environment

4.2.1 Petroleum Hydrocarbons

Petroleum, a naturally occurring substance, is one of the most complex mixtures known (Rodgers et al. 2005). It is mainly composed of hydrocarbons with minor proportions of oxygen, sulfur, and nitrogen heteroatom-containing compounds, with the presence of metals such as nickel, vanadium, iron, and copper, and other trace elements (Speight 2011). Four major groups of compounds are recognized in petroleum, when it is fractioned by solubility in different solvents with solid-phase absorbents: saturates or aliphatics (comprising linear, branched, and cyclic alkanes), aromatics [comprising BTEXs, monoaromatics, and polycyclic aromatic hydrocarbons (PAHs)], resins (or polar fraction), and asphalthenes, defined as the crude oil portion insoluble in *n*-heptane or *n*-pentane (Speight 2011). The complex composition of petroleum precludes the isolation and identification of all its individual constituents; nevertheless, conservative estimations indicate that the number of chemically distinct species in crude oil may be greater than 50,000 (Rodgers et al. 2005).

Many petroleum-derived products are manufactured by means of oil refining, from light distillates (e.g., solvents, jet fuels, and naphtha for planes and cars), medium distillates (such as diesel oil, with widespread use in transportation, agriculture, and industry), and heavy products (e.g., lubricant oils, bunker fuel destined for shipping combustibles, and industrial usages), coke, and asphalt (Speight 2011). Although oil-derived products contain only a fraction of the original crude oil, according to the manufacturing processes which they undergo, the composition is still complex. For example, a straight-run diesel fuel has more than 30,000 individual different compounds (Vendeuvre et al. 2007). Moreover, a broad variety of additives are usually added to petroleum products to enhance its properties (Auffret et al. 2009; Rizvi 2009). During their use, petroleum-derived products undergo physicochemical changes, mainly decomposition and oxidation (Rizvi 2009). From a bioremediation perspective, oil hydrocarbons, additives, and their decomposition/oxidation products are the target substrates to be biodegraded in spills of hydrocarbon wastes or wastewater.

4.2.2 *Hydrocarbon-Degrading Metabolic Capabilities of Microorganisms*

Because petroleum is naturally occurring, many microorganisms have evolved with the ability to utilize its constituents as sources of carbon and energy (Prince et al. 2010). Besides oil seeps entering naturally into the oceans, biosynthesis of some hydrocarbons is carried out by organisms such as bacteria, plants, and phytoplankton (Damsté et al. 2004; Ladygina et al. 2006). Thus, low levels of such compounds are distributed in the biosphere, being able to subtend hydrocarbon-degrading microorganisms in most environmental compartments. Consequently, it is not surprisingly that hydrocarbon-degrading microorganisms are found to be ubiquitous. In the marine environment, more than 175 genera of Bacteria, as well as some members of Archaea and Eukarya, have been described as hydrocarbon degraders (Head et al. 2006; Prince et al. 2010; Rojo 2010; McGenity et al. 2012). Predominantly, hydrocarbon-degrading bacteria belong to the phyla Actinobacteria, Bacteroidetes, Firmicutes, and Proteobacteria (Prince et al. 2010).

Among bacteria, two groups may be distinguished from the aspect of hydrocarbon metabolism. First are those that are strict hydrocarbonoclastics, which utilize almost exclusively a narrow range of hydrocarbons as carbon sources. For example, members of the genera *Alcanivorax*, *Marinobacter*, *Thalassolituus*, *Oleiphilus*, and *Oleispira* are strict alkane degraders, whereas *Cycloclasticus* is an aromatic hydrocarbon degrader (Harayama et al. 2004; Yakimov et al. 2007). These microorganisms are believed to be important in the biodegradation of hydrocarbons in the marine environment (McKew et al. 2007; Yakimov et al. 2007). Another group of diverse generalist microorganisms has a more versatile metabolism, utilizing hydrocarbons as one of its carbon and energy sources. Members of the genera *Marinobacter*, *Pseudomonas*, *Rhodococcus*, *Vibrio*, *Marinomonas*, and *Pseudoalteromonas*, among many others, are found in the generalist group (Harayama et al. 2004; Prince et al. 2010).

The metabolic pathways of hydrocarbon microbial degradation are well established for many microorganisms. A distinguishable feature of hydrocarbon-degrading microorganisms is the presence of key catabolic enzymes involved in the activation of alkane and aromatic hydrocarbons, identified as the most critical step of hydrocarbon catabolism (Habe and Omori 2003; Pérez-Pantoja et al. 2010; Rojo 2010). Thus, once activated by the aerobic insertion of oxygen or anaerobic addition of succinate, for example, hydrocarbons enter the inherent biochemistry of living organisms and are readily oxidized to carbon dioxide and water (Prince et al. 2010). Most frequently, the aerobic degradation of *n*-alkanes starts with the oxidation of a terminal or subterminal methyl group with subsequent transformation to primary or secondary alcohol, with further oxidation to the corresponding aldehyde or ketone, and finally conversion into fatty acids. Then, fatty acids are metabolized to acetyl coA by β-oxidation (Rojo 2010). This pathway is catalyzed by a broad range of monooxygenase enzymes. AlkB-like enzymes are widely distributed in *α*-, *β*-, and *γ-Proteobacteria* and the Actinomicetales (van Beilen et al. 2003; van Beilen and Funhoff 2007); cytochrome P450 enzymes of the CYP153 family have been identified in many genera, including *Sphingomonas*, *Rhodococcus*, *Marinobacter*, *Halomonas*, and *Alcanivorax* (Sabirova et al. 2006; van Beilen et al. 2006; Wang et al. 2010). AlkB-like and cytochrome P450 CYP153 enzymes have specificity in the alkane range (e.g., medium-chain-length alkanes, C5–C16; or long-chain-length alkanes, >C12). In addition, methane monooxygenases (pMMO) are found in microorganisms that are able to degrade short and volatile alkanes aerobically (C1–C4) (McDonald et al. 2006) and genes encoding enzymes responsible for degradation of alkanes longer than C15 (e.g., *almA* in *Acinetobacter* sp. and *ladA* in *Geobacillus thermodenitrificans* NG80-2) (Wentzel et al. 2007; Li et al. 2008). Many bacterial strains contain only one alkane hydroxylase system; however, it is rather common to find strains that contain more than one of such systems, which seems to give them selective advantages (van Beilen et al. 2006; Rojo 2010; Wang et al. 2010). Regarding aromatic hydrocarbons, the enzymes involved in the aerobic aromatic degradation are oxygenases that catalyze the initial hydroxylation and ring cleavage of PAHs and the intermediates metabolites into a few central intermediate compounds such as catechols, protocatechuates, gentisates, and (hydroxy)benzoquinols. Further, these compounds are oxidized by extradiol and intradiol cleavage, funneling into the Krebs cycle (Pérez-Pantoja et al. 2010). The PAH oxygenase prototype is the naphthalene dioxygenase encoded by the *nahAaAbAcAd* gene of *Pseudomonas putida* G7, which is able to degrade naphthalene, anthracene, and phenanthrene through the catechol meta-cleavage pathway into pyruvate and acetaldehyde. Similar PAH-degrading genes have been described in many strains (Abbasian et al. 2015). The aerobic and anaerobic aliphatic and aromatic hydrocarbon catabolic pathways have been reviewed in recent articles (Kanaly and Harayama 2000, 2010; Habe and Omori 2003; Pérez-Pantoja et al. 2010; Rojo 2010; Abbasian et al. 2015).

4.2.3 Biodegradation of Complex Substrates by Microbial Communities

Biodegradation of petroleum-derived hydrocarbons, both under environmental conditions such as oil spills, or in man-made scenarios, such as wastewater treatment plants, is an extremely complex process. Here, a multicomponent substrate, containing thousands of individual compounds, is metabolized by microbial communities, with multiple coexisting microbial species having specific hydrocarbon capabilities. Altogether, multiple parallel and interrelated biochemical reactions occur simultaneously, also constrained by physical and chemical variables such as oxygen and nutrient availability, temperature, and mixing conditions (Röling et al. 2004; Garcia-Blanco et al. 2007; Shokrollahzadeh et al. 2008; Ibarbalz et al. 2013). Thus, bioremediation processes are site dependent and also difficult to predict precisely (Röling et al. 2004; Gertler et al. 2012).

Hydrocarbon-degrading microorganisms are usually present in nature in very low numbers. However, a pollution event, such as an oil spill or a periodic oily wastewater discharge, induces a relatively rapidly growth of hydrocarbon-degrading microorganisms, becoming dominant during the first period after contamination (Harayama et al. 2004). Then, physical, chemical and biochemical changes take place, being relevant dissolution, photooxidation, emulsification, dispersion, and biodegradation (NRC 2003; Stout and Wang 2007; Gong et al. 2014). A succession of hydrocarbon-degrading specialist microorganisms during natural biodegradation processes, termed natural attenuation, has been observed as a general trait. Thus, at first aliphatic hydrocarbon degraders seem to be dominant, for example, members of the genus *Alcanivorax*, and then the abundance of aromatics hydrocarbon degraders increases, such as *Cycloclasticus* or members of *α-Proteobacteria* (Head et al. 2006; McGenity et al. 2012; Rodriguez-R et al. 2015). These changes are generally consistent with the type of hydrocarbons that remained available for biodegradation.

Multispecies microbial communities are essential for the successful biodegradation of complex petroleum derivates, not only because a broad pool of hydrocarbon enzymatic capabilities is needed, but also because positive interaction may be carried out by other members of the microbial community, not necessarily hydrocarbon degraders (McGenity et al. 2012). Some examples of such interspecies positive relationships are cooperative hydrocarbon consumption, co-metabolism of the more recalcitrant hydrocarbons (for example, benzo[*a*]pyrene), provision of nutrients, and biosurfactant or bioemulsifier production becoming hydrocarbons available to the water phase (Kanaly and Harayama 2000; Head et al. 2006; Musat et al. 2010; McGenity et al. 2012).

4.3 Hydrocarbon Remediation by Patagonian Microorganisms

4.3.1 Petroleum Hydrocarbons in Coastal Patagonia

The sea and coastal zones are particularly sensitive environments subjected to hydrocarbon exposure as a global trend (NRC 2003; Lucas and MacGregor 2006; Díez et al. 2007; Er-Raioui et al. 2009; Gong et al. 2014; Rodriguez-R et al. 2015), the Patagonian coast not being an exception. Argentinean Patagonian region, the east-southernmost zone of South America, is limited at the east with the South Atlantic Ocean with more than 3,000 km of coastline (from 39°42′S to 55°S). The region includes a vast semiarid coastal zone where two petroleum basins are actually exploited, extending onshore and offshore: the Golfo San Jorge Basin (44°–47°S, near 170,000 km^2) (Sylwan 2001) and the Austral Basin (46°30′–54°30′S in NNW–SSE direction, near 170,000 km^2), which continues the offshore edging Malvinas Basin (Rossello et al. 2008). The former, located in central Patagonia, is actually the most productive zone from Argentina, which has been exploited for more than a century. The complete oil production from the onshore Golfo San Jorge Basin, nearly 15 billion cubic meters (m^3) in 2014 (MMyERA 2016), is tanker-transported by sea from two oil terminals named Caleta Cordova (45°46′23″S, 67°19′25″W) and Caleta Olivia (46°25′32″S; 67°28′42″W). Offshore oil and gas are produced in Austral Basin, where the Río Cullen, San Sebastian, and Punta Loyola oil terminals are located. The maritime transportation route from the Austral and Golfo San Jorge basins parallels the coast of Argentinean Patagonia from south Argentine to the north with crude oil, returning with refined products, also by sea shipping. Some crude oil spill episodes have taken place on the Argentinean Patagonian coast, the most recent being a release of nearly 300 m^3 in Caleta Cordova in 2007 (Mercopress 2007). Even though minor, these quite frequent episodes contribute to chronically pollute certain zones, particularly those in the surroundings of crude oil terminals.

The region also possesses many ports, where commercial and fishery activities take place. The use of fossil fuels as well as the generation of hydrocarbons wastes (i.e., bilge wastes) from shipping activities are reported as a diffuse source of hydrocarbon pollution in the marine environment (NRC 2003; Kostianoy et al. 2005; Lucas and MacGregor 2006; Er-Raioui et al. 2009). Although the Argentinean Patagonian coast is a relatively nonpolluted region, periodic hydrocarbon pollution has been described, generally associated with port activities and petroleum exploitation zones (Commendatore et al. 2000, 2012, 2015; Commendatore and Esteves 2004, 2007; Esteves et al. 2006; Massara Paletto et al. 2008). Therefore, bioremediation technologies for the marine environment in the Argentinean Patagonia are attractive subjects for petroleum operators, governmental environmental decision makers, and harbour hydrocarbon-waste managers.

4.3.2 Bioprospection of Hydrocarbon-Degrading Microorganisms from Argentinean Patagonia Coastal Sediments

The Argentinean Patagonia maritime zone holds numerous niches for bioprospecting microorganisms, with potential biotechnological application adapted to the particularly harsh marine environment (Dionisi et al. 2012). Economic activities developed in Patagonia such as offshore oil exploitation, maritime oil transportation, fishing, marine transport, and port activities may result in hydrocarbon spillages that negatively affect coasts and waters, converting this region into an attractive reservoir of microorganisms from less explored sites with a certain history of hydrocarbon pollution. Marine sediments usually act as a sink for hydrophobic pollutants, which adsorb on organic matter as well as in sediment fine particles, exposing autochthonous microbial communities to a chemical selection pressure (Gong et al. 2014). Coastal marine sediments constitute particular ecosystems, specially intertidal zones where environmental conditions are daily modified according to tide level, for example, with drastic changes in redox condition, irradiation, and water availability, that in turn drive microbial degradation processes (Garcia-Blanco et al. 2007; Cravo-Laureau and Duran 2014; Rodriguez-R et al. 2015). Moreover, microorganisms in sediments are much more concentrated than in the water column, usually forming biofilms, subjected to changing gradients of nutrients and oxygen (McGenity et al. 2012). Consequently, a great biodiversity of microorganisms is found in coastal sediments with potential for bioremediation.

Recently, the Patagonian Austral Interjurisdictional Marine Coastal Park located in the north zone of the San Jorge Gulf (44°50′–45°11′S, 65°32′–66°42′W), the first marine park in Argentine, was designated a Biosphere Reserve in the frame of the Man and Biosphere Program of UNESCO (Jun 2015). This zone is characterized by a great biodiversity of organisms and microenvironments. However, a petroleum hydrocarbon accumulation zone was described in this Park, in Faro

Aristizábal (Commendatore et al. 2000). Here, Olivera and coworkers (2009) surveyed marine sediments, searching for hydrocarbon-degrading microorganisms with biosurfactant-producing capabilities. Isolation was performed in a marine medium using crude oil with and without nutrient addition and on phenanthrene with nutrient addition, as carbon sources. These authors recovered 96 different morphological strains. Among them, 17 strains evidenced surface tension reduction and 8 showed emulsifying activities; all were recovered in media with nutrient amendment. Further characterization of the most promising strains based on the 16S rRNA gene sequence associated them with *Alcanivorax borkumensis* SK2^T (similarities of 98.9 % and 99.6 %), *Alcanivorax venustensis* ISO4^T (similarities of 98.9 % and 99.6 %), *Cobetia marina* ATCC25374^T, *Halomonas salaria* M27^T (98.6 % similarity), and *Pseudomonas putida* (100 % similarity). Except for one strain related to *Cobetia marina* that only grew on phenanthrene-supplemented medium, all the strains were able to use hydrocarbons from bilge wastes and *n*-hexadecane to grow. *A. borkumensis* is a cosmopolitan marine bacterium that blooms in oil-contaminated areas, where it can constitute up to 80 % of the bacterial population (Golyshin et al. 2003; Röling et al. 2004). *Alcanivorax* strains were dominant in microcosms containing coastal waters from the Mediterranean Sea, crude oil, and nutrients (Cappello et al. 2007; Gertler et al. 2012). In addition, *Alcanivorax* strains were found in similar experiments with water from the Pacific Ocean in East Asia (Harayama et al. 2004) and in surface seawater in a survey in the Atlantic Ocean from Kingston to Cape Town (13°09′N–34°06′S) (Wang et al. 2010). The success of *A. borkumensis* in becoming dominant is believed to be underlain by its multiple alkane catabolism. This species has two AlkB-like alkane hydroxylases and three cytochrome P450-dependant alkane monooxygenases with different alkane specificity (Rojo 2010). To our knowledge, isolation of *Alcanivorax* strains from the South Atlantic Ocean at the mid-Patagonian coast was reported for the first time by Olivera and coworkers. The strain *Alcanivorax* sp. PA2, closely related to *A. borkumensis,* shows high hydrophobicity (~80 %) during the exponential growth phase on hydrocarbons. PA2 hydrophobicity decreases from the exponential to the stationary growth phase; it also produces a bioemulsifier and is able to form biofilms (Olivera et al. 2009; Sepúlveda et al. 2012). Also, it was isolated associated to another strain, *P. putida*, forming a naturally occurring consortium. As expected, *Alcanivorax* sp. PA2 exhibited some of the characteristic traits reported for *A. borkumensis*. On the Argentinean Patagonia coast, the genus *Alcanivorax* was recently detected by culture-independent methods in hydrocarbon-polluted sediments of Ushuaia Bay (54°48′S, 68°18′W) and Caleta Cordova (45°46′23″S, 67°19′25″W) (Lozada et al. 2014; Guibert et al. 2016). This fact, in addition to isolation of *Alcanivorax* strains in the north zone of the San Jorge Gulf, suggest that this species may be widely distributed along the Atlantic Patagonian coast. The reported isolate *Alcanivorax* PA2 and related strains confirm the presence of this cosmopolitan species for the first time in Patagonia, reinforcing its broad distribution in the temperate marine environment and its presence in South Atlantic coasts.

4.3.3 Bilge Waste Bioremediation by Patagonian Autochthonous Microbial Communities

The chemical complexity of wastes from hydrocarbon-derived products requires microbial communities with broad enzymatic capabilities for bioremediation processes to be successful in in situ and ex situ scenarios. Hence, depending on the hydrocarbon waste composition and its physical characteristics, particular microbial communities and process operation conditions for bioremediation are needed. In that frame, in Argentinean Patagonia, the autochthonous microorganisms of bilge waste have been studied extensively in Puerto Madryn Harbour, aiming to provide information about hydrocarbon waste treatment and remediation of such wastes in the marine environment. Those studies focused on (1) the characterization of bilge waste, (2) the biodegradation performance of the bilge waste microbial community of the different classes of hydrocarbons present in bilge wastes assessed at various experimental scales, and (3) the characterization of members of the autochthonous microbial community.

4.3.3.1 Oily Wastes from Ships

Bilge wastes or bilge waters are oily wastes resulting from normal ship operations. All types of conveyors, from artisanal fishing vessels or sport boats to large fishing and cargo ships, cruises, and oil tankers, generate bilge wastes (IMO 2004; USEPA 2008). Such wastes consist of a mixture of water, oily fluids, lubricants, cleaning fluids, and other similar wastes that accumulate in the bilge, the lowest part of the vessel. These liquid streams get into the bilge from a variety of sources including the main and auxiliary engines, boilers, evaporators and related auxiliary systems and other mechanical and operational sources found throughout the machinery spaces of vessels (USEPA 2008). They also contain seawater entering the ship by infiltration. Thus, bilge wastes have an extremely variable composition, mostly depending on the age, type, and size of the ship, and the type of fuel, lubricants, and related hydrocarbon products used in the ship, as much as on the operation criteria of the ship's company owner and the crew. Bilge wastes need to be downloaded from ships as a requirement for navigation security.

Marine hydrocarbon pollution from ships is a longstanding global problem. Bilge wastes are regulated by the International Maritime Organization (IMO) agreement "International Convention for the Prevention of Pollution from Ships (MARPOL 73/78)" (IMO 2004). This treaty requires that the Party States ensure to provide ship waste reception facilities in ports, and also legislates national directives and regulations for vessels flying their flags to complain to MARPOL. Regarding bilge waste, onboard oil-waste separation systems are required for ships >400 gross tonnes, and only water with less than 15 ppm of hydrocarbon content is allowed to be0 discharged; otherwise, bilge wastes are required to be retained onboard for subsequent discharge on port reception facilities. MARPOL governs in international

Fig. 4.1 Bilge wastes in ports. **a** Fishing ships at Patagonian ports, Argentina. (Photograph by Diego González Zevallos.) **b** Accidental bilge waste spill at Puerto Madryn, Argentinean Patagonia

waters, so for ship navigation in the country's territorial waters and economic exclusive zone (12 and 200 nautical miles from the shoreline, respectively) the country laws apply (Lin et al. 2007). In Argentina, National Regulations adopted the same restrictions as MARPOL and also declared *Special Protection Zones*, most of them located in Patagonia, where any type of discharge is banned (PNA 1998). Despite regulations, illegal discharge is a quite common problem, and accidental bilge waste spills near shore may also pollute the water column and sediments (Fig. 4.1). The characteristics of the chemicals from both the aqueous and oily phases from bilge wastes confer hazardous waste status in these wastes (CEU 2006).

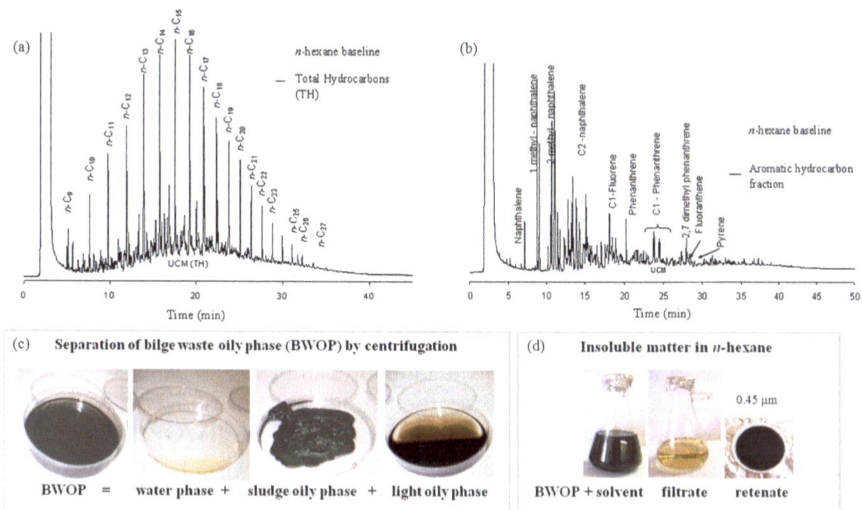

Fig. 4.2 Bilge waste oily phase (BWOP) from Puerto Madryn's bilge wastes, Argentinean Patagonia. **a** Total hydrocarbons gas chromatography-flame ionization detector (GC-FID) profile. **b** Aromatic hydrocarbon fraction GC-FID profile

4.3.3.2 Bilge Wastes at Puerto Madryn, Golfo Nuevo, Argentinean Patagonia

Onshore dumping pools were used as reception facilities for bilge wastes in Puerto Madryn city (42°46′S, 65°02′W) for about 10 years, accumulating large amounts of wastes downloaded from ships. These ponds operate as open storage tanks, where the phases of bilge wastes were separated by gravity. The composition of the accumulated residue reflected approximately an average of the physicochemical properties of bilge waste downloaded at the port (Nievas 2006). Puerto Madryn's harbour operates principally with fishing ships, typically smaller than 100 m length, mostly fishing in the Patagonian region, in addition to cargo ships and cruise ships (Fig. 4.1).

The bilge waste oily phase (BWOP), even after it is gravity decanted, contains some water in form of emulsion (Karakulski et al. 1998); thus, its resulting hydrocarbon content is quite variable. In a general trend, bilge waste from Puerto Madryn ports showed similar hydrocarbon content with time, when it was characterized by gas chromatography (GC) (Fig. 4.2a, b). Typical oil-derived product GC profiles in the range of n-C_9 to n-C_{30} were found (Morán et al. 2000; Olivera et al. 2003; Nievas et al. 2005, 2006). Depending on time, bilge waste showed unimodal and bimodal unresolved complex mixtures (UCM), in agreement with a mix of fuel and lubricant oils (Morán et al. 2000; Nievas et al. 2006). In the aromatic hydrocarbon fraction, parent and alkylated derivative PAHs with 2 to 4 rings were found; this fraction thus represents a minor proportion of the total hydrocarbon mass, usually less than 15 %

by weight (Olivera et al. 2003; Nievas et al. 2006). Typically, when the gravity-separated oily phase of bilge waste (density, 888 kg m^{-3} at 15 °C) undergoes further centrifugation, three phases may be identified: light oil (density, 862 kg m^{-3} at 15 °C), sludge oil, and a minor amount of water phase (Fig. 4.2c) (Nievas 2006). The BWOP had an 86 % wt of total organic carbon on a dry basis, according to the many hydrocarbon-derived products. Further, BWOP had typical carbonaceous deposits described for exhaust lubricant oils quantified as insoluble matter in *n*-hexane (2.1 wt%) and in toluene (1.6 wt%) (Fig. 4.2d) (Nievas 2006; Rizvi 2009). The kinematic viscosity of the BWOP (~20 mm^2 s^{-1} at 40 °C) was at least two times greater than those of marine fuel distillates, and nearly five times higher than those of medium diesel or fuel distillates, indicating a significant content of waste lubricant oil (Nievas et al. 2006). Notably, total hydrocarbons (TH), considered as the sum of aliphatic and aromatic hydrocarbons, although measured with different methodology, represent only a part of the total mass of the BWOP. For example, Olivera et al. (2003) reported that around 30 % of BWOP (wet basis) was constituted by aliphatic and aromatic hydrocarbons. Nievas and coworkers found that the total hydrocarbon fraction of BWOP was approximately 30 % [on a wet basis; this increases to 46 % when correcting by water content (Nievas et al. 2006), 39 % (wet basis; Nievas et al. 2005), and 46–56 % (Nievas et al. 2008a)]. These facts mean that polar compounds of bilge waste account for the remaining mass. The "no-hydrocarbon" portion of bilge waste may be composed of lubricant additives, which may represent up to 15 % of lubricant oils, polar compounds of fuel, and lubricants (Haus et al. 2001), and degradation products of lubricants and fuel produced during their lifetime use (Rizvi 2009).

4.3.3.3 Bilge Waste Biodegradation by Microbial Consortium

Biological treatment, within other technologies such as physically or chemically improved separation (Ghidossi et al. 2009; Rincón and La Motta 2014; Bilgili et al. 2016), chemical oxidation (Portela et al. 2003; Cazoir et al. 2012), and thermal destruction (Önenç et al. 2012), may be applied to manage bilge wastewater (Olivera et al. 2003; Nievas et al. 2006, 2008a, b; Sun et al. 2009, 2010; Mancini et al. 2012). Bioremediation treatments of hydrocarbon-impacted matrices have less environmental impact, and are generally cheaper and also less energy consuming than other processes, but usually they require longer treatment times and some compounds may remain recalcitrant for certain treatments (USEPA 2000; Khan et al. 2004; Fakhru'l-Razi et al. 2009, 2010; Reddy et al. 2011). Hence, waste biological treatments need to achieve a large extent of biodegradation in times as short as possible to be competitive. A classical approach to enhance hydrocarbon biodegradation is the addition of a surfactant or dispersant to overcome the low solubility of these compounds, making them available to microorganisms that develop in the aqueous phase. Biosurfactants are biodegradable and less toxic in environmental applications than synthetic surfactants and thus are preferred for bioremediation (Makkar and Rockne 2003; Olivera and Nievas 2010; Cappello et al. 2011; Rocha e Silva et al. 2014). Biosurfactants or bioemulsifiers can lower the water surface tension or oil–water interfacial tension, increasing the hydrocarbon apparent solubility and the surface interfacial area of oil–water systems by means of stabilizing emulsions (Ron and Rosenberg 2002; Olivera and Nievas 2010). Many types of biosurfactant are known; among them, surfactin is well characterized (Peypoux et al. 1999; Morán et al. 2002; Schaller et al. 2004). This lipopeptide low molecular weight biosurfactant, produced by *Bacillus subtilis* strains, is able to decrease the water surface tension to less than 30 N m^{-1} at concentrations greater then its critical micelle concentration (CMC) (Olivera and Nievas 2010). Olivera et al. (2000) and Morán et al. (2000) first studied the effect of surfactin addition and bioaugmentation on bilge waste hydrocarbon remediation (Table 4.1). The biosurfactant used was produced by a *Bacillus subtilis* strain, isolated from hydrocarbon-polluted sediments from Argentinean Patagonia (Morán et al. 2000). Surfactin added at supra-CMC concentration was able to enhance the biodegradation of the aliphatic and aromatic fractions of the BWOP when the native bilge waste microbial community was used as inoculum. Bilge waste degradation was also studied in microcosm aquaria simulating accumulation pond conditions for the bilge waste aqueous phase, which could be treated in situ (Olivera et al. 2000) (Table 4.1). Interestingly, in outdoor experiments surfactin crude extract added at sub-CMC level stimulated biodegradation of bilge waste hydrocarbons (Table 4.1) (Olivera et al. 2000). The same extent of biodegradation (<10 % hydrocarbon remaining) was reached in 10 days with biosurfactant, in comparison with control or bioaugmented experiments that required 20 days to reach similar hydrocarbon reduction levels. No preferential degradation of hydrocarbon was produced by the surfactin addition, nor was significant enhancement by bioaugmentation. The latter demonstrates that the native microbial consortium of bilge waste of the control treatment, added by the nonsterile oily phase, had adequate metabolic versatility to degrade the aliphatic and aromatic hydrocarbons, very similar to those of the enriched inoculum.

In addition, the native microbial communities from Puerto Madryn bilge wastes dumping pools harbor biosurfactant or bioemulsifier producer members. For example, we reported bioemulsifier production by the native microbial community of bilge wastes, in the early stationary growth phase, that concomitantly happened with the medium acidification during biodegradation of bilge wastes (Nievas et al. 2005). Two treatments for neutralizing the medium were tested, by intermittent addition of alkali and by buffering. For both, an enhancement in biodegradation occurs in comparison with a no controlled-pH treatment, which undergoes a pH decrease from 7.8 to 5.5 keeping constant at this value from day 6. Further, cells, supernatant, and the hydrocarbon phase of these cultures were assessed for emulsification activity. Some biosurfactants are produced by hydrocarbon-degrading microorganisms in an extracellular form, for example, surfactin in *Bacillus subtilis* or rhamnolipids in *Pseudomonas* sp. (Ron and Rosenberg 2002; Soberón-Chávez et al. 2005; Nitschke and Pastore 2006), whereas other surfactans/emulsifiers are produced bonded to the cell wall, for example, a glucose–lipid in *Alkanivorax borkumensis* and emulsan in *Acinetobacter calcoaceticus* (Abraham et al. 1998; Ron and Rosenberg 2002). In bilge waste hydrocarbon biodegradation assays, the bio-

Table 4.1 Bioremediation studies conducted in marine medium with autochthonous microbial community of bilge wastes from Puerto Madryn's bilge waste dumping pools

Assay configuration	Reactions conditions	Initial concentration[a]	Consortium[b]	Study focus	Study results	References
125-ml Erlenmeyer with 40 ml media	Reciprocal agitation at 80 strokes min^{-1} Temperature 25 °C by 7 days	0.60 % v/v	Enriched with glucose and bilge waste for 24 h	Addition of biosurfactant to enhance hydrocarbon biodegradation	Whole composition without surfactant (7 days): 17.4 % TAl reduction[c] 0 % TAr reduction[d] Whole composition with surfactant (7 days): 35.5 % TAl reduction[c] 41 % TAr reduction[d]	Morán et al. (2000)
20-l aquaria with 8-l media	Outdoors nonaseptic conditions temperature 2–33 °C Mean wind speed 17.3 km h^{-1} Aerated by air pump through diffuser	0.63 % v/v	Enriched with glucose and bilge waste for 24 h	Evaluation of bioaugmentation and biosurfactant addition to enhance hydrocarbon biodegradation	Compositions of the oily phase: At 20 days control and bioaugmented were closed 7 % initial content; biosurfactant reach the similar level at 10 days Same degradation extent for all treatment in 20 days	Olivera et al. (2000)

80-l batch bioreactor with 47.5 l media	Agitation at 200 rpm Indoors nonaseptic conditions Room temperature 12–22 °C Continuous aeration (dissolved oxygen over 60 % saturation)	5.0 % v/v	Enriched with glucose and bilge waste for 24 h	Biodegradation at bench scale with biosurfactant (surfactin)	Compositions of the oily phase (17 days): Control loss 12 % initial TAl and 20 % TAr Culture with biosurfactant: reduction 83 % TAl[c]; 76 % TAr[d], and PAHs in the range of 70–100 % Feasibility to apply bioremediation at bench scale and relatively high waste concentration	Olivera et al. (2003)
125-ml Erlenmeyer with 50 ml media	Orbital agitation at 160 rpm Temperature 25 °C	1.0 % v/v	Enrichment with bilge waste transferred each 4 days	Effect of pH on hydrocarbon biodegradation and bioavailability	Emulsifier production with pH decrease during biodegradation pH control in the range 6.5–7.0 enhance aliphatic and aromatic biodegradation	Nievas et al. (2005)
3-l batch bioreactor with 2-l media	Continuous agitation at 250 rpm Temperature 25 °C Continuous aeration 1 vvm	1.0 % v/v	Enrichment with bilge waste transferred each 4 days	Biodegradation at controlled reaction conditions Evaporation quantitation	Feasibility of applied bioremediation at 1 % v/v without external biosurfactant added Hydrocarbon evaporation rates[c]	Nievas et al. (2006)

(continued)

Table 4.1 (continued)

Assay configuration	Reactions conditions	Initial concentration[a]	Consortium[b]	Study focus	Study results	References
3-l batch bioreactor with 2-l media	Continuous agitation at 250 rpm Temperature 25 °C Continuous aeration 1 vvm pH controlled between 7 and 8	0.53 % v/v 0.18 % v/v	Enrichment with bilge waste transferred each 4 days	Obtain hydrocarbon biodegrading kinetics under controlled conditions in batch reactor Evaporation quantitation	Sequential biodegradation of alkanes and UCM; key role of bioemulsification in UCM availability and biodegradation: Total reduction of hydrocarbons *n*-alkanes >96 % in 2 days TH > 68 % in 6 days[c]	Nievas et al. (2008a)

TAl total aliphatic hydrocarbons, *TAr* total aromatic hydrocarbons, *TH* total hydrocarbons, *d* days
[a]Experiments were performed with the oily phase of bilge wastes as substrate
[b]Autochthonous bilge waste consortium
[c]Measured by GC-FID
[d]Measured by fluorescence

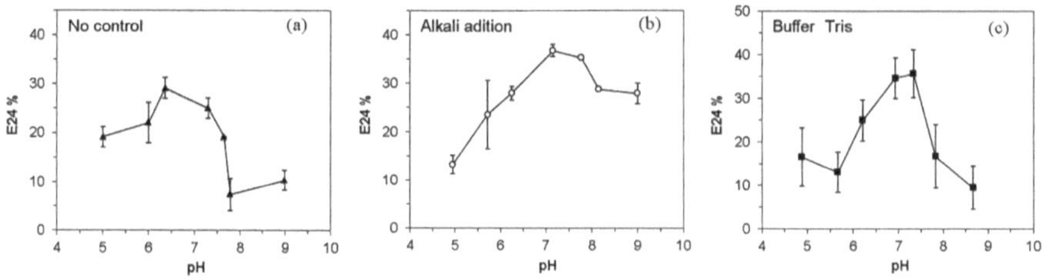

Fig. 4.3 Effect of pH on emulsification activity (E24%) of an emulsifier produced by the autochthonous microbial community of bilge waste (Nievas et al. 2005). E24% performed with a dispersion of the hydrocarbon phase of a 10-day culture, diluted 1:1 (v/v) in buffer solution of different pH. **a** Oily phase from a biodegradation assay without external modification of medium pH. **b** Oily phase from a biodegradation assay with intermittent alkali neutralization. **c** Oily phase from a biodegradation assay with 10 mM TRIS, pH 7.8

emulsifier was found associated to the oily phase. Moreover, the pH of the culture medium affected the emulsifier activity, being optimal for the range of 6.5–7.5 (Fig. 4.3) (Nievas et al. 2005). This observation means that enhanced biodegradation could be in part attributed to a better emulsification. Taking into account that the same microbial inoculum was used for all experiments and no significant differences were found in microbial biomass (assessed by microbial count), the enhancement in biodegradation could be better explained in terms of hydrocarbon accessibility to microorganisms than on the intrinsic metabolic capabilities of the microbial consortium. Thus, this study established neutral pH as the optimal condition to carry out bilge waste biodegradation.

The autochthonous microbial community showed a high biodegradation extent of aliphatic and aromatic hydrocarbons of bilge wastes. In outdoor aquaria experiments hydrocarbon reduction, assessing the composition of the oily phase, was 97 % and 94 % for the aliphatic and aromatic fractions in treatment with the native inoculum, whereas in a bioaugmented treatment reduction was near 98 % and 95 %, respectively, after 20 days. For the biosurfactant-added treatment, nearly 93 % reduction of the initial content for both hydrocarbon fractions was obtained in 10 days (Table 4.1) (Olivera et al. 2000). In the 80-l bioreactor batch experiment amendment with biosurfactant, 83 % and 76 % reduction for total aliphatic and total aromatic hydrocarbons of the initial content (5 % v/v BWOP) was found in 17 days (Table 4.1) (Olivera et al. 2000). Reduction in hydrocarbons in these experiments accounts for biodegradation and evaporation. Also, measurements were performed in the oily phase, which prevents closing a mass balance. Although some dilution of hydrocarbons in polar metabolites in the oily phase could exist, the high hydrocarbon diminutions are clear evidence of the potential of the autochthonous consortium to be applied in bioremediation of bilge wastes. Further, when bioremediation was assessed in 2-l batch bioreactors at 1 % v/v BWOP initial concentration without external surfactant added, reduction of the total mass in the reactor of the aliphatic and aromatic hydrocarbons and total *n*-hexane extractable were nearly 92 %, 99 %, and 85 %, respectively, in 14 days (Nievas et al. 2006). For experiments performed with another consortium, obtained from bilge wastes, reduction percentages were 96 % for *n*-alkanes in less than 3 days and 68–71 % for total hydrocarbons in 6 days, for initial waste concentrations of 0.18 % and 0.53 % v/v (Nievas et al. 2008a) (Table 4.1). Notably the greatest reduction in UCM, and in consequence in total hydrocarbons, in the culture with initial concentration of 0.53 % v/v BWOP happened in less than 24 h after complete emulsification of the reaction medium occurred by the self-biosurfactant-producing consortia. The sequential degradation of *n*-alkanes and further UCM after emulsifying production suggests that *n*-alkane depletion may be involved in emulsification triggering. Some hydrocarbon-degrading bacteria grow attached to hydrocarbon water droplets, and when droplets became exhausted of *n*-alkanes, cells detached from the interface, releasing the biopolymer, which remains attached to the hydrocarbons and acts as a potent emulsifier. In example, this occurs with *Acinetobacter calcaoceticus* producing Emulsan or *Alcanivorax* sp. strain growing on BWOP (Ron and Rosenberg 2002; Olivera et al. 2009). This fact also confirms the relevant role of emulsification during bilge waste biodegradation in a bioreactor, by a self-producing emulsifier consortium. In addition, a large extent of biodegradation was also obtained for individual 2- to 4-PAH parent compounds and alkylated derivatives in the range of 50–100 % assessed in the whole media in a 2-l bioreactor (Nievas et al. 2006) and of 71–100 % for the assessment in a 80-l bioreactor experiment in the oily phase (Olivera et al. 2003).

Hydrocarbon evaporation during biodegradation experiments was assessed in the gas off-stream from bioreactors aiming to provide information for dimensioning exhaust air treatment needs (Nievas et al. 2006, 2008a). Total hydrocarbon evaporation was in the range of 5–12 % of the initial hydrocarbon content for 0.18 %, 0.53 %, and 1 % v/v BWOP experiments carried out at 25 °C at 1 vvm aeration. The most important mass loss was observed in the first 2 days. Evaporated compounds were found in all hydrocarbon classes monitored up to volatilities similar to *n*-C16 when assessed by CG (Nievas et al. 2006, 2008a).

The studies summarized in Table 4.1 cannot be directly compared because they were performed with BWOP differing in composition, with autochthonous bilge waste obtained at different times and subsequently enriched under different conditions, and also analyzed with different methodologies. However, altogether these results constitute the basic knowledge to establish the biological treatment of bilge wastes as a viable option for the management of these wastes. Further investigation is needed to improve the process, and scale it up, as well also to evaluate new reaction configurations; for example, by providing a continuous treatment process. Combination of treatment technologies, such as biodegradation with a native microbial community step followed by separation by means of membrane filtration, is feasible and seems to be the best alternative for bioreactor-based processes, achieving short treatment times.

4.3.3.4 Bilge Waste Autochthonous Microbial Community

The native microbial community of bilge wastes from Puerto Madryn was studied through isolation of hydrocarbon-degrading microorganisms, assessing its metabolic hydrocarbon profile, and also by characterizing the naphthalene degradation potential. Olivera et al. (2003) isolated 14 strains from the oily phase of bilge wastes belonging to the genera *Pseudomonas* and *Vibrio*. The isolates identified with the API test resulted in *P. stutzeri, P. putrefaciens*, *Vibrio harveyi*, other unidentified *Pseudomonas* sp. and *Vibrio* sp. strains, and 4 unidentified strains. All the isolated strains were able to growth on naphthalene, except one *P. stutzeri* 78 that only grew on *n*-alkanes. The gene *nah*Ac was found ubiquitously distributed in the whole microbial community of the dumping pools at the water phase, oily phase, bottom, and surroundings. Moreover, 4 *Pseudomonas* isolates yield *nah*Ac gene amplification. This gene codifies the upper part of the catabolic pathways of naphthalene biodegradation to salicylate in many *Pseudomonas* (Habe and Omori 2003). Because some isolates that were able to degrade naphthalene showed no amplification of *nah*Ac, likely the bilge waste microbial community has a diversity of PAH catabolic genes divergent from the classical *nah*-like type (Olivera et al. 2003). In a later survey, 16 strains were isolated from the oily phase of bilge wastes (Nievas et al. 2006). Four of them were gram positive and showed no hydrocarbon degradation activity. The remaining 12 isolates, identified by 16S rRNA, belong to the genus *Pseudomonas* and were closely related to *P. stutzeri*. On the taxonomic level, members of *P. stutzeri* are grouped into at least 17 genomic variant groups termed genomovars (Lalucat et al. 2006). Five isolates from bilge wastes form a clade with *P. stutzeri* genomovar 3, where the *Pseudomonas* sp. NAP-3-1 strain isolated from PAH-polluted sediment and other isolates from the marine environment are grouped (Rockne et al. 2000). The bilge wastes isolates showed ability to degrade a broad range of hydrocarbon mixtures (diesel oil, petroleum, bilge wastes), aliphatics (*n*-hexane, *n*-hexadecane), and PAH hydrocarbons (naphthalene, phenanthrene, fluorene, pyrene) (Nievas et al. 2006). One of these *P. stutzeri* strains was further studied in the Microresp system with different hydrocarbons as the single carbon sources. It was able to mineralize the same PAHs and *n*-hexadecane, and it also formed biofilms (Sepúlveda et al. 2015). Many studies demonstrated the high metabolic versatility of *P. stutzeri* and its capacity to degrade hazardous compounds including petroleum hydrocarbons, particularly 2- to 4-ring PAH compounds (Stringfellow and Aitken 1995; Rockne et al. 2000; Röling et al. 2004; Santisi et al. 2015).

On the other hand, the microbial community enriched from the oily phase of bilge waste from Puerto Madryn was analyzed by polymerase chain reaction-denaturing gradient gel electrophoretic (PCR-DGGE), a culture-independent method, to identify the dominant taxons and to assess its population dynamics during bilge waste biodegradation (Nievas et al. 2010). An aerated 2-l batch bioreactor, with initial content of 0.53 % v/v BWOP in mineral medium inoculated with the autochthonous bilge waste microbial consortium, was evaluated at different times (Table 4.1) (Nievas et al. 2008a). Members of the genus *Marinobacter* were dominant in the enrichment, and *Pseudomonas*, *Shewanella*, and *Halomonas* were also detected. During the biodegradation experiment, the exponential growth phase agreed with *n*-alkane depletion and an increase in the prevalence of *Pseudomonas* and *Shewanella*. Only after emulsification happened, biodegradation of the more recalcitrant hydrocarbons found in an UCM occurred, associated with a predominance of *Marinobacter* and *Shewanella* (Nievas et al. 2010). Remarkably, strains belonging to the genera *Shewanella*, *Marinobacter*, *Halomonas*, *Pseudomonas*, and *Vibrio* can use diverse electron acceptors other than oxygen, most of them being able to reduce nitrate coupled with hydrocarbon degradation (Yakimov et al. 1998; Lalucat et al. 2006; Hau and Gralnick 2007; Grimaud 2010; Kaser and Coates 2010; Duran 2010). *P. stutzeri* is able to transform nitrate in nitrogen, and several strains have been isolated under nitrate-reducing conditions as degraders of dibenzothiophene and 2- to 4-ring PAHs (Lalucat et al. 2006). *Shewanella*, a genus of which most of the known isolates are from the marine environment, is a facultative anaerobe and has the ability to utilize a diverse array of final electronic acceptors such as nitrate and metals, which may explain its adaptation to extreme environmental conditions. Its broad metabolic capabilities include biodegradation of hydrocarbons and halogenated hydrocarbons (Hau and Gralnick 2007). *Vibrio* and *Marinobacter* are both nitrate utilizers and hydrocarbon degraders in the marine environment (Duran 2010; Kaser and Coates 2010; Rodriguez-R et al. 2015). *Halomonas* are aerobic chemo-organoheterotrophs, frequently found in hypersaline environments contaminated with oil, and have been identified also in halophilic oil-degrading consortia (McGenity 2010). Many *Halomonas* strains have the ability to grow by respiring nitrate (McGenity 2010). Bilge waste typically becomes anoxic, both in ships and also in storage tanks or dumping pools, because the oily floating phase prevents its aeration. This feature seem to be determinant in selecting typically marine hydrocarbon degraders, which are able to survive and even develop favourably under facultative/anoxic conditions, in the microbial community that colonizes bilge waste from Puerto Madryn dumpling pools.

4.4 Conclusions

The increasing demand of higher environmental standards at the global level requires efficient, safer, and low-cost biotechnological processes to recover impacted sites and to treat and dispose hazardous waste. Argentinean Patagonia, the microbial diversity of which is still mostly unknown, remains as an interesting reservoir of marine microorganisms with potential for biotechnological application related to environmental hydrocarbon bioremediation. This chapter briefly introduces the bioprospection of microorganisms and microbial communities recovered from some polluted sites and wastes from Patagonia, which revealed hydrocarbon biodegradation potential for bioremediation processes development. An *Alcanivorax* strain was isolated from sediments with a hydrocarbon pollution history, reinforcing the broad distribution in temperate marine environment of this hydrocarbonoclastic bacteria and its presence in the South Atlantic. The autochthonous microbial communities of bilge wastes from the port of Puerto Madryn harbor bacteria from the marine origin, described as versatile hydrocarbon degraders of both aliphatic and aromatic hydrocarbons and producers of biosurfactants and bioemulsifiers. Efficient aerobic hydrocarbon biodegradation from bilge waste was demonstrated by the autochthonous microbial community at different experimental scales and reactor configurations, denoting that further research focus on scale-up of hydrocarbon remediation processes, and its combination with separation technologies, would be worthwhile to establish feasible biological treatment for hydrocarbon-polluted saline waters. Moreover, bilge waste microbial communities are promising candidates to be applied in versatile bioreactor waste treatments where oxygen availability varies significantly, such as a biofilm reactor or sequencing batch reactor, because presumably such may be able also to utilize hydrocarbon under facultative/anoxic conditions.

Acknowledgments The authors gratefully acknowledge support from CONICET (PIP 112-200801-01856) and ANPCyT (PICT-2006-01876, PICT-2011-01242) of Argentina.

References

Abbasian F, Lockington R, Megharaj M, Naidu R (2015) A review on the genetics of aliphatic and aromatic hydrocarbon degradation. Appl Biochem Biotechnol 178:224–250

Abraham WR, Meyer H, Yakimov M (1998) Novel glycine containing glucolipids from the alkane using bacterium *Alcanivorax borkumensis*. Biochim Biophys Acta 1393:57–62

Auffret M, Labbé D, Thouand G, Greer CW, Fayolle-Guichard F (2009) Degradation of a mixture of hydrocarbons, gasoline, and diesel oil additives by *Rhodococcus aetherivorans* and *Rhodococcus wratislaviensis*. Appl Environ Microbiol 75:7774–7782

Bilgili MS, Ince M, Tari GT, Adar E, Balahorli V, Yildiz S (2016) Batch and continuous treatability of oily wastewaters from port waste reception facilities: a pilot scale study. J Electroanal Chem 760:119–126

Camilli R, Reddy CM, Yoerger DR, Van Mooy BAS, Jakuba MV, Kinsey JC, McIntyre CP, Sylva SP, Maloney JV (2010) Tracking hydrocarbon plume transport and biodegradation at deepwater horizon. Science 330(6001):201–204

Cappello S, Crisari A, Denaro R, Crescenzi F, Porcelli F, Yakimov M (2011) Biodegradation of a bioemulsificant exopolysaccharide (EPS2003) by marine bacteria. Water Air Soil Pollut 214:645–652

Cappello S, Denaro R, Genovese M, Giuliano L, Yakimov MM (2007) Predominant growth of *Alcanivorax* during experiments on "oil spill bioremediation" in mesocosms. Microbiol Res 162:185–190

Cazoir D, Fine L, Ferronato C, Chovelon JM (2012) Hydrocarbon removal from bilgewater by a combination of air-stripping and photocatalysis. J Hazard Mater 235-236:159–168

CEU (Council of the European Union) (2006) Regulation (EC) No 1013/2006 of the European Parliament and of the Council of 14 June 2006 on shipments of waste. Offic J Eur Commun 49N°L190, 1-98. ISSN 1725-2555

Commendatore MG, Esteves JL (2004) Natural and anthropogenic hydrocarbons in sediments from the Chubut River (Patagonia, Argentina). Mar Pollut Bull 48:910–918

Commendatore MG, Esteves JL (2007) An assessment of oil pollution in the coastal zone of Patagonia, Argentina. Environ Manag 40:814–821
Commendatore MG, Esteves JL, Colombo JC (2000) Hydrocarbons in coastal sediments of Patagonia, Argentina: levels and probable sources. Mar Pollut Bull 40:989–998
Commendatore MG, Franco MA, Gomes Costa P, Castro IB, Fillmann G, Bigatti G, Esteves JL, Nievas ML (2015) Butyltins, polyaromatic hydrocarbons, organochlorine pesticides, and polychlorinated biphenyls in sediments and bivalve mollusks in a mid-latitude environment from the Patagonian coastal zone. Environ Toxicol Chem 34:2750–2763
Commendatore MG, Nievas ML, Amin O, Esteves JL (2012) Sources and distribution of aliphatic and polyaromatic hydrocarbons in coastal sediments from the Ushuaia Bay (Tierra del Fuego, Patagonia, Argentina). Mar Environ Res 74:20–31
Cravo-Laureau C, Duran R (2014) Marine coastal sediments microbial hydrocarbon degradation processes: contribution of experimental ecology in the omics' era. Front Microbiol 5:39
Damsté JSS, Muyzer G, Abbas B, Rampen SW, Massé G, Allard WG, Belt ST, Robert J-M, Rowland SJ, Moldowan JM, Barbanti SM, Fago FJ, Denisevich P, Dahl J, Trindade LAF, Schouten S (2004) The rise of the rhizosolenid diatoms. Science 304:584–587
Díez S, Jover E, Bayona JM, Albaigés J (2007) Prestige oil spill. III. Fate of a heavy oil in the marine environment. Environ Sci Technol 41:3075–3082
Dionisi HM, Lozada M, Olivera NL (2012) Bioprospection of marine microorganisms: biotechnological applications and methods. Rev Argent Microbiol 44:49–60
Duran R (2010) *Marinobacter*. In: Timmis K (ed) Handbook of hydrocarbon and lipid microbiology. Springer, Berlin, pp 1725–1735
Er-Raioui H, Bouzid S, Marhraoui M, Saliot A (2009) Hydrocarbon pollution of the Mediterranean coastline of Morocco. Ocean Coast Manag 52:124–129
Esteves JL, Commendatore MG, Nievas ML, Paletto VM, Amín O (2006) Hydrocarbon pollution in coastal sediments of Tierra del Fuego Islands, Patagonia Argentina. Mar Pollut Bull 52:582–590
Fakhru'l-Razi A, Pendashteh A, Abdullah LC, Biak DRA, Madaeni SS, Abidin ZZ (2009) Review of technologies for oil and gas produced water treatment. J Hazard Mater 170:530–551
Fakhru'l-Razi A, Pendashteh A, Abidin ZZ, Abdullah LC, Biak DRA, Madaeni SS (2010) Application of membrane-coupled sequencing batch reactor for oilfield produced water recycle and beneficial re-use. Bioresour Technol 101:6942–6949
Garcia-Blanco S, Venosa AD, Suidan MT, Lee K, Cobanli S, Haines JR (2007) Biostimulation for the treatment of an oil-contaminated coastal salt marsh. Biodegradation 18:1–15
Gertler C, Näther DJ, Cappello S, Gerdts G, Quilliam RS, Yakimov MM, Golyshin PN (2012) Composition and dynamics of biostimulated indigenous oil-degrading microbial consortia from the Irish, North and Mediterranean Seas: a mesocosm study. FEMS Microbiol Ecol 81:520–536
Ghidossi R, Veyret D, Scotto JL, Jalabert T, Moulin P (2009) Ferry oily wastewater treatment. Sep Purif Technol 64:296–303
Golyshin PN, Martins Dos Santos VAP, Kaiser O, Ferrer M, Sabirova YS, Lünsdorf H, Chernikova TN, Golyshina OV, Yakimov MM, Pühler A, Timmis KN (2003) Genome sequence completed of *Alcanivorax borkumensis*, a hydrocarbon-degrading bacterium that plays a global role in oil removal from marine systems. J Biotechnol 106:215–220
Gong Y, Zhao X, Cai Z, O'Reilly SE, Hao X, Zhao D (2014) A review of oil, dispersed oil and sediment interactions in the aquatic environment: influence on the fate, transport and remediation of oil spills. Mar Pollut Bull 79:16–33
Grimaud R (2010) *Marinobacter*. In: Timmis K (ed) Handbook of hydrocarbon and lipid microbiology. Springer, Berlin, pp 1289–1296
Guibert LM, Loviso CL, Borglin S, Jansson JK, Dionisi HM, Lozada M (2016) Diverse bacterial groups contribute to the alkane degradation potential of chronically polluted subantarctic coastal sediments. Microb Ecol 71:100–112
Habe H, Omori T (2003) Genetics of polycyclic aromatic hydrocarbon metabolism in diverse aerobic bacteria. Biosci Biotechnol Biochem 67:225–243

Harayama S, Kasai Y, Hara A (2004) Microbial communities in oil-contaminated seawater. Curr Opin Biotechnol 15:205–214

Hau HH, Gralnick JA (2007) Ecology and biotechnology of the genus *Shewanella*. Annu Rev Microbiol 61:237–258

Haus F, German J, Junter G-A (2001) Primary biodegradability of mineral base oils in relation to their chemical and physical characteristics. Chemosphere 45:983–990

Head IM, Jones DM, Röling WF (2006) Marine microorganisms make a meal of oil. Nat Rev Microbiol 4:173–182

Ibarbalz FM, Figuerola ELM, Erijman L (2013) Industrial activated sludge exhibit unique bacterial community composition at high taxonomic ranks. Water Res 47:3854–3864

International Maritime Organization (IMO) (2004) Resolution MEPC.117(52): Amendments to the annex of the protocol of 1978 relating to the international convention for the prevention of pollution from ships, 1973, 2000

Kaiser MJ, Attrill MJ (2011) Marine ecology: processes, systems, and impacts. Oxford University Press, Oxford

Kanaly RA, Harayama S (2000) Biodegradation of high-molecular-weight polycyclic aromatic hydrocarbons by bacteria. J Bacteriol 182:2059–2067

Kanaly RA, Harayama S (2010) Advances in the field of high-molecular-weight polycyclic aromatic hydrocarbon biodegradation by bacteria. Microb Biotechnol 3:136–164

Karakulski K, Morawski WA, Grzechulska J (1998) Purification of bilge water by hybrid ultrafiltration and photocatalytic processes. Sep Purif Technol 14:163–173

Kaser F, Coates J (2010) Nitrate, perchlorate and metal respirers. In: Timmis K (ed) Handbook of hydrocarbon and lipid microbiology. Springer, Berlin, pp 2033–2047

Khan FI, Husain T, Hejazi R (2004) An overview and analysis of site remediation technologies. J Environ Manag 71:95–122

Kostianoy AG, Litovchenko KT, Lebedev SA, Stanichny SV, Soloviev DM, Pichuzhkina OE (2005) Operational satellite monitoring of oil spill pollution in the southeastern Baltic Sea. In: Oceans 2005—Europe, vol 181, pp 182–183. doi: 10.1109/OCEANSE.2005.1511706. Accessed 20–23 June 2005

Ladygina N, Dedyukhina EG, Vainshtein MB (2006) A review on microbial synthesis of hydrocarbons. Process Biochem 41:1001–1014

Lalucat J, Bennasar A, Bosch R, García-Valdés E, Palleroni NJ (2006) Biology *of Pseudomonas stutzeri*. Microbiol Mol Biol Rev 70:510–547

Li L, Liu X, Yang W, Xu F, Wang W, Wang L, Feng L, Bartlam M, Rao Z (2008) Crystal structure of long-chain alkane monooxygenase (LadA) in complex with coenzyme FMN: unveiling the long-chain alkane hydroxylase. J Mol Biol 376:453–465

Lin B, Lin CY, Jong TC (2007) Investigation of strategies to improve the recycling effectiveness of waste oil from fishing vessels. Mar Policy 31:415–420

Lozada M, Marcos MS, Commendatore MG, Gil MN, Dionisi HM (2014) The bacterial community structure of hydrocarbon-polluted marine environments as the basis for the definition of an ecological index of hydrocarbon exposure. Microbes Environ 29:269–276

Lucas Z, MacGregor C (2006) Characterization and source of oil contamination on the beaches and seabird corpses, Sable Island, Nova Scotia, 1996–2005. Mar Pollut Bull 52:778–789

Makkar RS, Rockne KJ (2003) Comparison of synthetic surfactants and biosurfactants in enhancing biodegradation of polycyclic aromatic hydrocarbons. Environ Toxicol Chem 22:2280–2292

Mancini G, Cappello S, Yakimov MM, Polizzi A, Torregrossa M (2012) Biological approaches to the treatment of saline oily waste (waters) originated from marine transportation. Chem Eng Trans 27:37–42, doi: http://citeseerx.ist.psu.edu/viewdoc/download?doi=10.1.1.461.9337&rep=rep1&type=pdf

Massara Paletto V, Commendatore M, Esteves J (2008) Hydrocarbon levels in sediments and bivalve mollusks from Bahía Nueva (Patagonia, Argentina): an assessment of probable origin and bioaccumulation factors. Mar Pollut Bull 56:2100–2105

McDonald IR, Miguez CB, Rogge G, Bourque D, Wendlandt KD, Groleau D, Murrell JC (2006) Diversity of soluble methane monooxygenase-containing methanotrophs isolated from polluted environments. FEMS Microbiol Lett 255:225–232
McGenity TJ (2010) Halophilic hydrocarbon degraders. In: Timmis KN (ed) Handbook of hydrocarbon and lipid microbiology. Springer, Berlin, pp 1939–1951
McGenity TJ, Folwell BD, McKew BA, Sanni GO (2012) Marine crude-oil biodegradation: a central role for interspecies interactions. Aquat Biosyst 8(10):10.1186. doi:10.1186/2046-9063-8-10
McKew BA, Coulon F, Osborn AM, Timmis KN, McGenity TJ (2007) Determining the identity and roles of oil-metabolizing marine bacteria from the Thames estuary, UK. Environ Microbiol 9:165–176
Mercopress (2007) Huge oil slick washes ashore north of Comodoro Rivadavia http://en.mercopress.com/2007/12/27/huge-oil-slick-washes-ashore-north-of-comodoro-rivadavia
Ministerio de Energía y Minería de la República Argentina (MMyERA) (2016) Información estadística de hidrocarburos. Producción de Petróleo y Gas. http://www.energia.gov.ar/home/hidrocarburos.php
Morán AC, Martínez MA, Siñeriz F (2002) Quantification of surfactin in culture supernatants by hemolytic activity. Biotechnol Lett 24:177–180
Morán AC, Olivera N, Commendatore M, Esteves JL, Siñeriz F (2000) Enhancement of hydrocarbon waste biodegradation by addition of a biosurfactant from *Bacillus subtilis* O9. Biodegradation 11:65–71
Musat F, Wilkes H, Behrends A, Woebken D, Widdel F (2010) Microbial nitrate-dependent cyclohexane degradation coupled with anaerobic ammonium oxidation. ISME J 4:1290–1301
National Research Council (NRC) (2003) Oil in the sea. III: Inputs, fates, and effects. The National Academies Press, Washington, DC
Nievas ML (2006) Diseño, simulación y optimización de un reactor biológico para el tratamiento de efluentes orgánicos provenientes de sentinas de buques. Ph.D. Thesis, Universidad Nacional del Sur, Bahía Blanca, Argentina
Nievas ML, Commendatore MG, Esteves JL, Bucalá V (2005) Effect of pH modification on bilge waste biodegradation by a native microbial community. Int Biodeter Biodegradr 56:151–157
Nievas ML, Commendatore MG, Esteves JL, Bucalá V (2008a) Biodegradation pattern of hydrocarbons from a fuel oil-type complex residue by an emulsifier-producing microbial consortium. J Hazard Mater 154:96–104
Nievas ML, Commendatore MG, Faleschini M, Sepúlveda MA, Esteves JL, Bucalá V (2008b) Evaluación de reactores de biofilm para el tratamiento de aguas de sentina de buques. Proceedings of II Argentinean Congress of SETAC, p 155
Nievas ML, Commendatore MG, Olivera NL, Esteves JL, Bucalá V (2006) Biodegradation of bilge waste from Patagonia with an indigenous microbial community. Bioresour Technol 97:2280–2290
Nievas ML, Ferrero M, Olivera NL, Dionisi HM, Commendatore MG, Esteves JL, Bucalá V (2010) Dynamics of a marine-microbial community during biodegradation of bilge waste hydrocarbons. Biocell 34:57
Nitschke M, Pastore GM (2006) Production and properties of a surfactant obtained from *Bacillus subtilis* grown on cassava wastewater. Bioresour Technol 97:336–341
Olivera N, Commendatore M, Morán A, Esteves J (2000) Biosurfactant-enhanced degradation of residual hydrocarbons from ship bilge wastes. J Ind Microbiol Biotechnol 25:70–73
Olivera NL, Commendatore MG, Delgado O, Esteves JL (2003) Microbial characterization and hydrocarbon biodegradation potential of natural bilge waste microflora. J Ind Microbiol Biotechnol 30:542–548
Olivera NL, Nievas ML (2010) Biosurfactants and their uses in the petroleum industry and hydrocarbon pollution remediation. In: Hagen ET (ed) Detergent: types, components and uses. Nova Science, Commack, pp 1–48

Olivera NL, Nievas ML, Lozada M, del Prado G, Dionisi HM, Siñeriz F (2009) Isolation and characterization of biosurfactant-producing *Alcanivorax* strains: hydrocarbon accession strategies and alkane hydroxylase gene analysis. Res Microbiol 160:19–26

Önenç S, Brebu M, Vasile C, Yanik J (2012) Copyrolysis of scrap tires with oily wastes. J Anal Appl Pyrol 94:184–189

Pérez-Pantoja D, González B, Pieper D (2010) Aerobic degradation of aromatic hydrocarbons. In: Timmis K (ed) Handbook of hydrocarbon and lipid microbiology. Springer, Berlin, pp 799–837

Peypoux F, Bonmatin J, Wallach J (1999) Recent trends in the biochemistry of surfactin. Appl Microbiol Biotechnol 51:553–563

Portela JR, Sanchez-Oneto J, López J, Nebot E, Martínez de la Ossa E (2003) Hydrothermal oxidation of oily wastes: an alternative to conventional treatment methods. Eng Life Sci 3:85–89

Prefectura Naval Argentina (PNA) (1998) Designación de zonas de protección especial en el litoral Argentino. Ordenanza N° 12/98 (DPMA), Tomo 6: Régimen para la Protección del Medio Ambiente, PNA, Buenos Aires

Prince R, Gramain A, McGenity TJ (2010) Prokaryotic hydrocarbon degraders. In: Timmis K (ed) Handbook of hydrocarbon and lipid microbiology. Springer, Berlin, pp 1669–1692

Reddy MV, Devi MP, Chandrasekhar K, Goud RK, Mohan SV (2011) Aerobic remediation of petroleum sludge through soil supplementation: microbial community analysis. J Hazard Mater 197:80–87

Rincón GJ, La Motta EJ (2014) Simultaneous removal of oil and grease, and heavy metals from artificial bilge water using electro-coagulation/flotation. J Environ Manag 144:42–50

Rizvi SQA (2009) Lubricant additives. Emissions in an internal combustion engine. In: Rizvi SQA (ed) A comprehensive review of lubricant chemistry, technology, selection, and design. ASTM, West Conshohocken, pp 100–208, 322–333

Rockne KJ, Chee-Sanford JC, Sanford RA, Hedlund BP, Staley JT, Strand SE (2000) Anaerobic naphthalene degradation by microbial pure cultures under nitrate-reducing conditions. Appl Environ Microbiol 66:1595–1601

Rocha e Silva NMP, Rufino RD, Luna JM, Santos VA, Sarubbo LA (2014) Screening of *Pseudomonas* species for biosurfactant production using low-cost substrates. Biocatal Agric Biotechnol 3:132–139

Rodgers RP, Schaub TM, Marshall AG (2005) Petroleomics: MS returns to its roots. Anal Chem 77:20A–27A. doi:10.1021/ac053302y

Rodriguez-R LM, Overholt WA, Hagan C, Huettel M, Kostka JE, Konstantinidis KT (2015) Microbial community successional patterns in beach sands impacted by the Deepwater Horizon oil spill. ISME J. doi:10.1038/ismej.2015.5

Rojo F (2010) Enzymes for aerobic degradation of alkanes. In: Timmis K (ed) Handbook of hydrocarbon and lipid microbiology. Springer, Berlin, pp 781–797

Röling WF, Milner MG, Jones DM, Fratepietro F, Swannell RP, Daniel F, Head IM (2004) Bacterial community dynamics and hydrocarbon degradation during a field-scale evaluation of bioremediation on a mudflat beach contaminated with buried oil. Appl Environ Microbiol 70:2603–2613

Ron EZ, Rosenberg E (2002) Biosurfactants and oil bioremediation. Curr Opin Biotechnol 13:249–252

Rossello EA, Haring CE, Cardinali G, Suárez F, Laffitte GA, Nevistic AV (2008) Hydrocarbons and petroleum geology of Tierra del Fuego, Argentina. Geol Acta 6:69–83

Sabirova JS, Ferrer M, Regenhardt D, Timmis KN, Golyshin PN (2006) Proteomic insights into metabolic adaptations in *Alcanivorax borkumensis* induced by alkane utilization. J Bacteriol 188:3763–3773

Santisi S, Gentile G, Volta A, Bonsignore M, Mancini G, Quatrini P, Cappello S (2015) Isolation and characterization of oil-degrading bacteria from bilge water. Growth 17:45–49

Schaller KD, Fox SL, Bruhn DF, Noah KS, Bala GA (2004) Characterization of surfactin from *Bacillus subtilis* for application as an agent for enhanced oil recovery. Appl Biochem Biotechnol 115:827–836

Sepúlveda M, Harris E, Olivera N, Ormazabal A, Bucalá V, Nievas M (2012) Interacción de microorganismos degradadores de hidrocarburos en la formación de biofilm sobre superficies sólidas. VIII Congreso Argentino de Microbiología General (SAMIGE), Mar del Plata, Julio de 2012

Sepúlveda M, Revuelta F, Olivera N, Nievas M (2015) Formación de biofilm de microorganismos marinos degradadores de hidrocarburos: Influencia del sustrato de crecimiento. III Congreso Argentino de Microbiología Agrícola y Ambiental. Buenos Aires, Noviembre de 2015

Shokrollahzadeh S, Azizmohseni F, Golmohammad F, Shokouhi H, Khademhaghighat F (2008) Biodegradation potential and bacterial diversity of a petrochemical wastewater treatment plant in Iran. Bioresour Technol 99:6127–6133

Soberón-Chávez G, Lépine F, Déziel E (2005) Production of rhamnolipids by *Pseudomonas aeruginosa*. Appl Microbiol Biotechnol 68:718–725

Speight JG (2011) Feedstocks. In: Speight JG (ed) The refinery of the future. Elsevier, Oxford, pp 1–22

Stout SA, Wang W (2007) Chemical fingerprinting of spilled or discharged petroleum: methods and factors affecting petroleum fingerprints in the environment. In: Stout SA, Wang W (eds) Oil spill environmental forensics. Fingerprinting and source identification. Elsevier, Amsterdam, pp 29–35

Stringfellow WT, Aitken MD (1995) Competitive metabolism of naphthalene, methylnaphthalenes, and fluorene by phenanthrene-degrading pseudomonads. Appl Environ Microbiol 61:1357–1362

Sun C, Leiknes T, Weitzenböck J, Thorstensen B (2009) The effect of bilge water on a biofilm-MBR process in an integrated shipboard wastewater treatment system. Desalination 236:56–64

Sun C, Leiknes T, Weitzenböck J, Thorstensen B (2010) Development of a biofilm-MBR for shipboard wastewater treatment: the effect of process configuration. Desalination 250:745–750

Sylwan CA (2001) Geology of the Golfo San Jorge Basin, Argentina (Geología de la Cuenca del Golfo San Jorge, Argentina). J Iberian Geol 27:123–158

USEPA (2000) Innovative remediation technologies: field-scale demonstration projects in North America (EPA 542-B-00-004), 2nd edn. USEPA, Washington, DC

USEPA (2008) Cruise ship discharge assessment report (EPA 842-R-07-005). USEPA, Washington, DC

van Beilen JB, Funhoff EG (2007) Alkane hydroxylases involved in microbial alkane degradation. Appl Microbiol Biotechnol 74:13–21

van Beilen JB, Funhoff EG, van Loon A, Just A, Kaysser L, Bouza M, Holtackers R, Röthlisberger M, Li Z, Witholt B (2006) Cytochrome P450 alkane hydroxylases of the CYP153 family are common in alkane-degrading eubacteria lacking integral membrane alkane hydroxylases. Appl Environ Microbiol 72:59–65

van Beilen JB, Li Z, Duetz WA, Smits THM, Witholt B (2003) Diversity of alkane hydroxylase systems in the environment. Oil Gas Sci Technol 58:427–440

Vendeuvre C, Ruiz-Guerrero R, Bertoncini F, Duval L, Thiébaut D (2007) Comprehensive two-dimensional gas chromatography for detailed characterisation of petroleum products. Oil Gas Sci Technol 62:43–55

Wang L, Wang W, Lai Q, Shao Z (2010) Gene diversity of CYP153A and AlkB alkane hydroxylases in oil-degrading bacteria isolated from the Atlantic Ocean. Environ Microbiol 12:1230–1242

Wentzel A, Ellingsen TE, Kotlar H-K, Zotchev SB, Throne-Holst M (2007) Bacterial metabolism of long-chain *n*-alkanes. Appl Microbiol Biotechnol 76:1209–1221

Yakimov MM, Golyshin PN, Lang S, Moore ERB, Abraham WR, Lünsdorf H, Timmis KN (1998) *Alcanivorax borkumensis* gen. nov., sp. nov., a new, hydrocarbon-degrading and surfactant-producing marine bacterium. Int J Syst Evol Microbiol 48:339–348

Yakimov MM, Timmis KN, Golyshin PN (2007) Obligate oil-degrading marine bacteria. Curr Opin Biotechnol 18:257–266

Zhu X, Venosa AD, Suidan MT, Lee K (2001) Guidelines for the bioremediation of marine shorelines and freshwater wetlands. US Environmental Protection Agency, Washington, DC

Chapter 5
Assessment of Microbial Patagonian Communities for Using in Heavy Metal Bioremediation

Ana Belén Segretin, María Alejandra Lima, Graciana Willis Poratti, María Sofía Urbieta, Cecilia Bernardelli, Alejandra Giaveno, and Edgardo Donati

Abstract Assessing microbial communities in extreme environments allows the search for novel extremophilic microorganisms that can be of use in the development or improvement of biotechnological processes. Most microbial communities developed in such harsh environments have different heavy metal resistance strategies of importance in some technological applications as biomining but also in the bioremediation of metal-polluted environments. In this chapter we describe the microbial diversity of an acidic, volcanic geothermal environment in Northern Patagonia of Argentina: the Copahue geothermal system, containing different geothermal manifestations with temperatures up to 90 °C and pH values from 2 to 7 under aerobic and anaerobic conditions, provoking an enormous biodiversity. In addition, we report some heavy metal applications of those communities, focusing mainly in the bioprecipitation of heavy metals using sulfate-reducing microorganisms.

5.1 Introduction

Assessment of prokaryotic biodiversity in extreme environments allows much more than a simple description of the microorganisms inhabiting a particular habitat. The knowledge of microbial diversity provides the chance to understand the species interrelationships, the influence of the environmental parameters on the community

A.B. Segretin • M.A. Lima • G. Willis Poratti • M.S. Urbieta • C. Bernardelli
CINDEFI (CCT La Plata-CONICET, UNLP), Facultad de Ciencias Exactas, Universidad Nacional de La Plata, Calle 47 y 115, 1900 La Plata, Argentina

E.R. Donati (✉)
Industrial Fermentation Research and Development Center (CINDEFI), CONICET-UNLP, School of Science, National University of La Plata, La Plata, Argentina
e-mail: donati@quimica.unlp.edu.ar

A. Giaveno
PROBIEN (CONICET, UNCo), Facultad de Ingeniería, Universidad Nacional del Comahue, Buenos Aires 1400, 8300 Neuquén, Argentina

N.L. Olivera et al. (eds.), *Biology and Biotechnology of Patagonian Microorganisms*, DOI 10.1007/978-3-319-42801-7_5

structure, and the role of those microorganisms in the geochemical cycles. But even more important, from such assessment the chances to isolate new species of microorganisms useful for different biotechnological applications are significantly improved (Rothschild and Mancinelli 2001).

Prokaryotic biodiversity is particularly attractive in thermal and/or acidic environments containing considerable heavy metal concentrations (Burgess et al. 2012). These extremophiles have developed mechanisms of resistance to thrive in the presence of heavy metals, and such strategies can be used for the bioremediation of those toxic chemical species. Within these extremophilic microorganisms, thermophiles can increase the rate of the processes because they can grow at moderate and high temperatures (Meyer-Dombard et al. 2005; Kublanov et al. 2009; Reigstad et al. 2010; Wemheuer et al. 2013; Urbieta et al. 2015a, b). In the same way, acidophilic microorganisms offer some additional advantages in processes working with heavy metals because most of them are soluble at low pH values.

Patagonia has many extreme environments and some geothermal zones very attractive for assessing polyextremophiles. Our research team has studied the prokaryotic diversity of an acidic, volcanic geothermal environment in Patagonia: the Copahue geothermal system, which is situated in the Cordillera Norpatagónica in the northwest of Neuquén Province, Argentina. The area is crowned by the Copahue volcano, whose activity produced diverse geothermal manifestations such as mud cones, hot springs, ponds, and pools grouped in five main areas named Copahue Thermal Centre, Las Máquinas, Las Maquinitas, Anfiteatro y Chancho-Co (over the Chilean side) (Varekamp et al. 2009). Most samples studied by our research team belong to Las Máquinas and Las Maquinitas. Las Máquinas has a large acidic (pH 3.2) moderate temperature (36 °C) pool with sulfur and pyrite deposits. Las Maquinitas shows hot springs with the most extreme conditions (temperature 90 °C and pH 2). Some photographs of these places are shown in Fig. 5.1.

Close to the crater of the Copahue volcano, there are two acidic (pH 0.3–2.3) thermal springs, whose streams meet forming the Rio Agrio. Rio Agrio runs down the volcano slope forming various cascades and finally discharges in Caviahue Lake at 13.5 km; along the way, the temperature of the water decreases greatly but remain very acidic, with pH values close to 1. Figure 5.2 shows some images from Rio Agrio.

In this chapter we describe the microbial diversity detected in several samples taken from the Copahue geothermal system. In addition, we report some heavy metal potential applications of those communities, focusing on the bioprecipitation of heavy metals and bioreduction of hexavalent chromium. Hyperresistance to arsenic manifested by some microbial communities is also reported.

5.2 Biodiversity Assessment

Our assessment strategy implies three different approaches, enrichment of the samples to the isolation of autochthonous extremophilic microorganisms, and two culture-independent alternatives using microbial ecology techniques such as

Fig. 5.1 Ponds and pools at Copahue geothermal field

Fig. 5.2 Cascades belong to Rio Agrio

amplification and sequencing of the complete 16S rRNA gene of the entire community and fluorescent in situ hybridization (FISH) for quantitative information on the species distribution and community structure.

5.2.1 Materials and Methods

Temperature, electrical conductivity, redox potential, and pH of the samples were measured using calibrated instruments. Water samples were collected in sterile plastic jars. Samples were fixed for FISH in the field using paraformaldehyde to achieve a 4 % final concentration. Samples were incubated for 4–12 h, then diluted in sterile water and filtered through a 0.22-μm membrane. Filters were washed and neutralized with phosphate-buffered saline (PBS) buffer. Different universal and specific CY3-labeled probes were used on the filters, and 4′,6′-diamidino-2-phenylindole (DAPI) stain was used in all hybridizations to evaluate total cell numbers. Vectashield mounting medium (Vector Laboratories, Burlingame, CA, USA) was added to preparations to avoid fluorescence fading. A Leica DM 2500 epifluorescence microscope was used to visualize hybridization. Images were taken using a Leica DFC 300 FX camera and its corresponding software (Leica Microscopy Systems, Heerbrugg, Switzerland). Total cell density was calculated as the average of at least 50 DAPI-stained fields. Hybridization percentages for universal probes were calculated as the quotient of the average recount of 20 hybridized fields divided by the average recount of those same fields with DAPI stain. Hybridization percentages for specific probes were calculated considering EUB338-based cell counts as 100 %.

A portion of each original sample was inoculated into specific media using different energy sources under physicochemical conditions similar to the environments where they were taken (more details can be found following). Previous results have been published in Chiacchiarini et al. (2010).

Other portions of the samples were filtered through 0.22-μm membranes. The filtrates were used for chemical analysis. Different metal concentrations were determined by atomic absorption spectrophotometry. The material retained on the membranes, after washing with sterile water and TE buffer, was exposed to DNA extraction using the Fast DNA Spin Kit. Cells were disrupted using a mixture of ceramic and silica beads in a vortex at maximum speed. 16S rRNA genes were amplified by polymerase chain reaction (PCR) using forward primers 8F: 5′-AGAGTTTGATC(A/C)TGGC-3′ for Bacteria and 25F: 5′-TCYGGTTGATCCYGCCRG-3′ for Archaea (Achenbach and Woese 1999), whereas the reverse primer for both was 1492r: 5′-TACCTTGTTACGACTT-3′. PCR conditions were as follows: initial denaturation at 95 °C for 5 min, followed by 38 cycles of denaturation at 95 °C for 1 min, an annealing temperature of 46 °C for the Bacteria domain primers and 50 °C for Archaea domain primers maintained for 1 min, and final extension at 72 °C for 1 min. Amplification reactions contained 20–30 ng DNA per 50-ml reaction volume, 1× PCR buffer (Promega Biotech), 2.5 mM of each of the deoxynucleotides, 2.5 mM $MgCl_2$, 500 mM of the forward and reverse primers, and 0.025 U ml^{-1} Taq DNA polymerase (Promega Biotech). PCR amplification was checked by 1.2 % agarose gel electrophoresis stained with ethidium bromide. Amplified 16S rRNA gene products (>1400 bp) were cloned using the Topo Ta Cloning Kit (Invitrogen, Carlsbad, CA, USA) and sequenced using a BigDye Sequencing Kit (Applied Biosystems) following the manufacturer's instructions. Sequences were checked for potential chimeras using the Bellerophon Chimera Check programme (http://greengenes.lbl.gov/cgi-bin/nph-bel3_interface.cgi) and Mallard software. Operational taxonomic units (OTUs) were defined at 0.03 average distance between sequences (equivalent to 97 % similarity) using the ARB software package (http://www.arb-home.de). A distance matrix generated using the Greengenes on-line tool (http://greengenes.lbl.gov/cgi-bin/nph-distance_matrix.cgi) was used as the input file to distance-based OTU and richness (DOTUR) software, which assigns sequences to OTUs for every possible distance. Rarefaction analysis and the Chao1 nonparametric diversity estimator were applied to the clone library to estimate how completely the library had been sampled and to extrapolate to total sequence diversity. Further analysis of one representative of each OTU was carried out using the Classifier and Taxomatic online tools of the Ribosomal Database Project (http://rdp.cme.msu.edu). Phylogenetic trees were constructed using ARB tools on a database constructed with more than 53,000 16S rRNA sequences updated with BLAST closest matches for Copahue sequences. Neighbor-joining and Jukes–Cantor correlation were used. The rRNA alignments were corrected manually, and bootstrap values were calculated for 1000 replications to give statistical support to the results. Biodiversity indices were calculated using PAST software (version 2.14).

5.2.2 Results

Biodiversity in the Copahue geothermal field was approached by two complementary strategies (Urbieta et al. 2012, 2014, 2015a, b). The assessments on water samples as well as in the microbial biofilms were done using amplification and sequencing of 16S rRNA gene of Bacteria and Archaea (the set of primers used was 8F: 5′-AGAGTTTGATC(A/C)TGGC-3′ for Bacteria and 25 F: 5′-TCYGGTTGATCCYGCCRG-3′ for Archaea; reverse primer for both was 1492r: 5′-TACCTTGTTACGACTT-3′). Hybridizations (FISH technique) were done with general bacteria and archaea probes or more specific ones according to the phylogenetic classes or genera detected. Figures 5.3, 5.4, 5.5, 5.6, and 5.7 show the results for the five most representative points into the geothermal field (Las Máquinas, LMa; Laguna Verde, LVE; Baño 9, B9; Laguna Sulfurosa, LS; Las Maquinitas, LMi); Figs. 5.6 and 5.7 show the results for two points along Rio

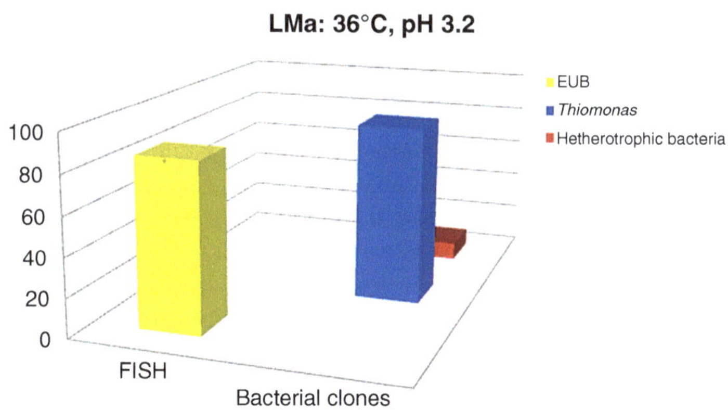

Fig. 5.3 Biodiversity at Las Máquinas pond

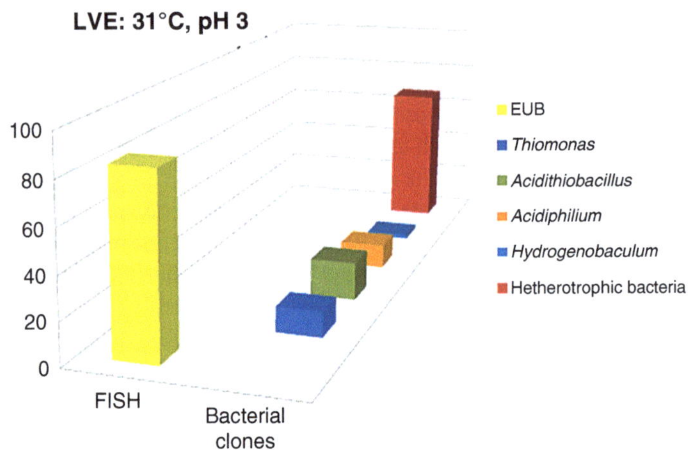

Fig. 5.4 Biodiversity at Laguna Verde pool

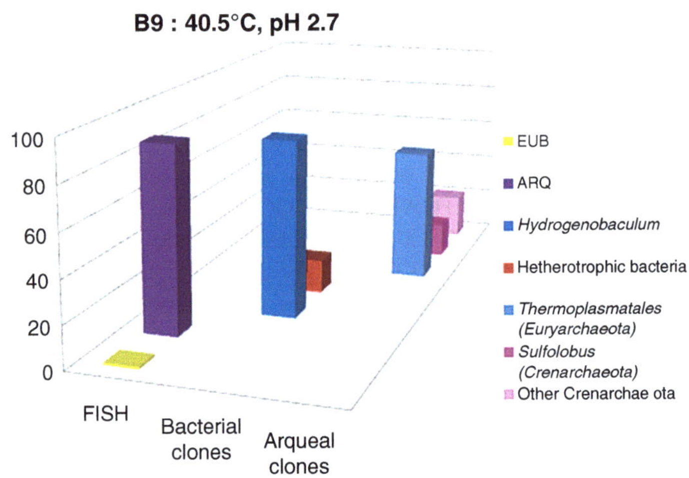

Fig. 5.5 Biodiversity at Baño 9 pond

Agrio (one of them, VA2, is close to the origin and the other, CV, is several kilometers downstream, just at the Cabellera de la Virgen cascade).

On the other hand diverse enrichment cultures and isolations from samples with specific environmental characteristics were carried out focusing on the isolation of microorganisms with certain metabolic features such as iron and/or sulfur compound oxidation, sulfate reduction, heavy metals transformation, and photosynthesis connected with lipid accumulation or other alternative energy sources production.

Ponds and hot springs selected from the thermal manifestations showed markedly different community structures, depending essentially on temperature but also on other factors such as anthropogenic intervention. The last influence can be checked

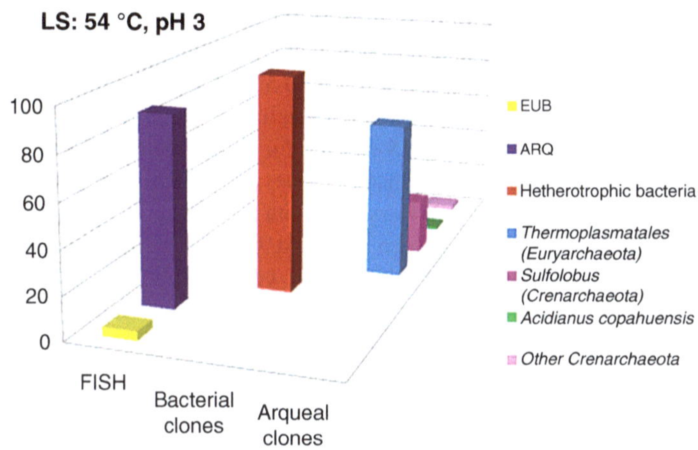

Fig. 5.6 Biodiversity at Laguna Sulfurosa pool

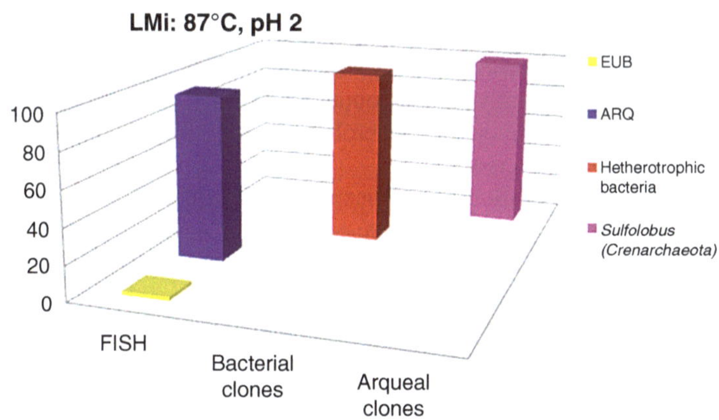

Fig. 5.7 Biodiversity at Las Maquinitas pond

in the different composition of the prokaryotic community found in two ponds (LMa and LVE) with similar acidic pH (~3) and moderate temperature conditions (around 30 °–35 °C). In both ponds, bacteria was dominant with more than 84 % hybridization with the probes EUBI-III that together cover most of the species in the domain Bacteria. However, LMa was colonized by acidophilic, mesophilic, autotrophic, or mixotrophic sulfur-oxidizing bacteria related to the genera *Thiomonas* (91 % of the clones analyzed) and *Acidithiobacillus* (5 %); at LVE, situated inside the healthcare facility at Copahue Thermal Centre, the number of species detected was higher, with a smaller representation of acidophiles and the presence of various species of heterotrophs, such as *Acinetobacter* and *Pseudomonas*, related to human activities. Archaea were not detected in these moderate temperature ponds by amplification of 16S rRNA genes.

It was interesting to probe that as the temperature of the ponds increased the prokaryotic community composition shifted toward Archaea. For instance, in the waters of the pond B9, with a temperature of 40.5 °C and a pH of 2.7, Archaea represented 84 % of all the microorganisms detected according to DAPI stain and hybridization with the probe ARQ915 specific for the domain Archaea.

Approximately 60 % of the archaeal clones were affiliated with the order *Thermoplasmatales* in the phylum Euryarchaeota (no further taxonomic classification was possible) and showed between 96 % and 99 % similarities with other uncultured archaeal clones detected in other acidic geothermal regions. The archaeal community was completed by two groups of Creanarchaeotas: one only classified as members of the class *Thermoprotei* and other associated to the genus *Sulfolobus*, which are aerobic, thermophilic, acidophilic sulfur-oxidizing Archaea very common in acidic high-temperature environments. Only one bacterial genus was detected in B9: *Hydrogenobaculum*, a group of acidophilic, thermophilic, chemolithotrophic, hydrogen- and sulfur-oxidizing species found in many sulfur-rich geothermal environments around the world.

In other acidic ponds in the Copahue Thermal Centre with temperature greater than 50 °C, *Thermoplasmatales* and *Sulfolobus* were still dominant and *Acidianus copahuensis* appeared. *A. copahuensis* is a novel thermophilic, acidophilic, facultative anaerobic, sulfur- and iron-oxidizing strain isolated, characterized, and reported by our group that is apparently autochthonous from Copahue (Giaveno et al. 2013). In these ponds, other thermophilic anaerobic Archaea, such as *Vulcanisaeta* and *Thermocladium*, were detected.

The most extreme conditions were found at Las Maquinitas with a temperature of 87 °C, close to the boiling temperature of water at Copahue's altitude, and a pH of 2. Such extreme conditions were reflected in the biodiversity found: Archaea represented 95 % of all microorganisms detected by DAPI and hybridization assays and were represented by only one sequence affiliated to the genus *Sulfolobus* that showed low sequence similarity (approximately 94 %) to uncultured archaeal clones retrieved from diverse acidic and high temperature thermal environments, which might indicate the existence of another novel autochthonous strain from Copahue. On the other hand, the bacteria detected in Las Maquinitas by amplification and sequencing of the 16S rRNA gene were all related to mesophilic and neutrophilic species typical of the soil or associated with human presence that were probably not metabolically active in the extreme environmental conditions of such hot springs.

Biofilms found at the borders of the pools and ponds were also sampled. The prokaryotic biodiversity of these biofilms, with temperatures from 30 ° to 36 °C, was completely different from what was found in the waters of the ponds. An important proportion of photosynthetic species and almost no presence of sulfur oxidizers were our main results. The nature of the photosynthetic species is apparently determined by the pH of the microbial biofilm: in the samples with pH lower than 4 the only photosynthetic species detected were related to the eukaryotes *Bacillariophyta* and *Chlorophyta*, whereas in the samples with higher pH values, *Cyanobacteria* related to the genera *Synechococcus*, *Leptolyngbya*, *Mastigocladus*, and *Fischerella*, and *Chloroflexi* from the genera *Roseiflexus* and *Chloroflexus*, were also detected.

Many of the sequences related to photosynthetic species presented low similarity with known species. Only a small percentage of sequences related to *Thiomonas* was detected in the samples with more acidic conditions.

The species found by cultivation from different ponds of the Copahue geothermal field are in agreement with the biodiversity detected in the molecular ecology assessments (Dopson and Johnson 2012). For instance, sulfur-oxidizing species related to *Acidithiobacillus (At.) thiooxidans* and *At. caldus* were isolated from various moderate temperature ponds from the Copahue geothermal system; Archaea related to *Sulfolobus* and *Acidianus* were isolated from ponds of more elevated temperature. Heterotrophic acidophilic species, mesophilic and thermophilic, related to *Acidiphilium*, *Sulfobacillus*, *Alicyclobacillus*, and *Mesoaciditoga* were isolated from all the samples tested. Surprisingly, although iron-oxidizing species were scarcely detected by molecular ecology techniques, they were isolated by cultivation from various samples, mainly species related to *At. ferrooxidans*, *Leptospirillum*, and *Sulfobacillus thermosulfidooxidans* (Chiacchiarini et al. 2010). Sulfate-reducing activity was found in enrichments carried out on anaerobic sediments. Several sulfate-reducing strains were isolated in overlay plates under anaerobic conditions. One of them presents 96 % similarity to the isolate *Peptococcaceae* bacterium CL4, an acidophilic sulfate-reducing bacteria isolated from the bottom layer of an acidic metal-rich stream in an abandoned mine in the Iberian Pyrite Belt in Spain, and the other is related to the genus *Desulfotomaculum*, which are anaerobic mesophilic and thermophilic spore-forming sulfate-reducing bacteria typical of subsurface environments. Photosynthetic species were also detected by cultivation; when samples from microbial biofilms or mats that developed near the ponds were inoculated in basal salt medium specific for enrichment of algae and cyanobacteria and cultivated at 25 °C with a light period of 12 h, species related to the cyanobacteria *Synechococcus elongates* and *Leptolyngbya* sp. and the microalgaes *Scenedesmus obliquus* and *Chlorella* sp. from the phylum *Chlorophyta* were identified.

In the case of acidic Río Agrio, we also approached the assessment of its biodiversity by culturing and non-culturing techniques. In the study using molecular ecology techniques it was found that the microbial community of Río Agrio consists of a relatively small number of acidophilic species commonly found in acid mine drainage or in natural acidic environments such as Rio Tinto. Our results (Figs. 5.8, 5.9) show that in spite of dilution that the river suffers from the input of tributary streams of snowmelt origin, the microbial community structure and the species found along its course do not change significantly. All through the river the same concentration of microorganisms was found: about 2.5×10^6 cells ml^{-1} according to DAPI stain recounts. An almost constant 40 % of *Archaea*, between 17 % and 31 % of *Gammaproteobacteria*, between 23 % and 31 % of *Alphaproteobacteria*, and between 0 % and 5 % of *Nitrospira* according to FISH hybridization, were detected. In the archaeal population, more than 94 % of the clones analyzed were related to *Ferroplasma*: these are acidophilic, lithoautotrophic or mixotrophic euryarchaeotas capable of iron and pyrite oxidation that are significant members of the microbial community in very acidic, heavy metal-rich environments. In the domain Bacteria, the *Gammaproteobacteria* were represented mostly by different

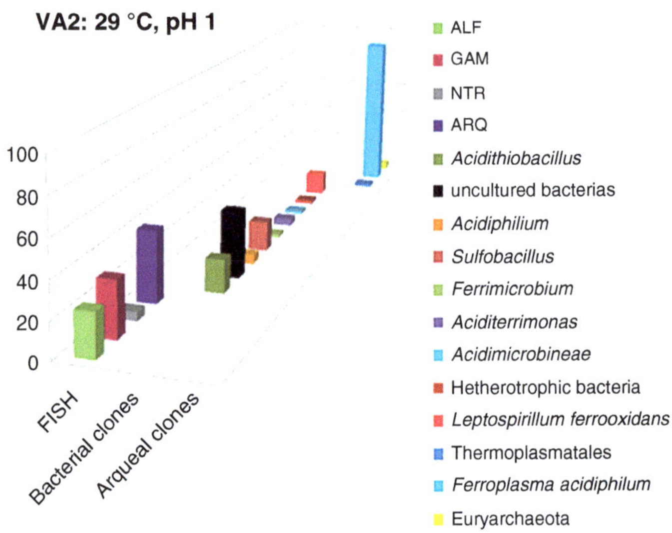

Fig. 5.8 Biodiversity at VA2 point (close to the origin of Río Agrio)

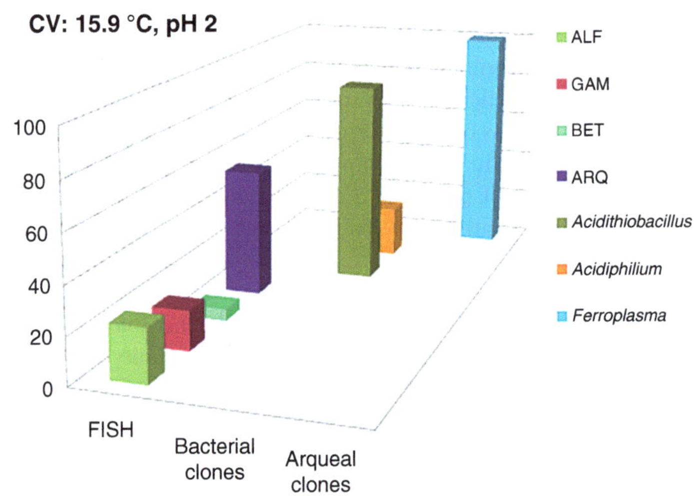

Fig. 5.9 Biodiversity at CV point (Río Agrio downstream)

Acidithiobacillus species such as *At. ferrivorans*, *At. Albertensis*, and *At. thiooxidans* and by a minor fraction of sequences that were not possible to classify beyond the class *Gammaproteobacteria*; however, they showed 99 % of sequence similarity to other sequences retrieved from diverse natural or mining-related acidic environments. In the class *Alphaproteobacteria*, the only species detected was *Acidiphilium*, an acidophilic chemoorganotrophic bacteria able to oxidize sulfur compounds that is generally found in acidic environments where chemolithotrophic species are found. In the *Nitrospira* group the only species found was *Leptospirillum*, mainly in the samples with higher concentrations of soluble ferrous iron, which is its only energy source. In the origin of the river where the pH is 1, *Leptospirillum* species would be metabolically favored for the oxidation of ferrous iron over *Acidithiobacillus* species whose enzymatic routes are inhibited at pH values lower than 1.3. Still, *Acidithiobacillus* species seemed to be more abundant than *Leptospirillum*, probably because sulfur compounds (that can be used as a source of energy by the first but not the second), products of the sustained volcanic activity, are much more abundant than ferrous iron. Close to the source of Río Agrio also were detected between 6 % and 8 % of sequences related to *Sulfobacillus*, a sulfur-oxidizing species that is a very common member of the microbial community of acid mine drainages, and between 6 % and 2 % of sequences related to *Ferrimicrobium*, an acidophilic heterotroph able to oxidize ferrous iron and to reduce ferric iron in anaerobic conditions.

Near to Caviahue Lake in the area of the waterfalls with pH values about 2, the number of species dropped, only *Acidithiobacillus*, *Acidiphilum*, and *Ferroplasma* being detected.

Mesophilic iron-oxidizing species related to *At. ferrooxidans* and *Leptospirillum* were isolated when samples from the area of the cascades were inoculated in basal salts solutions supplemented with ferrous iron and cultivated at 30 °C. On the other hand, moderately thermophilic iron oxidizers and Archaea related to *Sulfolobus* and *Acidianus* were isolated from the samples collected near the source of the river. Sulfur oxidizers, some of them related to *At. thiooxidans* and *At. caldus*, were cultivated from samples all through the river. No acidophilic heterotrophic microorganisms were isolated from Rio Agrio samples, emphasizing that the prevailing primary producers are lithoautotrophic species (Chiacchiarini et al. 2010).

5.3 Biotechnological Applications

Extremophiles are receiving great attention because of their mechanisms of biochemical adaptation to extreme conditions that can be used in different biotechnological applications. Not only can microorganisms isolated from extreme environments have a wide biotechnological potential, but also many of their biomolecules—and especially extremoenzymes—could be used to improve current processes or to develop new ones.

Although biotechnological applications for extremophiles are extremely diverse, in this chapter we focus on those capable of being used for heavy metal remediation. Many technologies have been developed to overcome the pollution produced by heavy metals, including reduction, filtration, electrochemical treatment, evaporation, reverse osmosis, ion exchange, and chemical precipitation (Hashim et al. 2011); most of these are expensive and/or inefficient at low metal concentrations and when pollution involves large areas or volumes. Biological treatments have many advantages from economical, environmental, and practical aspects. These treatments are related to the mechanisms the resistant microorganisms use to survive in hostile environments with high metal concentrations. The main metal resistance mechanisms include exclusion by permeability barrier, intra- or extracellular sequestration, active-transport efflux pumps, and enzymatic transformation to other, less toxic, species. The first step to detect suitable microorganisms for metal (and metalloids) bioremediation is analyzing the tolerance/resistance to different metals and metalloids. Thus, many environmental screenings to detect metal (and metalloids) resistance have been reported and are being daily reported.

In this way, samples from the Copahue geothermal field have been inoculated in increasing metal/metalloid concentration cultures, and we have been successful to detect microbial communities with high resistance to certain metal(loid)s, including some that are not present in the field. For example, we achieved a significant growth for some samples from Rio Agrio in heterotrophic cultures with very high arsenic concentrations: up to 1.4 and 33.75 g l^{-1} of As(III) and As(V), respectively (arsenic is present in the water samples but in very low concentrations). This hyperresistance to arsenic suggests the chance to use those microbial communities to remediate arsenic-contaminated effluents (Drewniak et al. 2008; Rawlings 2008).

The mechanisms for surviving in the presence of high metal concentration allow designing some strategies for metal bioremediation. These strategies can be classified in three different groups: immobilization of metals present in liquid effluents, mobilization of metals contained in solid wastes, residues, and spent and exhausted devices, and transformation of metals to less toxic chemical species.

5.3.1 Metal Mobilization

Many solid residues and wastes have a high content of metals that disallows their final disposal according to environmental regulations. In addition, such residues and wastes can be regarded as potential sources for metals. The biotechnologies to mobilize metals are based in leaching processes. Microorganisms are able to generate many different leaching products (oxidizing agents, reducing agents, mineral acids, organic acids, solvents, etc.) to extract metals from mineral wastes, contaminated sediments, electronic wastes, spent catalysts, etc. Acidophilic iron- and sulfur-oxidizing microorganisms are surely among those most used for those purposes. These microorganisms (Bacteria and Archaea) are able to generate sulfuric acid from the oxidation of elemental sulfur or reduced sulfur compounds under aerobic conditions; moreover, some generate ferric iron from the oxidation of ferrous iron. These acidic and oxidizing media can be used to solubilize metals from many insoluble compounds.

The main sulfur-oxidizing microorganisms are mesophilic and autotrophic bacteria belonging to the genus *Acidithiobacillus*. Within *Acidithiobacillus*, *Acidithiobacillus thiooxidans* and *Acidithiobacillus ferrooxidans* are the most important species. The last is also able to oxidize iron (II), which is also the process catalyzed by *Leptospirillum ferrooxidans*. Also, many Archaea are able to catalyze the same processes; the most important Archaea belong to the genera *Sulfolobus*, *Acidianus*, and *Ferroplasma*.

Our research group has published several papers using those microorganisms to mobilize metals from residues and wastes (Cerrutti et al. 1998; Viera and Donati 2004; Coto et al. 2008; Viera et al. 2008).

5.3.2 Metal Immobilization

The two main biotechnologies to immobilize metals are biosorption and bioprecipitation. Metal biosorption can occur either by metabolism-independent passive sorption or by intracellular, metabolism-dependent active uptake. Both processes may occur in the same organism. The terms sorption and adsorption are usually used when passive accumulation is being considered whereas uptake and accumulation are used when metabolism-dependent intracellular transport is implied. The non-metabolic biosorption processes comprise different mechanisms: chemisorption, ion exchange, complexation, coordination, chelation, and physical adsorption. Although there are many reports on the higher biosorption capacity of living biomass in comparison with dead biomass, the first case has some disadvantages; the process is irreversible, and in addition substrates must be added to cultivate the biomass, and it is necessary to avoid possible inhibition caused by metals.

The second biotechnology to immobilize metals is bioprecipitation; in this process some microorganisms generate metabolic products that readily precipitate with heavy metals. The most important case within bioprecipitation is that provoked by sulfate-reducing microorganisms (SRM). These microorganisms (mainly bacteria) are able to obtain energy for cell synthesis and growth by coupling the oxidation of low molecular weight organic substrates (lactate, acetate, propionate, ethanol, etc.), under anaerobic conditions, to the reduction of sulfate to sulfide.

SRM are a complex physiological prokaryote group with a variety of essential functions in many anaerobic environments. To date, there are four groups of SRM: gram-negative mesophilic SRM, gram-positive spore-forming SRM, thermophilic bacterial SRM, and thermophilic archaeal SRM. The main families of gram-negative SRM are *Desulfovibrionaceae* and *Desulfobacteriaceae*. The group of gram-positive spore-forming SRM is dominated by the genus *Desulfotomaculum*. Within the group of bacterial thermophilic SRM, the main species are *Thermodesulfobacterium commune* and *Thermodesulfovibrio yellowstonii*. Finally, the archaeal group is characterized by optimal growth temperatures above 80 °C. Two species have been completely described: *Archaeoglobus fulgidus* and *Archaeoglobus profundus*.

The sulfide produced by SRM activity can be used for metal precipitation (Kikot et al. 2010; Kieu et al. 2011). This ion reacts with metal ions present in contaminated waters and precipitates as insoluble metal sulfide. The solubility products of most heavy metal sulfides are very low, which is why even low concentrations of sulfide can remove metals to levels permitted in the environment. In addition to precipitating metal sulfides, sulfate reduction utilizes protons (especially when the initial strong acid is converted in a weak acid) and provokes an elevation in pH that can also contribute to precipitation of certain metals. The main advantage of bioprecipitation over the chemical treatment of effluents with heavy metals is the reduction of the sludge volume that is generated when carbonates are used for precipitation. In addition, metal sulfides are more stable under anaerobic conditions than hydroxides or carbonates because of their lower solubility product.

The major limitation in the application of this technology is the sensitivity of SRM to acidity and high metal concentration. The inhibition at low pH values is related to the use of small organic acids as carbon sources; these molecules are almost not dissociated at such pH values and can cross the cell membranes to dissociate into the cells where the pH is almost neutral. To avoid such inhibition, organic acids must be replaced by other nonacidic carbon sources (such as glycerol); in addition, adaptation and/or isolation from acidic environmental samples or a soft previous neutralization allow obtaining a suitable growth at pH values below 4.5.

5.3.2.1 Case Study

In this section we describe some experiments on metal bioprecipitation using neutrophilic and acidophilic SRM isolated from the Copahue geothermal zone.

To obtain neutrophilic SRM communities, samples from Copahue were enriched using Postgate B medium, with lactate as the electron and carbon source and sulfate as the terminal electron acceptor. The basal medium contains the following (in l^{-1} distilled water): KH_2PO_4 0.5 g, NH_4Cl 1 g; $CaSO_4$ 1 g, $MgSO_4{\bullet}7H_2O$ 2 g, sodium lactate 3.5 g, yeast extract 1 g, $FeSO_4{\bullet}7H_2O$ 0.5 g, ascorbic acid 0.1 g, thioglycolic acid 0.1 g. The pH was adjusted to 7.0 with NaOH 5 M. The medium was divided into 50-ml flasks, sealed with rubber stoppers, and sterilized by autoclaving at 121 °C for 20 min. After that, the flasks were opened under sterile conditions, sediment samples added, and the flasks immediately closed. Flasks containing the medium and sediment samples were incubated at 30 °C for 15–20 days. Growth of cultures was monitored periodically by measuring the remaining sulfate concentration using a turbidimetric method with $BaCl_2$ and by observing the formation of black precipitates (FeS). Positive enrichments were used in bioprecipitation assays.

Samples for acidophilic communities were enriched in a SRM medium with glycerol 3 mM, yeast extract 0.01 % (w/v), zinc 4 mM, $FeSO_4$ 100 μM, K_2SO_4 0.87 g l^{-1}, basal salts, and trace elements (initial pH 3.0). Nitrogen was bubbled through the media to displace oxygen to create an environment for the growth of anaerobic microorganisms. After that, the medium was sterilized by autoclaving and rapidly divided into 50-ml flasks under sterile conditions. Then, sediment samples were added to the flasks containing anaerobic medium. Flasks were incubated in sealed anaerobic jars at 30 °C. After 1 month, sulfate and glycerol were measured using ionic chromatography and zinc using atomic absorption spectrophotometry. Positive enrichments were used in the bioprecipitation studies.

In bioprecipitation assays, 10 ml of the growth enrichment was added to 90 ml of sterile SRM medium (initial pH 3.0) or Postgate C medium and incubated anaerobically. Postgate C medium contains the following (in l^{-1} distilled water): KH_2PO_4 0.5 g, NH_4Cl 1 g, Na_2SO_4 4.5 g, $CaCl_2{\bullet}6H_2O$ 0.06 g, $MgSO_4{\bullet}7H_2O$ 0.06 g, sodium lactate 6 g, yeast extract 1 g, $FeSO_4{\bullet}7H_2O$ 0.004, sodium citrate 0.3 g, pH 7.0). For acidophilic SRM inoculum the medium was sparked with N_2 before sterilization to remove traces of dissolved oxygen. In the case of Postgate C medium, 50 μl anaerobic solution (0.2 g ascorbic acid, 200 μl thioglycolic acid, 10 ml distilled water) was added to obtain anoxic conditions. The concentrations of glycerol, zinc, sulfate, and pH were measured at the end of growth.

Samples for bioprecipitation tests were prepared in 10-ml flasks. Each of them was filled under sterile conditions with 9 ml SRM or Postgate C medium (prepared as before) at pH 3.0 or pH 7.0, respectively. In each case the medium was prepared without the addition of $ZnSO_4$ or $FeSO_4$, respectively, to test the precipitation of other heavy metals. Different volumes of a 1000 mg l^{-1} stock solution of each heavy metal were added to reach different concentration (5, 10, and 25 mg l^{-1}). The metals tested were Cr(III) and Ni(II). Finally, 1 ml of grown SRM enrichment cultures was added to each flask. The flasks were then sealed with rubber stoppers and incubated anaerobically at 30 °C for 1 month. Experiments were performed in duplicate. One extra tube was prepared in the same conditions and was used for the initial measurements, for example, to determine the initial precipitation of metals as a result of the dissolved sulfide found in the inocula. Control tests without SRM inoculation were also performed with both media to distinguish the amount of heavy metals removed by biological mechanisms (bioprecipitation) from those removed by chemical precipitation. Samples for analytical determinations from the flasks were previously filtered through a 0.22-μm cellulosic acetate membrane to remove any possible precipitates. The initial and final concentrations of heavy metals were determined by atomic absorption spectrophotometry. Previously, samples were diluted in 0.14 N nitric acid. For both enrichment cultures bioprecipitation percentage (% BP) was calculated considering final (30 days) and initial (0 days) soluble metal concentrations.

Figure 5.10 shows the results of nickel bioprecipitation. It is clear that neutrophilic SRM were able to precipitate nickel more than an abiotic control at the lower metal concentration. When nickel concentration was higher (25 mg l^{-1}), microorganisms were fully inhibited. However, acidophilic SRM showed an excellent performance for nickel precipitation at all metal concentrations used in the experiments, and even at this pH value where the abiotic controls reached low percentages of precipitation.

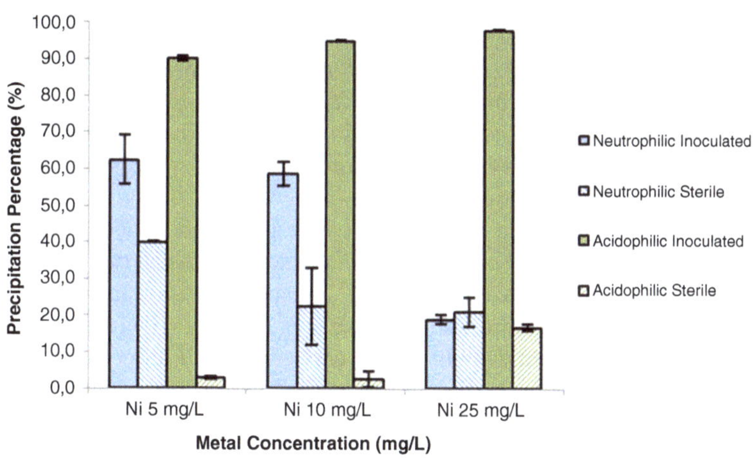

Fig. 5.10 Nickel bioprecipitation by neutrophilic and acidophilic sulfate-reducing microorganisms (SRM)

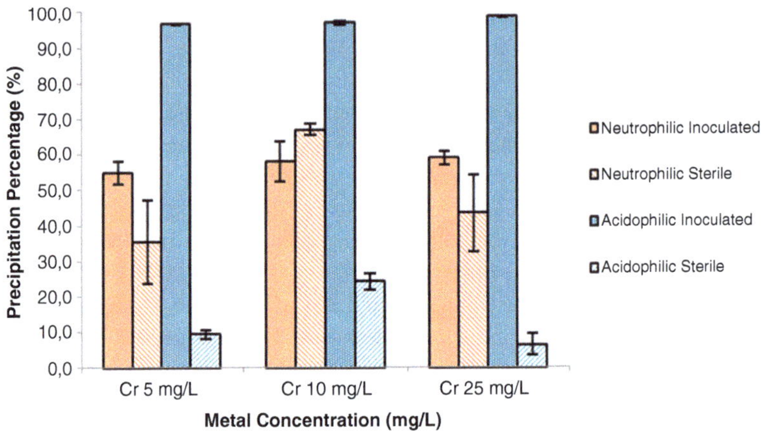

Fig. 5.11 Chromium bioprecipitation by neutrophilic and acidophilic SRM

This behavior was also observed for chromium (Fig. 5.11). In this case, the neutrophilic microbial community is only able to overpass the precipitation reached in the abiotic controls at higher metal concentration because abiotic chromium precipitation is important even at pH values close to 7. As was observed for nickel, the acidophilic microbial community achieved high percentages of chromium precipitation at all metal concentrations used in the experiments.

5.3.3 Metal Transformation

Microorganisms can transform metals, metalloids, and organometallic compounds by reduction and oxidation processes. In some cases, the changes in the oxidation states allow the mobilization/immobilization of the metal(loid)s. For example, the oxidation of Fe(II) to Fe(III) or Mn(II) to Mn(IV) reduces the solubility of those metals whereas the inverse processes increase their mobility; both processes can be catalyzed by different microbial species and should be included in the previous sections.

In other cases, changes in the oxidation state provoked a modification in the toxicity of the metal(loid)s. For example, some microorganisms can detoxify arsenic by efficiently oxidizing As(III) to As(V). In nature, arsenic appears as trivalent arsenic and pentavalent arsenic. Although both are toxic, As(III) is relatively more toxic than As(V) because it can bind sulfhydryl groups and dithiol groups of proteins. As(III) removal is more difficult because of its relatively higher solubility; As(V) is less soluble and less bioavailable. Thus, some microorganisms have developed the ability to oxidize As(III) to As(V) as a mechanism for detoxification (Rahman et al. 2014).

Chromium is a similar case. Hexavalent chromium compounds are highly water soluble and adsorb poorly to soil and organic matter, so they are mobile in soil and groundwater. Cr(VI) can be reduced to Cr(III), reducing mobility, bioavailability, and toxicity. Many microorganisms are able to catalyze such processes in a very efficient way (Allegretti et al. 2006; Cabrera et al. 2007; Cheung and Gu 2007; Murugavelh and Mohanty 2012; Thatoi et al. 2014; Singh et al. 2015). Next, we describe this ability in microbial communities from the Copahue geothermal system.

5.3.3.1 Case Study

Enrichment culture in nutrient broth (pH 7) was made from a sample taken from the *Las Maquinas* site (LMa: 68.5 °C and pH 7.5). The culture was incubated overnight at 50 °C with agitation. The enrichment obtained was tested for Cr(VII) reduction activity. Experiment was carried out in duplicate with a concentration of 20 mg l^{-1} Cr(VI) ($K_2Cr_2O_7$) with the respective sterile controls. Total chromium and Cr(VI) were measured at different times during 24 h using atomic absorption spectroscopy and 1,5-diphenylcarbazide colorimetric method techniques for identification of total chromium and Cr(VI), respectively. Additionally, absorbance at 600 nm was used to measure bacterial growth. As shown in Fig. 5.12, after 24 h, 91 % of the chromium added was reduced in the enrichment (LMa) whereas only a slight decrease in total chromium concentration was observed. In the same figure, microbial growth is shown through optical density at 600 nm (OD_{600}), which is correlated with the decrease in Cr(VI) concentration.

Additional controls were also made to understand in more detail the process of Cr(VI) reduction. A well-grown culture was centrifuged at 8000 rpm for 20 min.

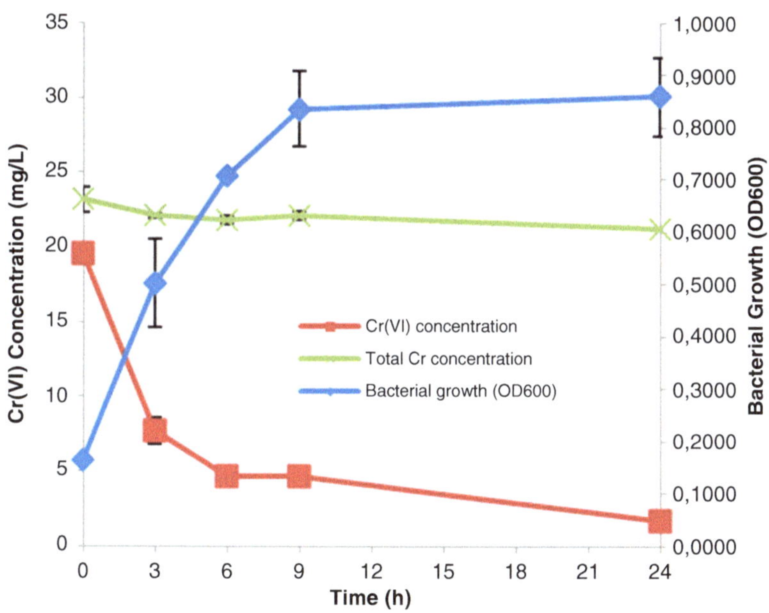

Fig. 5.12 Chromate reduction by a microbial community from Las Máquinas

The supernatant was filtrated by a 0.22-μm filter to obtain a suspension free of cells. Both filtered supernatant and the resuspended pellet were phased to a 20 mg l^{-1} Cr(VI) concentration and total chromium, Cr(VI), and OD_{600} were measured. After 24 h only 25 % of Cr(VI) reduction was detected in the supernatant systems although complete reduction was accomplished in the resuspended pellet systems. This result shows that the reductive activity has a place on or inside the biomass and is not the result of a soluble compound produced by the microorganisms. Because chromium reduction was also observed (45 % in 24 h) in sterile controls, another control in minimal culture media (M9) was made to see if the reduction seen in the sterile controls in nutritive broth was caused by oxidation of the contained organic compounds. No reduction took place in this system, as the oxidation of the organic components of the nutrient broth was the reason for Cr(VI) decrease in the sterile controls.

Summarizing the results presented in this section demonstrates the potential of this community isolated from Copahue to reduce Cr(VI) to Cr(III).

5.4 Conclusions

The Copahue geothermal system, including pools, ponds, and the Rio Agrio, presents a particular prokaryotic biodiversity. According to our results, temperature was one of the main factors that define the biodiversity: in moderate temperature sites the predominance of bacteria [particularly acidophilic species related to acid and metalliferous drainage (AMD)] was found whereas those sites with high temperature (ponds) were dominated by Archaea. In the water samples, the dominant prokaryotes were chemolithoautotrophic or mixotrophic, mainly sulfur oxidizing, whereas in microbial biofilms photosynthetic species were the most important primary producers. In the case of Rio Agrio, also mainly sulfur-oxidizing bacteria and Archaea were found although some genera of iron-oxidizing microorganisms were also detected. Some microbial communities isolated from those samples showed the ability to precipitate metals even at very low pH values and to reduce hexavalent chromium to a less toxic trivalent chromium.

Acknowledgments The authors gratefully acknowledge support from ANPCyT (PICT 2012 0623 and PICT 2013 0630).

References

Achenbach L, Woese C (1999) 16S and 23S rRNA-like primers. In: Sower KR, Schreier HJ (eds) Archaea: a laboratory manual. Cold Spring Harbor Laboratory Press, Cold Spring Harbor, NY, pp 521–523

Allegretti P, Furlong J, Donati E (2006) The role of higher polythionates in the reduction of chromium (VI) by *Acidithiobacillus* and *Thiobacillus* cultures. J Biotechnol 22:55–61

Burgess EA, Unrine JM, Mills GL, Romanek CS, Wiegel J (2012) Comparative geochemical and microbiological characterization of two thermal pools in the Uzon Caldera, Kamchatka, Russia. Microb Ecol 63:471–489

Cabrera G, Viera M, Gómez JM, Cantero D, Donati E (2007) Bacterial removal of chromium (VI) and (III) in a continuous system. Biodegradation 18:505–513

Cerrutti C, Curutchet G, Donati E (1998) Bio-dissolution of spent nickel-cadmium batteries using *Thiobacillus ferrooxidans*. J Biotechnol 62:209–219

Cheung KH, Gu J-D (2007) Mechanism of hexavalent chromium detoxification by microorganisms and bioremediation application potential: a review. Int Biodeter Biodegr 59:8–15

Chiacchiarini P, Lavalle L, Giaveno A, Donati E (2010) First assessment of acidophilic microorganisms from geothermal Copahue-Caviahue system. Hydrometallurgy 104:334–341

Coto O, Galizia F, Hernández I, Marrero J, Donati E (2008) Cobalt and nickel recoveries from laterite tailings by organic and inorganic bioacids. Hydrometallurgy 94:18–22

Dopson M, Johnson DB (2012) Biodiversity, metabolism and applications of acidophilic sulphur-metabolizing microorganisms. Environ Microbiol 14:2620–2631

Drewniak L, Styczek A, Majder-Lopatka M, Sklodowska A (2008) Bacteria, hypertolerant to arsenic in the rocks of an ancient gold mine, and their potential role in dissemination of arsenic pollution. Environ Pollut 156:1069–1074

Giaveno MA, Urbieta MS, Ulloa R, González-Toril E, Donati ER (2013) Physiologic versatility and growth flexibility as the main characteristics of a novel thermoacidophilic *Acidianus* strain isolated from Copahue geothermal area in Argentina. Microb Ecol 65:336–346

Hashim MA, Mukhopadhyay S, Sahu JN, Sengupta B (2011) Remediation technologies for heavy metal contaminated groundwater. J Environ Manag 92:2355–2388

Kieu HTQ, Müller E, Horn H (2011) Heavy metal removal in anaerobic semi-continuous stirred tank reactors by a consortium of sulfate-reducing bacteria. Water Res 45:3863–3870

Kikot P, Viera M, Mignone C, Donati E (2010) Study of the effect of pH and dissolved heavy metals on the growth of sulfate-reducing bacteria by a fractional factorial design. Hydrometallurgy 104:494–500

Kublanov IV, Perevalova AA, Slobodkina GB, Lebedinsky AV, Bidzhieva SK, Kolganova TV, Bonch-Osmolovskaya EA (2009) Biodiversity of thermophilic prokaryotes with hydrolytic activities in hot springs of Uzon Caldera, Kamchatka (Russia). Appl Environ Microbiol 75:286–291

Meyer-Dombard DR, Shock EL, Amend JP (2005) Archaeal and bacterial communities in geochemically diverse hot springs of Yellowstone National Park, USA. Geobiology 3:211–227

Murugavelh S, Mohanty K (2012) Bioreduction of hexavalent chromium by free cells and cell free extracts of *Halomonas* sp. Chem Eng J 203:415–422

Rahman S, Kim K-H, Saha SK, Swaraz AM, Paul DK (2014) Review of remediation techniques for arsenic (As) contamination: a novel approach utilizing bio-organisms. J Environ Manag 134:175–185

Rawlings DE (2008) High level arsenic resistance in bacteria present in biooxidation tanks used to treat gold-bearing arsenopyrite concentrates: a review. Trans Nonferrous Metals Soc China 18:1311–1318

Reigstad LJ, Jorgensen SL, Schleper C (2010) Diversity and abundance of Korarchaeota in terrestrial hot springs of Iceland and Kamchatka. ISME J 4:346–356

Rothschild LJ, Mancinelli RL (2001) Life in extreme environments. Nature (Lond) 409:1092–1101

Singh R, Dong H, Liu D, Zhao L, Marts AR, Farquhar E, Tierney DL, Almquist CB, Briggs BR (2015) Reduction of hexavalent chromium by the thermophilic methanogen *Methanothermobacter thermautotrophicus*. Geochim Cosmochim Acta 148:442–456

Thatoi H, Das S, Mishra J, Rath BP, Das N (2014) Bacterial chromate reductase, a potential enzyme for bioremediation of hexavalent chromium: a review. J Environ Manag 146:383–399

Urbieta MS, González-Toril E, Giaveno MA, Aguilera A, Donati E (2012) First prokaryotic biodiversity assessment using molecular techniques of an acidic river in Neuquén, Argentina. Microb Ecol 64:91–104

Urbieta MS, González-Toril E, Aguilera A, Giaveno MA, Donati E (2014) Archaeal and bacterial diversity in five different hydrothermal ponds in the Copahue region in Argentina. Syst Appl Microbiol 37:429–441
Urbieta MS, Donati ER, Chan K-G, Shahar S, Sin LL, Goh KM (2015a) Thermophiles in the genomic era: biodiversity, science, and applications. Biotechnol Adv 33:633–647
Urbieta MS, González-Toril E, Aguilera A, Giaveno MA, Donati E (2015b) Comparison of the microbial communities of hot springs waters and the microbial biofilms in the acidic geothermal area of Copahue (Neuquén, Argentina). Extremophiles 9:437–450
Varekamp JC, Ouimette A, Herman S, Flynn KS, Bermúdez AH, Delpino DH (2009) Naturally acid waters from Copahue volcano, Argentina. Appl Geochem 24:208–220
Viera M, Donati E (2004) Microbial processes to metal recovery from waste products. Curr Top Biotechnol 1:117–127
Viera M, Donati E, Bosio V (2008) Integrated bacterial process for the treatment of a spent nickel catalyst. J Hazard Mater 154:804–810
Wemheuer B, Taube R, Akyol P, Wemheuer F, Daniel R (2013) Microbial diversity and biochemical potential encoded by thermal spring metagenomes derived from the Kamchatka Peninsula. Archaea 2013:136714. doi:10.1155/2013/136714

Chapter 6
Microbiological and Biochemical Indicators for Assessing Soil Quality in Drylands from Patagonia

Magalí S. Marcos and Nelda Lila Olivera

Abstract Soil monitoring programs aimed to diagnose the state of the soils and to prevent their deterioration are needed to make use of this resource in a sustainable manner. The development of such programs requires the identification of indicators that can be used to detect changes in soil quality. Although several chemical, physical, and biological parameters serve this purpose, microbiological properties have the advantages of being highly sensitive and responding quickly to changes in soil quality. In Argentina, the information of soil quality indicators is fragmented and mainly directed toward the analysis of physicochemical parameters in agricultural lands. In this chapter, we review microbiological/biochemical indicators of soil quality measured in water-limited natural ecosystems of Patagonia. This baseline information contributes to the future design of site-specific monitoring programs devoted to prevent the deterioration of lands in Patagonia.

6.1 Definition and Importance of Soil Quality

Soils are the naturally occurring mineral and organic material at the Earth's surface that provide an environment for living organisms, and they have critical importance affecting life sustainability through their role in controlling the Earth's environment (Voroney 2007). They provide habitat, shelter, water, and food to living organisms, and offer essential ecosystem services such as maintenance of the atmospheric gas composition, the regulation of populations of human, animal, and plant pathogens, the bioremediation of toxic chemicals, the cycling of nutrients, and the filtering and buffering of potential pollutants (Wall 2012, www.nrcs.usda.gov). Furthermore,

M.S. Marcos (✉) • N.L. Olivera
Patagonian Institute for the Study of Continental Ecosystems (IPEEC-CENPAT), National Scientific and Technical Research Council (CONICET), Argentina
e-mail: magali@cenpat-conicet.gob.ar

N.L. Olivera et al. (eds.), *Biology and Biotechnology of Patagonian Microorganisms*, DOI 10.1007/978-3-319-42801-7_6

soils are a reservoir of yet unexplored genetic diversity (Wall 2012). Because soils are an invaluable resource for life, it is fundamental to use them in a rational manner that guarantees their preservation for future generations (Karlen et al. 1997). Such is the importance of soils that 2015 was declared the "International Year of Soils" by the United Nations in an effort to recognize their relevance and to promote their conservation (United Nations 2014). In particular, water-limited soils cover more than 41 % of the Earth's land surface, provide habitat to more than 2 billion people, and support rangelands and croplands in 90 % of their extension (Niemeijer et al. 2005). Therefore, their preservation is a subject of global concern.

Soil quality was defined by the S-581 Ad Hoc Committee convoked by the Soil Science Society of America as "the capacity of a specific kind of soil to function, within natural or managed ecosystem boundaries, to sustain plant and animal productivity, maintain or enhance water and air quality, and support human health and habitation" (Karlen et al. 1997). There is no clear difference between the terms 'soil health' and 'soil quality'. The latter term has been used to refer to soil fitness for a specific use, whereas 'soil health' has been used in a broader sense, meaning soil capacity to function sustainably (Doran and Zeiss 2000). However, both terms are generally used as synonyms. Another term related to the previous concepts is 'soil fertility', defined as "the capacity of a soil to provide physical, chemical and biological requirements for the growth of plants for productivity, reproduction and quality relevant to plant type, soil type, land use, and climatic conditions" (Abbott and Murphy 2007). While this concept has been associated to agricultural production systems (Abbott and Murphy 2007), soil quality has been related to both agricultural and natural ecosystems.

Healthy soils are characterized by high biodiversity and resilience and have the capability to function at their highest potential, whereas poor-quality/unhealthy soils manifest the opposite symptoms and function below their full potential (Karlen et al. 1997; van Bruggen and Semenov 2000; Lehman et al. 2015). Because soil quality has an inherent component determined by soil-forming processes, it is not possible to define absolute soil quality indicators and reference values to be universally applied in different soils and land use practices (Karlen et al. 2001; Bloem et al. 2006). However, soil quality can be compared in similar soils (in terms of their inherent properties) subjected to different land management practices, or in the same soil at different times (Schloter et al. 2006). Overall, the monitoring of soil quality is a valuable tool that can contribute to our understanding of ecosystem sustainability and to help us evaluate if land use decisions and practices ensure the conservation and sustainable use of this resource (Karlen et al. 2001).

6.2 Soil Quality Indicators

Soil quality indicators can be classified into physical, chemical, and biological; the last parameter includes microbiological, biochemical, and soil faunal indicators as well (Fig. 6.1). Physicochemical parameters are more frequently used than biological measures to assess soil quality, despite the fact that they are less sensitive

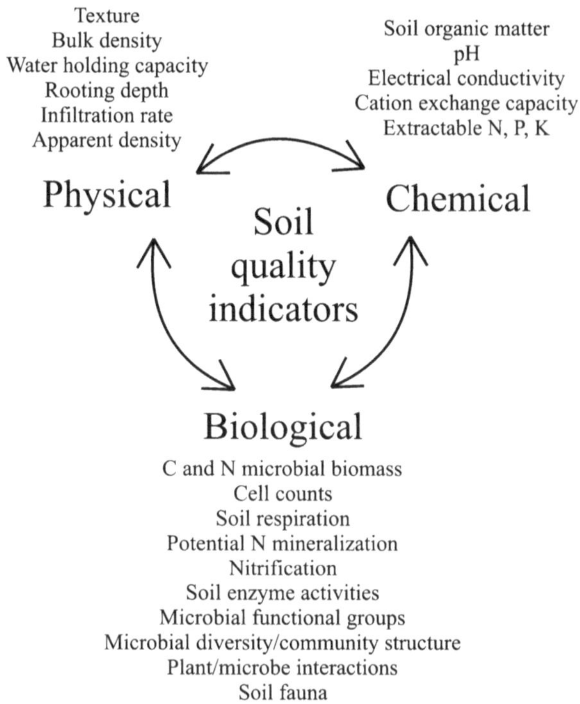

Fig. 6.1 Soil quality indicators. *Arrows* represent the interrelationship among physical, chemical, and biological properties

because they only respond to drastic changes in soil (Gil-Sotres et al. 2005). In contrast, microorganisms respond faster to changes in soil properties and environmental conditions, and together with soil enzymes are responsible for key processes associated with soil quality, such as nutrient cycling and organic matter decomposition (Burns et al. 2006). Soil microorganisms also combine the following characteristics of good soil quality indicators: they are sensitive enough to detect long-term changes in soil quality resulting from land management and climate, but not too sensitive to vary in response to short-term conditions; well correlated with beneficial soil functions; useful to predict whether a soil will provide a beneficial function and why (or why not); easy to comprehend and useful to land managers; and easy and inexpensive to measure (Doran and Zeiss 2000). In addition, indicators should be reproducible and representative of the study site (Burns et al. 2006).

Microbiological/biochemical indicators can be classified in four groups according to the information they provide (Benedetti and Dilly 2006):

- Microbial biomass and number: include measures of microbial biomass carbon (C) and nitrogen (N), viable cell counts in cultures or direct counts using a microscope.
- Microbial activity: indicators of soil respiration, N mineralization, nitrification, enzyme activities, and abundance/activity of specific functional groups of microorganisms, such as nitrifiers. Includes both actual and potential activities.

- Microbial diversity and community structure: nowadays assessed using molecular techniques based on the analysis of nucleic acids [e.g., sequence analysis of clone libraries, fingerprinting analysis of polymerase chain reaction (PCR) products], phospholipid fatty acids, and community-level physiological profiles. Although modern high-throughput molecular techniques (e.g., small-subunit rDNA/rRNA pyrosequencing, metagenomics, and other "omics") have the enormous potential to be used as well, they have not yet been used very often as soil quality indicators, probably because they require complex data analysis and are expensive compared to traditional microbiological measures (Bastida et al. 2008; Lehman et al. 2015).
- Plant–microbe interactions: biomass/activity/diversity of microorganisms in the rhizosphere, nodulating potential of nitrogen fixers, arbuscular mycorrhizae.

Individual properties have been often combined into simple or multiparametric indexes, depending on whether they relate two or more individual parameters, respectively. Further information about these indexes was reviewed in detail by Bastida et al. (2008) and Paz-Ferreiro and Fu (2016).

6.3 The Arid Lands of Patagonia: Soil, Climate, and Vegetation Description

There is a wide variety of soil types in Patagonia, although not evenly distributed. More than 50 % of Patagonia is covered by Aridisols, which are characteristic of dry areas, specially deserts (del Valle 1998; Coronato et al. 2008). These are followed by Entisols (22 %), most of which have a poor development of pedogenic horizons (del Valle 1998), and by Mollisols (13 %), which are dark soils usually located in slopes in the western part of Patagonia (del Valle 1998). Six other soil types cover small areas along the subhumid to humid western region. In general, soils are poorly developed, probably as a consequence of several interruptions of the pedogenetic processes (Coronato et al. 2008).

As described in detail in the Introduction chapter of this book, climatic conditions are very particular in Patagonia. The extra-Andean region has an arid to semiarid climate, with low rainfall (<250 mm per year in most of the area) that decreases from west to east and shows high intra- and interannual variation, strong dry winds blowing from the west, and temperate to cold temperatures decreasing in a north–south direction (Ares et al. 1990; Coronato et al. 2008).

The distribution of plant forms developed under the climate and soil characteristics of the arid extra-Andean Patagonia delimits two phytogeographic provinces. The Patagonian Monte (the so-called southern portion of the Monte Phytogeographic province, which also extends northward from Patagonia) is a temperate semidesert shrubland with low grass cover, no trees, and more than 50 % bare soil (León et al. 1998; Coronato et al. 2008). Vegetation consists of shrub-grass plant patches surrounded by bare soil areas. Some of the most frequent shrubs include species of the genera *Larrea*, *Lycium*, *Chuquiraga*, and *Prosopis*; among the most frequent perennial grasses are species of *Stipa* and *Poa* (León et al. 1998). In contrast, the Patagonian Phytogeographic Province has different types of vegetation, including grass steppes dominated by *Festuca* spp., shrub-grass steppes dominated by the shrubs *Mulinum spinosum*, *Senecio filaginoides*, and *Adesmia volckmannii*, and the grasses *Stipa speciosa*, *Pappostipa* spp., and *Poa ligularis*, and semideserts of dwarf (*Nassauvia glomerulosa*, *N. ulicina*, and *Chuquiraga aurea*) or other (*Chuquiraga avellanedae*, *Colliguaja integerrima*, *Nardophyllum obtusifolium*, and *Mulguraea tridens*) shrubs (Paruelo et al. 2006). It also includes biomes called "mallines", humid grasslands associated with permanent water supplies such as rivers and creeks (Coronato et al. 2008).

Adverse climatic conditions in arid environments can cause soil degradation through wind or water erosion, leading to declined soil fertility and desertification (Chartier et al. 2013), a problem that already affects more than 90 % of Patagonia (del Valle 1998). In addition, human economic activities developed in this region, such as raising sheep or hydrocarbon exploitation, could contribute to increased land and vegetation degradation (Mazzoni and Vázquez 2010). Fortunately, the early detection of land degradation processes is possible through the implementation of tools for monitoring soil quality, therefore allowing landowners and decision makers to adopt policies that could prevent further land degradation in timely fashion. In the next section, we describe studies that included microbiological measures to detect changes in soil quality in arid lands from Patagonia exposed to either sheep overgrazing or hydrocarbon pollution.

6.4 Microbiological and Biochemical Indicators Measured in Drylands from Patagonia

Argentina is compromised in soil conservation and has been part of several national and international projects to fight desertification and monitor arid lands degradation, including its cooperation with five other countries, FAO and UNEP in the already completed, GEF-founded Land Degradation Assessment in Drylands (LADA) Project (Kellner et al. 2011). A product of this project was the creation of the National Observatory of Land Degradation and Desertification, which produced maps and indicators of erosion, aridity, land cover, and degradation. In addition, the Agricultural Technology National Institute (INTA) has been monitoring soil quality in several agricultural lands from humid and semiarid regions of Argentina. However, these measures were taken in fertile and productive lands, and less attention has been paid to the less fertile natural ecosystems of Patagonia. In these ecosystems, most of the studies that used microbiological/biochemical indicators analyzed grazed lands, and only a few studies assessed the effects of contamination on soil properties (Table 6.1).

Table 6.1 Microbiological indicators measured in Patagonian arid lands

Site	Indicator used	Soil conditions/land use practices	Reference
Monte phytogeographic province, southern portion	Potential N-mineralization N-microbial biomass	Soil from grass-shrub patches, incipient grass-shrub patches, grass patches, and bare soil	Mazzarino et al. (1996, 1998)
Patagonian phytogeographic province	Microbial biomass C β-Glucosidase	Soil associated with shrubs, grasses, mosses, and bare soil	Gonzalez-Polo and Austin (2009)
Monte and Patagonian phytogeographic provinces	Potential N-mineralization Microbial-N flush	Soil underneath perennial grasses and shrubs	Bertiller et al. (2006)
Monte phytogeographic province, southern portion	N-Mineralization Microbial-N immobilization C-Respiration	Soil associated to perennial grasses and evergreen shrubs	Vargas et al. (2006)
Monte phytogeographic province, southern portion	Potential N-mineralization Microbial-N flush	Soil underneath perennial grasses and evergreen shrubs	Carrera et al. (2003, 2005)
Monte phytogeographic province, southern portion	Potential N-mineralization	Soil at low and high grazing disturbance sites	Carrera and Bertiller (2013)
Patagonian phytogeographic province	C-mineralization Microbial biomass Fungi/bacteria	Soil with and without biocide addition Soil with different UV intensities (full sun, UV-B filtered, blocked total)	Austin and Vivanco (2006)
Monte phytogeographic province, southern portion	Microbial biomass-C Total heterotrophic fungi/total heterotrophic bacteria Dehydrogenase/β-glucosidase/protease Alkaline phosphatase/acid phosphatase	Soil from patches and interpatches at sites under low and high grazing intensity pressures	Prieto et al. (2011)
Monte phytogeographic province, southern portion	Microbial biomass-C Dehydrogenase/β-glucosidase Alkaline phosphatase/acid phosphatase Protease	Soil from patches and interpatches at sites under low and high grazing intensity pressures	Olivera et al. (2014)
Monte phytogeographic province, southern portion	Bacterial community structure by means of PCR-DGGE of 16S rRNA genes	Soil from patches and interpatches	Olivera et al. (2016)
Patagonian phytogeographic province	Soil respiration	Soil in a vegetation gradient and soil under two different grazing pressures in a dry steppe	Peri et al. (2015)
Monte phytogeographic province, southern portion	Nitrifying enzyme activity Abundance of *amoA* genes from AOB and AOA by qPCR	Soil amended with plant litter from low and high grazing disturbed sites, under both low and high moisture conditions	Marcos et al. (2016)

Patagonian phytogeographic province	Net nitrification	Nitrification inhibited vs. control, at two different years	Austin et al. (2006)
Patagonian phytogeographic province	Potential soil respiration Potential soil nitrification Potential soil ammonification	Soil associated to shrubs and grasses	Flombaum and Sala (2012)
Patagonian phytogeographic province	Net N-mineralization Net nitrification Net ammonification	Effects of irrigation treatments on litter decomposition and soil N-mineralization	Yahdjian and Sala (2008)
Patagonian phytogeographic province	Potential microbial respiration Potential net N-mineralization Microbial biomass-N	Nontreated soil (control) and soil treated with inorganic fertilization, biosolids compost, or municipal solid waste compost	Kowaljow and Mazzarino (2007)
Patagonian phytogeographic province	Potential microbial respiration Potential net N-mineralization	Nontreated soil (control) and soil treated with inorganic fertilization, biosolids compost, or municipal solid waste compost	Kowaljow et al. (2010)
Patagonian phytogeographic province	Microbial biomass-C Respiration β-Glucosidase/acid phospho-monoesterase/leucine-aminopeptidase Phenol oxidase	Nontreated soil (control) and soil treated with inorganic fertilization, biosolids compost, or municipal solid waste compost	González Polo et al. (2015)
Patagonian phytogeographic province	Total heterotrophic bacteria Hydrocarbon-degrading bacteria Metabolic diversity (number of substrate BIOLOG microplates)	Crude oil-polluted soil and unpolluted soil	Pucci et al. (2000)
Patagonian phytogeographic province	Total heterotrophic bacteria Hydrocarbon-degrading bacteria	Crude oil-polluted soil and unpolluted soil	Peressutti et al. (2003)
Patagonian phytogeographic province	CFU Total cell counts by flow cytometry Metabolic diversity (BIOLOG substrates) Community composition	Soil at different concentrations of NaCl	Kleinsteuber et al. (2006)
Specific location not reported	Total heterotrophic bacteria Hydrocarbon-degrading bacteria	Soil under natural attenuation, or amended with diesel oil, diesel oil + N, benzoate (an inductor of aromatic hydrocarbons mineralization), and benzoate + N	Acuña et al. (2012)
Monte phytogeographic province, southern portion	Dehydrogenase activity	Near-root soil after 15 days of different plant growth (tomato, sunflower, soybean, alfalfa, and unplanted soil)	Mitton et al. (2014)

6.4.1 Microbiological and Biochemical Indicators in Grazed Lands

Sheep became the predominant domestic herbivores in Patagonia after their introduction at the beginning of the 1900s (Bertiller et al. 2002). Reductions of total plant cover, and replacements of species within and between functional groups are produced by long-term grazing disturbance (Bertiller and Bisigato 1998), and these changes in vegetation, in turn, could affect soil properties and processes. Mazzarino et al. (1996) studied availability, mineralization and immobilization of N in different vegetation patches (resulting from exposure to grazing disturbances) and bare soil of the Patagonian Monte. In their study, soil underneath undisturbed patches had higher concentrations of inorganic-N (mainly ammonium) and N retained in the microbial biomass than bare soil. Although N-mineralization was low in all conditions, undisturbed patches constitute an important source of N (inorganic and mineralizable), and therefore the authors suggested that they should be conserved and used as a tool to monitor the conservation state of the ecosystem (Mazzarino et al. 1996). In a subsequent study, the same authors confirmed that the flush of microbial-N increased in undisturbed patches of vegetation, and in addition, they observed that net N-mineralization depended on water availability (Mazzarino et al. 1998). Gonzalez-Polo and Austin (2009) demonstrated that microenvironmental and biochemical soil conditions associated with different plant life forms influence biotic soil characteristics. These authors observed that shrubs appear to create hotspots of microbial activity by the establishment of protected microsites with high organic matter. They also highlighted the importance of heterogeneous vegetation distribution on the distribution of soil organic-C available to microorganisms (Gonzalez-Polo and Austin 2009). In contrast, Bertiller et al. (2006) found that neither N-mineralization nor microbial-N varied in soils underneath shrubs and perennial grasses from arid lands, although both characteristics increased when trees from humid ecosystems of Patagonia were included in the analysis.

Because plants produce litter with different chemical characteristics, changes in plant species composition induced by grazing alter the chemistry of the plant litter mixture that is produced, and indirectly, the input of nutrients that the soil receives. A decomposition experiment with leaf litter from perennial grasses and evergreen shrubs showed that different species of perennial grasses produced litter of similar characteristics that induced similar C and N dynamics in soil microcosms (Vargas et al. 2006). In contrast, shrub species varied to a greater extent in their plant litter recalcitrance and in their effects on C and N processes. Therefore, changes in plant species composition induced by grazing could have different effects on soil biogeochemistry depending on the species replaced (Vargas et al. 2006). In accordance, the effects of N-conservation strategies of perennial grasses and evergreen shrubs on soil N-dynamics were studied in the Patagonian Monte (Carrera et al. 2003). Perennial grasses with high N-resorption efficiencies (low N concentration in plant litter) decreased soil N, microbial-N flush, and N-mineralization rates; in contrast, shrubs contributed to increasing soil N concentration by producing N-rich litter that mineralized faster (Carrera et al. 2003). Potential N-mineralization and soil C and N concentrations were higher in soils under the influence of plant litter from shrubs than from perennial grasses, which in turn was associated with the higher decomposability of shrub tissues (Carrera et al. 2005). In a different study, changes in canopy structure induced by grazing lead to the production of leaf litter of higher recalcitrance than low grazing-disturbed sites. These changes in canopy structure and plant litter characteristics lead to reduced litter decomposition and soil N-mineralization rates in high grazing- compared to low grazing-disturbed sites (Carrera and Bertiller 2013). The authors suggested that the amendment of soils from the highly grazed site with nonrecalcitrant litter may stimulate soil decomposer activity and nutrient release, which in turn would contribute to the reestablishment of plant cover (Carrera and Bertiller 2013). Interestingly, not only soil decomposers govern litter decomposition in these environments. Austin and Vivanco (2006) observed that photodegradation exerted a major control on plant litter decomposition in soils from the Patagonian Steppe, and concluded that changes in radiation interception (e.g., because of changes in vegetation cover or cloudiness) could have important effects on C balance in this ecosystem.

Prieto et al. (2011) used a polyphasic approach to study the soil quality in the Patagonian Monte, as they measured the physical (bulk density), chemical (organic-C, total N, and pH), microbiological (microbial biomass-C, microorganism counts), and enzymatic (dehydrogenase, β-glucosidase, protease, alkaline and acid phosphatase activities) properties of soils under contrasting grazing intensities. Grazing reduced the cover of perennial grasses, soil organic-C, and the activity of several enzymes (β-glucosidase, phosphatase, protease), which could produce a negative effect on processes related with C, P, and N cycles (Prieto et al. 2011). On the other hand, grazing was associated with increased microbial biomass-C and microorganism counts in highly grazed sites, probably resulting from nutrient input from urine and feces depositions (Prieto et al. 2011). Overall, grazing affected soil properties through direct (depositions input, trampling) and indirect (litter-mediated) effects (Prieto et al. 2011). A seasonal study in the same region showed that long-term continuous grazing induced a reduction of plant litter quantity and quality, which negatively affected some soil enzyme activities (alkaline and acid phosphatase and β-glucosidase) and microbial biomass-C (Olivera et al. 2014). However, in line with Prieto et al. (2011), there were localized positive effects induced by grazing in interpatched areas from highly grazed sites (enhanced soil-N, alkaline phosphatase and dehydrogenase activities, microbial biomass-C), presumably as a response to nutrient redistribution through animal excreta, which promotes belowground microsites with high microbial activity (Olivera et al. 2014). More recently, denaturing gradient gel electrophoresis (DGGE) analysis of the 16S rRNA phylogenetic marker showed that grazing intensity also affected the structuring of bacterial communities from patched and interpatched areas of the Patagonian Monte, including those in biological soil crusts (Olivera et al. 2016). Soil respiration also changed in semiarid grasslands from southern Patagonia in response to different grazing intensities, land uses, and climate and vegetation gradients (Peri et al. 2015). According to the authors, this information is essential to estimate the C balance in a range of ecosystems from Patagonia (Peri et al. 2015).

We recently analyzed the effects of soil moisture and grazing-induced changes in plant litter quality on the nitrifying activity and abundance of bacterial (AOB) and archaeal (AOA) ammonia-oxidizing genes in arid soils from the Patagonian Monte (Marcos et al. 2016). Changes in plant litter induced by grazing negatively influenced nitrification by providing poorer substrates with higher inhibitory compounds. In contrast, plant litter from lightly grazed sites combined with high soil moisture promoted nitrification and AOB abundance, while AOA remained constant at lower abundances (Marcos et al. 2016). Nitrifying microorganisms have a major role in ecosystem functioning, and their inhibition may not only affect the N but also the C cycles (Austin et al. 2006). In an experiment in the Patagonian Steppe where nitrification was inhibited, plant cover decreased and litter decomposition was reduced (Austin et al. 2006). The reduction in plant cover appeared to be associated with a reduction of nitrate and an increment of ammonium concentrations in soils, suggesting that plants from this ecosystem may prefer nitrates over ammonium (Austin et al. 2006). Furthermore, potential soil nitrification together with rooting depth and soil thermal amplitude were the traits that had the highest influence on ecosystem functioning in soils from the Patagonian Steppe (Flombaum and Sala 2012). In a different study, the effects of precipitation on litter decomposition and soil-N mineralization were examined in the Patagonian steppe (Yahdjian and Sala 2008). Decomposition was affected by drought but not by increases in precipitation, indicating that wet and dry years may not have the same effect on decomposition. Nitrification increased in response to moisture increments, but ammonification did not, suggesting that microorganisms involved in these processes may have different sensitivities to soil moisture (Yahdjian and Sala 2008).

In an arid ecosystem of Patagonia affected by wildfires, organic composts of different quality significantly improved soil properties after 1 year of application (Kowaljow and Mazzarino 2007). Organic composts did not improve soil moisture, but they did improve soil organic C, microbial respiration, and potential net N-mineralization, and together with inorganic fertilizers they also improved microbial biomass-N (Kowaljow and Mazzarino 2007). Inorganic fertilizers did not contribute to soil restoration, but they lead to an increase in inorganic-N (the main limiting nutrient in arid ecosystems after water) and produced a recovery of plant cover, which in turn contributed to decreased soil degradation (Kowaljow and Mazzarino 2007). However, the authors proposed that this could be a short-term effect of inorganic fertilizers. Similar results were obtained in a subsequent study in the same environment, where potential microbial respiration, net N-mineralization, organic matter, and nutrients remained higher in organic fertilizer-amended soils than in nonfertilized controls after 3 years of compost application (Kowaljow et al. 2010). These effects of a single dose of organic composts and inorganic fertilizers on soil quality were followed in the long term (6 years after the application of the fertilizers). Organic compost application significantly stimulated total soil C and N, and enhanced potential respiration and the activities of β-glucosidase, acid phosphomonoesterase, and phenol oxidase enzymes, although they did not affect microbial biomass (González Polo et al. 2015). In contrast, inorganic fertilizers did not stimulate any of the mentioned parameters.

The authors of this study proposed that although different organic amendments can induce different changes on soil properties and enzyme activities and these changes should be studied before any management decision, the application of organic composts is a plausible restoration measure for ecosystems in semiarid Patagonia (González Polo et al. 2015).

6.4.2 Microbiological Indicators in Contaminated Soils

Few studies have been performed in drylands from Patagonia to measure changes in soil quality indicators in response to exposure to pollutants, and these studies were mostly directed to analyzing hydrocarbon-contaminated soils. Pucci et al. (2000) compared the abundance of total heterotrophic and hydrocarbon-degrading bacteria and the BIOLOG substrate utilization in soils impacted compared to not impacted by crude oil contamination. Total heterotrophic bacteria were one order of magnitude higher in the unpolluted than in the polluted soil soon after the spill occurred, although this was reversed 7 months later, presumably by an increase of hydrocarbon degraders (Pucci et al. 2000). In addition, the bacterial community from the contaminated soil substantially increased the amount of used substrates from the BIOLOG plate in comparison to the bacteria from the nonpolluted site (Pucci et al. 2000). The authors concluded that those semiarid soils hold bacterial communities adapted to degrading high concentrations of hydrocarbons (Pucci et al. 2000). A similar response of the total heterotrophic and hydrocarbon-degrading bacteria was observed in a crude oil biodegradation experiment, in which these groups were about three and seven orders of magnitude higher in contaminated than in noncontaminated samples after 4 months of incubation, respectively (Peressutti et al. 2003).

To establish detection tools capable of determining hydrocarbon degradation potential in natural environments, Kleinsteuber et al. (2006) analyzed the diversity and dynamics of a bacterial community from a diesel fuel-contaminated soil and identified active hydrocarbon degraders. The analyzed soils held a complex bacterial community capable of degrading diesel fuel and shifting in response to different soil salinities (Kleinsteuber et al. 2006). Acuña et al. (2012) analyzed hydrocarbon-polluted soil microcosms with or without N-fertilization to study N effect on aliphatic and aromatic hydrocarbon degradation. N-fertilization appeared to favor aliphatic hydrocarbon degradation, although unfertilized soils with N deficiency enhanced aromatic hydrocarbon degradation (Acuña et al. 2012).

There is only one report of arid soils contaminated with nonhydrocarbon pollutants, in which the authors analyzed plant species potential to phytoremediate dichloro diphenyl trichloroethane (DDT)-polluted soils from the Patagonian Monte (Mitton et al. 2014). Dehydrogenase activity was used to assess soil microbial activity in soils seeded with four different plants and in unplanted control soil (Mitton et al. 2014). From the different tested plants, tomato appeared to be the best phytoremediator candidate, showing higher accumulation potential, no phytotoxicity effects, and enhanced soil dehydrogenase activity (Mitton et al. 2014).

6.5 Interpretation of Microbiological and Biochemical Measures and Selection of Suitable Soil Quality Indicators

Although, as mentioned in the previous section, several studies from Patagonian arid lands have measured microbiological/biochemical parameters that are generally used as indicators of soil quality, the results of these studies so far suggest that only a few of these parameters (namely, microbial-N flush and some enzyme activities) could potentially be used as indicators of soil quality in this region. These variables took different values at contrasting land uses, and these differences were consistent in various studies, allowing the identification of possible threshold values to differentiate disturbed (possibly lower quality) from undisturbed (higher quality) soils. For example, the microbial-N flush was $\geq$19 mg N kg soil^{-1} in soil from plant patches at ungrazed sites and sites where grazing was excluded for several years, whereas lower values were reported in soils from grazed patches and in soils burned after a wildfire (Fig. 6.2) (Mazzarino et al. 1996, 1998; Carrera et al. 2003; Kowaljow and Mazzarino 2007). Similarly, β-glucosidase and alkaline and acid phosphatase activities showed values higher than 200, 300, and 90 μg pNP g soil^{-1} h^{-1}, respectively, in soils from plant patches at conserved sites, whereas lower values were observed at degraded soils (Fig. 6.2) (Prieto et al. 2011; Olivera et al. 2014). These enzyme values are similar to those in semiarid uncultivated natural rangelands from Iran, semiarid shrublands and pine forests from Spain, and semiarid undisturbed shrublands of the Western United States (Bastida et al. 2006; Blecker et al. 2013; Raiesi and Beheshti 2014; Hedo et al. 2015), except for alkaline phosphatase, that showed similar although more dispersed values in non-Patagonian soils. Interestingly, the Geometric Mean of enzyme activity (GMea), an integrative combination of enzyme measures into a unique index, has been proposed to assess soil quality (Hinojosa et al. 2004; García-Ruiz et al. 2008). We calculated this index based on enzyme measures in published studies from Patagonian soils and found an overall decrease in enzyme activities, between 26 % and 31 %, in soils from plant patches (Prieto et al. 2011; Olivera et al. 2014) and of 25 % in bare soil areas (Prieto et al. 2011) in response to increased grazing pressure. In other soils from Patagonia, the GMea index increased 17–62 % in response to amendment with inorganic and organic fertilizers (González Polo et al. 2015). Although GMea is just one of the several existing microbiological/biochemical indexes, it was the only one that we could calculate with the information available from Patagonian soils. Overall, this index seems to be suitable to detect changes in soil quality; however, more studies should be performed in undisturbed high-quality soils to determine and validate reference GMea values for Patagonia.

Other microbiological variables showed ambiguous results that overlapped in soils from different studies or land uses, hindering the finding of a clear limit to distinguish soils of different qualities (Fig. 6.2). Potential N mineralization in Patagonian arid soils (0.09–0.98 mg N kg soil^{-1} day^{-1}) was slightly lower than in semiarid soils from an undisturbed shrubland in the western United States (3.5–15.2 mg N kg soil^{-1} day^{-1})

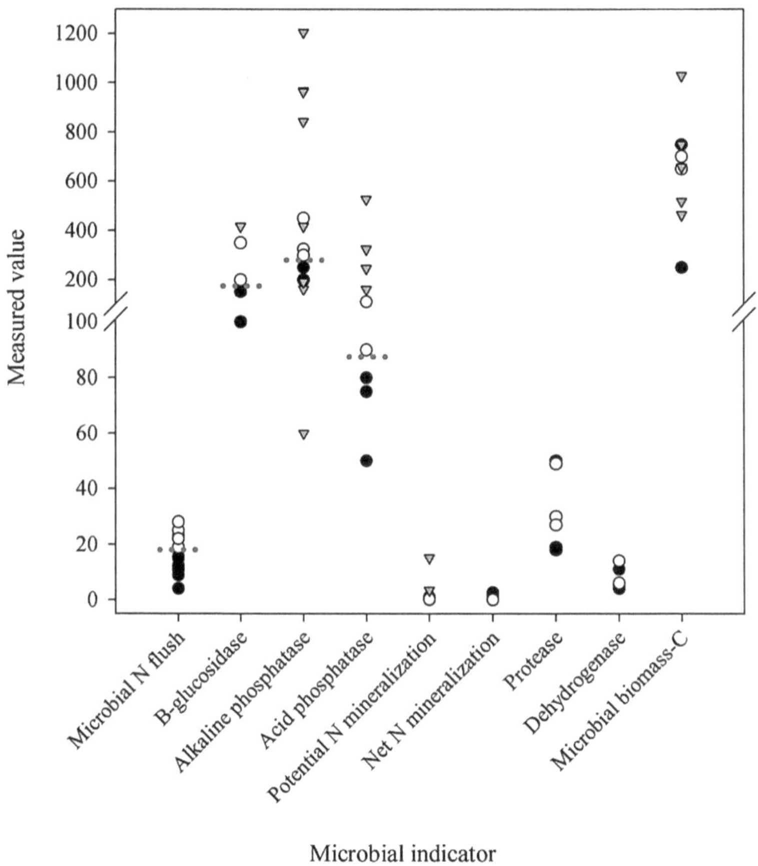

Fig. 6.2 Some microbiological indicators measured in arid soils from Patagonia and in soil quality studies performed in semiarid lands from other countries. *Circles* represent values obtained in studies from Patagonia (*white circles*, undisturbed/lightly disturbed soils; *black circles*, highly disturbed soils) and *triangles* represent values obtained in undisturbed lands from other countries (Iran (Raiesi and Beheshti 2014), Spain (Bastida et al. 2006; Hedo et al. 2015) and Western USA (Blecker et al. 2013)). Indicator units are mg N kg soil^{-1} (microbial N flush), µg pNP g soil^{-1} h^{-1} (β-glucosidase, alkaline and acid phosphatase enzymes), mg N kg soil^{-1} day^{-1} (potential and net N mineralization), µg Tyr g soil^{-1} h^{-1} (proteases), µg TPF g soil^{-1} h^{-1} (dehydrogenase), and µg C g soil^{-1} (microbial biomass-C). Only microbial biomass-C values obtained by the fumigation-extraction method were included in the figure. *Dashed grey lines* in some of the indicators represent suggested threshold values that could be used to differentiate low- from high-quality soils

(Blecker et al. 2013), and did not differ in response to grazing pressure, plant life forms, or fertilizer amendment (Carrera et al. 2003, 2005; Bertiller et al. 2006; Kowaljow and Mazzarino 2007; Kowaljow et al. 2010; Carrera and Bertiller 2013). Similarly, net N-mineralization and protease enzymes did not differ in soils covered by different plant life forms or under different grazing pressures (Mazzarino et al. 1996, 1998; Vargas et al. 2006; Prieto et al. 2011; Olivera et al. 2014). Contrarily to the previously described enzymes, dehydrogenase values overlapped in soil from plant patches at grazing disturbed and conserved sites (Fig. 6.2). However, this enzyme activity could be useful to identify land degradation at bare soil areas, where differential values were observed in response to different grazing pressures (values < 2.5 μg TPF g $soil^{-1}$ h^{-1} at lightly grazed sites and above that value in highly grazed sites) (Prieto et al. 2011; Olivera et al. 2014).

By using the fumigation-extraction method, microbial biomass-C values between 200 and 750 μg C g $soil^{-1}$ were observed in Patagonian soils associated with plant patches from disturbed and undisturbed sites (Fig. 6.2) (Gonzalez-Polo and Austin 2009; González Polo et al. 2015), and these values were in accordance with those in semiarid forests and shrublands from Murcia, Spain (464–1030 μg C g $soil^{-1}$) (Bastida et al. 2006; Hedo et al. 2015). In addition, values between 300 and 600 μg C g $soil^{-1}$ were obtained in Patagonian soils under different grazing pressures using the substrate-induced respiration technique (Prieto et al. 2011; Olivera et al. 2014). Nevertheless, as these studies measured microbial biomass-C with different techniques, it is not possible to make a straight comparison. Total heterotrophic fungi and bacteria were also counted using different media and culture conditions, and therefore no further generalizations are performed based on these measures.

6.6 Conclusions

In this chapter, we reviewed studies performed in arid lands from Patagonia that included different soil biochemical/microbiological measures to assess their suitability as soil quality indicators in arid environments. Mainly, we identified four microbiological/biochemical parameters (microbial-N flush, β-glucosidase, alkaline and acid phosphatase activities) that could potentially serve as indicators of soil quality in vegetated areas; as well as one indicator (dehydrogenase activity) for its use in bare soil areas. The flush of N is the N mineralized by a recolonizing population of soil bacteria after fumigation and can be used to estimate microbial biomass-N in soil (Shen et al. 1984). β-Glucosidases are involved in the degradation of carbohydrates, releasing important energy sources for soil microorganisms (Eivazi and Tabatabai 1988). Alkaline and acid phosphatases hydrolyze phosphoric monoesters, releasing phosphate, and are therefore important in soil organic-P mineralization (Tabatabai 1982). Overall, these indicators are involved in the cycling of three of the most important chemical elements that influence soil fertility: N, C, and P. Dehydrogenase activity is used as a measure of the overall microbial activity of soil. We found that these variables took different values at disturbed and undisturbed Patagonian soils, and therefore suggest that they might be useful to detect changes in soil quality produced as a consequence of human activities. However, more studies are still needed to clarify how other factors such as climatic conditions could affect these indicators. Moreover, so far, in the published literature there is not enough information to calculate multiparametric quality indexes for Patagonia, and because the use of a single property as a soil quality indicator is not recommended

(Gil-Sotres et al. 2005), it is desirable to expand the knowledge on this topic. Additionally, further studies in undisturbed high-quality soils are needed for the identification of reference values of any biochemical/microbiological indicator to be used in this region. There is also a need to standardize field and laboratory protocols of the indicator measures for comparison purposes.

The record of microbiological and biochemical data in drylands from Patagonia has some missing information that precludes us from achieving generalizations about their meaning. For example, several indicators (potential soil respiration, microbial N-immobilization, net and potential nitrification, soil ammonification, ammonia monooxygenase, acid phosphomonoesterase, leucine-aminopeptidase phenol oxidase, the abundance of bacterial and archaeal *amoA* genes, and the abundance of hydrocarbon-degrading bacteria) have been measured in only one study, and/or under only one state of disturbance (i.e., in nondisturbed/high-quality soils or in disturbed/low-quality soils, but not in both states). Further evidence of this missing information is that most microbiological data were assessed by indirect measures (e.g., microbial biomass-C, N-mineralization/immobilization/nitrification), few studies targeted microbial populations with specific functions, and microbial diversity is almost entirely unknown. Besides, studies have been much localized; with most ecosystems within this vast territory still remaining unexplored. Altogether, these gaps of information need to be filled to determine which of these indicators are suitable for this region and hold sufficient information to discriminate soils of different quality. We hope that the information presented in this chapter might help researchers in the design of future projects aimed to bridge these gaps.

Acknowledgments The authors gratefully acknowledge support from ANPCyT (PICT 2013-1505), Argentina.

References

Abbott LK, Murphy DV (2007) What is soil biological fertility? In: Abbott LK, Murphy DV (eds) Soil biological fertility: a key to sustainable land use in agriculture. Springer, Dordrecht, pp 1–15

Acuña AJ, Pucci OH, Pucci GN (2012) Effect of nitrogen deficiency in the biodegradation of aliphatic and aromatic hydrocarbons in Patagonian contaminated soil. Int J Res Rev Appl Sci 11:470–476

Ares J, Beeskow AM, Bertiller M, Rostagno M, Irisarri M, Anchorena J, Defossé G, Merino C (1990) Structural and dynamic characteristics of overgrazed lands of northern Patagonia, Argentina. In: Breymeyer A (ed) Managed grasslands. Elsevier, Amsterdam, pp 149–175

Austin AT, Vivanco L (2006) Plant litter decomposition in a semi-arid ecosystem controlled by photodegradation. Nature (Lond) 442:555–558

Austin AT, Sala OE, Jackson RB (2006) Inhibition of nitrification alters carbon turnover in the Patagonian Steppe. Ecosystems 9:1257–1265

Bastida F, Moreno JL, Hernández T, García C (2006) Microbiological degradation index of soils in a semiarid climate. Soil Biol Biochem 38:3463–3473

Bastida F, Zsolnay A, Hernández T, García C (2008) Past, present and future of soil quality indices: a biological perspective. Geoderma 147:159–171

Benedetti A, Dilly O (2006) Introduction. In: Bloem J, Hopkins DW, Benedetti A (eds) Microbiological methods for assessing soil quality. CABI Publishing, Wallingford, pp 3–14
Bertiller MB, Bisigato A (1998) Vegetation dynamics under grazing disturbance. The state-and-transition model for the Patagonian steppes. Ecol Austr 8:191–199
Bertiller MB, Ares JO, Bisigato AJ (2002) Multiscale indicators of land degradation in the Patagonian Monte, Argentina. Environ Manag 30:704–715
Bertiller MB, Mazzarino MJ, Carrera AL, Diehl P, Satti P, Gobbi M, Sain CL (2006) Leaf strategies and soil N across a regional humidity gradient in Patagonia. Oecologia 148:612–624
Blecker SW, Stillings LL, Amacher MC, Ippolito JA, DeCrappeo NM (2013) Development and application of a soil organic matter-based soil quality index in mineralized terrane of the Western US. Environ Earth Sci 68:1887–1901
Bloem J, Schouten AJ, Sørensen SJ, Rutgers M, van der Werf A, Breure AM (2006) Monitoring and evaluating soil quality. In: Bloem J, Hopkins DW, Benedetti A (eds) Microbiological methods for assessing soil quality. CABI Publishing, Wallingford, pp 23–49
Burns RG, Nannipieri P, Benedetti A, Hopkins DW (2006) Defining soil quality. In: Bloem J, Hopkins DW, Benedetti A (eds) Microbiological methods for assessing soil quality. CABI Publishing, Wallingford, pp 15–22
Carrera AL, Bertiller MB (2013) Combined effects of leaf litter and soil microsite on decomposition process in arid rangelands. J Environ Manag 114:505–511
Carrera AL, Bertiller MB, Sain CL, Mazzarino MJ (2003) Relationship between plant nitrogen conservation strategies and the dynamics of soil nitrogen in the arid Patagonian Monte, Argentina. Plant Soil 255:595–604
Carrera AL, Vargas DN, Campanella MV, Bertiller MB, Sain CL, Mazzarino MJ (2005) Soil nitrogen in relation to quality and decomposability of plant litter in the Patagonian Monte, Argentina. Plant Ecol 181:139–151
Chartier MP, Rostagno CM, Videla LS (2013) Selective erosion of clay, organic carbon and total nitrogen in grazed semiarid rangelands of northeastern Patagonia, Argentina. J Arid Environ 88:43–49
Coronato AMJ, Coronato F, Mazzoni E, Vázquez M (2008) The physical geography of Patagonia and Tierra del Fuego. In: Rabassa J (ed) The late Cenozoic of Patagonia and Tierra del Fuego, vol 11. Elsevier, Amsterdam, pp 13–55
del Valle HF (1998) Patagonian soils: a regional synthesis. Ecol Austr 8:103–123
Doran JW, Zeiss MR (2000) Soil health and sustainability: managing the biotic component of soil quality. Appl Soil Ecol 15:3–11
Eivazi F, Tabatabai MA (1988) Glucosidases and galactosidases in soils. Soil Biol Biochem 20:601–606
Flombaum P, Sala OE (2012) Effects of plant species traits on ecosystem processes: experiments in the Patagonian steppe. Ecology 93:227–234
García-Ruiz R, Ochoa V, Hinojosa MB, Carreira JA (2008) Suitability of enzyme activities for the monitoring of soil quality improvement in organic agricultural systems. Soil Biol Biochem 40:2137–2145
Gil-Sotres F, Trasar-Cepeda C, Leirós MC, Seoane S (2005) Different approaches to evaluating soil quality using biochemical properties. Soil Biol Biochem 37:877–887
González Polo M, Kowaljow E, Castán E, Sauzet O, Mazzarino MJ (2015) Persistent effect of organic matter pulse on a sandy soil of semiarid Patagonia. Biol Fertil Soils 51:241–249
Gonzalez-Polo M, Austin AT (2009) Spatial heterogeneity provides organic matter refuges for soil microbial activity in the Patagonian steppe, Argentina. Soil Biol Biochem 41:1348–1351
Hedo J, Lucas-Borja ME, Wic C, Andrés-Abellán M, de Las Heras J (2015) Soil microbiological properties and enzymatic activities of long-term post-fire recovery in dry and semiarid Aleppo pine (*Pinus halepensis* M.) forest stands. Solid Earth 6:243–252
Hinojosa MB, García-Ruíz R, Viñegla B, Carreira JA (2004) Microbiological rates and enzyme activities as indicators of functionality in soils affected by the Aznalcóllar toxic spill. Soil Biol Biochem 36:1637–1644

Karlen DL, Mausbach MJ, Doran JW, Cline RG, Harris RF, Schuman GE (1997) Soil quality: a concept, definition, and framework for evaluation (a guest editorial). Soil Sci Soc Am J 61:4–10
Karlen DL, Andrews SS, Doran JW (2001) Soil quality: current concepts and applications. Adv Agron 74:1–40
Kellner K, Risoli C, Metz M (2011) Terminal evaluation of the UNEP/FAO/GEF project 'Land Degradation Assessment in Drylands (LADA)'. United Nations Environment Programme
Kleinsteuber S, Riis V, Fetzer I, Harms H, Müller S (2006) Population dynamics within a microbial consortium during growth on diesel fuel in saline environments. Appl Environ Microbiol 72:3531–3542
Kowaljow E, Mazzarino MJ (2007) Soil restoration in semiarid Patagonia: chemical and biological response to different compost quality. Soil Biol Biochem 39:1580–1588
Kowaljow E, Mazzarino MJ, Satti P, Jiménez-Rodríguez C (2010) Organic and inorganic fertilizer effects on a degraded Patagonian rangeland. Plant Soil 332:135–145
Lehman RM, Acosta-Martinez V, Buyer JS, Cambardella CA, Collins HP, Ducey TF, Halvorson JJ, Jin VL, Johnson JMF, Kremer RJ, Lundgren JG, Manter DK, Maul JE, Smith JL, Stott DE (2015) Soil biology for resilient, healthy soil. J Soil Water Conserv 70:12A–18A
León RJC, Bran D, Collantes M, Paruelo JM, Soriano A (1998) Grandes unidades de vegetación de la Patagonia extra andina. Ecol Austr 8:125–144
Marcos MS, Bertiller MB, Saraví Cisneros H, Olivera NL (2016) Nitrification and ammonia-oxidizing bacteria shift in response to soil moisture and plant litter quality in arid soils from the Patagonian Monte. Pedobiol J Soil Ecol 59:1–10
Mazzarino MJ, Bertiller MB, Sain CL, Laos F, Coronato FR (1996) Spatial patterns of nitrogen availability, mineralization, and immobilization in northern Patagonia, Argentina. Arid Soil Res Rehabil 10:295–309
Mazzarino MJ, Bertiller MB, Sain C, Satti P, Coronato F (1998) Soil nitrogen dynamics in northeastern Patagonia steppe under different precipitation regimes. Plant Soil 202:125–131
Mazzoni E, Vázquez M (2010) Desertificación en la Patagonia. In: Latrubesse EM (ed) Developments in Earth surface processes. Elsevier, Amsterdam, pp 351–377
Mitton FM, Miglioranza KSB, Gonzalez M, Shimabukuro VM, Monserrat JM (2014) Assessment of tolerance and efficiency of crop species in the phytoremediation of DDT polluted soils. Ecol Eng 71:501–508
Niemeijer D, Puigdefabregas J, White R, Lal R, Winslow M, Ziedler J, Prince S, Archer E, King C (2005) Dryland systems. UNEP, London, pp 623–662
Olivera NL, Prieto L, Carrera AL, Saraví Cisneros H, Bertiller MB (2014) Do soil enzymes respond to long-term grazing in an arid ecosystem? Plant Soil 378:35–48
Olivera NL, Prieto L, Bertiller MB, Ferrero MA (2016) Sheep grazing and soil bacterial diversity in shrublands of the Patagonian Monte, Argentina. J Arid Environ 125:16–20
Paruelo JM, Golluscio RA, Jobbágy EG, Canevari M, Aguiar MR (2006) Situación ambiental en la Estepa Patagónica. In: Brown A, Martinez Ortiz U, Acerbi M, Corcuera J (eds) La situación ambiental Argentina 2005. Fundación Vida Silvestre Argentina, Buenos Aires, pp 303–313
Paz-Ferreiro J, Fu S (2016) Biological indices for soil quality evaluation: perspectives and limitations. Land Degrad Dev 27:14–25
Peressutti SR, Alvarez HM, Pucci OH (2003) Dynamics of hydrocarbon-degrading bacteriocenosis of an experimental oil pollution in Patagonian soil. Int Biodeter Biodegr 52:21–30
Peri PL, Bahamonde H, Christiansen R (2015) Soil respiration in Patagonian semiarid grasslands under contrasting environmental and use conditions. J Arid Environ 119:1–8
Prieto LH, Bertiller MB, Carrera AL, Olivera NL (2011) Soil enzyme and microbial activities in a grazing ecosystem of Patagonian Monte, Argentina. Geoderma 162:281–287
Pucci OH, Bak MA, Peressutti SR, Klein I, Härtig C, Alvarez HM, Wünsche L (2000) Influence of crude oil contamination on the bacterial community of semiarid soils of Patagonia (Argentina). Acta Biotechnol 20:129–146
Raiesi F, Beheshti A (2014) Soil C turnover, microbial biomass and respiration, and enzymatic activities following rangeland conversion to wheat-alfalfa cropping in a semi-arid climate. Environ Earth Sci 72:5073–5088

Schloter M, Munch JC, Tittarelli F (2006) Managing soil quality. In: Bloem J, Hopkins DW, Benedetti A (eds) Microbiological methods for assessing soil quality. CABI Publishing, Wallingford, pp 50–62
Shen SM, Pruden G, Jenkinson DS (1984) Mineralization and immobilization of nitrogen in fumigated soil and the measurement of microbial biomass nitrogen. Soil Biol Biochem 16:437–444
Tabatabai MG (1982) Soil enzymes. In: ASA-SSSA (ed) Methods of soil analysis. Part 2. Microbiological and biochemical properties. ASA-SSSA, Madison, WI, pp 903–947
United Nations (2014) Resolution A/RES/68/232
van Bruggen AHC, Semenov AM (2000) In search of biological indicators for soil health and disease suppression. Appl Soil Ecol 15:13–24
Vargas DN, Bertiller MB, Ares JO, Carrera AL, Sain CL (2006) Soil C and N dynamics induced by leaf-litter decomposition of shrubs and perennial grasses of the Patagonian Monte. Soil Biol Biochem 38:2401–2410
Voroney RP (2007) The soil habitat. In: Paul EA (ed) Soil microbiology, ecology and biochemistry. Elsevier, Amsterdam, pp 25–49
Wall DH (2012) Introduction. In: Wall DH (ed) Soil ecology and ecosystem services. Oxford University Press, Oxford, pp 1–2
Yahdjian L, Sala OE (2008) Do litter decomposition and nitrogen mineralization show the same trend in the response to dry and wet years in the Patagonian steppe? J Arid Environ 72:687–695

Part II
Patagonian Microorganisms for Industrial and Sanitary Applications

Chapter 7
Molecular Ecology of Class 1 Integrons in Patagonia as Model System for Understanding the Rise of Antibiotic Resistance Isolates Around the World

Verónica Elizabeth Álvarez, María Paula Quiroga, Gabriel Alejandro Castro, Marcelo Hernán Cassini, and Daniela Centrón

Abstract The understanding of the molecular basis of the flux of antimicrobial resistance genes (ARG) is a fundamental requirement to develop a comprehensive explanation of how pathogenic strains adapt to extreme resistance phenotypes at a global scale. However, it is difficult to assess how ARG interact with human activities. This constraint led us to seek a suitable biological model system, which has two components: class 1 integrons as the main molecular element of bacterial resistance to antibiotics, and Tierra del Fuego Island as the study area. Both clinical and non-clinical strains were studied from this island for the presence of class 1 integrons. The 30 % of clinical strains and the 11 % of strains isolated from the open environment were *intI1* positive, respectively, sharing 3 *intI1* "clinical" alleles previously described in nosocomial strains from Europe, Africa, and Asia, depicting an interchange of genes among both habitats, the hospital and the sites close to human activities in Ushuaia City. It is likely that once the *intI1*-positive clinical strains are released from the hospital, the *intI1* genes can be transferred through the mechanisms of lateral genetic transfer to other bacterial species from the open environment where they can be maintained over time.

V.E. Álvarez • M.P. Quiroga • D. Centrón (✉)
Laboratorio de Investigaciones de los Mecanismos de Resistencia a Antibióticos, Instituto de Microbiología y Parasitología Médica (IMPaM, UBA-CONICET), Facultad de Medicina, Universidad de Buenos Aires, Ciudad Autónoma de Buenos Aires, Argentina
e-mail: dcentron@gmail.com

G.A. Castro
Hospital Regional de Ushuaia, Ushuaia, Argentina

M.H. Cassini
Grupo GEMA, Departamento de Ciencias Básicas, Universidad Nacional de Luján, Luján, Buenos Aires, Argentina

Laboratorio de Biología del Comportamiento, IBYME, Ciudad Autónoma de Buenos Aires, Argentina

N.L. Olivera et al. (eds.), *Biology and Biotechnology of Patagonian Microorganisms*, DOI 10.1007/978-3-319-42801-7_7

7.1 The Global Threat of Increased Resistance to Antibiotics

The discovery of antibiotics was one of the greatest scientific achievements of the twentieth century. Unfortunately, their effectiveness and easy access led to overuse. As a result of that, during the past 80 years, there has been a remarkable evolution of antibiotic-resistant strains of bacteria, leading to the appearance of pathogens with resistance to almost all families of antibiotic agents (Davies and Davies 2010; Arduino et al. 2012). Nowadays, this progressive increase of multidrug resistance in all geographic regions is considered a main global threat that has been identified as a public health priority according to several organizations such as the US Centers for Disease Control and Prevention (CDC), the European Centre for Disease Prevention and Control (ECDC), and the World Health Organization (WHO) (http://www.who.int/drugresistance).

The guiding force behind the increasing rates of resistance is the abuse, misuse, and overuse of antibiotic agents, used not only in human patients but also at a large scale in livestock, as growth promotion in food animals, and also released into the environment (Roca et al., 2015). The emergence of resistant microorganisms is a natural phenomenon that takes place either by mutations when bacteria replicate themselves erroneously or by the exchange of antibiotic resistance genes (ARG) between them. Just after these intracellular molecular processes, in the nosocomial niche, the exposure to antibiotic agents provides the necessary selective pressure for the acceleration and dissemination of resistant pathogens. In this scenario, new resistance mechanisms can emerge and spread globally.

Although there is no worldwide information concerning the mortality associated to resistant bacteria, in the recent report published by WHO (http://www.cdc.gov/drugresistance/pdf/ar-threats-2013-508.pdf), *Antibiotic Resistance Threats in the United States* (2013), there are some data of the hypothetically calamitous significances of inaction. This year the CDC estimates that in the United States, more than 2 million people are sickened every year with antibiotic-resistant infections, with at least 23,000 dying as a result. Regarding the level of concern of the resistant bacterial species causing severe human infections, CDC ranked bacterial species in this report into one of three categories: urgent, serious, and concerning. Besides, it must be assumed that antibiotic resistance endangers healthcare advances to society, because without effective antibiotics not only prevention and treatment of infections, but also the success of organ transplantation, chemotherapy, surgery, and general medical clinics, may be endangered.

7.2 Origin of the Antibiotic Resistance Genes

The source of ARG has been a matter for deep investigation. In some cases, the origin of ARG in human pathogens has been demonstrated as proceeding from bacteria from the open environment, such as the extended spectrum β-lactamases CTX-M-like genes (Arduino et al. 2002; Hawkey 2008; Nordmann et al. 2008; Rossolini et al. 2008). This is an example of how the resistome works. This term, "resistome," has been used to identify the collection of all genes that contributes, direct or indirectly, to a phenotype of antibiotic resistance (D'Costa et al. 2006). It is a suitable concept, because it emphasizes the role of environmental bacteria as the source and origin of the ARG. More than 20,000 genes have been collected in a database of ARG (Liu and Pop 2009), but they represent only an minor fraction of the resistome, because novel genomes and metagenomes are revealing huge amounts of resistance genes capable of handling multiple families of antibiotic agents (Gillings 2014). It has been shown that ARG are ancient components of the pangenome, such as the transposon Tn*5045* that has a class 1 integron with the *aadA2* gene cassette isolated from permafrost dating from 10,000 years past (Petrova et al. 2011), or a diverse collection of genes encoding resistance to β-lactam, tetracycline, and glycopeptide antibiotics identified in metagenomes from permafrost dating from 30,000 years ago (D'Costa et al. 2011), or a macrolide kinase gene related to a known family of kinases circulating in modern drug-resistant pathogens identified in bacteria from a cave microbiome that has been inaccessible for 4 million years (Bhullar et al. 2012). Although these data reveal that ARG were common in environmental bacteria independently of anthropogenic activities, the role of environmental bacteria in driving the expansion of antibiotic resistance in the clinic is still a matter of debate.

Understanding of the molecular and environmental basis of the flux of ARG from the environment to the clinic and vice versa is a fundamental requirement to develop a comprehensive explanation of how pathogenic strains adapt to extreme resistance phenotypes at a global scale. However, it is difficult to assess the directionality of the flux of genes between the natural environment and human habitats to identify how genes, or genetic platforms, interact with human activities. This constraint led us to seek a suitable and appropriate biological model system, where fewer variables intervene, and able to make inferences regarding the flux of genes. Our model system is characterized by two main components: class 1 integrons as the main molecular element of bacterial resistance to antibiotics, and Tierra del Fuego Island as study area.

7.3 What Are Class 1 Integrons?

Integrons are genetic platforms that promote bacterial diversity. Nowadays, the accepted definition of an integron is that formulated by Hall and Collis, in which they defined it as an element that contains all the genetic determinants to mediate site-specific recombination to capture mobile gene cassettes. Basically, integrons are composed of three elements: (1) a gene that encodes an integron integrase (*intI*), (2) a recombination site (*attI*), and (3) a promoter that allows the expression of the gene cassettes inserted in the variable region of the integrons (Hall and Collis 1995) (Fig. 7.1). Generally, the *intI* gene and the *attI* recombination site are called 5′-conserved segment (5′-CS) (Stokes and Hall 1989). Each class of integron harbours a specific *attI* recombination site located on the last 40–70 bp of the 3′-end of 5′-CS (Recchia et al. 1994; Collis et al. 1998; Hall et al. 1999).

Bacteria harbouring integrons can capture and express genes found inserted in the chromosome or free in the bacterial cytoplasm as circular gene cassettes; the

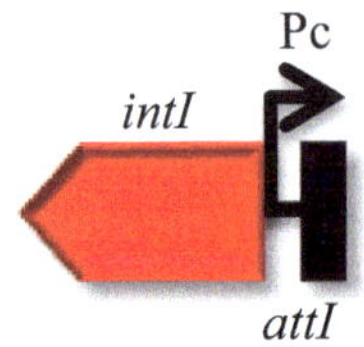

Fig. 7.1 Integron structure. The three basic components of integrons are the *intI1* gene, which encodes an integron integrase; the *attI* recombination site; and Pc, the promoter that allows the expression of the gene cassettes inserted in the variable region of the integron

gene cassettes are composed of a gene and a recombination site called *attC*. The *attC* recombination sites are imperfect palindromic sequences, which vary in sequence and length from 57 to 141 bp and harbour four union domains to the integrase (Cameron et al. 1986; Recchia and Hall 1995; Stokes et al. 1997). From 5′ to 3′, those domains are 1L, also called inverse core site (ICS; RYYYAAC; R = purine and Y = pyrimidine); 2L (GTTCRARY); 2R (RYTYAAC); and last 1R, also called core site (CS; GTTRRRY) (Hall et al. 1991; Collis and Hall 1992; Stokes et al. 1997). The *attC* folds in a secondary structure generating a stem loop and the integrase recognizes the 1L-1R and 2L–2R sites and cleaves between C and A in the bottom strand of 1R (Stokes et al. 1997; Collis et al. 2001; Biskri et al. 2005; Bouvier et al. 2005). When the gene cassette is on its lineal conformation inserted in the variable region of an integron, the last six nucleotides of 1L (CS) G′TTRRRY are upstream of the gene cassette ORF, and downstream from it is the remaining sequence of the *attC* recombination site, ending at the first nucleotide of the CS G'TTRRRY (the ends of the cassette are underlined and the apostrophe shows the cleavage site) (Recchia and Hall 1995). By site-specific recombination, the integron integrase mediates the integration and excision of gene cassettes by recognizing the *attI* and *attC* recombination sites, leading to a variety of gene cassettes arrays. Gene cassettes could encode antibiotic resistance determinants or different metabolic functions among other diverse functions, which allows bacteria to better adapt to the rapidly changing environment (Partridge et al. 2009).

Integrons are not mobile elements per se, but instead they are usually linked to different genetic platforms such as transposons, insertion sequences, and/or conjugative plasmids. This association allows integrons to be mobilized within a genome or to another bacterial species, placing them in the lateral gene transfer scenario and also as one of the major contributors to the dissemination of antibiotic resistance.

Initially, integrons were classified in classes according to the nucleotide sequence of the integrase gene (Hall and Collis 1995). Nowadays, more than 100 integron integrases have been identified in different niches and bacterial genomes. For practical purposes they can be separated into two large groups: (1) those called mobile integrons or antibiotic resistance integrons, which were first and mainly identified in bacteria from clinical settings and are inserted in transposons, plasmids, and/or genomic islands that mobilize them and are responsible for the dissemination of antibiotic resistance; and (2) those localized in bacterial chromosomes, which are called chromosomal integrons. and when they harbour more than 20 gene cassettes in the variable region they are named superintegrons (Mazel 2006; Hall 2012).

Class 1 integrons were the first class of integrons identified, and they were found in clinical settings harbouring more than 130 antibiotic resistance gene cassettes (Stokes and Hall 1989; Partridge et al. 2009). Several genetic structures have been described at the 3′-end of the variable region of class 1 integrons (Nardelli et al. 2012). There are three genetic platforms spreading class 1 integrons in clinical samples from Argentina (Fig. 7.2): (1) the most common one harbours the well-known

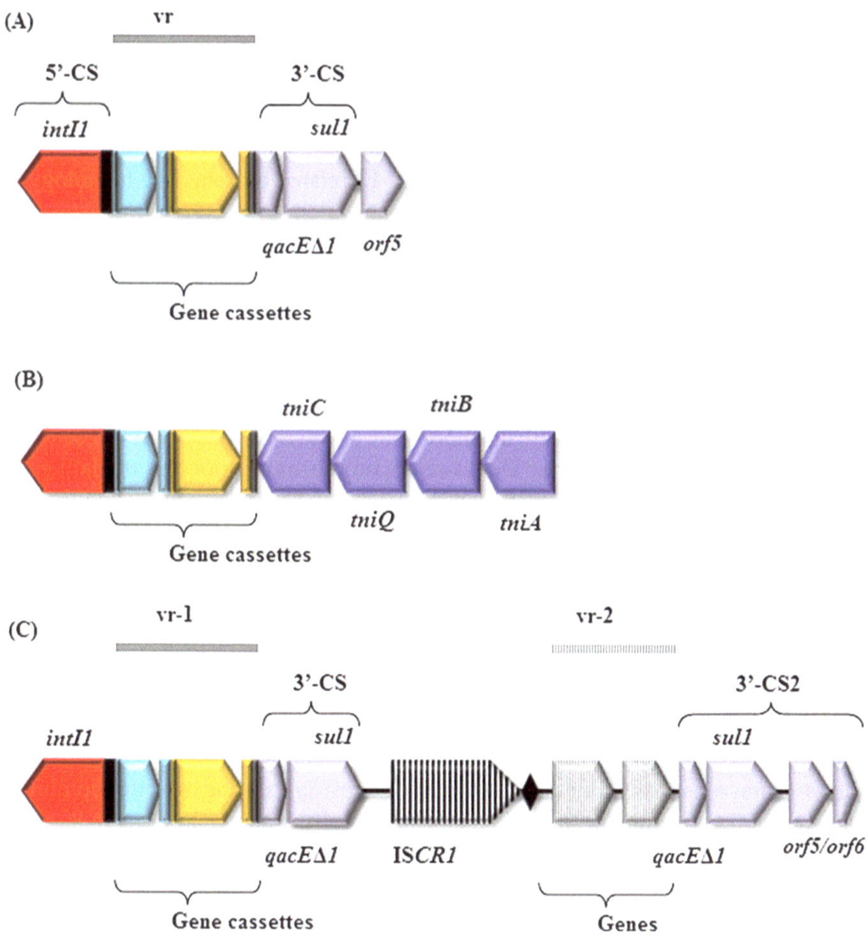

Fig. 7.2 Genetic platforms of class 1 integrons circulating in Argentina. **a** General genetic organization of class 1 integrons. The type 1 integrase gene within the 5′-conserved segment (5′-CS) is shown by a *red arrow box*; the 3′-conserved segment (3′-CS) is shown by *grey arrow boxes*; the *attI1* recombination site is shown by a *black rectangle*; the gene cassettes are shown in *green* and *purple* by *arrow boxes* with the *attC*s represented by *thick* and *thin rectangles*, where the *thick rectangles* shows from the ICS to the first position of the CS and the *thin rectangle* shows the remaining bases of the CS. **b** General genetic organization of class 1 integrons with the complete module of Tn*402*; *tniC-tniQ-tniB-tniA* genes are shown by *dark grey arrow boxes*. **c** General genetic organization of complex class 1 integrons. The 3′-CS and the second copy of this element (3′-CS2) are shown by *grey arrow boxes*; genes of the variable region 2 (vr-2) are shown with *striped arrow boxes*; and the putative origin of replication of IS*CR1* is represented by a *black diamond*

3′-conserved segment (3′-CS) downstream of the variable region, containing the *qacEΔ1* gene that is a deleted form of *qacE*, the quaternary ammonium compounds resistance gene cassette, followed by the *sul1* gene that confers resistance to sulphonamides, and the *orf5* of unknown function (Orman et al. 2002; Quiroga et al. 2013) (Fig. 7.2a); (2) the complete or incomplete module of Tn*402*, *tniC-tniQ-tniB-tniA* (Marchiaro et al. 2010) (Fig. 7.2b); and (3) the so-called unusual class 1 integrons, complex class 1 integron, complex *sul1*-type integrons, or IS*CR1* elements because they share a common region (CR) that includes the *orf513* gene and has been identified as the insertion sequence common region 1 (IS*CR1*) (Verdet et al. 2000; Arduino et al. 2002; Toleman et al. 2006). When several complex class 1 integrons are aligned, the general organization can be identified as a composite of three genetic structures (Fig. 7.2c): (1) the typical class 1 integron harbouring the 5′-CS with the variable region 1 (vr-1) and a first copy of the 3′-CS; (2) the *orf513* gene, which codifies the putative recombinase Orf513 related by sequence homology to the IS*CR* family proteins and by its genetic organization to the IS*91*-like transposases (Partridge and Hall 2003; Toleman et al. 2006), and (3) the variable region 2 (vr-2) coding for a remarkable variety of ARG followed by a duplication of the 3′-CS (Toleman et al. 2006; Quiroga et al. 2007). In Argentina, nosocomial isolates harbouring complex class 1 integrons with $bla_{\mathrm{CTX\text{-}M\text{-}2}}$, $bla_{\mathrm{DHA\text{-}1}}$, *qnrB2*, or *qnrB10* genes have been described since 2002 (Arduino et al. 2002; Di Conza et al. 2002; Quiroga et al. 2007, 2013; Albornoz et al. 2012; Cejas et al. 2012; Andrés et al. 2013).

Although class 1 integrons are the more widespread integrons in clinical settings, it is also worth noting that they were found chromosomally located predating the association with Tn*402*-like transposon in non-clinical samples, suggesting that the ancestor of the clinical class 1 integron was more like a typical chromosomal integron (Stokes et al. 2006).

7.4 Class 1 Integrons as Paradigm of Multidrug Resistance in Gram-Negative Bacilli

Class 1 integrons with more than 130 antibiotic resistance gene cassettes combined in diverse arrays within the variable region are usually found in 20 %, up to 80 %, of *Enterobacteriaceae* and *Pseudomonas aeruginosa* clinical isolates around the world (Domingues et al. 2015). A previous analysis of class 1 integrons in 867 nonrepeated isolates comprising 8 species of *Enterobacteriaceae* from 23 European hospitals showed a significant relationship between multidrug resistance isolates and the presence of the *intI1* gene, independent of species or origin (Leverstein-van Hall et al. 2003). Of major concern, the ability of certain *Enterobacteriaceae* species, in particular *Enterobacter cloacae*, *Serratia marcescens*, and *Klebsiella pneumoniae*, have been documented as causing nosocomial outbreaks through the dispersion of a multidrug-resistant epidemic clone carrying class 1 integrons (Merkier et al. 2013).

In the report published by WHO (http://www.cdc.gov/drugresistance/pdf/ar-threats-2013-508.pdf), most of the clinical strains of bacterial species identified as "urgent" (carbapenem-resistant *Enterobacteriaceae*) and "serious" (multidrug-resistant *Acinetobacter*, drug-resistant *Campylobacter*, extended-spectrum β-lactamase producing *Enterobacteriaceae*, multidrug-resistant *Pseudomonas aeruginosa*, drug-resistant non-typhoidal *Salmonella*, and drug-resistant *Shigella*) threats for causing severe human infections, harbour more than one class 1 integron, evidencing the relevant role that this genetic element plays in the evolution to multidrug and extreme resistance.

7.5 Why Study Samples from Tierra del Fuego Island to Understand the Flux of Antibiotic Resistance Genes at a Global Scale?

The relevant role of natural communities as a reservoir and original source of class 1 integrons was recently identified (Gillings et al. 2008; Nardelli et al. 2012). Since then, their distribution has been reported in environments with different degrees of human disturbance. The molecular features of class 1 integrons as genetic tool for studying the flux of antimicrobial resistance genes is based on the different alleles of the *intI1* gene found from both habitats, the clinic and the open environment, which led to the identification of the sources of both "environmental" and "clinical" class 1 integrons (Gillings et al. 2008). About this conception, the type of allele of *intI1* gene was used as a footprint to delineate the directionality of strains (Gillings et al. 2008; Nardelli et al. 2012).

Environmental alleles are those variants originated in bacteria that grow in soil, water, wildlife, or other substrates that are is not directly affected by antibiotics. In the open environment with a medium or low degree of urbanization, only environmental alleles are expected to be found because the influence of antibiotics should be null or low. This presumption was clearly established recently with clinical and nonclinical samples from Tierra del Fuego Island (Nardelli et al. 2012). In contrast, in clinical, anthropic or urban environments such as cities, mostly clinical alleles were found (Nardelli et al. 2012). In turn, clinical alleles are circulating in the hospital niche and associated with clinical samples (Nardelli et al. 2012).

The other component of our model system of study is Tierra del Fuego, which is a Patagonian island from Argentina and Chile recognized as one of the last places on Earth that contains land areas that may still be considered wild or "clean," given its great expanses of intact natural vegetation and large vertebrate assemblages, along with a low human population density (Mittermeier et al. 2002). The area that lies within the Sub-Antarctic Deciduous Beech Forest, characterised by two species of southern beech, *Nothofagus pumilio* (Lenga) and *Nothofagus betuloides* (Guindo) (Gutiérrez et al. 1991). Its climate belongs to the sub polar oceanic type. Temperatures are cold all year round, with an average annual temperature of 5.7 °C and low annual temperature variation, ranging from −0.3 °C in July to 9.4 °C in January.

The study was conducted in the southeast portion of Tierra del Fuego Island that contains large extensions of natural and seminatural environments (Mittermeier et al. 2002). There are also two urban sites: Ushuaia city with 80,000 inhabitants, on the southern coast of the island, bathed in the Beagle Channel, and Tolhuin, a town of about 8000 inhabitants. The geography and the history of this region provided a suitable mosaic of areas with different degrees of urbanization, which is an ideal model system for the study of role of the environment on the rise of antibiotic resistance. Another advantage provided by the area is that it permits the collection of environmental and nonpathogenic strains that grow from 4 ° to 25 °C. With these conditions, during the collection of strains in the field, the probability of contamination with *intI1*-positive strains from human activities can be considered very rare.

7.6 Multidrug Resistance Class 1 Integrons from Hospitals and Environmental Class 1 Integrons from the Open Environment

Both clinical and non-clinical strains were studied in Tierra del Fuego Island for the presence of class 1 integrons. The clinical strains ($n=32$) were from the Regional Hospital of Ushuaia collected in April 2007 and June 2010 (Table 7.1). The non-clinical strains ($n=127$) were collected from nine sites of Tierra del Fuego with different degrees of urbanisation in February 2006 (Nardelli et al. 2012) and January 2009 (Table 7.2). The 30 % of clinical strains and the 11 % of non-clinical strains isolated from the open environment were *intI1* positive, respectively. The variable region of the nine *intI1*-positive clinical strains was within a *sul*-type class 1 integron with antibiotic resistance gene cassettes arrays previously described worldwide (*aadA1*, *aadB-aaddA1*, *dfrA1-aadA1*, *aac(6′)-Ib-aadA1*, and *aac(6′)-IId*) (Orman et al. 2002). The isolates from the environment harbouring clinical alleles were also studied for the content of antibiotic resistance gene cassettes as previously described (Orman et al. 2002) (Table 7.1). However, negative results were

Table 7.1 Features of the clinical strains isolated from the Regional Hospital of Ushuaia

Isolate	Year of isolation	Variable region	Alleles of *intI1*
Klebsiella pneumoniae	2007	*aac(6′)-Ib-aadA1*	KJ701247
Escherichia coli	2007	*aac(6′)-IId*	KJ701247
Escherichia coli	2007	*dfrA1-aadA1*	JN870908
Escherichia coli	2010	*aadA1*	JN870911
Escherichia coli	2010	*dfrA1-aadA1*	JN870908
Escherichia coli	2010	*aadB-aaddA1*	KJ701247
Escherichia coli	2010	*aac(6′)-IId*	JN870911
Escherichia coli	2010	*aac(6′)-IId*	KJ701247
Escherichia coli	2010	*aadB-aaddA1*	KJ701247

All strains were from urine samples isolated from community patients. The accession number of each allele is also shown. *aac(6′)-Ib* confers resistance to amikacin and tobramycin, *aadA1* confers resistance to streptomycin, *aac(6′)-IId* confers resistance to gentamicin, *dfrA1* confers resistance to trimethoprim, and *aadB* confers resistance to gentamicin

found, suggesting that in the open environment the role of class 1 integrons is associated to different activities not related to antibiotic agents usually used in the clinic. Because a high frequency of class 1 integrons was found in non-clinical samples from Tierra del Fuego Island, including sites far from urbanised villages (Nardelli et al. 2012), it is likely that they also represent a valuable tool for the adaptation of bacteria to different niches in response to environmental stresses.

7.7 Discharge, Maintenance, and Dissemination of *intI1* Genes into the Open Environment

The *intI1*-positive clinical strains from the hospital of Ushuaia harboured three *intI1* "clinical" alleles previously described in nosocomial strains from Europe, Africa, and Asia (Tables 7.1, 7.2; Fig. 7.3) (Nardelli et al. 2012). Interestingly, although "environmental" alleles were identified in sites with a medium and low degree of urbanisation from Tierra del Fuego Island (Nardelli et al. 2012), the three "clinical" alleles were only found in isolates from sites with a high degree of urbanisation from Tierra del Fuego Island, also close to the hospital of Ushuaia, depicting an

Table 7.2 Features of the non-clinical strains isolated from sites with high level of urbanization from Ushuaia in Tierra del Fuego Island

Isolate	Year of isolation	Source	Allele of *intI1*	Site	Accession number
Pseudomonas sp. 1SL5	2006	Soil	Clinical	S54u 499 5899 W68u 219 0599	JN870908
Aeromonas media 1AC2	2006	Water of stream	Clinical	S54u 499 5899 W68u 219 0599	JN870911
Vibrio sp. 1AC4	2006	Water of stream	Non-clinical	S54u 499 5899 W68u 219 0599	JN870903
Pseudomonas sp. 10AN4	2006	Water of stream	Clinical	S54u 479 3799 W67u 159 3699	KJ701247
Enterobacter sp. 10AL1	2009	Water of stream	Non-clinical	S54u 479 3799 W67u 159 3699	JN870907
Pantoea dispersa 10FZSS14	2009	Fox Feces	Clinical	S54u 479 3799 W67u 159 3699	KJ701247
Enterobacter sp. 1SN9	2009	Soil	Clinical	S54u 499 5899 W68u 219 0599	KJ701247
Pseudomonas sp. 1AN2	2009	Water of stream	Clinical	S54u 499 5899 W68u 219 0599	JN870908
Serratia sp. 10AN5	2009	Water of stream	Clinical	S54u 479 3799 W67u 159 3699	KJ701247

All strains were isolated from sites with high level of urbanisation corresponding to site 1 (S54u 499 5899 W68u 219 0599) and to site 2 (S54u 479 3799 W67u 159 3699) of the manuscript of Nardelli et al. (2012). The accession number of each *intI1*gene as well as the corresponding type of allele (clinical or environmental) is also shown

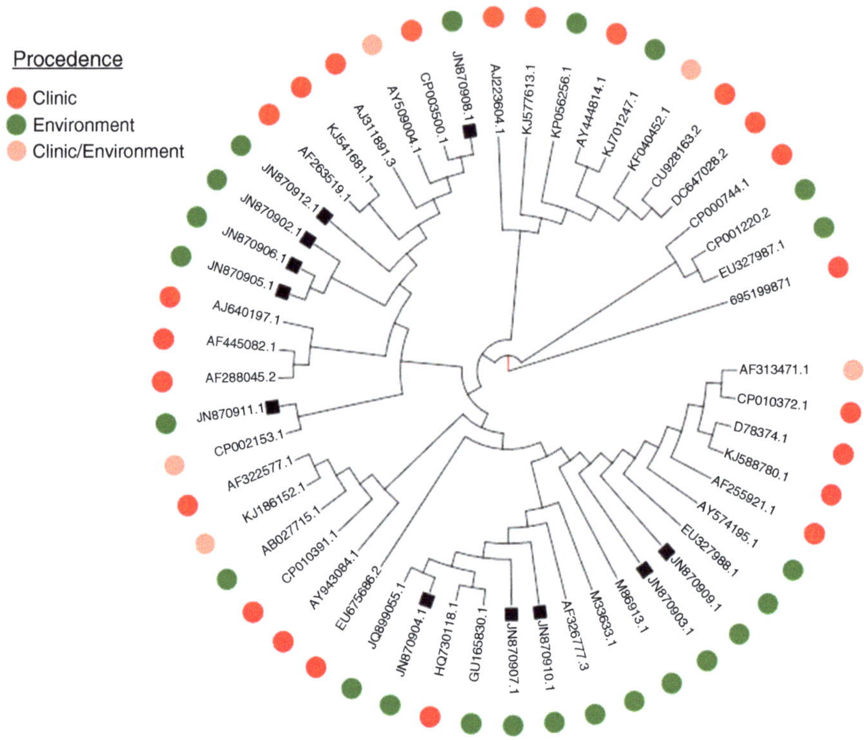

Fig. 7.3 Phylogenetic tree of strains representing the most frequent alleles found in the environment, the clinic, or both. A total of 52 *intI1* genes harbouring different alleles were selected from a BLASTn query using AF313471 as reference (April 2015). The phylogenetic tree was constructed using the maximum likelihood algorithm implemented in MEGA v6.0

interchange of genes among the two habitats, the hospital and the sites close to human activities in Ushuaia City (Fig. 7.3). The same *intI1* clinical alleles were found in the non-clinical strains during the 4 years of bacterial sampling in the open environment. It is likely that once the *intI1*-positive clinical strains are released from the hospital, the *intI1* genes can be transferred through the mechanisms of lateral genetic transfer (LGT) to other bacterial species from the open environment (*Vibrio* sp., *Aeromonas media*, *Pseudomonas* non-*aeruginosa*) (Table 7.2) where they can be maintained over time. These data represent the first study performed at the same time on both isolates from the open environment with different degrees of urbanisation and isolates from hospitals in situ that allows identifying an active flux of *intI1* genes, at least in one direction, from the hospital to the environment (Fig. 7.4). Also, the clinical *intI1* allele most common in clinical and non-clinical strains, which probably has been discharged from nosocomial settings, was found in the microbiota of a wild fox, evidencing how successful these genetic elements can be for the wide spread of antibiotic resistance gene cassettes through animals that are used as bacterial genetic transporters.

Fig. 7.4 Flux of class 1 integrons between the environment and the clinic. The *arrow* indicates the direction of the flux of *intI1* genes as analysed in this study. The *intI1* gene is shown in *red* and the antimicrobial resistance gene cassettes are shown in *orange*, *violet*, *blue*, *grey*, and *yellow*. The gene cassettes of the variable region of class 1 integrons from the environment are shown in *green* because the typical antibiotic resistance gene cassettes found in the clinic including those from the Regional Hospital from Ushuaia were not found

7.8 What About Complex Class 1 Integrons in the Open Environment?

The IS*CR1* gene was also found in non-clinical samples from Patagonia in γ-and β-*Proteobacteria*, in Actinobacteria, and in *Pseudomonaceae* isolates. The sequence of 475 bp from the 11 IS*CR1*-positive isolates of that study showed 100 % identity with the clinical allele (EU722351) (Nardelli et al. 2012). The distribution of these elements has been extensively studied in clinical isolates but very little in the environment (Quiroga et al. 2013). The data collected from both studies, clinical and environmental, are very interesting, revealing that some genes associated with class 1 integrons are very variable in nature such as we discussed earlier for the great number of alleles of *intI1*, and others very little such as the case of IS*CR1*, that is spreading with the same allele in clinical and non-clinical samples. This information can be vital to understand how biological systems allow pathogens to accumulate multiple ARG over time.

The high frequency of IS*CR1*, as well as its distribution in several taxa in Patagonian samples, supports also the important role of the open environment as a reservoir of this gene in Argentina. The *ISCR1* gene was found in bacteria from the same sampling sites where *intI1*, *qacE1/qacE*Δ*1* and/or *sul1* were also identified, which ensures an encounter among these bacteria and a putative transfer of genes. If this genetic transfer is followed by recombination events and selection, the whole processes can result in the formation of mosaic structures as usually detected in multidrug-resistant isolates from the clinic.

7.9 Flux of Antibiotic Resistance Genes and Mobile Elements in Different Degrees of Urbanisation in Patagonia

The pangenome contains a huge amount of genes that are mobilized very frequently between bacterial species by the mechanisms of LGT. In theory, any part of the pangenome can be mobilized by LGT mechanisms along evolutionary time. Some genetic elements are specialized on moving DNA within and between bacterial genomes, such as plasmids, transposons, integrons, insertion sequences, MITEs, and integrative conjugative elements are known as the "mobilome." The mobilome is responsible for most of the vast genome diversification within bacterial genomes aided by homologous and nonhomologous recombination events.

It is expected that ARG will respond differently than LGT of mobile elements (LGTME) to the geographic variation of urbanization because of the different functions of these elements. If LGTME occurs exclusively as a response to the use of antibiotics, significantly high frequencies of occurrence are expected in habitats contaminated with antibiotics (typically urbanized areas) than in natural and rural areas that are free of antibiotics. In contrast, if LGTME serves as a general mechanism of response to environmental stress, they should be present in both "clean" (where antibiotics have not been dumped) and urbanised habitats as far as bacteria meet stressful conditions in the former (for example, extreme seasonal or daily variations in weather conditions). Thus, they should present a weak relationship with the degree of urbanisation. In contrast, ARG should be closely related to antibiotic pressure, and thus will present a strong link to geographic variations in urbanisation and human presence.

An evaluation of two types of genetic determinant ARG associated to class 1 integrons (*sul1* and *qacE1*/*qacE*Δ*1* genes) and lateral genetic elements (LGE) (*intI1*, IS*CR1* and *tniC* genes) in a model of a culture-based method without antibiotic selection was conducted in a gradient of anthropogenic disturbances in Tierra del Fuego Island (Nardelli et al. 2012). On one hand, each ARG showed different ecological and molecular behaviours in environmental samples. Although the *sul1* gene frequency was associated with urbanization, the *qacE1*/*qacE*Δ*1* gene showed an adaptive role to several habitats. On the other hand, the *intI1* and IS*CR1* genes and *intI1* pseudogenes that were found widespread throughout natural communities were not associated with urbanisation ($p>0.05$). The abundance and widespread occurrence of IS*CR1* and *intI1* throughout Patagonian sites with different degrees of urbanization, and within different taxa, may be considered as one of the causes of the increasing frequency of multidrug-resistant isolates that have characterized Argentina for decades.

References

Albornoz E, Lucero C, Rodriguez C, WHONET-Argentina Group, Galas M, Centrón D, Corso A, Petroni A (2012) Prevalence of plasmid mediated quinolone resistance mechanisms (PMQR) in Argentina, 52th ICAAC. ICAAC, San Francisco, CA

Andrés P, Lucero C, Soler-Bistué A, Guerriero L, Albornoz E, Tran T, Zorreguieta A, Galas M, Corso A, Tolmasky ME, Petroni A (2013) Differential distribution of plasmid mediated quinolone resistance genes in clinical enterobacteria with unusual phenotypes of quinolone susceptibility from Argentina. Antimicrob Agents Chemother 57:2467–2475

Arduino SM, Roy PH, Jacoby GA, Orman BE, Pineiro SA, Centrón D (2002) $bla_{CTX\text{-}M\text{-}2}$ is located in an unusual class 1 integron (In35) which includes Orf513. Antimicrob Agents Chemother 46:2303–2306

Arduino SM, Quiroga MP, Ramírez MS, Merkier AK, Errecalde L, Di Martino A, Smayevsky J, Kaufman S, Centrón D (2012) Transposons and integrons in colistin-resistant clones of Klebsiella pneumoniae and Acinetobacter baumannii with epidemic or sporadic behaviour. J Med Microbiol 61:1417–1420

Bhullar K, Waglechner N, Pawlowski A, Koteva K, Banks ED, Johnston MD, Barton HA, Wright GD (2012) Antibiotic resistance is prevalent in an isolated cave microbiome. PLoS One 7(4), e34953

Biskri L, Bouvier M, Guerout AM, Boisnard S, Mazel D (2005) Comparative study of class 1 integron and *Vibrio cholerae* superintegron integrase activities. J Bacteriol 187(5):1740–1750

Bouvier M, Demarre G, Mazel D (2005) Integron cassette insertion: a recombination process involving a folded single strand substrate. EMBO J 24:4356–4367

Cameron FH, Groot Obbink DJ, Ackerman VP, Hal RM (1986) Nucleotide sequence of the *aad(2")* aminoglycoside adenylyltransferase determinant *aadB* evolutionary relationship of this region with those surrounding *aadA* in R538-1 and *dhfrII* in R388. Nucleic Acids Res 14(21):8625–8635

Cejas D, Fernandez Canigia L, Quinteros M, Giovanakis M, Vay C, Lascialandare S, Mutti D, Pagniez G, Almuzara M, Gutkind G, Radice M (2012) Plasmid-encoded AmpC (pAmpC) in *Enterobacteriaceae*: epidemiology of microorganisms and resistance markers. Rev Argent Microbiol 44:182–186

Collis CM, Hall RM (1992) Site-specific deletion and rearrangement of integron insert genes catalyzed by the integron DNA integrase. J Bacteriol 174:51574–51585

Collis CM, Kim MJ, Stokes HW, Hall RM (1998) Binding of the purified integron DNA integrase Intl1 to integron- and cassette-associated recombination sites. Mol Microbiol 29(2):477–490

Collis CM, Recchia GD, Kim MJ, Stokes HW, Hall RM (2001) Efficiency of recombination reactions catalyzed by class 1 integron integrase IntI1. J Bacteriol 183(8):2535–2542

Davies J, Davies D (2010) Origins and evolution of antibiotic resistance. Microbiol Mol Biol Rev 74:417–433

D'Costa VM, McGrann KM, Hughes DW, Wright GD (2006) Sampling the antibiotic resistome. Science 311:374–377.

D'Costa VM, King CE, Kalan L, Morar M, Sung WWL, Schwarz C, Froese D, Zazula G, Calmels F, Debruyne R, Golding GB, Poinar HN, Wright GD (2011) Antibiotic resistance is ancient. Nature (Lond) 477:457–461

Di Conza J, Ayala JA, Power P, Mollerach M, Gutkind G (2002) Novel class 1 integron (InS21) carrying blaCTX-M-2 in *Salmonella enterica* serovar Infantis. Antimicrob Agents Chemother 46:2257–2261

Domingues S, da Silva GJ, Nielsen KM (2015) Global dissemination patterns of common gene cassette arrays in class 1 integrons. Microbiology 161:1313–1337

Gillings MR, Krishnan S, Worden PJ, Hardwick SA (2008) Recovery of diverse genes for class 1 integron-integrases from environmental DNA samples. FEMS Microbiol Lett 287:56–62

Gillings MR (2014) Integrons: past, present, and future. Microbiol Mol Biol Rev 78:257–277.

Gutiérrez E, Vallejo VR, Romanya J, Fons J (1991) The subantarctic *Nothofagus* forests of Tierra del Fuego: distribution, structure and production. Oecol Aquat 10:351–366

Hall RM (2012) Integrons and gene cassettes: hotspots of diversity in bacterial genomes. Ann N Y Acad Sci 1267:717–718

Hall RM, Collis CM (1995) Mobile gene cassettes and integrons: capture and spread of genes by site-specific recombination. Mol Microbiol 15(4):593–600

Hall RM, Brookes DE, Stokes HW (1991) Site-specific insertion of genes into integrons: role of the 59-Base element and determination of the recombination cross-over point. Mol Microbiol 5(8):1941–1959

Hall RM, Collis CM, Kim MJ, Partridge SR, Recchia GD, Stokes HW (1999) Mobile gene cassettes and integrons in evolution. Ann N Y Acad Sci 870:68–80

Hawkey PM (2008) Prevalence and clonality of extended-spectrum beta-lactamases in Asia. Clin Microbiol Infect 1:159–165.

Leverstein-van Hall MA, Blok HE, Donders AR, Paauw A, Fluit AC, Verhoef J (2003) Multidrug resistance among *Enterobacteriaceae* is strongly associated with the presence of integrons and is independent of species or isolate origin. J Infect Dis 187:251–259

Liu B, Pop M (2009) ARDB-antibiotic resistance genes database. Nucleic Acids Res 37:D443–447.

Marchiaro P, Viale AM, Ballerini V, Rossignol G, Vila AJ, Limansky A (2010) First report of a Tn*402*-like class 1 integron carrying $bla_{\text{VIM-2}}$ in *Pseudomonas putida* from Argentina. J Infect Dev Ctries 4:412–416

Mazel D (2006) Integrons: agents of bacterial evolution. Nat Rev Microbiol 4(8):608–620

Merkier AK, Rodríguez C, Togneri A, Brengi S, Osuna C, Pichel M, Cassini MH, Centrón D, Serratia marcescens Argentinean Collaborative Group (2013) Outbreak of a cluster with epidemic behavior due to *Serratia marcescens* after colistin administration in a hospital setting. J Clin Microbiol 51(7):2295–2302

Mittermeier RA, Mittermeier CG, Robles Gil P, Pilgrim JD, Konstant WR et al (2002) Wilderness: Earth's last wild places. CEMEX, Mexico City

Nardelli M, Scalzo PM, Ramírez MS, Quiroga MP, Cassini MH, Centrón D (2012) Class 1 integrons in environments with different degrees of urbanization. PLoS One 7(6):e39223

Nordmann P, Lartigue MF, Poirel L (2008) Beta-lactam induction of ISEcp1B-mediated mobilization of the naturally occurring bla(CTX-M) beta-lactamase gene of Kluyvera ascorbata. FEMS Microbiol. Lett 288:247–249.

Orman BE, Pineiro SA, Arduino S, Galas M, Melano R, Caffer MI, Sordelli DO, Centrón D (2002) Evolution of multiresistance in nontyphoid *Salmonella* serovars from 1984 to 1998 in Argentina. Antimicrob Agents Chemother 46:3963–3970

Partridge SR, Hall RM (2003) In34, a complex In5 family class 1 integron containing orf513 and dfrA10. Antimicrob Agents Chemother 47:342–349

Partridge SR, Tsafnat G, Coiera E, Iredell JR (2009) Gene cassettes and cassette arrays in mobile resistance integrons. FEMS Microbiol Rev 33(4):757–784

Petrova M, Gorlenko Z, Mindlin S (2011) Tn5045, a novel integron-containing antibiotic and chromate resistance transposon isolated from a permafrost bacterium. Res Microbiol 162:337–345

Quiroga MP, Arduino SM, Merkier AK, Quiroga C, Petroni A, Argentinian Integron Study Group, Roy PH, Centrón C (2013) Distribution and functional identification of complex class 1 integrons. Infect Genet Evol 19:88–96

Quiroga MP, Andres P, Petroni A, Soler Bistue AJ, Guerriero L, Vargas LJ, Zorreguieta A, Tokumoto M, Quiroga C, Tolmasky ME, Galas M, Centron D (2007) Complex class 1 integrons with diverse variable regions, including *aac(6')-Ib-cr*, and a novel allele, *qnrB10*, associated with IS*CR1* in clinical enterobacterial isolates from Argentina. Antimicrob Agents Chemother 51:4466–4470

Recchia GD, Hall RM (1995) Gene cassettes: a new class of mobile element. Microbiology 141:3015–3027

Recchia GD, Stokes HW, Hall RM (1994) Characterisation of specific and secondary recombination sites recognised by the integron DNA integrase. Nucleic Acids Res 22(11):2071–2078

Roca I, Akova M, Baquero F, Carlet J, Cavaleri M, Coenen S, Cohen J, Findlay D, Gyssens I, Heure OE, Kahlmeter G, Kruse H, Laxminarayan R, Liébana E, López-Cerero L, MacGowan A, Martins M, Rodríguez-Baño J, Rolain JM, Segovia C, Sigauque B, Taconelli E, Wellington

E, Vila J (2015) The global threat of antibiotic resistance: science for intervention. New Microb New Infect 6:22–29

Rossolini GM, D'Andrea MM, Mugnaioli C (2008) The spread of CTX-M-type extended-spectrum beta-lactamases. Clin Microbiol Infect 1:33–41.

Stokes HW, Hall RM (1989) A novel family of potentially mobile DNA elements encoding site-specific gene-integration functions: integrons. Mol Microbiol 3(12):1669–1683

Stokes HW, O'Gorman DB, Recchia GD, Parsekhian M, Hall RM (1997) Structure and function of 59-base element recombination sites associated with mobile gene cassettes. Mol Microbiol 26(4):731–745

Stokes HW, Nesbo CL, Holley M, Bahl MI, Gillings MR, Boucher Y (2006) Class 1 integrons potentially predating the association with Tn*402*-like transposition genes are present in a sediment microbial community. J Bacteriol 188:5722–5730

Toleman MA, Bennett PM, Walsh TR (2006) Common regions, e.g., orf513 and antibiotic resistance: IS91-like elements evolving complex class 1 integrons. J Antimicrob Chemother 58:1–6

Verdet C, Arlet G, Barnaud G, Lagrange PH, Philippon A (2000) A novel integron in *Salmonella enterica* serovar Enteritidis, carrying the $bla_{\mathrm{DHA\text{-}1}}$ gene and its regulator gene ampR, originated from *Morganella morganii*. Antimicrob Agents Chemother 44:222–225

Chapter 8
Novel Sources of Antimicrobials from Pristine and Poorly Explored Environments. The Patagonia Microbiota Case

Gonzalo V. Arnau, Mariana E. Danilovich, Leandro A. Sánchez, Federico Acosta, and Osvaldo Delgado

Abstract Antimicrobial compounds are molecules widespread in life forms to mediate competition, and their industrial production could be important for potential use as preservatives in the food, cosmetic, and pharmaceutical industries. Pathogen resistance causes high mortality rates in hospitals and important economic losses in health institutions. Pathogens such as methicillin-resistant *Staphylococcus aureus*, penicillin-resistant *Streptococcus pneumoniae*, and beta-lactam-resistant *Pseudomonas aeruginosa* are the most common examples. This worldwide problem requires the discovery of new molecules with antibiotic activity, effective therapeutics strategies, and research of promising targets. Most antibiotics come from screening programs of natural sources, including the isolation of new microorganisms. Extremophile microorganisms are a valuable source for novel biomolecules with unusual properties, including antimicrobial activity. Because of their harsh conditions, the Patagonian, sub-Antarctic, and Antarctic environments are ideal places for bioprospecting. Patagonia is a sparsely populated region located at the southern end of South America, shared by Argentina and Chile. The overall climate is cool and dry; the east coast is warmer than the west, especially in summer, because a branch of the Southern Equatorial Current reaches its shores, whereas the west coast is washed by a cold current. Cold environments could be a suitable source of microorganisms with ability to produce cold-active antimicrobial compounds with potential use in biotechnology.

G.V. Arnau • M.E. Danilovich • L.A. Sánchez
Planta Piloto de Procesos Industriales Microbiológicos PROIMI-CONICET,
Av. Belgrano y Pje. Caseros, 4000 San Miguel de Tucumán, Tucumán, Argentina

F. Acosta • O. Delgado (✉)
Centro de Investigaciones y Trasferencia de Catamarca CITCA-CONICET,
Prado 366, 4700 San Fernando del Valle de Catamarca, Catamarca, Argentina

Facultad de Ciencias Exactas y Naturales, Universidad Nacional de Catamarca
FACEN-UNCa, Av. Belgrano 300, 4700 San Fernando del Valle de Catamarca,
Catamarca, Argentina
e-mail: odelgado@proimi.org.ar

N.L. Olivera et al. (eds.), *Biology and Biotechnology of Patagonian Microorganisms*, DOI 10.1007/978-3-319-42801-7_8

8.1 Patagonian Environments

Patagonia is a sparsely populated region located at the southern area of South America, shared by Argentina and Chile. The overall climate is cool and dry. The east coast is warmer than the west, especially in summer, as a branch of the Southern Equatorial Current reaches its shores, whereas the west coast is washed by a cold current.

With an area of about 260,000 square miles (673,000 km^2), Patagonia constitutes a vast area of steppe and desert that extends from 37° to 51° south latitude and is considered a pristine and oligotrophic ecosystem (Carrasco et al. 2002). The area is limited by the Patagonian Andes to the west, the Colorado River to the north (except where the region extends into the Andean borderlands), the Atlantic Ocean to the east, and the Strait of Magallanes to the south. The southern portion of the strait (Tierra del Fuego) is divided between Argentina and Chile and often included in Patagonia.

Patagonian environments are known by their low temperatures, reaching values below 0 °C during winter periods (July), features mainly determined by air masses coming from the Pacific Ocean and also the presence of the Cordillera de los Andes. Precipitation is concentred in winter periods, where 46 % of annual precipitations occur, increasing the water deficit in summer periods and causing impacts in phenology (Paruelo et al. 1998; Carrasco et al. 2002). Other relevant features are the subpolar low-pressure belt at approximately 60°S (Prohaska 1976) and the presence of winds from the west region with low humidity (Beltrán 1997).

8.2 Cold-Living Microorganisms

More than three quarters of our planet is occupied by cold ecosystems, including deep oceans, polar regions, and the high peaks. However, despite the low temperatures, these natural spaces have been colonized by extremophile organisms named *psychrophiles*. Psychrophilic microorganisms are not only adapted to survive at low temperatures, but in some cases, their adaptation is often accompanied by other conditions such as high levels of pressure, salinity, or ultraviolet radiation.

To survive these stressful conditions, microorganisms have developed genetic and physiological mechanisms and thus survive and colonize under these extreme and unfavorable conditions (Beales 2004). Some examples involving changes in membrane lipid composition to ensure membrane performance or regarding enzyme activity and solute transport are well known (Brown and Minnikin 1973; Russell 1984; Tsuchiya et al. 1987; Russell et al. 1995; Mastronicolis et al. 1998).

Microorganisms have a great capacity for adaptation in environments with extreme pH values, salinity, pressure, radiation, and temperature. These characteristics are caused by the versatility that bacteria have to modify their biological structures through natural selection in relatively short periods of time (Edwards 1990). In fact, many polyextremophiles have been described as being able to live in environments that include two or more extreme conditions such as *thermoacidophiles*, *thermobarophiles*, *haloalcalophiles*, or UV radiation-resistant *oligotrophs* (Dib et al. 2008; Bowers et al. 2009).

Psychrophilic microorganisms possess a high biotechnological potential because their enzymes and molecules have high activity at low temperatures, which might mean, for example, a great advantage over the energy costs that are involved in industrial processes. A classic example is the use of cold-active proteases in the leather industry, which avoids the need to heat water at 37 °C during processing (Kobayashi 1989). Cold-active enzymes may also be helpful for domestic uses, for example, in detergent production. In food industry, a cold-active beta-galactosidase has been patented with the purpose of removing the lactose from milk during its storage at low temperatures. On the other hand, cold-active pectinases may decrease the viscosity and clarify fruit juices at low degrees (Feller and Gerday 2003).

8.3 Bioprospection at Sub-Antarctic Environments

The Convention on Biological Diversity (CBD) defines bioprospecting as "the exploration of biodiversity for commercially valuable genetic and biochemical resources"; however, this definition varies among different authors. Other authors define it as a systematic research for natural compounds, genes, designs, and organisms in their natural ecosystem with potential application, using biological, biophysics, biochemical, and genetic methods without disruption to nature (Mateo et al. 2001). However, searching and application of biological resources is as old as humankind and has been a key to the survival, adaptation, and evolution of the human species.

Because psychrophilic microorganisms are adapted to unusual living conditions, this fact forced them to acquire some specific adaptations to survive and grow in such ecosystems (Cavicchioli et al. 2002). Thus, some microorganisms gained the ability to produce antimicrobial compounds as a defense mechanism. Psychrotolerant organisms are defined as organisms that grow well at temperatures close to the freezing point of water, but have optimal growth rates above 20 °C, whereas psychrophilic organisms grow faster at temperatures near to 15 °C or lower; however, they are unable to grow above 20 °C (Cavicchioli et al. 2002). Synthesis of substances with antibacterial effects is considered a strategy to inhibit the growth of neighbour microorganisms, acquiring advantages in colonization and nutrient competition in environments where they are scarce (Grossart et al. 2003; O'Brien et al. 2004; Lo Giudice et al. 2007).

In the area of drug discovery, there are two sources for antimicrobial leads: natural products and synthetic compounds. During the so-called golden age of antibiotics (from 1940 to 1960), most of the successful antibiotic drugs or chemical scaffolds for chemical modification and development of novel drugs were derived from natural products, mainly isolated from bacteria and fungi (Singh and Barrett 2006). For the past few decades, there seems to be a "void" in the antimicrobial drug pipelines because of the difficult of introducing new classes of antibiotics. Mining of natural sources frequently retrieves the same known chemical scaffolds; however, this fact does not indicate the exhaustion of natural resources but rather basic flaws in discovery methodology (Davies 2011).

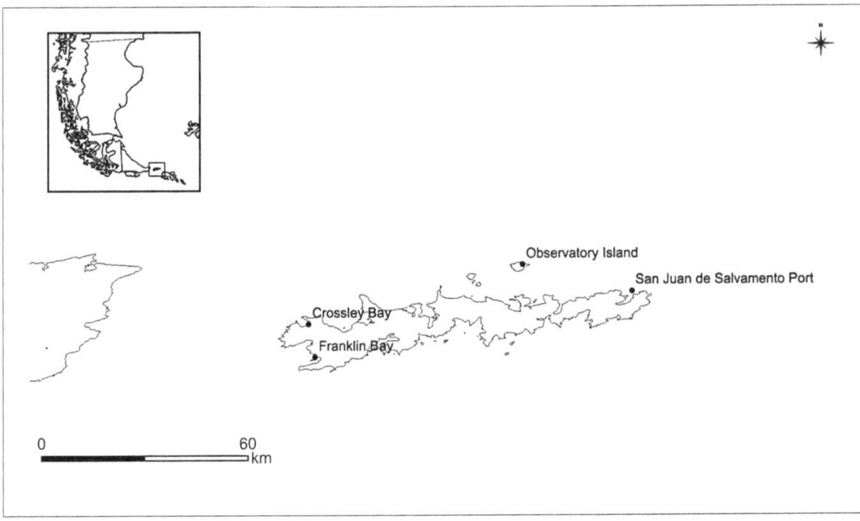

Fig. 8.1 Isla de los Estados Reservation (Ushuaia) sampling sites. Four locations including Observatory Island (54°39′ S, 64°08′ W) Crossley Bay (54°48′ S, 64°41′ W), Franklin Bay (54°53′ S, 64°40′ W), and San Juan del Salvamento Port (54°43′ S, 63°51′ W) were sampled by Sánchez et al. (2009) where cold-adapted microorganisms producing antimicrobials were isolated

Argentinean environments are valuable sources for bioprospecting for microorganisms and other natural compounds with biotechnological potential. Patagonia, a semiarid scrub plateau that covers nearly the entire southern portion of mainland Argentina, represents one of those exploitable natural environments with its microbial biodiversity.

At southern Patagonia, the Argentinian island, Isla de los Estados, lies 29 km of the eastern portion of Tierra del Fuego, Ushuaia (Fig. 8.1). The island is 65 km long (east–west) and 15 km wide, with an area of 534 km^2. The highest point is 823 m above sea level, and it is considered the last prominence of the Cordillera de los Andes. It has a cold climate with an average temperature in the warmest month below 10 °C and abundant precipitation (Ponce and Fernández 2014). This sub-Antarctic region has been described in the literature as a valuable ecosystem and as a novel source of antibacterial compounds (Sánchez et al. 2009).

Cold-adapted microorganisms isolated from Isla de los Estados soil samples were described according to their ability to produce novel antimicrobials (Sánchez et al. 2009). Most of the isolates showed a wide spectrum of activity against both gram-positive and gram-negative enteropathogenic bacteria. Antimicrobials were suspected to be microcin-like compounds and exhibited high activity even after freezing at −20 ° and −80 °C. Those cold-living microorganisms with the ability to produce cold-active compounds have potential application in chilled food preservation (Sánchez et al. 2009).

A *Serratia proteamaculans* strain isolated from Isla de los Estados was described as a producer of a novel cold-active antimicrobial substance. This compound was purified, partially characterized, and temporarily named as Serraticin A (Sánchez et al. 2010). The low molecular weight bacteriocin showed a wide inhibition spectrum and an interesting biotechnological potential as a control agent against pathogenic bacteria. Further studies allowed us to purify and to elucidate the structure by LC-MS/MS and $^{1}H/^{13}C$ NMR approaches. The elucidated structure was in agreement with those ^{1}H and ^{13}C NMR data that Needham et al. (1994) described. The antimicrobial compound was identified as Andrimid (**1**; Fig. 8.2) (Sánchez et al. 2013),

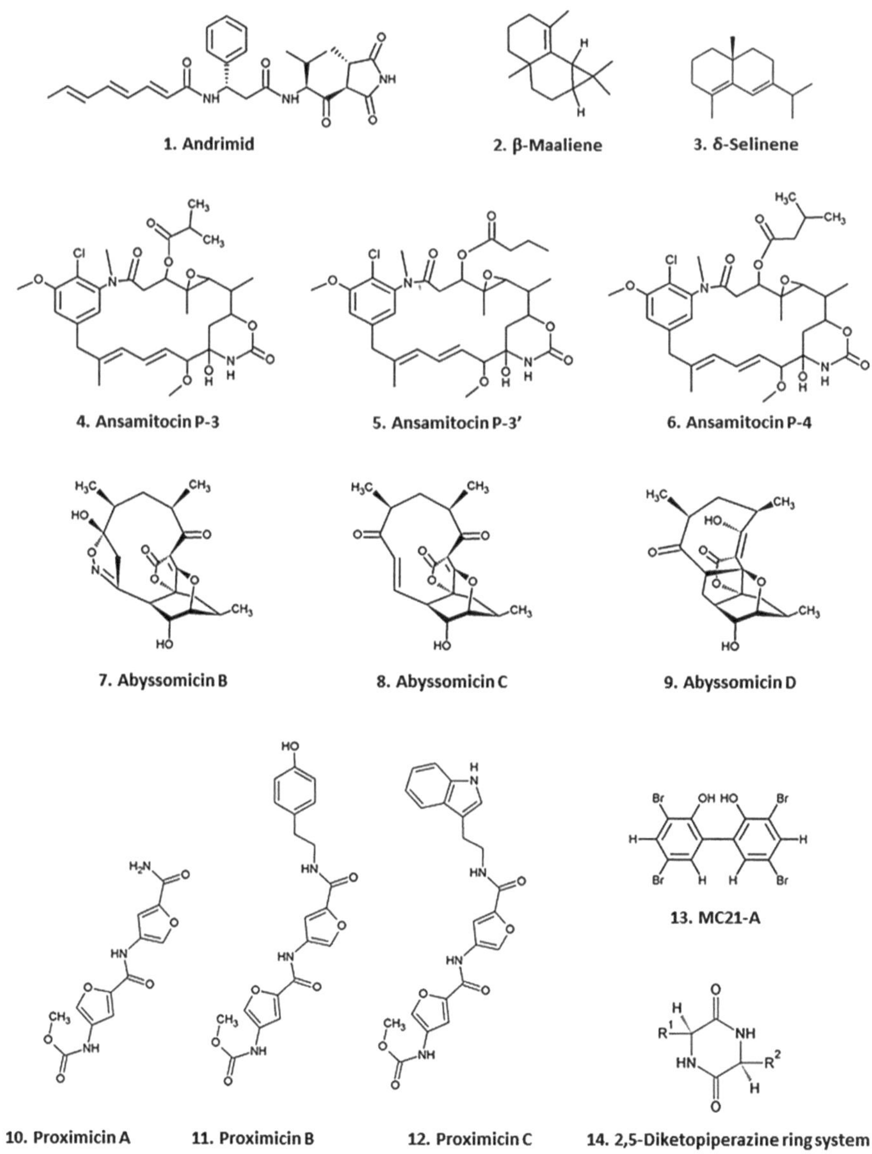

Fig. 8.2 Antimicrobial structures *1–14*

a substance with antibacterial activity at low temperatures that has gained attention for its potential ability to be applied in cold-dependent processes such as cold chain maintenance, cryopreservation, and cosmetics. However, not only Argentinean Patagonia has been studied as a source of novel antimicrobial compounds: Chilean Patagonia has also been reported as a pristine natural environment that gained attention in recent years. The ability of three psychrotrophic gram-negative bacilli strains isolated from cold-water rivers to produce bioactive metabolites with antimicrobial activity has been reported by Barros et al. (2013). The strains were identified based on 16S rDNA sequence analysis as *Pseudomonas* sp. and *Yersinia* sp. with 99.6 % and 99.5 % identity, respectively. The extracts with antibacterial activity obtained showed inhibition against both gram-positive and gram-negative bacteria; however, *Listeria monocytogenes* was not inhibited. The thermal and proteolytic resistances of the antibacterial metabolite fractions were also reported. To gain insight into chemical structures, a GC-MS analysis was conducted with active extracts suggesting antibacterial activity could be related to compounds such as sesquiterpenes derivatives from β-maaliene (**2**) or δ-selinene (**3**) (Fig. 8.2).

Biologically active substances produced by *Streptomyces* spp. isolated from several plant species have been described by Castillo et al. (2007). These substances presented antifungal activity toward major plant pathogens such as *Phytophthora erythroseptica*, *Pythium ultimum*, *Sclerotinia sclerotiorum*, and *Mycosphaerella fijiensis*, but none of the isolates showed antibacterial properties.

Other authors have described bacteriocins produced by *Enterococcus faecium* isolated from ewes' milk and cheese samples from the Argentine Patagonian region, by mechanical milking of animals belonging to dairy farms located in Chubut Province. These bacteriocins showed antimicrobial activity against diverse strains of *Listeria monocytogenes* and *Staphylococcus aureus*. They exhibited a bactericidal mode of action, thermostability, and pH resistance, becoming potential candidates to aid in food preservation (Rivas et al. 2012).

In addition, marine bacteria represent a significant reservoir of antimicrobial active substances of commercial interest. Antimicrobial compounds isolated from marine bacteria have gained attention in various research centres throughout the world and several of them are in clinical trials (Debnath et al. 2007).

Bioprospecting of marine environments has achieved relevant developments in antimicrobial yield. Bacteriocins produced by the marine bacterium *Vibrio vulnificus* have been described as a tool to preserve seafood because of their bactericidal effect against closely related species (Shehane and Sizemore 2002). Ansamitocins (**4–6**; Fig. 8.2), a group of novel maytansinoid antibiotics with antifungal and antitumoral activity produced by *Nocardia* sp., were also obtained from marine bacteria (Higashide et al. 1977; Asai et al. 1979). Abyssomicins B–D (**7–9**; Fig. 8.2), polycyclic polyketide antibiotics from the marine actinomycete *Verrucosispora maris* AB-18-032, were isolated from sediment samples collected from the Sea of Japan at a depth of 289 m, and their structure was elucidated in the year 2007 (Riedlinger et al. 2004; Keller et al. 2007). One year later, Fiedler et al. (2008) described the isolation of another *Verrucosispora* strain named MG-37 from sediment samples of Raune Fjord (Norway) at a depth of 250 m that produces a family of structurally related peptide metabolites named proximicin A, B, and C (**10–12**; Fig. 8.2). *Pseudoalteromonas phenolica* O-BC30^T isolated from the marine environment was reported to produce the bactericidal antibiotic *MC21-A* (**13**; Fig. 8.2) with special activity against methicillin-resistant *Staphylococcus aureus* (Isnansetyo and Kamei 2003). Argentinian authors have reported several psychrotolerant microorganisms as producers of promising antimicrobials with a wide spectrum of activity against pathogenic bacteria. Bacteria were isolated from soil samples collected in sites of *Isla de los Estados* Reservation (South Atlantic Argentina) that are phylogenetically related to *Serratia proteamaculans* (Sánchez et al. 2009). Recently, Arnau et al. (2015) reported the novel species *Pseudomonas yamanorum* 8H1^T isolated from Isla de los Estados Reservation with antagonistic activity against both gram-positive and gram-negative pathogenic bacteria. The novel species was proposed based on a polyphasic approach, including a multilocus sequences analysis of 16S rRNA, *gyr*B, *rpo*B, and *rpo*D genes, DNA–DNA hybridization, and biochemical and chemotaxonomic features. GC-MS analysis of the active methanolic extract of freeze-dried supernatant suggests that activity could be related to 2,5-diketopiperazine (**14**; Fig. 8.2) compounds (2,5-DKPs) (unpublished data). 2,5-Diketopiperazines are cyclodipeptides formed by the condensation of two α-amino acids, representing an abundant class of biologically active natural compounds. They have been found in bacteria, fungi, plants, and mammals, and they represent interesting compounds because of their variety of biological activities, such as antibacterial, antifungal, antiviral, antitumor, plant growth promotion, and quorum-sensing signaling among many others (Huang et al. 2010; Borthwick 2012; Elkahoui et al. 2013). 2,5-DKPs represent attractive scaffolds for drug discovery because of their small size, rigidity, chirality, stability to proteolysis, and diversity of substituent groups, which can be introduced at up to six positions, and stereochemistry controlled at up to four positions (Borthwick 2012).

The Patagonian Sea includes an extensive continental shelf (approximately 1,000,000 km^2) of relief and scarce slope, where the majority does not exceed 100 m deep. It widens progressively to the south, reaching a maximum width of 860 km at 51°S. The continental shelf ends to the east in a sharp break, where depth increases to 160–200 m, and from this point, the slope grows at a rate of 1 m every 1000 m in the so-called continental slope (west to east). The continental rise presents a gentle slope crossed by various canyons and valleys, located between 3200 and 5000 m water depth. To the east, the continental rise is connected to the abyssal plain, which reaches a maximum water depth of 6212 m. The Patagonian Sea is fed by two boundary currents: Brazil and Malvinas. The confluence of the warm, nutrient-poor, and salty waters of Brazil with the cold, nutrient-rich, and relatively fresh waters of Malvinas creates a stratified region of high variability of water mass properties. The Malvinas current is responsible for the high level of biological activity found in the Patagonia region, which is considered a class I marine ecosystem (Falabella et al. 2009; Dionisi et al. 2012b).

The microbial communities inhabiting Argentinian marine environments remain mostly unexplored and unexploited. Natural products with antimicrobial activity isolated from marine invertebrates have been reported. For instance, the novel antifungal disulfated triterpene glycoside, patagonicoside A (**15**; Fig. 8.3), was isolated

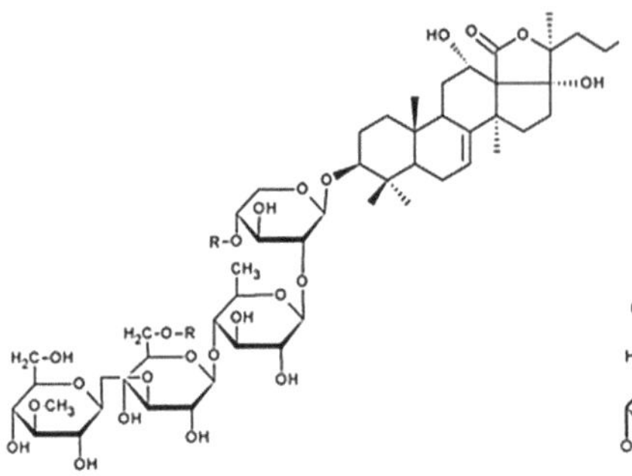

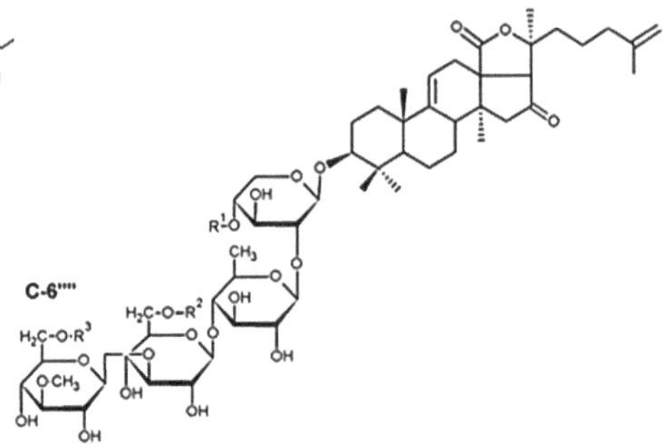

15. Patagonicoside A; R= $SO_3^-Na^+$
16. ds-Patagonicoside A; R= H

17. Hemoiedemoside A; R^1, R^2= $SO_3^-Na^+$ R^3= H
18. Hemoiedemoside B; R^1, R^2, R^3= $SO_3^-Na^+$

Fig. 8.3 Antimicrobial structures *15–18*

from the sea cucumber *Psolus patagonicus* collected from Bahía Ensenada, Ushuaia (Murray et al. 2001). The glycosidic part of patagonicoside A is a linear disulfated tetrasaccharide, similar to those found in cucumechinosides A and C isolated from *Cucumarina echinata* (Miyamoto et al. 1990). However, the aglycon moiety of patagonicoside A represents the first example of a holothurin with a Δ^7, 3β, 12α, 17α trihydroxy holostane-type aglycon. Patagonicoside A exhibited antifungal activity against the pathogenic fungus *Cladosporium cucumerinum*, although its desulfated derivative (**16**) did not (Fig. 8.3), suggesting that the presence of sulfated groups in the oligosaccharide chain is important in the antifungal activity of these triterpene glycosides. Other authors also reported the isolation of two new sulfated triterpene glycosides, hemoiedemosides A (**17**) and B (**18**) (Fig. 8.3), from the sea cucumber *Hemoiedema spectabilis* collected from the Patagonian coast of Argentina (Chludil et al. 2002a). Hemoiedemoside A showed an antifungal activity higher than patagonicoside A and even higher than hemoiedemoside B. Hemoiedemoside B differs from hemoiedemoside A in the presence of a third sulfate group at C-6″″. These results suggest that both the structure of the triterpenoidal aglycon and the presence and number of sulfate groups at the oligosaccharide chain may be essential for the antifungal activity of these saponins. Cytotoxic activity of these compounds was also observed, thus making them potentially interesting as anticancer drugs. Same authors also described another two new sulfated steroidal hexaglycosides, anasterosides A and B (**19–20**; Fig. 8.4), with antifungal activity isolated from the Patagonian starfish *Anasterias minuta* (Chludil et al. 2002b).

Besides macroorganisms, bacteria were also reported as antimicrobial producers. Several Argentinian research efforts focus on lactic acid bacteria (LAB) isolated from the intestinal tract, tegument, and gills of marine fish, marine sediment, and seawater (Vallejo et al. 2009; Sica et al. 2010; Marguet et al. 2011). LAB produce antimicrobial compounds (e.g., *bacteriocins*) that can be used as biological control agents in marine and freshwater aquaculture (Sica et al. 2012; Sahoo et al. 2014; Garcés et al. 2015), as food biopreservatives (Gálvez et al. 2007; Calo-Mata et al. 2008), and as functional starter cultures for the development of anchovy-based products (Belfiore et al. 2010).

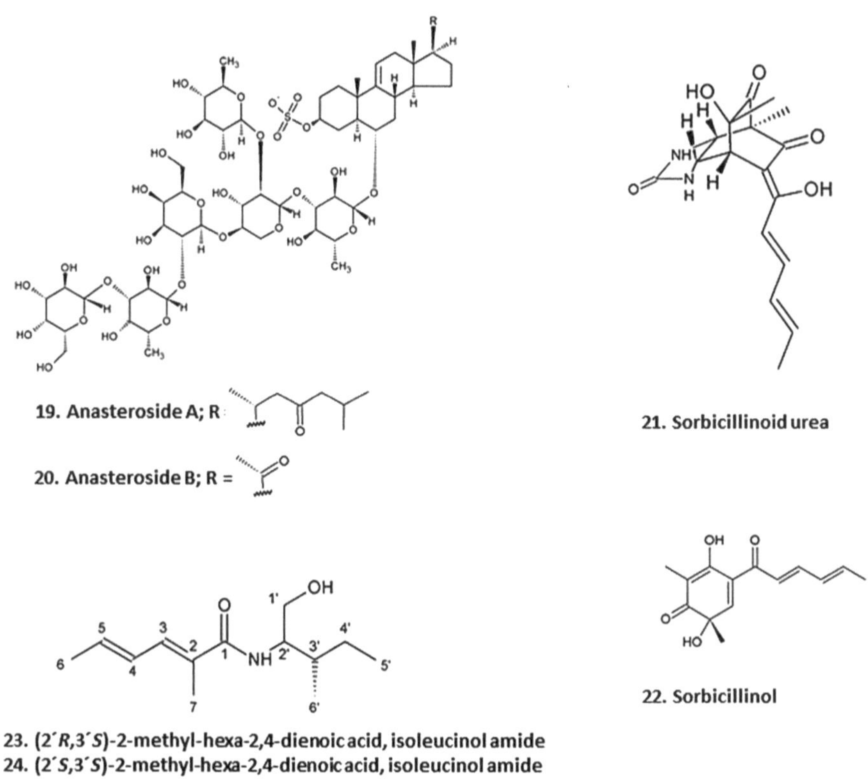

Fig. 8.4 Antimicrobial structures *15–24*

Sequeiros and coworkers (2010) have isolated a bacteriocin-producing bacterium with potential use as probiotics in aquaculture. The strain was isolated from the intestinal tract of the Patagonian fish *Odontesthes platensis* collected on the northeast coast of Chubut Province, Patagonia, Argentina and identified as *Lactococcus lactis* TW34 (Sequeiros et al. 2010). Bacteriocin was later identified as nisin Z (Sequeiros et al. 2015), and it was optimally produced at 15 °C, showing strong in vitro inhibitory activity against the closely related *Lactococcus garvieae* (causal agent of lactococcosis), and also against other gram-positive fish pathogens. However, gram-negative pathogens tested were not inhibited. Recently, Schelegueda et al. (2015) reported the isolation of a bacteriocin-producing strain, *Enterococcus mundtii* Tw56, from the fish *Odontesthes platensis*. Bacteriocin was identified by PCR as mundticin KS, and it was able to inhibit a wide spectrum of both gram-positive and gram-negative pathogens, and also the psychrophilic flora of *Odontesthes platensis*. Inhibitory activity against gram-negative bacteria is an unusual characteristic for bacteriocins produced by lactic acid bacteria.

Screening programs for fungal metabolites with antimicrobial activities from terrestrial (Levy et al. 2000, 2003; Cabrera et al. 2002) and marine (Cabrera and Seldes 1997; Gallo et al. 2004) isolates have been carried out. Cabrera et al. (2006) reported a new sorbicillinoid compound with a urea group (**21**; Fig. 8.4) and a known bioactive diketopiperazines from the fungus *Paecilomyces marquandii* (Massee) Hughes, isolated from an intertidal marine sediment sample. The diketopiperazines *cyclo*(L-Phe-L-Val) and *cyclo*(L-Phe-L-Leu) were responsible for the weak antimicrobial activity observed against *Bacillus subtilis* ATCC 6633 and *Escherichia coli* ATCC 25922. Although sorbicillinoid urea (**21**) was not active against target strains; it contributes to increase the number of new chemical scaffolds that could serve for the development of new drugs; this is the first report of the isolation of a Diels-Alder product of sorbicillinol (**22**, Fig. 8.4) with a nitrogen-containing moiety.

Another fungal strain, *Acremonium furcatum* isolated from intertidal marine sediment, was reported to produce two new amino alcohol derivatives (**23**, **24**; Fig. 8.4) with antibacterial and antifungal activities (Gallardo et al. 2006). The structure of the compounds was assigned as 2-methyl-hexa-2,4-dienoic acid, isoleucinol amide, and they were biosynthesized as a 3:1 mixture of C-2 epimers, with absolute configuration being 2′*R*, 3′*S* for the major epimer (**23**) and 2′*S*, 3′*S* for the minor compound (**24**). The synthetic compounds (**23** and **24**) were bioassayed against the bacteria *Bacillus subtilis*, *Staphylococcus aureus*, and *Escherichia coli* (at 50 μg of sample/6 mm disk; agar diffusion method) and the phytopathogenic fungi *Fusarium virguliforme*, *Colletotrichum truncatum*, *Macrophomina phaseolina*, *Botrytis cinerea*, and *Aspergillus fumigatus* (at 50 μg of sample/spot; direct bioautography on TLC). Interestingly, the L-isoleucine derivative (**24**) showed moderate antibiotic activity and low antifungal activity, whereas the D-*allo* derivative (**23**) showed antifungal activity. Another worthy observation is that antimicrobial compounds were not produced in culture medium without artificial seawater, even at trace levels, although the strain grew well under this condition. It has been reported that metabolite production by fungi is sometimes dependent on media salinity (Christophersen et al. 1998).

8.4 Problematics

Despite advances in therapy of bacterial infections, the frequency of bacterial resistance to common antibiotics continues rising. Among them, pathogens such as methicillin-resistant *Staphylococcus aureus* (Reynolds et al. 2004), penicillin-resistant *Streptococcus pneumoniae* (Karchmer 2004), and beta-lactam-resistant *Pseudomonas aeruginosa* (Paterson 2006) are the most common examples. Consequently, diseases that were easily healed are becoming a relevant problem affecting the Health System, causing important economics losses. The inappropriate uses of antibiotics, the detriment in pharmaceutical companies involved in novel antimicrobial discovery, and the emergence of new pathogenicity highlight the risks and impel scientists to become involved with this problem (Sánchez et al. 2009). Therefore, the development of novel antimicrobial compounds has turned into a worldwide priority (Wenzel 2004; Spížek et al. 2010).

Common infectious diseases could once again be the first cause of worldwide death. Antibiotics to prevent and to treat bacterial infections have been used for seven decades. Bacteria respond to the use of these antagonistic substances, turning into resistant forms by a different process. Nowadays, bacterial resistance causes high mortality rates in hospitals and important economic losses in healthcare institutions.

Some key factors of this worldwide problem are the inappropriate use of antibiotics (societal factor), the emergence of new pathogenicity or resistance (biological factor), and the restricted investment of pharmaceutical companies in the research and development of new antimicrobial drugs (economic factor). The urgent race against ever-evolving resistant pathogens requires the discovery of new molecules with antibiotic activity, effective therapeutics strategies, and research of promising targets. The development of novel antimicrobial compounds has become in a priority since the discovery of penicillin by Alexander Fleming in 1980. Nowadays, of 5000 antimicrobial compounds analysed or discovered, only 1 of them reaches the customers (Wratschko 2009) and actinomycetes are the main producers of antibiotics in current use. Traditional isolation and culturing procedures have been used as a typical technique to screen for novel products or to discover biotechnological capabilities from microbial communities. Isolation and identification of producer microorganisms are vital in applied research work (Dionisi et al. 2012a). However, the limitations of the common isolation techniques mean the antimicrobial detection rates are lower than expected, depending on the criteria used. The limited number of environmental microorganisms isolated in a given culture media or even using enrichment methods usually select fast-growing or opportunistic microorganisms (Zengler et al. 2002) although the majority of producers of secondary metabolites are often slow-growing cells. In addition, those culture media are not suitable for microbial growth, and the biotechnological potential could be not extensively exploited.

The rate of antimicrobial compound detection or discovery depends on the criteria used for selection, but generally are low values in the commonly used isolation procedures. For antimicrobial compounds active at low temperatures, the detection rates oscillate between 0.16 % (Sánchez et al. 2009) and 0.29 % (O'Brien et al. 2004). From dairy and meat sources, a detection rate of 0.20 % was reported for bacteriocin producers using direct plating methods and 3.4 % from fish and vegetables sources (Coventry et al. 1997). However, in other studies focused in general antibacterial activities, without considering the nature of the antagonistic molecule, the detection rate was higher (Hentschel et al. 2001).

Marine organisms are on the top of microorganisms that are not cultivable with traditional isolation methods. To overcome these limitations, deep studies in novel isolation techniques of the marine ecosystem were done to increase the ratio of cultured microorganisms (Joint et al. 2010). Selective isolation employing dispersion and differential centrifugation techniques (DDC) (Hopkins et al. 1991) applied to sediment samples led to the isolation of *Actinomycetes* previously reported only in terrestrial sediments (Maldonado et al. 2005).

New isolation techniques have opened the possibility for isolation of novel products to enhance the efficiency of drug discovery. Different improvements and changes in some traditional approaches have been applied to improve the recovery of microorganisms in culture. Combination of high-throughput screening with the use of the natural site environment as culture media represented the first important progress in marine bacteria cultivation. Other approaches such as microdroplet encapsulation in an agarose matrix (Zengler et al. 2002, 2005), diffusion chamber (Kaeberlein et al. 2002), and co-cultivation with "helper strains" (Nichols et al. 2008; Lewis et al. 2010) have increased cell recovery in culture. Nowadays, automatized methods have been reported using cultivation devices. One example is a design of the isolation chip (iChip). Based on several hundred diffusion chambers, each inoculated with one cell, this platform allows the cultivation and isolation of uncultivable microorganisms from various environments and also allows to access to different unexplored biological compounds (Nichols et al. 2010). This novel high-throughput method led to the isolation of new organisms that are sources of new antimicrobials. With this novel technology, a new antibiotic named teixobactin has been discovered in a screening of uncultured bacteria (Ling et al. 2015). This new antimicrobial compound did not show resistance in *Staphylococcus aureus* or *Mycobacterium tuberculosis* strains, and the inhibition mechanisms were described as inhibition of cell-wall synthesis by binding in conserved and specific motif of lipid II (precursor of peptidoglycan) and lipid III precursor of cell-wall teichoic acid.

Thus, innovation, research, and development of novel techniques would allow the isolation of novel uncultivable microorganisms and would allow the discovery of new biologically active compounds with antimicrobial activity. Currently, such compounds remain unexplored in different valuable environments such as the Patagonia region or Antarctica. Those advances would contribute to the problems associated with pathogenic resistance and would enhance clinical treatments.

Regarding the food industry, use of bacteriocins as food preservatives is still limited mainly because of low activity. Nisin is the only bacteriocin used as a food preservative and, although its application is authorized, it is restricted by its low activity at neutral or alkaline pH (Gálvez et al. 2007). The need of natural additives for chilled-food preservation highlights the need to discover novel active compounds.

The majority of antibiotics have been isolated from natural sources (Gootz 1990) by specific programs using microbiological techniques (D'Souza et al. 1982). Recently, Howe et al. (2015) reported the first drug candidate that is directed against a riboswitch or RNA capable of recognizing essential molecules (vitamins, metabolites, coenzymes) and regulating translation of RNA into proteins. This riboswitch recognizes riboflavin, which is an essential molecule for metabolism of bacteria, enabling its translation. The new drug candidate, called ribocil, competes effectively with vitamin B_2 for binding to the riboswitch blocking translation rather than activating it, killing the bacteria. Ribocil is therefore the first member of a new generation of antibiotics against bacteria that currently lack resistance. Researchers found ribocil by a conventional screening method that screened a library of 57,000 small synthetic molecules. The innovation remained on the method of selection, which is specifically directed to molecules that block the synthesis of vitamin B_2.

However, when bacteria in culture are exposed to prolonged ribocil use (in sublethal concentrations), ultimately they generate drug resistance.

8.5 Bioprospecting from Antarctic Environments

In addition to the Argentinian Patagonian region, described as a valuable source of antimicrobial compounds, another important sampling site described in the literature as a relevant ecosystem valuable as a source of novel antibacterial compounds is the Argentine Antarctic. The exploration of this environment with its unusual habitats and poorly explored areas would be one strategy for the discovery of new biologically active substances (Kennedy et al. 2008; Rojas et al. 2009; Liu et al. 2013; Chávez et al. 2015). Generally, new strains are thought to be sources of new compounds with diverse biological activities. Antarctica is the southernmost continent and is surrounded by the Southern Ocean. It is the coldest, driest, and windiest continent where approximately 98 % of its surface is covered by ice. The average temperature of coastal areas is about 0 °C in summer (up to 15 °C at the Antarctic Peninsula) and around −10 ° to −30 °C in winter. The polar plateau presents more extreme conditions, with average temperatures of −30 °C in summer and −70 °C in winter (López-Martínez 2009). There is a great diversity of aquatic habitats in Antarctica, such as the benthos, water columns, fluctuating sea-ice cover by the circumpolar Southern Ocean, glaciers, and lakes. These habitats range in salinity from ultra-oligotrophic glacial lakes on Signy Island (South Orkney Islands) to the hypersaline and permanently ice-covered lakes of the McMurdo dry valley region and Vestfold Hills. In the same way, terrestrial habitats range from the most eutrophic brown earth soils at maritime Signy Island to endolithic rock communities of cold deserts at the edge of the Polar Plateau (Wynn-Williams 1996). Despite the harsh conditions found in Antarctica, including cold temperatures, freeze–thaw cycles, hypersalinity, low nutrient availability, extreme pH, high UV radiation, and dryness, the use of molecular techniques has revealed abundant microbial communities (Franzmann 1996; Tindall 2004; Niederberger et al. 2008; Margesin and Miteva 2011).

Argentine maintains a territorial claim over Antarctica in an area located from south of the 60°S latitude to the South Pole, and from 25° to 74°W, which includes the Antarctic Peninsula, within 13 stations in this area. Because of global warming, the Antarctic Peninsula experiences a 2 °C increase in the annual mean temperature, being one of the most rapidly warming regions on Earth. The west coast of the Antarctic Peninsula is characterized by deep embayments interconnected by channels that facilitate the transport of heat and nutrients. During the ice-free season, the Antarctic Peninsula Coastal Current sets a favourable environment for the biological production of this marine ecosystem. South of the Weddell Sea, the Filchner-Ronne Ice Shelf is floating over the continental shelf and usually releases ice fragments (icebergs) from the ice front. These free-drifting icebergs on the Weddell Sea might represent hotspots of constant micronutrient release that could serve as areas of increased production and sequestration of organic carbon (Dionisi et al. 2012b).

The Argentinian marine environments, with their notable extension, complex topography, and water circulation patterns, as well as high biological productivity and biodiversity, hold numerous valuable niches for the bioprospection of microorganisms with biotechnological potential. In particular, sub-Antarctic and Antarctic marine regions, remote and mostly pristine, contain microbial communities adapted to extreme conditions that are remarkably suitable for bioprospection (Dionisi et al. 2012b).

Thus, Antarctic marine and terrestrial habitats are an excellent resource of extremophiles, which often produce small molecules belonging to unusual structure classes with novel biological activities (Wilson and Brimble 2009; Giddings and Newman 2015a, b).

The antimicrobial activity of microorganisms from the Antarctic environs (including the sub-Antarctic) has been investigated (O'Brien et al. 2004; Sánchez et al. 2009; Encheva-Malinova et al. 2014; Tomova et al. 2015). A soil-derived *Streptomyces griseus* strain NTK 97 isolated from Terra Nova Bay at Edmunson Point, Antarctica, was reported to produce a new angucyclinone antibiotic, frigocyclinone (**25**, Fig. 8.5) (Bruntner et al. 2005). Frigocyclinone consists of a tetrangomycin moiety attached

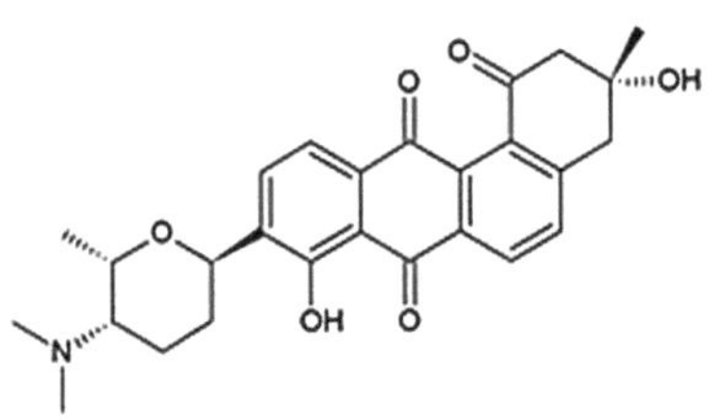

25. Frigocyclinone

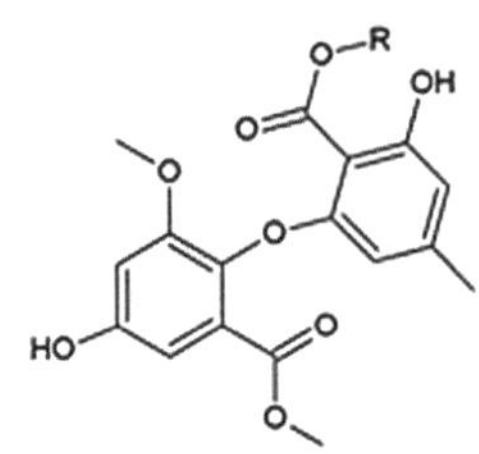

26. Asterric acid; R= H
27. Ethyl asterrate; R= CH_2CH_3
28. *n*-Butyl asterrate; R= $(CH_2)_3CH_3$

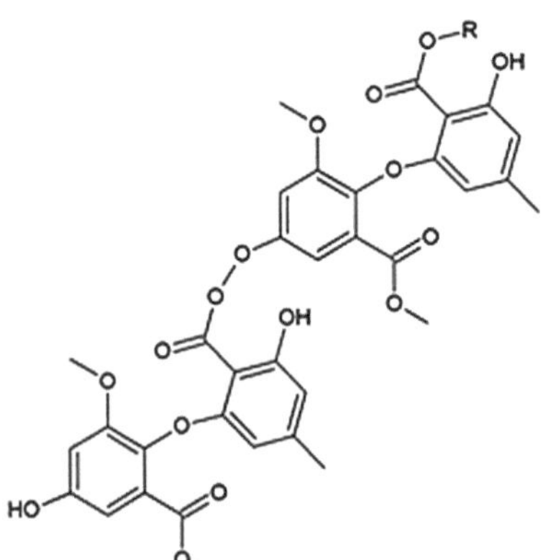

29. Geomycin A; R= CH_3
30. Geomycin B; R= H

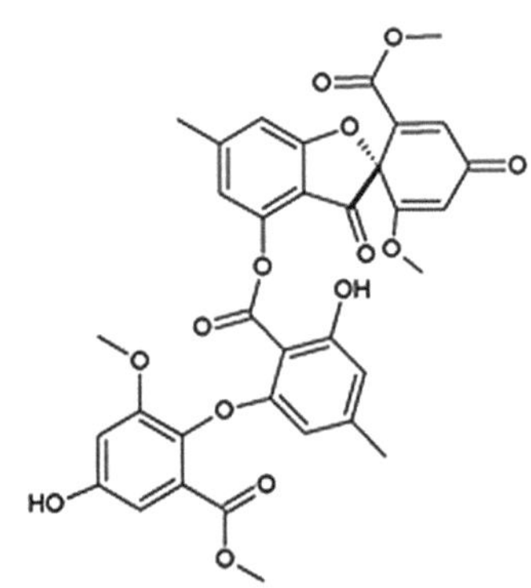

31. Geomycin C

Fig. 8.5 Antimicrobial structures *25–31*

through a C-glycosidic linkage with the aminodeoxysugar ossamine. Its UV-visible spectrum was highly similar to that of urdamycin B produced by *Streptomyces fradiae* (Drautz et al. 1986). Frigocyclinone exhibited good antibacterial activity against gram-positive eubacteria, such as *Bacillus subtilis* DSM 10 (MIC, 4.6 μg ml^{-1}) and *Staphylococcus aureus* DSM 21231 (MIC, 15 μg mL^{-1}). The minimal inhibition concentration (MIC) of frigocyclinone was significantly higher in comparison to that of vancomycin and erythromycin; however, frigocyclinone revealed a more potent activity than urdamycin B. No inhibitory activity against some gram-negative bacteria (*Escherichia coli, Pseudomonas fluorescens, Proteus mirabilis*), filamentous fungi (*Botrytis cinerea, Aspergillus viridinutans, Penicillium notatum, Paecilomyces variotii*), and yeasts (*Saccharomyces cerevisiae, Candida albicans*) was observed. In addition, other terrestrial psychrophiles producing antimicrobial compounds have been isolated from Antarctic soils. Li et al. (2008) isolated the psychrotolerant fungus *Geomyces* sp. from the Foldes Peninsula and King George Island, which produced five new asterric acid derivatives (**26**), ethyl asterrate (**27**), *n*-butyl asterrate (**28**), and geomycins A–C (**29**–**31**) (Fig. 8.5). Ethyl asterrate and *n*-butyl asterrate did not exhibit noticeable in vitro antimicrobial activity, indicating that ester chain length is not responsible for the antimicrobial activity of asterric acids. Asterric acid derivatives have been claimed as useful in the treatment of myocardial infarction and renal insufficiency; however, the antifungal activity had not been reported until now. Furthermore, geomycin B exhibited antifungal activity against *Aspergillus fumigatus* (MIC, 20 μg ml^{-1}) whereas geomycin C exhibited antimicrobial activity against the gram-positive bacteria *Staphylococcus aureus* (MIC, 51 μg ml^{-1}) and *Streptococcus pneumoniae* (MIC, 103 μg ml^{-1}) and the gram-negative bacterium *Escherichia coli* (MIC, 20 μg ml^{-1}).

References

Arnau VG, Sánchez LA, Delgado OD (2015) *Pseudomonas yamanorum* sp. nov., a psychrotolerant bacterium isolated from a subantarctic environment. Int J Syst Evol Microbiol 65:424–431

Asai M, Mizuta E, Izawa M, Haibara K, Kishi T (1979) Isolation, chemical characterization and structure of ansamitocin, a new antitumor ansamycin antibiotic. Tetrahedron 35:1079–1085

Barros J, Becerra J, González C, Martínez M (2013) Antibacterial metabolites synthesized by psychrotrophic bacteria isolated from cold-freshwater environments. Folia Microbiol 58:127–133

Beales N (2004) Adaptation of microorganisms to cold temperatures, weak acid preservatives, low pH, and osmotic stress: a review. Compr Rev Food Sci Food Saf 3:1–20

Belfiore C, Björkroth J, Vihavainen E, Raya R, Vignolo G (2010) Characterization of *Leuconostoc* strains isolated from fresh anchovy (*Engraulis anchoita*). J Gen Appl Microbiol 56:175–180

Beltrán A (1997) Caracterización microclimática del Distrito Occidental de la estepa patagónica. Magister thesis, Universidad de Buenos Aires, Buenos Aires

Borthwick AD (2012) 2, 5-Diketopiperazines: synthesis, reactions, medicinal chemistry, and bioactive natural products. Chem Rev 112:3641–3716

Bowers KJ, Mesbah NM, Wiegel J (2009) Biodiversity of poly-extremophilic bacteria: does combining the extremes of high salt, alkaline pH and elevated temperature approach a physico-chemical boundary for life? Saline Syst 5:9

Brown CM, Minnikin DE (1973) Effect of growth temperature on the fatty acid composition of some psychrophilic marine pseudomonads. J Gen Microbiol 75:R9

Bruntner C, Binder T, Pathom-aree W, Goodfellow M, Bull AT, Potterat O, Puder C, Horer S, Schmid A, Bolek W, Wagner K, Mihm G, Fiedler H-P (2005) Frigocyclinone, a novel angucyclinone antibiotic produced by a *Streptomyces griseus* strain from Antarctica. J Antibiot (Tokoyo) 58:346–349

Cabrera GM, Seldes AM (1997) Citrinin derivatives from an intertidal marine *Penicillium.* Ann Asoc Quím Arg 85:193–196

Cabrera GM, Roberti MJ, Wright JE, Seldes AM (2002) Cryptoporic and isocryptoporic acids from the fungal cultures of *Polyporus arcularius* and *P. ciliatus*. Phytochemistry 61:189–193

Cabrera GM, Butler M, Rodriguez MA, Godeas A, Haddad R, Eberlin MN (2006) A sorbicillinoid urea from an intertidal *Paecilomyces marquandii.* J Nat Prod 69:1806–1808

Calo-Mata P, Arlindo S, Boehme K, de Miguel T, Pascoal A, Barros-Velazquez J (2008) Current applications and future trends of lactic acid bacteria and their bacteriocins for the biopreservation of aquatic food products. Food Bioproc Technol 1:43–63

Carrasco JF, Casassa G, Rivera A (2002) Meterological and climatological aspects of the Southern Patagonia Icefield. In: Casassa G, Sepulveda V, Sinclair RM (eds) The Patagonian icefields: a unique laboratory for environmental and climate change studies. Springer, New York, pp 29–41

Castillo UF, Browne L, Strobel G, Hess WM, Ezra S, Pacheco G, Ezra D (2007) Biologically active endophytic Streptomycetes from *Nothofagus* spp. and other plants in Patagonia. Microb Ecol 53:12–19

Cavicchioli R, Siddiqui KS, Andrews D, Sowers KR (2002) Low-temperature extremophiles and their applications. Curr Opin Biotechnol 13:253–261

Chávez R, Fierro F, García-Rico RO, Vaca I (2015) Filamentous fungi from extreme environments as a promising source of novel bioactive secondary metabolites. Front Microbiol 6:903

Chludil HD, Muniaín CC, Seldes AM, Maier MS (2002a) Cytotoxic and antifungal triterpene glycosides from the Patagonian sea cucumber *Hemoiedema spectabilis.* J Nat Prod 65:860–865

Chludil HD, Seldes AM, Maier MS (2002b) Antifungal steroidal glycosides from the Patagonian starfish *Anasterias minuta*: structure–activity correlations. J Nat Prod 65:153–157

Christophersen C, Crescente O, Frisvad JC, Gram L, Nielsen J, Nielsen PH, Rahbæk L (1998) Antibacterial activity of marine-derived fungi. Mycopathologia 143:135–138

Coventry MJ, Gordon JB, Wilcock A, Harmark K, Davidson BE, Hickey MW, Hillier AJ, Wan J (1997) Detection of bacteriocins of lactic acid bacteria isolated from foods and comparison with pediocin and nisin. J Appl Microbiol 83:248–258

Davies J (2011) How to discover new antibiotics: harvesting the parvome. Curr Opin Chem Biol 15:5–10

Debnath M, Paul AK, Bisen PS (2007) Natural bioactive compounds and biotechnological potential of marine bacteria. Curr Pharm Biotechnol 8:253–260

Dib J, Motok J, Zenoff VF, Ordonez O, Farias ME (2008) Occurrence of resistance to antibiotics, UV-B, and arsenic in bacteria isolated from extreme environments in high-altitude (above 4400 m) Andean wetlands. Curr Microbiol 56:510–517

Dionisi HM, Lozada M, Olivera NL (2012a) Bioprospection of marine microorganisms: biotechnological applications and methods. Rev Argent Microbiol 44:49–60

Dionisi HM, Lozada M, Olivera NL (2012b) Bioprospection of marine microorganisms: potential and challenges for Argentina. Rev Argent Microbiol 44:122–132

Drautz H, Zähner H, Rohr J, Zeeck A (1986) Metabolic products of microorganisms. 234. Urdamycins, new angucycline antibiotics from *Streptomyces fradiae*. I. Isolation, characterization and biological properties. J Antibiot (Tokyo) 39(12):1657–1669

D'Souza SF, Kaul R, Nadkarni GB (1982) Immobilization of microbial cells in hen egg white. Biotechnol Bioeng 24:1701–1704

Edwards C (1990) Thermophiles. In: Microbiology of extreme environments. Open University Press, Milton Keynes, pp 1–32

Elkahoui S, Abdel Rahim H, Tabbene O, Shaaban M, Limam F, Laatsch H (2013) *Cyclo*-(His, Leu): a new microbial diketopiperazine from a terrestrial *Bacillus subtilis* strain B38. Nat Prod Res 27:108–116

Encheva-Malinova M, Stoyanova M, Avramova H, Pavlova Y, Gocheva B, Ivanova I, Moncheva P (2014) Antibacterial potential of streptomycete strains from Antarctic soils. Biotechnol Biotec Eq 28:721–727

Falabella V, Campagna C, Croxall J (eds) (2009) Atlas del Mar Patagónico. Especies y Espacios. Wildlife Conservation Society, Buenos Aires

Feller G, Gerday C (2003) Psychrophilic enzymes: hot topics in cold adaptation. Nat Rev Microbiol 1:200–208

Fiedler HP, Bruntner C, Riedlinger J, Bull AT, Knutsen G, Goodfellow M, Jones A, Maldonado L, Pathom-aree W, Beil W, Schneider K, Keller S, Sussmuth RD (2008) Proximicin A, B and C, novel aminofuran antibiotic and anticancer compounds isolated from marine strains of the actinomycete *Verrucosispora*. J Antibiot (Tokyo) 61:158–163

Franzmann PD (1996) Examination of Antarctic prokaryotic diversity through molecular comparisons. Biodivers Conserv 5:1295–1305

Gallardo GL, Butler M, Gallo ML, Rodríguez MA, Eberlin MN, Cabrera GM (2006) Antimicrobial metabolites produced by an intertidal *Acremonium furcatum*. Phytochemistry 67:2403–2410

Gallo ML, Seldes AM, Cabrera GM (2004) Antibiotic long-chain and α, β-unsaturated aldehydes from the culture of the marine fungus *Cladosporium* sp. Biochem Syst Ecol 32:545–551

Gálvez A, Abriouel H, López RL, Omar NB (2007) Bacteriocin-based strategies for food biopreservation. Int J Food Microbiol 120:51–70

Garcés ME, Sequeiros C, Olivera NL (2015) Marine *Lactobacillus pentosus* H16 protects *Artemia franciscana* from *Vibrio alginolyticus* pathogenic effects. Dis Aquat Organ 113:41–50

Giddings LA, Newman DJ (2015a) Bioactive compounds from terrestrial Extremophiles. In: Tiquia-Arashiro SM, Mormile M (eds) Extremophilic bacteria. Springer briefs in microbiology. Springer, New York, pp 1–75

Giddings LA, Newman DJ (2015b) Bioactive compounds from marine Extremophiles. In: Tiquia-Arashiro SM, Mormile M (eds) Extremophilic bacteria. Springer briefs in microbiology. Springer, New York, pp 1–124

Gootz TD (1990) Discovery and development of new antimicrobial agents. Clin Microbiol Rev 3:13–31

Grossart HP, Kiørboe T, Tang K, Ploug H (2003) Bacterial colonization of particles: growth and interactions. Appl Environ Microbiol 69:3500–3509

Hentschel U, Schmid M, Wagner M, Fieseler L, Gernert C, Hacker J (2001) Isolation and phylogenetic analysis of bacteria with antimicrobial activities from the Mediterranean sponges *Aplysina aerophoba* and *Aplysina cavernicola*. FEMS Microbiol Ecol 35:305–312

Higashide E, Asai M, Ootsu K, Tanida S, Kozai Y, Hasegawa T, Kishi T, Sugino Y, Yoneda M (1977) Ansamitocin, a group of novel maytansinoid antibiotics with antitumour properties from *Nocardia*. Nature (Lond) 270:721–722

Hopkins DW, MacNaughton SJ, O'Donnell AG (1991) A dispersion and differential centrifugation technique for representatively sampling microorganisms from soil. Soil Biol Biochem 23:217–225

Howe JA, Wang H, Fischmann TO, Balibar CJ, Xiao L, Galgoci AM, Malinverni JC, Mayhood T, Villafania A, Nahvi A, Murgolo N, Barbieri CM, Mann PA, Carr D, Xia E, Zuck P, Riley D, Painter RE, Walker SS, Sherborne B, de Jesus R, Pan W, Plotkin MA, Wu J, Rindgen D, Cummings J, Garlisi CG, Zhang R, Sheth PR, Gill CJ, Tang H, Roemer T (2015) Selective small-molecule inhibition of an RNA structural element. Nature (Lond) 526:672–677

Huang R, Zhou X, Xu T, Yang X, Liu Y (2010) Diketopiperazines from marine organisms. Chem Biodivers 7:2809–2829

Isnansetyo A, Kamei Y (2003) MC21-A, a bactericidal antibiotic produced by a new marine bacterium, *Pseudoalteromonas phenolica* sp. nov. O-BC30^{T}, against methicillin-resistant *Staphylococcus aureus*. Antimicrob Agents Chemother 47:480–488

Joint I, Mühling M, Querellou J (2010) Culturing marine bacteria—an essential prerequisite for biodiscovery. Microb Biotechnol 3:564–575
Kaeberlein T, Lewis K, Epstein SS (2002) Isolating "uncultivable" microorganisms in pure culture in a simulated natural environment. Science 296:1127–1129
Karchmer AW (2004) Increased antibiotic resistance in respiratory tract pathogens: PROTEKT US—an update. Clin Infect Dis 39:142–150
Keller S, Nicholson G, Drahl C, Sorensen E, Fiedler HP, Süssmuth RD (2007) Abyssomicins G and H and atrop-abyssomicin C from the marine *Verrucosispora* strain AB-18-032. J Antibiot (Tokyo) 60:391–394
Kennedy J, Marchesi JR, Dobson AD (2008) Marine metagenomics: strategies for the discovery of novel enzymes with biotechnological applications from marine environments. Microb Cell Fact 7:27
Kobayashi H (1989) Liquid leather cleaners. Jpn Patent 1:225–700
Levy LM, Cabrera GM, Wright JE, Seldes AM (2000) Indole alkaloids from a culture of the fungus *Aporpium caryae*. Phytochemistry 54:941–943
Levy LM, Cabrera GM, Wright JE, Seldes AM (2003) 5H-Furan-2-ones from fungal cultures of *Aporpium caryae*. Phytochemistry 62:239–243
Lewis K, Epstein S, D'Onofrio A, Ling LL (2010) Uncultured microorganisms as a source of secondary metabolites. J Antibiot (Tokyo) 63:468–476
Li Y, Sun B, Liu S, Jiang L, Liu X, Zhang H, Che Y (2008) Bioactive asterric acid derivatives from the Antarctic ascomycete fungus *Geomyces* sp. J Nat Prod 71:1643–1646
Ling LL, Schneider T, Peoples AJ, Spoering AL, Engels I, Conlon BP, Mueller A, Schäberle TF, Hughes DE, Epstein S, Jones M, Lazarides L, Steadman VA, Cohen DR, Felix CR, Fetterman KA, Millett WP, Nitti AG, Zullo AM, Chen C, Lewis K (2015) A new antibiotic kills pathogens without detectable resistance. Nature (Lond) 517:455–459
Liu J-T, Lu X-L, Liu X-Y, Gao Y, Hu B, Jiao B-H, Zheng H (2013) Bioactive natural products from the Antarctic and Arctic organisms. Mini Rev Med Chem 13:617–626
Lo Giudice A, Bruni V, Michaud L (2007) Characterization of Antarctic psychrotrophic bacteria with antibacterial activities against terrestrial microorganisms. J Basic Microbiol 47:496–505
López-Martínez J (2009) Antártida. Introducción a un continente remote. Rodolfo Sánchez Albatros. Buenos Aires, p 96
Maldonado LA, Stach JEM, Pathom-aree W, Ward AC, Bull AT, Goodfellow M (2005) Diversity of cultivable actinobacteria in geographically widespread marine sediments. Antonie Van Leeuwenhoek 87:11–18
Margesin R, Miteva V (2011) Diversity and ecology of psychrophilic microorganisms. Res Microbiol 162:346–361
Marguet ER, Vallejo M, Sierralta Chichisola V, Quispe JL (2011) Actividad antagonista de bacterias lácticas aisladas del medio marino contra cepas de *Listeria*. Acta Bioquim Clin Latinoam 45:305–310
Mastronicolis SK, German JB, Megoulas N, Petrou E, Foka P, Smith GM (1998) Influence of cold shock on the fatty-acid composition of different lipid classes of the food-borne pathogen *Listeria monocytogenes*. Food Microbiol 15:299–306
Mateo N, Nader W, Tamayo G (2001) Bioprospecting. Encyclopedia of biodiversity, vol 1. Academic Press, Cambridge, pp 471–488
Miyamoto T, Togawa K, Higuchi R, Komori T, Sasaki T (1990) Six newly identified biologically active triterpenoid glycoside sulfates from the sea cucumber *Cucumaria echinata*. Liebigs Ann Chem 1990:453–460
Murray AP, Muniaín C, Seldes AM, Maier MS (2001) Patagonicoside A: a novel antifungal disulfated triterpene glycoside from the sea cucumber *Psolus patagonicus*. Tetrahedron 57:9563–9568
Needham J, Kelly MT, Ishige M, Andersen RJ (1994) Andrimid and moiramides A-C, metabolites produced in culture by a marine isolate of the bacterium *Pseudomonas fluorescens*: structure elucidation and biosynthesis. J Org Chem 59:2058–2063

Nichols D, Lewis K, Orjala J, Mo S, Ortenberg R, O'Connor P, Zhao C, Vouros P, Kaeberlein P, Epstein SS (2008) Short peptide induces an "uncultivable" microorganism to grow in vitro. Appl Environ Microbiol 74:4889–4897

Nichols D, Cahoon N, Trakhtenberg EM, Pham L, Mehta A, Belanger A, Epstein SS (2010) Use of ichip for high-throughput *in situ* cultivation of "uncultivable" microbial species. Appl Environ Microbiol 76:2445–2450

Niederberger TD, McDonald IR, Hacker AL, Soo RM, Barrett JE, Wall DH, Cary SC (2008) Microbial community composition in soils of Northern Victoria Land, Antarctica. Environ Microbiol 10:1713–1724

O'Brien A, Sharp R, Russell NJ, Roller S (2004) Antarctic bacteria inhibit growth of food-borne microorganisms at low temperatures. FEMS Microbiol Ecol 48:157–167

Paruelo JM, Beltrán A, Jobbágy E, Sala OE, Golluscio RA (1998) The climate of Patagonia: general patterns and controls on biotic processes. Ecol Austr 8:85–101

Paterson DL (2006) The epidemiological profile of infections with multidrug-resistant *Pseudomonas aeruginosa* and *Acinetobacter* species. Clin Infect Dis 43:43–48

Ponce JF, Fernández M (2014) Climatic and environmental history of Isla de los Estados, Argentina. In: Lohmann G, Rabassa J, Notholt J, Mysak LA, Unnithan V (eds) South America and the Southern Hemisphere. Springer briefs in Earth system sciences. Springer, New York

Prohaska F (1976) The climate of Argentina, Paraguay and Uruguay. Clim Cent South Am 22:13–112

Reynolds R, Potz N, Colman M, Williams A, Livermore D, MacGowan A (2004) Antimicrobial susceptibility of the pathogens of bacteraemia in the UK and Ireland 2001–2002: the BSAC Bacteraemia Resistance Surveillance Programme. J Antimicrob Chemother 53:1018–1032

Riedlinger J, Reicke A, Zähner H, Krismer B, Bull AT, Maldonado LA, Ward AC, Goodfellow M, Bister B, Bischoff D, Süssmuth RD, Fiedler HP (2004) Abyssomicins, inhibitors of the *para*-aminobenzoic acid pathway produced by the marine *Verrucosispora* strain AB-18-032. J Antibiot (Tokyo) 57:271–279

Rivas FP, Castro MP, Vallejo M, Marguet E, Campos CA (2012) Antibacterial potential of *Enterococcus faecium* strains isolated from ewes' milk and cheese. LWT Food Sci Technol 46:428–436

Rojas JL, Martín J, Tormo JR, Vicente F, Brunati M, Ciciliato I, Losib D, Van Trappenc S, Mergaertc J, Swingsc J, Marinellid F, Genilloud O (2009) Bacterial diversity from benthic mats of Antarctic lakes as a source of new bioactive metabolites. Mar Genomics 2:33–41

Russell NJ (1984) Mechanisms of thermal adaption in bacteria: blueprints for survival. Trends Biochem Sci 9:108–112

Russell NJ, Evans RI, Ter Steeg PF, Hellemons J, Verheul A, Abee T (1995) Membranes as a target for stress adaption. Int J Food Microbiol 28:255–261

Sahoo TK, Jena PK, Patel AK, Seshadri S (2014) Bacteriocins and their applications for the treatment of bacterial diseases in aquaculture: a review. Aquaculture research. Wiley, New York

Sánchez LA, Gómez FF, Delgado OD (2009) Cold-adapted microorganisms as a source of new antimicrobials. Extremophiles 13:111–120

Sánchez LA, Hedström M, Delgado MA, Delgado OD (2010) Production, purification and characterization of serraticin A, a novel cold-active antimicrobial produced by *Serratia proteamaculans* 136. J Appl Microbiol 109:936–945

Sánchez LA, Sierra MG, Siñeriz F, Delgado OD (2013) Andrimid production at low temperature by a psychrotolerant *Serratia proteamaculans* strain. World J Microbiol Biotechnol 29:1773–1781

Schelegueda LI, Vallejo M, Gliemmo MF, Marguet ER, Campos CA (2015) Synergistic antimicrobial action and potential application for fish preservation of a bacteriocin produced by *Enterococcus mundtii* isolated from *Odontesthes platensis*. LWT Food Sci Technol 64:794–801

Sequeiros C, Vallejo M, Marguet ER, Olivera NL (2010) Inhibitory activity against the fish pathogen *Lactococcus garvieae* produced by *Lactococcus lactis* TW34, a lactic acid bacterium isolated from the intestinal tract of a Patagonian fish. Arch Microbiol 192:237–245

Sequeiros C, Garcés ME, Vallejo M, Marguet ER, Olivera NL (2015) Potential aquaculture probiont *Lactococcus lactis* TW34 produces nisin Z and inhibits the fish pathogen *Lactococcus garvieae*. Arch Microbiol 197:449–458
Shehane SD, Sizemore RK (2002) Isolation and preliminary characterization of bacteriocins produced by *Vibrio vulnificus*. J Appl Microbiol 92:322–328
Sica MG, Olivera NL, Brugnoni LI, Marucci PL, López-Cazorla AC, Cubitto MA (2010) Isolation, identification and antimicrobial activity of lactic acid bacteria from the Bahía Blanca Estuary. Rev Biol Mar Oceanogr 45:389–397
Sica MG, Brugnoni LI, Marucci PL, Cubitto MA (2012) Characterization of probiotic properties of lactic acid bacteria isolated from an estuarine environment for application in rainbow trout (*Oncorhynchus mykiss*, Walbaum) farming. Antonie Van Leeuwenhoek 101:869–879
Singh SB, Barrett JF (2006) Empirical antibacterial drug discovery: foundation in natural products. Biochem Pharmacol 71:1006–1015
Spížek J, Novotná J, Řezanka T, Demain AL (2010) Do we need new antibiotics? The search for new targets and new compounds. J Ind Microbiol Biotechnol 37:1241–1248
Tindall BJ (2004) Prokaryotic diversity in the Antarctic: the tip of the iceberg. Microb Ecol 47:271–283
Tomova I, Stoilova-Disheva M, Lazarkevich I, Vasileva-Tonkova E (2015) Antimicrobial activity and resistance to heavy metals and antibiotics of heterotrophic bacteria isolated from sediment and soil samples collected from two Antarctic islands. Front Life Sci 8(4):348–357
Tsuchiya H, Sato M, Kanematsu N, Kato M, Hoshino Y, Takagi N, Namikawa I (1987) Temperature-dependent changes in phospholipid and fatty acid composition and membrane lipid fluidity of *Yersinia enterocolitica*. Lett Appl Microbiol 5:15–18
Vallejo M, Olivera N, Sequeiros C, Marguet E (2009) Actividad antilisteria de bacterias ácido lácticas aisladas de peces marinos. Analecta Vet 29:19–23
Wenzel RP (2004) The antibiotic pipeline: challenges, costs, and values. N Engl J Med 351:523–526
Wilson ZE, Brimble MA (2009) Molecules derived from the extremes of life. Nat Prod Rep 26:44–71
Wratschko K (2009) Empirical setting: the pharmaceutical industry. In: Strategic orientation and alliance portfolio configuration. Gabler, Wissenchaft, pp 87–96
Wynn-Williams DD (1996) Antarctic microbial diversity: the basis of polar ecosystem processes. Biodivers Conserv 5:1271–1293
Zengler K, Toledo G, Rappé M, Elkins J, Mathur EJ, Short JM, Keller M (2002) Cultivating the uncultured. Proc Natl Acad Sci USA 99:15681–15686
Zengler K, Walcher M, Clark G, Haller I, Toledo G, Holland T, Mathur EJ, Woodnutt G, Short JM, Keller M (2005) High-throughput cultivation of microorganisms using microcapsules. Methods Enzymol 397:124–130

Chapter 9
Bioprospecting for Bioactive Actinomycetes from Patagonia

María Soledad Vela Gurovic

Abstract Actinomycetes have been and still are the most promising bacterial source of antibiotic and antifungal substances. It is known that the genetic potential of these microorganisms has been underexplored. Up to date, almost all continents have been subject to bioprospecting for bioactive actinomycetes. Reports on the screening of actinomycetes come mainly from Europe, India, and countries of Africa. In contrast, some biologically rich regions, particularly Patagonia and other regions from South America, have been scarcely explored. Bioprospecting based on the exploration of new environments, such as the deep sea and marine invertebrates, led to the discovery of new and unique bioactive metabolites. Patagonia offers a vast diversity of such potential environments and sources of bioactive strains, including autochthonous marine invertebrates, endemic plants and lichens, wide unoccupied desert areas with high-saline environments exposed to high temperatures, pristine environments, and ancient forests. The latest most successful approaches in bioprospecting for bioactive actinomycetes are reviewed in this chapter, together with a discussion of the available reports on bioprospecting for actinomycetes from Patagonia.

9.1 Introduction

Actinomycetes comprise a large group of gram-positive bacteria. Although they are typically abundant in soil as saprophytic microorganisms, actinomycetes also inhabit freshwater and deep sea. Actinomycetes are the source of many clinically successful antibiotics. For decades, the isolation of bioactive metabolites from actinomycetes provided new hits for the treatment of infections and cancer, among other diseases. The discovery of many bioactive compounds from actinomycete cultures has revolutionized human medicine and other disciplines such as veterinary pharmacology and the food industry. Ivermectin is the best example to illustrate

M.S. Vela Gurovic (✉)
CERZOS UNS-CONICET CCT-Bahía Blanca,
Camino La Carrindanga Km7 (B8000FWB), Bahía Blanca, Argentina
e-mail: svela@uns.edu.ar

N.L. Olivera et al. (eds.), *Biology and Biotechnology of Patagonian Microorganisms*, DOI 10.1007/978-3-319-42801-7_9

this. The Nobel laureate in physiology or medicine 2015, Prof. Dr. Satoshi Omura, isolated the producing microorganism *Streptomyces avermitilis* from the soil near a golf course.

Although the research on actinomycete metabolites was very successful in the past century, the discovery of new bioactive entities from natural sources declined over the years, although the frequency of rediscovering known metabolites increased (Baltz 2007). However, the prevailing criterion suggests that natural sources have not been depleted. Indeed, genomic and metagenomic studies showthat the metabolic potential of bacteria has been underestimated. In the past years, sequencing and annotation of bacterial genomes showed that many strains harbour tens of biosynthetic gene clusters related to secondary metabolism. This biosynthetic potential probably remained hidden during routine bioassays because of the lack of proper media and conditions required to trigger the expression of these gene clusters. In addition, metagenomic studies revealed the existence of a large number of uncultured bacteria with underestimated capabilities for the production of bioactive metabolites.

Nowadays, there is concern about the emerging resistance to antibiotics. The need of new antibiotics and chemotherapeutics with better pharmacological profiles boosted the development of novel tools and approaches for uncovering the biosynthetic potential of microorganisms. Figure 9.1 summarizes some of the issues that should be taken into account when searching for novel microbial bioactive compounds.

New methods are needed for isolating and growing uncultured bacteria. These achievements would allow the isolation of bacteria growing in underexplored environments, such as the deep sea and deserts. Genomic-guided screening of isolated microorganisms becomes useful to identify potential strains. The amplification and sequencing of biosynthetic genes related to secondary metabolism is now fast and easy to perform. It is therefore possible to compare partial or total sequences with those already deposited in public databases. This comparison serves as preliminary data about the novelty of the resulting metabolite. For those strains with annotated genomes, genome mining allows the identification of biosynthetic gene clusters related to secondary metabolism and the prediction of the putative structure of the

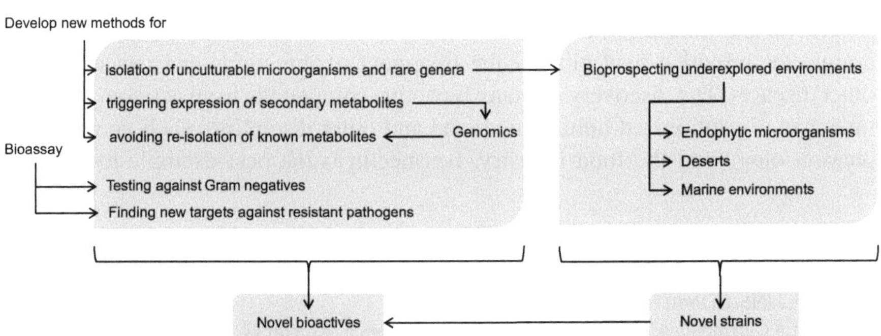

Fig. 9.1 Bottlenecks in bioprospecting for bioactive Actinomycetes

metabolite. However, if the optimal growing conditions for the production of the metabolite are not found, it is not possible to prove its structure and activity. If this is the case, heterologous expression is the only alternative.

Most of the antibiotics in clinical use are active against gram-positive strains. Testing against gram negatives, as well as finding new targets and bioassays against resistant pathogens and other diseases, would improve the chance of clinical success from the beginning and would also decrease the rate of reisolation of known metabolites (Baltz 2007).

Regarding the hypothesis "everything is everywhere," a study showed that samples from distant regions harbour unique bacterial populations (Wawrik et al. 2007). The authors concluded that the distribution of bacteria depended on factors such as soil, pH, and ecosystem type. Actinomycetes and their secondary metabolite genes were patchily distributed, suggesting that actinomycetes could be endemic: this supports the screening of new actinomycetes and new metabolic capabilities from underexplored environments. When considering bioprospecting for actinomycetes, such environments currently include marine environments such as marine sediments and marine invertebrates, deserts, and endophytic microorganisms. It is worth here to remember that the discovery of a novel strain is not always associated with the discovery of new metabolites.

9.2 Underexplored Environments

Active bacteria from marine environments comprise those associated with marine invertebrates (20 %), seaweeds (11 %), seawater (7 %), and sediments (5 %) (Gupta et al. 2013). It is of interest to note that many compounds released in seawater could be of ecological relevance. For example, a protective role of the microbial secondary metabolites has been associated to the epibiotic bacteria in seaweed. This kind of ecological association is highly appreciated when looking for bioactive substances that could act as chemical messengers between living organisms. The isolation of actinomycetes associated with marine plants and animals lead to a high rate of bioactive isolates, mainly belonging to the *Streptomyces* and *Micromonospora* genera (Zheng et al. 2000). Forthcoming research will certainly focus on arid environments, endophytic microorganisms, and extreme environments (Peeters et al. 2011).

9.3 The Isolation Method

The successful exploration of particular environments depends on the isolation conditions. Special media and physicochemical factors count for the isolation of representative microorganisms. For example, the recovery of cultivable actinomycetes from sandy low-nutrient soils is highly dependent on soil moisture and zinc concentration (Vreulink et al. 2007). Because the genus *Streptomyces* is highly ubiquitous and has been intensively studied, some screening programs aim to find novel taxa by the selective isolation of rare genera (Hayakawa et al. 2004). The use of antibiotics and chemotactic and other known methods tend to achieve this selectivity (Takahashi and Omura 2003).

Further, the composition of the isolation media can be manipulated to isolate bioactive-producing strains (Basilio et al. 2003). The OSMAC concept (one strain many compounds) refers to the capability of one single strain of producing up to 20 different metabolites belonging to diverse product families when growing conditions such as media composition and aeration are changed. It has been shown that factors such as metal stress could be successfully used to induce antibiotic production when screening actinomycetes (Haferburg and Kothe 2013).

9.4 Genomics-Guided Approach

The increasing genomic information now accessible from public databases allows the design of screening methods based on the presence of specific biosynthetic genes in the genomes of the isolates. These target genes encode a biosynthetic enzyme specific for certain type of metabolites. The method consists of the isolation of DNA from each isolate, and the use of polymerase chain reaction (PCR) with specific primers for the detection of such genes. The primer design is based on the available sequences for that gene. Some of the primers described in the literature (Fig. 9.2) include primers for the screening of type I polyketide synthases (PKS) and nonribosomal peptide synthases (NRPS; Ayuso-Sacido and Genilloud 2005), type II PKS (Metsä-Ketelä et al. 1999), cytochrome P450 hydroxylase (CYP; Hwang et al. 2007), and 3-amino-5-hydroxybenzoic acid (AHBA) synthase (Huitu et al. 2009).

The genomic tools help to find novel compounds, but there is a need of special isolation and cultivation media for the production of the metabolite. Some authors agree that a combination of both genomics and OSMAC approaches is the best choice for bioprospecting (Zotchev 2012; Zazopoulos et al. 2003).

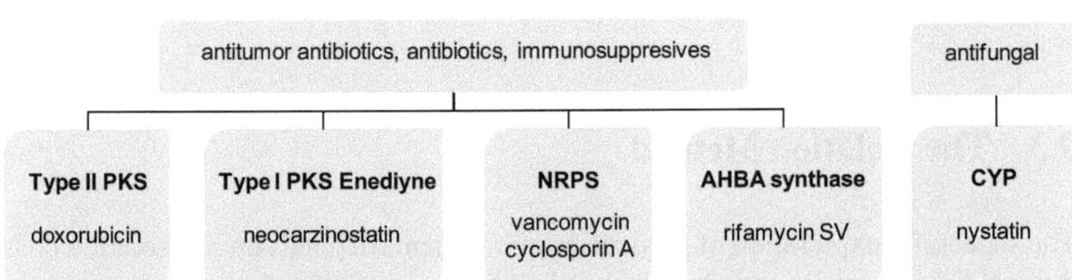

Fig. 9.2 Biosynthetic enzymes targeted for genomic-guided screening

9.5 Combined Approaches: Ansamycins

Ansamycins are antitumor antibiotics with a particular mechanism of action. Perhaps the most famous members of this group are the antibiotics against mycobacteria, the rifamycins. Studies on bioprospecting for ansamycin producers combined bioprospecting in underexplored environments, a genomic approach and the use of selective producing media for the isolation of metabolites. Strains belonging to *Streptomyces* were isolated from the Atacama Desert in Chile, an underexplored ecosystem. Six selective media were used for the isolation. The genomic assay for ansamycins was based on the detection of 3-amino-5-hydroxybenzoic acid (AHBA) synthase of actinomycetes. They found one positive strain among 21 isolates. After growing the positive isolate in special media for the production of the metabolites, ansamycin-type antibiotics were detected (Huitu et al. 2009). They modified ISP-2 media (International Streptomyces Project Medium 2, Yeast Extract-Malt Extract) by replacing glucose with glycerol to elicit the production of two of the isolated ansamycins. A total of four novel ansamycin antibiotics were finally identified (Okoro et al. 2009; Rate et al. 2011).

9.6 Actinomycetes from Patagonia

The information concerning this topic is currently scarce. Actinomycetes from Patagonia were the subject of bioremediation and biocontrol screening programs. To date, there exists only one report on the screening of antibiotic-producing actinomycetes from Patagonia. Endophytic streptomycetes have been isolated and characterized from several species of *Nothofagus* and other plants growing in the southern reaches of Patagonia. The authors report the isolation of six streptomycetes, all of them being active against plant pathogenic fungi (Castillo et al. 2007).

Researchers from the University of Comahue in northern Patagonia have been studying actinomycetes for many years. They contributed most of the information available about distribution of actinomycetes in Patagonia (Vobis et al. 2001). In one study, they describe cultivable actinomycetes associated with lichenized microecosystems (*Pseudocyphellaria berberina*) from the temperate Valdivian rainforests. The isolates belonged to the genera *Actinoplanes*, *Dactylosporangium*, *Pseudonocardia*, *Micromonospora*, *Streptomyces*, and *Streptosporangoium* (Scervino et al. 2014). In another study, a total number of 122 strains of actinomycetes were isolated from both rhizosphere and rhizoplane of the plant *Discaria trinervis* (Solans and Vobis 2003). Regarding bioactivity, the antifungal potential against plant pathogenic fungi of three saprophytic strains from this microbial collection has been evaluated (Solans and Vobis 2013).

Bioprospecting at CENPAT CONICET in Chubut province afforded a total of 60 isolates (Olivera and Vela Gurovic 2013; Vela Gurovic and Olivera 2014). This screening, which is the first report on antibacterial actinomycetes from Patagonia, covered regions from east Patagonia, specifically the departments of Biedma, Rawson, and Gaiman of Chubut Province, Patagonia. Samples were taken from the rhizosphere, soil, and plants. About a third of the isolates displayed bioactivity against bacteria. The isolates were grown in different media to trigger the secondary metabolism. Different supernatants of the same isolate where then tested against gram positives and gram negatives. The supernatants of many isolates were active in special media, but displayed lower or no activity in media commonly used for cultivation of actinomycetes such as ISP-2 (Fig. 9.3).

It was possible to detect a higher rate of active isolates by using different cultivation media. Studies are being conducted to identify the active compounds of the isolate that showed the best bioactivity profile (Fig. 9.4).

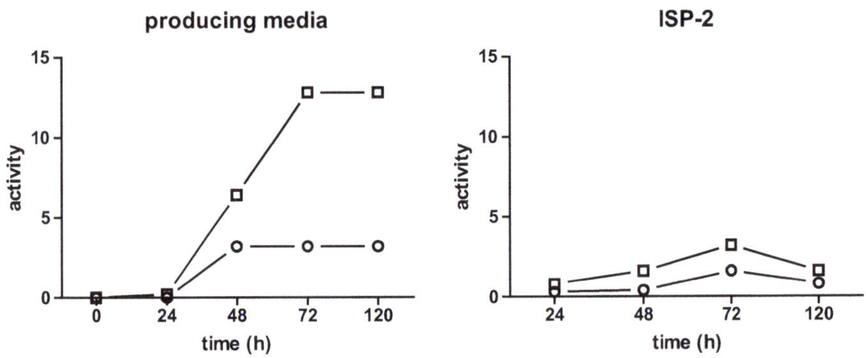

Fig. 9.3 Activity profiles of supernatants of a Patagonian streptomycete over time. *Right*: activity profile of the isolate growing in special media; *left*: the same isolate growing in ISP-2. *Squares* gram positive, *circles* gram negative

Fig. 9.4 *Streptomyces* sp., isolated from Patagonian soil, growing on ISP-2 agar (*left*). Liquid cultures of the same isolate in different media ISP-2 (*left*) and production media for triggering secondary metabolism (*right*)

9.7 Conclusions

Patagonia is an underexplored region with high potential for the discovery of novel microbial strains with novel metabolic potential. As the bioprospecting for actinomycetes turns to specific ecological niches, where bacteria associate with other living systems such as fungi, invertebrates, plants, and sponges, novel bacterial species associated with endemic higher organisms are found at a higher rate. The most successful screening programs include a combination of different approaches to find novel strains and novel metabolites. The use of genomics, selective isolation and growing media, and proper criteria when deciding where to screen would guarantee the discovery of novel bioactives for the development of therapeutic agents.

Acknowledgments This work was supported by ANPCyT (Agencia Nacional de Promoción Científica y Tecnológica, PICT 2012-2361). M. S. Vela Gurovic is a research member of the National Scientific and Technical Research Council of Argentina (CONICET).

References

Ayuso-Sacido A, Genilloud O (2005) New PCR primers for the screening of NRPS and PKS-I systems in Actinomycetes: detection and distribution of these biosynthetic gene sequences in major taxonomic groups. Microbial Ecol 49:10–24

Baltz RH (2007) Antimicrobials from Actinomycetes: back to the future. Microbe 2:125–131

Basilio A, González I, Vicente MF, Gorrochategui J, Cabello A, González A, Genilloud O (2003) Patterns of antimicrobial activities from soil Actinomycetes isolated under different conditions of pH and salinity. J Appl Microbiol 95:814–823

Castillo UF, Browne L, Strobel G, Hess WM, Ezra S, Pacheco G, Ezra D (2007) Biologically active endophytic Streptomycetes from *Nothofagus* spp. and other plants in Patagonia. Microb Ecol 53:12–19

Gupta AP, Pandotra P, Sharma R, Kushwaha M, Gupta S (2013) Marine resource: a promising future for anticancer drugs (Chapter 8). In: Atta-ur-Rahman (ed) Studies in natural products chemistry, vol 40. Elsevier, Amsterdam

Haferburg G, Kothe E (2013) Activation of silent genes in Actinobacteria by exploiting metal stress. In: Amoroso MJ, Benimeli CS, Cuozzo SA (eds) Actinobacteria. Application in bioremediation and production of industrial enzymes. CRC Press, Boca Raton, pp 56–73

Hayakawa M, Yoshida Y, Iimura Y (2004) Selective isolation of bioactive soil Actinomycetes belonging to the *Streptomyces violaceusniger* phenotypic cluster. J Appl Microbiol 96:973–981

Huitu Z, Linzhuan W, Aiming L, Guizhi S, Feng H, Qiuping L, Yuzhen W, Huanzhang X, Qunjie G, Yiguang W (2009) PCR screening of 3-amino-5-hydroxybenzoic acid synthase gene leads to identification of ansamycins and AHBA-related antibiotic producers in Actinomycetes. J Appl Microbiol 106:755–763

Hwang Y-B, Lee M-Y, Park H-J, Han K, Kim E-S (2007) Isolation of putative polyene-producing Actinomycetes strains via PCR-based genome screening for polyene-specific hydroxylase genes. Process Biochem 42:102–107

Metsä-Ketelä M, Salo V, Halo L, Hautala A, Hakala J, Mäntsälä P, Ylihonko K (1999) An efficient approach for screening minimal PKS genes from *Streptomyces*. FEMS Microbiol Lett 180:1–6

Okoro CK, Brown R, Jones AL, Andrews BA, Asenjo JA, Goodfellow M, Bull AT (2009) Diversity of culturable Actinomycetes in hyper-arid soils of the Atacama Desert, Chile. Antonie van Leeuwenhoek J Microb 95:121–133

Olivera NL, Vela Gurovic MS (2013) Bioprospección de actinomicetes nativos de la región patagónica con actividad antimicrobiana contra patógenos de peces. Rev Argent Microbiol 45(S1):173

Peeters K, Ertz D, Willem A (2011) Culturable bacterial diversity at the Princess Elisabeth Station (Utsteinen, Sør Rondane Mountains, East Antarctica) harbours many new taxa. Syst Appl Microbiol 34:360–367

Rate ME, Houssen WE, Arnold M, Abdelrahman MH, Deng H, Harrison WTA, Okoro CK, Asenjo JA, Andrews Barbara A, Ferguson G, Bull AT, Goodfellow M, Ebel R, Jaspars M (2011) Chaxamycins AD, bioactive Ansamycins from a hyper-arid desert *Streptomyces* sp. J Nat Prod 74:1491–1499

Scervino JM, Messuti MI, Solans M, Vobis G (2014) Actinomicetes cultivables asociados a micro-ecosistemas liquénicos de la selva templada valdiviana, Argentina. Bol Soc Argent Bot 49:441–445

Solans M, Vobis G (2003) Actinomicetes saprofíticos asociados a la rizósfera y rizoplano de *Discaria trinervis*. Ecol Austr 13:97–107

Takahashi Y, Omura S (2003) Isolation of new actinomycete strains for the screening of new bioactive products. J Gen Appl Microbiol 49:141–154

Vela Gurovic MS, Olivera NL (2014) Characterization of the bioactive diffusible pigment of an actinomycete isolated from Patagonian soil. In: Conference proceedings of the 3rd international meeting on Pharmaceutical Sciences, 18–19 Sept, Córdoba, Argentina. International Journal of Pharmaceutical Sciences and Research. http://ijpsr.com/wp-content/uploads/2015/05/Proceedings-RICIFa2014-Cordoba-ARGENTINA.pdf

Solans M, Vobis G (2013) Biology of actinomycetes in the rhizosphere of nitrogen-fixing plants. In: Amoroso MJ, Benimeli CS, Cuozzo SA (eds) Actinobacteria. Application in bioremediation and production of industrial enzymes. CRC Press, Boca Raton, pp 1–25

Vobis G, Solans M, Chaia E (2001) Géneros raros o pocos conocidos de Actinomicetes en Argentina. In: Abstracts of the XXVIII Jornadas Argentinas de Botánica, Santa Rosa, La Pampa, 21–25 Oct 2001

Vreulink J-M, Esterhuyse A, Jacobs K, Botha A (2007) Soil properties that impact yeast and actinomycete numbers in sandy low nutrient soils. Can J Microbiol 53:1369–1374

Wawrik B, Kutliev D, Abdivasievna UA, Kukor JJ, Zylstra GJ, Kerkhof L (2007) Biogeography of Actinomycetes communities and type II polyketide synthase genes in soils collected in New Jersey and Central Asia. Appl Environ Microbiol 73:2982–2989

Zazopoulos E, Huang K, Staffa A, Liu W, Bachmann BO, Nonaka K, Ahlert J, Thorson JS, Shen B, Farnet CM (2003) A genomics-guided approach for discovering and expressing cryptic metabolic pathways. Nat Biotechnol 21:187–190

Zheng Z, Zeng W, Huang Y, Yang Z, Li J, Cai H, Su W (2000) Detection of antitumor and antimicrobial activities in marine organism associated Actinomycetes isolated from the Taiwan Strait, China. FEMS Microbiol Lett 188:87–91

Zotchev SB (2012) Marine Actinomycetes as an emerging resource for the drug development pipelines. J Biotechnol 158:168–175

Chapter 10
Microorganisms from Patagonia and Antarctica and Their Cold-Active Skills for Using Polymeric Materials

Fiorella F. Gomez, Jaime D. Babot, Gustavo A. Lovrich, Faustino Siñeriz, and Osvaldo Delgado

Abstract Both aquatic and soil ecosystems are environments rich in polymeric material from the different structures of the organisms belonging to the three domains of life that exist on our planet: Archaea, Bacteria, and Eukarya. Moreover, microbial communities play important roles in food webs, contributing to the regeneration of nutrients within these systems. The ecosystems of Patagonia and Antarctica could be taken into account as important reservoirs of microorganisms with potential biotechnological interest because of the cold-active hydrolytic enzyme activities produced there and also because of the diversity of enzyme-producing microbes. Cold-active amylase, pectinase, cellulase, carboxymethylcellulase, xylanase, galactosidase, glucosidase, chitinase, α-rhamnosidase, and protease activities have been detected so far in bacterial isolates from the sub-Antarctic region. The same activities, plus lipase, urease, and esterase activities, have been reported for fungal isolates obtained from Patagonia and Antarctica.

F.F. Gomez
Planta Piloto de Procesos Industriales Microbiológicos PROIMI-CONICET,
Av. Belgrano y Pje. Caseros, 4000 San Miguel de Tucumán, Tucumán, Argentina

Facultad de Ciencias Exactas y Naturales, Universidad Nacional de Catamarca
FACEN-UNCa, Av. Belgrano 300, 4700 San Fernando del Valle de Catamarca,
Catamarca, Argentina

J.D. Babot • F. Siñeriz
Planta Piloto de Procesos Industriales Microbiologicos PROIMI-CONICET,
Av. Belgrano y Pje. Caseros, 4000 San Miguel de Tucumán, Tucumán, Argentina

G.A. Lovrich
Centro Austral de Investigaciones Científicas CADIC-CONICET,
Bernardo Houssay 200, 9410 Ushuaia, Tierra del Fuego, Argentina

O. Delgado (✉)
Facultad de Ciencias Exactas y Naturales, Universidad Nacional de Catamarca FACEN-UNCa,
Av. Belgrano 300, 4700 San Fernando del Valle de Catamarca, Catamarca, Argentina

Centro de Investigaciones y Transferencia de Catamarca CITCA-CONICET,
Prado 366, 4700 San Fernando del Valle de Catamarca, Catamarca, Argentina
e-mail: odelgado@proimi.org.ar

N.L. Olivera et al. (eds.), *Biology and Biotechnology of Patagonian Microorganisms*, DOI 10.1007/978-3-319-42801-7_10

10.1 Microbial Diversity

Although microorganisms are abundant in nature and ubiquitously distributed, only a small percentage of them (0.1–0.5 %) has been studied by means of culture-dependent techniques (Klappenbach et al. 2001). This property has seriously hindered the analysis of their biodiversity, structure, and roles in microbial communities, among other parameters. Studies focused on these parameters are now possible following the design of molecular techniques. Among these approaches, it is worthy to mention denaturing gradient gel electrophoresis (DGGE) (Ercolini 2004), real-time polymerase chain reaction (RT-PCR) (Vitali et al. 2010), and also pyrosequencing and next-generation sequencing (NGS) (Metzker 2010). The microbial communities in several environments have been assessed through these techniques.

10.1.1 Sub-Antarctic Region Bacterial Diversity

Cutting-edge marine ecosystems ecology studies have been performed by 16S rDNA sequence analysis, and thus bacterial populations are now known. Among pelagic bacteria, most of them belong to *α-Proteobacteria*, an abundant and widely distributed group in marine environments, and aggregating bacteria would be mainly represented by members of the subclass *γ-Proteobacteria* and group CFB (*Cytophaga-Flavobacterium-Bacteroides*). On the other hand, cultivable microorganisms have been classified as strict aerobes or facultative anaerobes and seem to have importance in the degradation of macromolecules, such as chitin, DNA, or cellulose (Reichenbach 1992).

Patagonia is located in the southernmost part of South America. Currently, research is taking place in different parts of this region. Several culture-dependent and culture-independent studies have been performed to isolate microorganisms and analyze their phylogeny and production of cold-active enzymes of biotechnological potential.

As the sole easily accessible sub-Antarctic coastal region, the Beagle Channel (55°S, 66, 71°W; Fig. 10.1) is particularly relevant as a source of novel extremophile microbes, but also for any research focused in the sub-Antarctic region. Average temperature ranges between 4 °C and 10 °C in this area, allowing the growth of psychrophilic and psychrotolerant microorganisms. Along the channel, samples have been taken from various environments, including water, sediment, invertebrates (crustaceans, bivalves, ascidians, isopods, salps, amphipods, polychaetes, starfishes), sea vertebrates (different fishes), and seaweed (Fernandes et al. 2002; Belchior and Vacca 2006; Olivera et al. 2007; Orrillo et al. 2007; Prabagaran et al. 2007; Cristóbal et al. 2009; Pucci et al. 2009; Tropeano et al. 2012). Multiple isolates and clones were obtained and characterized as members of the classes *α-Proteobacteria*, *γ-Proteobacteria*, *ε-Proteobacteria*, and group CFB. Other noncultivable microorganisms were detected but could not be identified by means

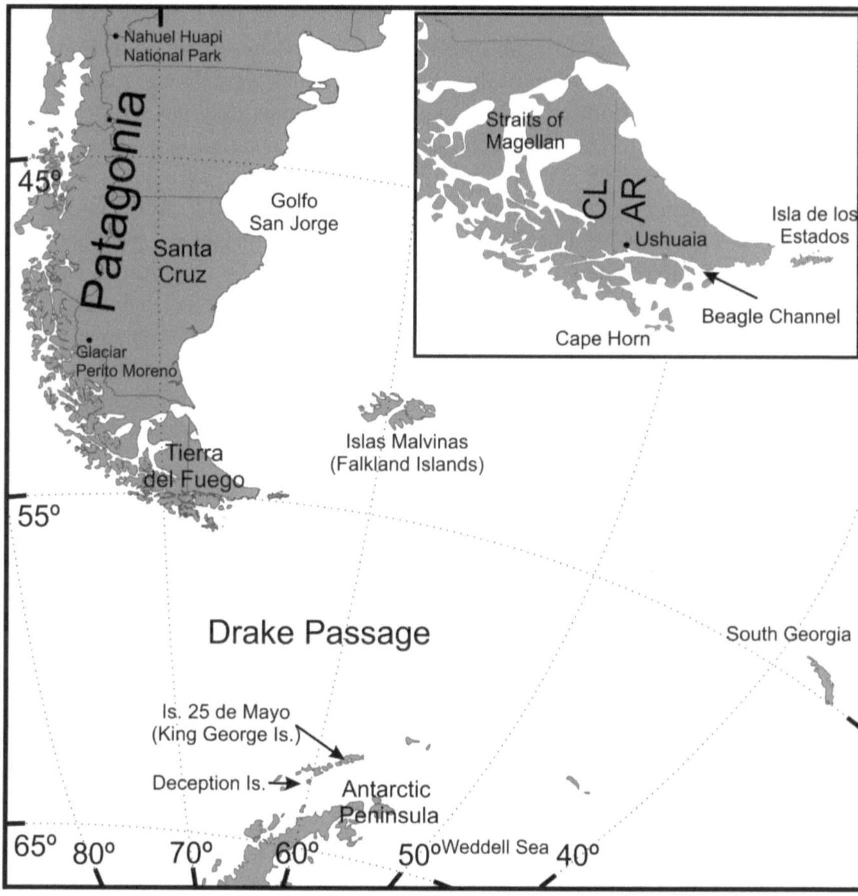

Fig. 10.1 Patagonia, nearer Sub-Antarctic region, and Antarctic Peninsula

of 16S rDNA sequence analysis (Prabagaran et al. 2007). Bacteria belonging to the CFB group are abundant in coastal areas, but little is known about their populations in the open sea. As far as it is known, they do not seem to be involved in cellulose degradation (Reichenbach 2006). The following genera belonging to the aforementioned classes were identified: *Marinomonas*, *Shewanella*, *Pseudoalteromonas*, *Roseobacter*, *Sulfitobacter*, *Glaciecola*, and *Psychrobacter* (Prabagaran et al. 2005, 2007; Orrillo et al. 2007; Cristóbal et al. 2009, 2011, 2014). Recently, the new species *Marinomonas ushuaiensis* was isolated, identified, and characterized by Prabagaran et al. (2005) from seawater sampled from the coast of Ushuaia. This species can use glucose, mannose, melibiose, and other carbohydrates as carbon sources, which in turn result from the decomposition of organic material.

Concerning other sampling areas near Beagle Channel, the bacterial populations would be similar to those of the channel. However, members of the genera *Colwellia* and *Planococcus*, and even one isolate of the family *Flavobacteriaceae*, were detected from sediment samples from the nearby Isla de los Estados (54°47′ S, 64°15′ W; Fig. 10.1) (Sánchez et al. 2009). *Colwellia* sp. showed an optimal growth temperature around 15 °C, whereas the other two isolates were able to grow best at 20 °C–25 °C; thus, the former was characterized as moderately psychrophilic and the latter were characterized as psychrotolerant. Other putative new species were detected because their 16S rDNA sequences showed a similarity index lower than 97 % within sequences available in databases (Olivera et al. 2007). At the Golfo San Jorge (45°–47°S, 65°40′W; Fig. 10.1), 53 lactic acid bacteria isolates were obtained from the intestinal contents of *Odontesthes platensis*, and their inhibitory activity was tested against a set of both gram-positive and gram-negative fish pathogens (Sequeiros et al. 2010). *Lactococcus lactis* TW34, an isolate able to grow at 10 °C, showed inhibitory activity against *Lactococcus garvieae* 03/8460 and other gram-positive bacteria, but not against gram-negative pathogens. The antimicrobial compound produced by this strain was reported to be a highly thermostable bacteriocin active between pH 3.0 and 11.0.

In a study performed in Potter Cove (Isla 25 de Mayo/King George Island, Antarctica) (62°14′S, 58°40′W; Fig. 10.1), 189 aerobic heterotrophic bacterial isolates were obtained throughout samples of seawater, marine sediment, algae, and different marine animals, and they were identified as members of the phyla Proteobacteria, Bacteroidetes, Actinobacteria, and Firmicutes, *γ-Proteobacteria* being the predominant class (Tropeano et al. 2012, 2013).

10.1.2 Fungal Diversity in Patagonia and Antarctica

Knowledge is limited about fungal diversity from Patagonia and Antarctica. In a worldwide study, Tedersoo et al. (2014) concluded that the overall richness of soil fungi increases toward the Equator. However, the diversity of saprotrophic fungi, parasites, and pathogens increased at high latitudes, such as those of Patagonia. Most studies on fungal diversity were performed in forests dominated by trees of the genus *Nothofagus* in northwestern Argentine Patagonia; however, the diversity of *Epichloë*, a plant endophytic fungus in Southern Patagonia (Santa Cruz and Tierra del Fuego provinces), was studied by Mc Cargo et al. (2014). Three species were found: *Epichloë tembladerae*, *Epichloë typhina* var. *aonikenkana*, and the new species *Epichloë cabralii*. Based on morphotyping and rDNA sequence analysis, the ectomycorrhizal fungi from *Nothofagus* forests in northwestern Patagonia were assessed by Nouhra et al. (2013). They found that this community was dominated by Basidiomycota, mainly from the genera *Cortinarius* and *Inocybe*, but there was higher diversity among Agaricales. Several *Tubeufiaceae* species were detected on the surface of *Nothofagus* trees in northwestern Patagonia by Sánchez and Bianchinotti (2010), including two species so far described only in the Northern Hemisphere: *Acanthostigma minutum* and *Tubeufia cerea*. A new species belonging to the genus *Acanthostigma*, *A. patagonica*, was described by Sanchez et al. (2012). The specimens were isolated from the bark of *Nothofagus* trees in northwestern

Patagonia. In another study performed in a similar forest, four new species of the genus *Cystangium* were described (Trierveiler-Pereira et al. 2015).

The prevailing harsh environmental conditions in Antarctica constitute significant limiting factors for plant and animal life. Therefore, the biology of Antarctica is dominated by microorganisms (Friedmann and Thistle 1993). The endophytic fungi associated with leaves of *Deschampsia antarctica* in Admiralty Bay at Isla 25 de Mayo/King George Island (Antarctica) (62°09′ S, 58°28′ W; Fig. 10.1) were characterized by Rosa et al. (2009). *Alternaria* and *Phaeosphaeria* were reported as the most frequent genera associated to plants, and other fungal isolates were identified as *Entrophospora* sp. and several undescribed *Ascomycete* species. Fungal abundance and diversity were studied in 18 distinct ice-free locations in Antarctica by Arenz and Blanchette (2011). Ascomycetes was found to be the predominant phylum, while Antarctic Peninsula and the Ross Sea Region had the highest Zygomycetes and Basidiomycetes abundances, respectively. The main genera found were *Geomyces*, *Cadophora*, *Rhodotorula*, and *Cryptococcus*. The same genera were reported as the most abundant in wood from historic expedition huts left by early explorers to the South Pole (Blanchette et al. 2004; Arenz et al. 2006).

Alike filamentous fungi diversity studies, most yeast diversity studies were also focused in Northwestern Patagonia. In a recent study by Fernández et al. (2012), morphological and molecular methods were used to assess the fungal community on the surface of seeds and noncommercial fruits of *Nothofagus nervosa* trees. A total of 171 isolates corresponding to 17 species were recovered, most of which belong to the phylum Ascomycota, but also several isolates from genera *Cryptococcus* and *Rhodotorula* were reported. In another study, Brandão et al. (2011) evaluated the diversity of yeasts in the Nahuel Huapi Lake, Nahuel Huapi National Park (41°05′ S, 71°20′ W; Fig. 10.1): a total of 149 isolates distributed in 14 genera and 34 species were obtained. The most frequently isolated species were *Rhodotorula mucilaginosa* and *Cryptococcus victoriae*. Soil yeasts from two forests dominated by *Nothofagus* trees were evaluated by Mestre et al. (2014). Each forest site showed a particular arrangement of species. *Cryptococcus podzolicus* was most frequently isolated in nutrient-rich soils, *Trichosporon porosum* dominated cold mountain forests with low nutrient and water availability in soil, and capsulated yeasts such as *Cryptococcus phenolicus* dominated forest sites with low precipitation. Recently, different strains of the psychrotolerant species *Saccharomyces eubayanus* and *S. uvarum* were isolated from the seeds and bark of *Araucaria araucana* trees (Rodríguez et al. 2014). In another study performed in the same region, the diversity of yeasts inhabiting the bulk soil, rhizosphere, and ectomycorrhizosphere of a *Nothofagus pumilio* forest was assessed. A total of 126 yeast isolates were obtained; basidiomycetous yeasts were predominant in all soil fractions, and the most frequently isolated species was *Cryptococcus podzolicus*. Many of the recovered yeast species were associated with lignocellulose compound degradation, which suggests that yeasts are important as decomposer in those forest soils (Mestre et al. 2011b). Two strains of the new yeast species *Lindnera rhizosphaerae* were isolated from rhizospheric soil samples of the same forest (Mestre et al. 2011a). In a study performed in southwestern Patagonia, 153 yeast isolates were obtained from melted glacial ice by de Garcia et al. (2012a). Ninety percent of the isolates were basidiomycetous; 16 genera and 29 species were identified, mainly *Dioszegia crocea* and *Cryptococcus victoriae*. Twenty-five percent of total isolates corresponded to psychrophilic yeasts, whereas 75 % were psychrotolerant yeasts. In a study of yeast diversity in soil and water from Isla 25 de Mayo/King George Island, Antarctica (Fig. 10.1), *Mrakia* and *Cryptococcus* genera contained the highest species diversity, and *Sporidiobolus salmonicolor* was the most ubiquitous species. Most of the yeasts were psychrotolerant (Carrasco et al. 2012). In another study performed in Antarctica, 97 yeast isolates belonging to 21 different species, 8 Ascomycota and 13 Basidiomycota, were recovered from marine and terrestrial samples. Encapsulated yeasts belonging to genera *Rhodotorula* and *Cryptococcus* were recovered from seven different samples. Moreover, *Candida glaebosa*, *Cryptococcus victoriae*, *Meyerozyma guilliermondii*, *Rhodotorula mucilaginosa*, and *R. laryngis* were the most abundant yeast species recovered (Duarte et al. 2013). In a similar study, 35 yeasts were isolated from soil, rocks, wood, and bones collected in the Isla 25 de Mayo/King George Island, Antarctica, and classified in the genera *Leucosporidiella*, *Rhodotorula*, *Cryptococcus*, *Bullera*, and *Candida*. *Cryptococcus victoriae* was by far the most ubiquitous species (Rovati et al. 2013).

10.2 Cold-Active Hydrolytic Enzymatic Activities

Microorganisms have an important role in marine trophic networks. They degrade dissolved organic matter, both that generated in the sea and that incorporated from other sources such as rivers, city wastes, and contaminant compounds such as hydrocarbons. (Pucci et al. 2009). By doing so, microbes regenerate chemical elements such as carbon and nitrogen and incorporate them into their respective biogeochemical cycles.

The sub-Antarctic region is dominated by temperatures ranging between 4 °C and 10 °C, depending on the season. As result of those low temperatures, enzyme-catalyzed chemical reactions are either diminished or stopped (D'Amico et al. 2002). However, psychrophilic and psychrotolerant microorganisms have successfully colonized this area (Fernandes et al. 2002), possible by a series of both physiological and structural adaptations, including the synthesis of cold-adapted enzymes with optimal performance at low temperature. Microorganisms are classified as psychrophilic or psychrotolerant according not only to their optimal growth temperature but also to the range of temperatures at which they are able to grow. Psychrophilic microorganisms are those with a minimum growth temperature of 0 °C or lower, an optimum growth temperature of 15 °C or lower, and a maximum growth temperature of 20 °C (Morita 1975).

Cold-adapted enzymes show promising characteristics that make them appealing for their use as biotechnological tools. The use of these enzymes in industrial processes would not only highly reduce costs related to heating, but even more would decrease the loss of volatile compounds and the risk of contamination in industrial processes (Tropeano et al. 2012). Such enzymes with proteolytic, amylolytic, lipolytic, and cellulolytic activities are already being incorporated into detergents; their properties allow also their use in the food industry for the elaboration of heat-sensitive products (Demain and Adrio 2008; Tropeano et al. 2012). Currently, cold-adapted enzymes are employed in the textile industry, in the manufacture of paper, and also in bioremediation, among other uses (Kirk et al. 2002).

10.2.1 Cold-Active Enzymatic Activities from Marine Bacteria in the Sub-Antarctic Region

So far, several cold-adapted enzymes with potential biotechnological applications have been detected around the sub-Antarctic region, for example, galactosidases, glucosidases, cellulases, and chitinases (Fernandes et al. 2002; Olivera et al. 2007; Cristóbal et al. 2009; Tropeano et al. 2013). The expression of most of these enzymes is triggered by their substrates, which are used as a carbon source by marine microbes and also by microorganisms inhabiting the gastrointestinal tract of marine animals.

Fernandes et al. (2002) characterized a β-galactosidase produced by the bacterial Antarctic isolate TAE 79b, phylogenetically related to the genus *Pseudoalteromonas*. This enzyme displayed an optimal activity at pH 9.0 and 26 °C; however, 28 % of its maximal activity was retained at 5 °C. The industrial interest of this enzyme lies in its use for lactose fermentation, because lactose intolerance affects two thirds of the world's population (Hoyoux et al. 2001).

On the other hand, benthic invertebrates such as *Munida gregaria* (morphotype "subrugosa") were collected from Beagle Channel and their intestinal contents were screened for hydrolytic enzyme-producing bacteria (Cristóbal et al. 2009). Isolate *Shewanella* sp. G5 showed remarkable potential by its psychrotolerant character (growth temperature ranging between 4 °C and 37 °C, with an optimal growth temperature of 20 °C) and its production of three intracellular β-glucosidase isoenzymes. Production of these isoenzymes was observed on glucose and cellobiose, with a maximal activity at pH 8.0 for both substrates.

β-Glucosidases, α-rhamnosidases, and α-arabinofuranosidases can be employed for acidity reduction of fruit juices and improvement of wine flavor by the release of volatile terpenes. In this context, psychrotolerant bacteria inhabiting both seawater and the gastrointestinal tract of marine invertebrates were isolated from the coastal area near Ushuaia, Argentina (54°80′ S, 68°31′ W; Fig. 10.1) and screened for α-rhamnosidase production by Orrillo et al. (2007). Of 140 isolates, all of them isolated from seawater, only 10 evidenced α-rhamnosidase activity. The highest α-rhamnosidase activity at 4 °C was detected for the isolate 005NJ, belonging to the *Pseudoalteromonas* genus. This thermosensitive activity (half-life of 4 min at 50 °C) could be utilized for food processing at low temperatures (Orrillo et al. 2007).

Other enzymatic activities also reported from this sub-Antarctic region included proteases (Olivera et al. 2007; Tropeano et al. 2013), amylases, pectinases, cellulases, carboxymethyl-cellulases, and xylanases (Tropeano et al. 2012, 2013). Proteases were detected from bacterial isolates belonging to the genera *Colwellia* and *Shewanella* obtained from Isla de los Estados sediments. *Colwellia* sp. showed thermosensitive proteolytic activity whereas that of *Shewanella* sp. was reported as thermostable. These results suggest important variations in the thermal properties of the proteases that can coexist in such environments (Olivera et al. 2007). Amylolytic, pectinolytic, cellulolytic, carboxymethyl-cellulolytic, and xylanolytic activities were detected in a set of bacteria isolated from Potter Cove (Isla 25 de Mayo/King George Island, Antarctica; Fig. 10.1). Each isolated bacterial species was able to produce one or more of these enzymatic activities. Interestingly, there seemed to be a certain association between the expression of enzymes when the production of more than one hydrolytic activity by the same strain occurred: this was observed mainly for the associated production of pectinases and carboxymethyl-cellulases. Most enzymatic activities were produced by isolates of the genera *Pseudoalteromonas* and *Pseudomonas* (Tropeano et al. 2013). These results suggest that Potter Cove is a promising source of biotechnological relevant biomolecules (Tropeano et al. 2012).

10.2.2 Cold-Active Hydrolytic Enzymes from Fungi in Patagonia and Antarctica

Few studies have focused on the study of fungal hydrolytic enzymes of microorganisms isolated from Patagonia and Antarctica. Elíades et al. (2015) studied four strains of *Humicolopsis cephalosporioides*, a soil fungus associated with *Nothofagus pumilio* forests. These strains were isolated from a *N. pumilio* forest that was harvested by shelter-wood cutting practiced 50 years ago in Tierra del Fuego, Argentina (Fig. 10.1). They found intense proteolytic, cellulolytic, and pectinolytic activities in all strains at temperatures below 25 °C. Pectinolytic and proteolytic activities were highest at 5 °C and 15 °C for all strains, respectively. Both cellulolytic and amylolytic activities were highest at 5 °C for two isolates, although the highest lipolytic and chitinolytic activities were observed at this temperature for one strain. The production of hydrolytic enzymes by *Tolypocladium cylindrosporum*, an entomopathogenic fungus proposed as a biological control agent against insects, was studied for a strain isolated from soil samples collected in a forest in Ushuaia, Argentina by Scorsetti et al. (2012). Proteolytic and urease activities were higher at 4 °C than at 12 °C and 24 °C. There were no differences between chitinolytic activity at 12° and 24 °C; however, activity was not detected at 4 °C. Similarly, lipolytic activity was higher at 12° and 24 °C than at 4 °C.

On the other hand, the study of hydrolytic enzymes produced by yeasts isolated from Patagonia and Antarctica has been more extensive, especially in northwestern Patagonia. The extracellular enzymatic activities (amylolytic, proteolytic, lipolytic, esterasic, pectinolytic, chitinolytic, and cellulolytic activities) of 91 basidiomycetaous yeasts isolated from glacial and subglacial waters of northwest Patagonia (41°10′ S, 71°50′ W, Argentina; Fig. 10.1) was investigated by Brizzio et al. (2007).

Three or more different enzymatic activities were reported at 4 °C for more than 15 % of strains, and more than 63 % of the strains showed two enzymatic activities at similar temperatures. On the other hand, almost 10 % of strains exhibited three or more enzymatic activities and 32 % of them showed two activities. The number of strains exhibiting amylolytic, proteolytic, and lipolytic activities was higher at 4 °C; more strains with pectinolytic activity were detected at 20 °C. Remarkably, no cellulolytic or chitinolytic activities were detected at 4 °C and 20 °C. In another study, 148 yeast isolates were obtained from subsurface water from the Nahuel Huapi Lake in northwestern Argentinean Patagonia (Fig. 10.1) by Brandão et al. (2011). Of these, 82 % showed at least one extracellular enzymatic activity at 5° or 20 °C. Esterase activity was the most widely expressed extracellular enzyme activity (positive for 71.8 % of the isolates), followed by cellulotytic (53.0 %), pectinolytic (42.9 %), amylolytic (26.8 %), and protease (22.1 %) activities. The number of strains showing amylolytic activity at 5 °C was higher than at 20 °C; however, the remnant enzymatic activities were more distributed among yeasts growing at 20 °C. Libkind et al. (2008) studied the occurrence of amylolytic, proteolytic, esterase, cellulolytic, pectinolytic, and pectate-lyase activities in two strains of *Xanthophyllomyces dendrorhous* (*Phaffia rhodozyma*) previously isolated from stromata of the ascomycetous fungus *Cyttaria hariotii* and from water samples of Lake Ilón in northwestern Patagonia, Nahuel Huapi National Park (71°56′ S, 41°11′ W; Fig. 10.1). No enzymatic activities were detected at 20 °C, and amylase was the only hydrolytic activity observed at 5 °C, suggesting that the synthesis of such enzymes in *X. dendrorhous* is favored by low temperatures. This enzyme has potential application in the production of maltose syrup and also in the beer industry. Concerning Southern Patagonia, several *Cryptococcus* strains were isolated from melted water, ice, and seawater from glaciers in Mount Tronador, Nahuel Huapi National Park (41°11′ S, 71°50′ W), Perito Moreno glacier (49°15′ S, 73°51′ W), and Horn Cape (57°25′ S, 66°34′ W) (Fig. 10.1), respectively (de Garcia et al. 2012b). The strains were tested for their ability to degrade starch, protein, pectin, cellulose, and fatty acids at 5 °C and 20 °C. *Cryptococcus victoriae* and related strains showed cellulolytic and esterase activities at both tested temperatures. Although amylolytic and proteolytic activities were not detected at any strain, two new species described in the study, *Cryptococcus tronadorensis* and *C. fonsecae*, showed the six enzymatic activities tested at both assayed temperatures. In Antarctica, Vaz et al. (2011) isolated 89 yeast strains from samples of the rhizosphere of *Deschampsia antarctica*, ornithogenic soil, soil, marine and lake sediments, marine water, and freshwater from lakes of Admiralty Bay (Isla 25 de Mayo/ King George Island, Shetland Islands; 62°09′ S, 58°28′ W) and Port Foster, Deception Island (62°55.5′ S, 60°37′ W; Fig. 10.1), during the austral summer season between November 2006 and January 2007. All strains were tested for their ability to degrade starch, protein, lipids, pectin, and cellulose at 4 °C and 20 °C. Cellulolytic and esterase activities were the most frequent, being present in 76 % of the isolates. Higher levels of esterase activity were detected at 4 °C than 20 °C. On the other hand, no significant differences between temperatures were observed for the other tested enzymatic activities. Nevertheless, most isolated yeasts showed enzymatic activity at 4 °C, providing evidence of their metabolic adaptation to cold environments. In another study, eight extracellular hydrolytic activities were tested on 78 yeast isolates obtained from both water and soil samples from the same area by Carrasco et al. (2012). All species displayed at least one of the eight extracellular enzymatic activities tested. Lipase, amylase, and esterase activities dominated, and chitinase and xylanase were less common. Two isolates identified as *Leuconeurospora* sp. and *Dioszegia fristingensis* displayed six enzyme activities. Finally, lipase, xylanase, and protease activities were tested in 97 isolates obtained from water and soil samples from continental Antarctica by Duarte et al. (2013). They found that 46.4 %, 37.1 %, and 14.4 % of the evaluated isolates were able to produce lipases (at 15 °C), xylanase (at 15 °C), and protease (at 25 °C), respectively. Interestingly, most lipolytic, proteolytic, and xylanolytic strains were distributed within the phylum Basidiomycota.

10.3 Conclusions

Throughout Patagonia, various microorganisms colonizing different environments were detected, all of them able to hydrolyze polymeric material, belonging to a wide variety of phyla. Among bacteria, members of *Proteobacteria*, mainly α-*Proteobacteria* and γ-*Proteobacteria*, abounded. On the other hand, Basidiomycota and Ascomycota were the predominant fungal phyla from Patagonia and Antarctica, respectively. Among yeasts, *Cryptococcus* and *Rhodotorula* seem to be the most widely distributed genera in the sub-Antarctic region. Because of the extreme conditions found in this region, these microorganisms naturally produce cold-active hydrolytic enzymes with potential applications in multiple industrial processes.

Given all the foregoing, the sub-Antarctic marine ecosystem could be considered as an important reservoir of microorganisms with potential biotechnological interest, as the result of their hydrolytic enzyme activities and also because of the diversity and fluctuation in microbial population density and the presence of different enzyme-producing microorganisms.

References

Arenz BE, Blanchette RA (2011) Distribution and abundance of soil fungi in Antarctica at sites on the Peninsula, Ross Sea Region and McMurdo Dry Valleys. Soil Biol Biochem 43:308–315

Arenz BE, Held BW, Jurgens JA, Farrell RL, Blanchette RA (2006) Fungal diversity in soils and historic wood from the Ross Sea Region of Antarctica. Soil Biol Biochem 38:3057–3064

Belchior SGE, Vacca G (2006) Fish protein hydrolysis by a psychrotrophic marine bacterium isolated from the gut of hake (*Merluccius hubbsi*). Can J Microbiol 52:1266–1271

Blanchette RA, Held BW, Jurgens JA, McNew DL, Harrington TC, Duncan SM, Farrell RL (2004) Wood-destroying soft rot fungi in the historic expedition huts of Antarctica. Appl Environ Microbiol 70:1328–1335

Brandão LR, Libkind D, Vaz AB, Santo LCE, Moliné M, de García V, van Broock M, Rosa CA (2011) Yeasts from an oligotrophic lake in Patagonia (Argentina): diversity, distribution and synthesis of photoprotective compounds and extracellular enzymes. FEMS Microbiol Ecol 76:1–13

Brizzio S, Turchetti B, De Garcia V, Libkind D, Buzzini P, Van Broock M (2007) Extracellular enzymatic activities of basidiomycetous yeasts isolated from glacial and subglacial waters of northwest Patagonia (Argentina). Can J Microbiol 53:519–525

Carrasco M, Rozas JM, Barahona S, Alcaíno J, Cifuentes V, Baeza M (2012) Diversity and extracellular enzymatic activities of yeasts isolated from King George Island, the sub-Antarctic region. BMC Microbiol 12:251–259

Cristóbal HA, Benito J, Lovrich GA, Abate CM (2014) Phylogenentic and enzymatic characterization of psychrophilic and psychrotolerant marine bacteria belong to γ-Proteobacteria group isolated from the sub-Antarctic Beagle Channel, Argentina. Folia Microbiol 60:183–198

Cristóbal HA, López MA, Kothe E, Abate CM (2011) Diversity of protease-producing marine bacteria from sub-Antarctic environments. J Basic Microbiol 51:590–600

Cristóbal HA, Schmidt A, Kothe E, Breccia J, Abate CM (2009) Characterization of inducible cold-active β-glucosidases from the psychrotolerant bacterium *Shewanella* sp. G5 isolated from a sub-Antarctic ecosystem. Enzyme Microb Technol 45:498–506

D'Amico S, Claverie P, Collins T, Georlette D, Gratia E, Hoyoux A, Meuwis M-A, Feller G, Gerday C (2002) Molecular basis of cold adaptation. Philos Trans R Soc Lond B Biol Sci 357:917–925

de Garcia V, Brizzio S, van Broock MR (2012a) Yeasts from glacial ice of Patagonian Andes, Argentina. FEMS Microbiol Ecol 82:540–550

de Garcia V, Zalar P, Brizzio S, Gunde-Cimerman N, van Broock M (2012b) *Cryptococcus species* (Tremellales) from glacial biomes in the southern (Patagonia) and northern (Svalbard) hemispheres. FEMS Microbiol Ecol 82:523–539

Demain AL, Adrio JL (2008) Contributions of microorganisms to industrial biology. Mol Biotechnol 38:41–55

Duarte A, Dayo-Owoyemi I, Nobre F, Pagnocca F, Chaud L, Pessoa A, Felipe M, Sette L (2013) Taxonomic assessment and enzymes production by yeasts isolated from marine and terrestrial Antarctic samples. Extremophiles 17:1023–1035

Elíades LA, Cabello MN, Pancotto V, Moretto A, Rago MM, Saparrat MC (2015) Preliminary data on growth and enzymatic abilities of soil fungus *Humicolopsis cephalosporioides* at different incubation temperatures. Rev Iberoam Micol 32:40–45

Ercolini D (2004) PCR-DGGE fingerprinting: novel strategies for detection of microbes in food. J Microbiol Methods 56:297–314

Fernandes S, Geueke B, Delgado O, Coleman J, Hatti-Kaul R (2002) β-Galactosidase from a cold-adapted bacterium: purification, characterization and application for lactose hydrolysis. Appl Microbiol Biotechnol 58:313–321

Fernández NV, Mestre MC, Marchelli P, Fontenla SB (2012) Yeast and yeast-like fungi associated with dry indehiscent fruits of *Nothofagus nervosa* in Patagonia, Argentina. FEMS Microbiol Ecol 80:179–192

Friedmann EI, Thistle AB (1993) Antarctic microbiology. Wiley, New York

Hoyoux A, Jennes I, Dubois P, Genicot S, Dubail F, François J-M, Baise E, Feller G, Gerday C (2001) Cold-adapted β-galactosidase from the Antarctic psychrophile *Pseudoalteromonas haloplanktis*. Appl Environ Microbiol 67:1529–1535

Kirk O, Borchert TV, Fuglsang CC (2002) Industrial enzyme applications. Curr Opin Biotechnol 13:345–351

Klappenbach JA, Saxman PR, Cole JR, Schmidt TM (2001) rrndb: the ribosomal RNA operon copy number database. Nucleic Acids Res 29:181–184

Libkind D, Moliné M, de García V, Fontenla S, van Broock M (2008) Characterization of a novel South American population of the astaxanthin producing yeast *Xanthophyllomyces dendrorhous* (*Phaffia rhodozyma*). J Ind Microbiol Biotechnol 35:151–158
Mc Cargo PD, Iannone LJ, Vignale MV, Schardl CL, Rossi MS (2014) Species diversity of *Epichloe* symbiotic with two grasses from southern Argentinean Patagonia. Mycologia 106:339–352
Mestre MC, Fontenla S, Rosa CA (2014) Ecology of cultivable yeasts in pristine forests in northern Patagonia (Argentina) influenced by different environmental factors. Can J Microbiol 60:371–382
Mestre MC, Rosa CA, Fontenla SB (2011a) *Lindnera rhizosphaerae* sp. nov., a yeast species isolated from rhizospheric soil. Int J Syst Evol Microbiol 61:985–988
Mestre MC, Rosa CA, Safar SV, Libkind D, Fontenla SB (2011b) Yeast communities associated with the bulk-soil, rhizosphere and ectomycorrhizosphere of a *Nothofagus pumilio* forest in northwestern Patagonia, Argentina. FEMS Microbiol Ecol 78:531–541
Metzker ML (2010) Sequencing technologies–the next generation. Nat Rev Genet 11:31–46
Morita RY (1975) Psychrophilic bacteria. Bacteriol Rev 39:144–167
Nouhra E, Urcelay C, Longo S, Tedersoo L (2013) Ectomycorrhizal fungal communities associated to *Nothofagus* species in Northern Patagonia. Mycorrhiza 23:487–496
Olivera NL, Sequeiros C, Nievas ML (2007) Diversity and enzyme properties of protease-producing bacteria isolated from sub-Antarctic sediments of Isla de Los Estados, Argentina. Extremophiles 11:517–526
Orrillo AG, Ledesma P, Delgado OD, Spagna G, Breccia JD (2007) Cold-active α-l-rhamnosidase from psychrotolerant bacteria isolated from a sub-Antarctic ecosystem. Enzyme Microb Technol 40:236–241
Prabagaran S, Suresh K, Manorama R, Delille D, Shivaji S (2005) *Marinomonas ushuaiensis* sp. nov., isolated from coastal sea water in Ushuaia, Argentina, sub-Antarctica. Int J Syst Evol Microbiol 55:309–313
Prabagaran SR, Manorama R, Delille D, Shivaji S (2007) Predominance of *Roseobacter*, *Sulfitobacter*, *Glaciecola* and *Psychrobacter* in seawater collected off Ushuaia, Argentina, Sub-Antarctica. FEMS Microbiol Ecol 59:342–355
Pucci GN, Acuña AJ, Llanes ML, Tiedemann MC, Pucci OH (2009) Identificación de bacterias marinas cultivables de la ciudad costera Comodoro Rivadavia, Argentina. Rev Biol Mar Oceanogr 44:49–58
Reichenbach H (1992) The order *Cytophagales*. The prokaryotes. Springer, New York, pp 3631–3675
Reichenbach H (2006) The order *Cytophagales*. The prokaryotes. Springer, New York, pp 549–590
Rodríguez ME, Pérez-Través L, Sangorrín MP, Barrio E, Lopes CA (2014) *Saccharomyces eubayanus* and *Saccharomyces uvarum* associated with the fermentation of *Araucaria araucana* seeds in Patagonia. FEMS Yeast Res 14:948–965
Rosa LH, Vaz AB, Caligiorne RB, Campolina S, Rosa CA (2009) Endophytic fungi associated with the Antarctic grass *Deschampsia antarctica* Desv. (Poaceae). Polar Biol 32:161–167
Rovati JI, Pajot HF, Ruberto L, Mac Cormack W, Figueroa LI (2013) Polyphenolic substrates and dyes degradation by yeasts from 25 de Mayo/King George Island (Antarctica). Yeast 30:459–470
Sánchez L, Gómez F, Delgado O (2009) Cold-adapted microorganisms as a source of new antimicrobials. Extremophiles 13:111–120
Sánchez RM, Bianchinotti M (2010) New records in the *Tubeufiaceae* from Andean Patagonian forests of Argentina. Mycotaxon 111:131–141
Sanchez RM, Miller AN, Bianchinotti MV (2012) A new species of *Acanthostigma* (Tubeufiaceae, Dothideomycetes) from the southern hemisphere. Mycologia 104:223–231
Scorsetti AC, Elíades LA, Stenglein SA, Cabello MN, Pelizza SA, Saparrat MC (2012) Pathogenic and enzyme activities of the entomopathogenic fungus *Tolypocladium cylindrosporum* (Ascomycota: Hypocreales) from Tierra del Fuego, Argentina. Int J Trop Biol Conserv 60:833–841

Sequeiros C, Vallejo M, Marguet ER, Olivera NL (2010) Inhibitory activity against the fish pathogen *Lactococcus garvieae* produced by *Lactococcus lactis* TW34, a lactic acid bacterium isolated from the intestinal tract of a Patagonian fish. Arch Microbiol 192:237–245

Tedersoo L, Bahram M, Põlme S, Kõljalg U, Yorou NS, Wijesundera R, Ruiz LV, Vasco-Palacios AM, Thu PQ, Suija A (2014) Global diversity and geography of soil fungi. Science 346:1078–1088

Trierveiler-Pereira L, Smith ME, Trappe JM, Nouhra ER (2015) Sequestrate fungi from Patagonian *Nothofagus* forests: *Cystangium* (Russulaceae, Basidiomycota). Mycologia 107:90–103

Tropeano M, Coria S, Turjanski A, Cicero D, Bercovich A, Mac Cormack W, Vazquez S (2012) Culturable heterotrophic bacteria from Potter Cove, Antarctica, and their hydrolytic enzymes production. Polar Res 31:18507–18517

Tropeano M, Vázquez S, Coria S, Turjanski A, Cicero D, Bercovich A, Mac Cormack W (2013) Extracellular hydrolytic enzyme production by proteolytic bacteria from the Antarctic. Pol Polar Res 34:253–267

Vaz AB, Rosa LH, Vieira ML, Garcia VD, Brandão LR, Teixeira LC, Moliné M, Libkind D, Van Broock M, Rosa CA (2011) The diversity, extracellular enzymatic activities and photoprotective compounds of yeasts isolated in Antarctica. Braz J Microbiol 42:937–947

Vitali B, Ndagijimana M, Cruciani F, Carnevali P, Candela M, Guerzoni ME, Brigidi P (2010) Impact of a synbiotic food on the gut microbial ecology and metabolic profiles. BMC Microbiol 10:4–16

Chapter 11
Alkaline Proteases from Patagonian Bacteria

Nelda Lila Olivera, Martín S. Iglesias, and Cynthia Sequeiros

Abstract In addition to their ecological importance in the acquisition of nitrogen-rich organic compounds, extracellular proteases also have interesting biotechnological applications. Particularly, alkaline proteases represent one of the most important groups of commercial enzymes. First, we introduce the classification and catalytic mechanisms of proteases. Then, this chapter reviews the advances in the bioprospection of alkaline proteases produced by bacteria adapted to selective conditions from different environments of Patagonia (Argentina). Among them, the arid soils of the Patagonian Monte are propitious for the development of alkaliphilic microorganisms. Thus, we focus on the description of the species *Bacillus patagoniensis* and the biochemical and catalytic properties of its alkaline protease. Then, we discuss investigations about alkaline protease-producing bacteria from the southern Patagonian coast, the prevalence of psychrophilic and psychrotolerant strains, and the response of their extracellular proteases to temperature.

11.1 Introduction

Enzymes that hydrolyze peptide bonds of polypeptides or proteins rendering peptides or amino acids are called proteases, although they are also termed peptidases, proteolytic enzymes, or proteinases (Rawlings et al. 2014). Microbial proteases take part in many important physiological processes of microorganisms, including processing of signal peptides and propeptides, sporulation, cell division, adsorption of proteinaceous nutrients, cleavage of cell-wall proteins, bacterial pathogenesis,

N.L. Olivera (✉) • M.S. Iglesias
Patagonian Institute for the Study of Continental Ecosystems (IPEEC-CENPAT), National Scientific and Technical Research Council (CONICET), Argentina
e-mail: olivera@cenpat-conicet.gob.ar

C. Sequeiros
Centro para el Estudio de Sistemas Marinos (CESIMAR-CCT CENPAT-CONICET), Blvd. Brown 2915, Puerto Madryn, Chubut, Argentina

N.L. Olivera et al. (eds.), *Biology and Biotechnology of Patagonian Microorganisms*, DOI 10.1007/978-3-319-42801-7_11

and defense (Rao et al. 1998; Wu and Chen 2011; Molière and Turgay 2013). They also have a major function in the global recycling of carbon and nitrogen through the depolymerization of proteins in the environment (Folse and Allison 2012). Particularly, protein degradation is associated with the first stages of nitrogen mineralization and release, regulating its availability to primary producers and microorganisms in the environment (Alkorta et al. 2003).

Commercially, proteases constitute one of the largest groups of the global market of industrial enzymes, which reached nearly 4.8 billion dollars in 2013 as per estimates by BCC Research group (http://www.bccresearch.com/market-research/ biotechnology/enzymes-industrial-applications-bio030h.html). Particularly, alkaline proteases are widely used in the detergent and leather industries, in food and feed processing, in medical applications, and in organic waste treatments. Microorganisms are the preferred source of alkaline proteases as they grow rapidly and in reduced spaces, and they can be genetically manipulated for protease engineering.

The screening of protease-producing microorganisms in unexplored environments, especially those showing selective conditions, is an important strategy for the isolation of robust enzymes with valuable technical characteristics (e.g., high specificity and stability toward temperature, pH, salt concentration, and chemical agents). In this chapter, we focus on the bioprospection of alkaline protease-producing bacteria from Patagonia. First, we review the classification and catalytic mechanisms of proteases and the main characteristics of alkaline proteases. Then, we discuss the occurrence and particularities of alkaline protease-producing bacteria from the arid soils of the Patagonian Monte and from cold marine ecosystems off the southern Patagonian coast.

11.2 Classification and Catalytic Mechanisms of Proteases

11.2.1 Classification and Nomenclature of Proteases

Two systems of classification and nomenclature of proteases are currently in use: the EC System recognized by the Nomenclature Committee of the International Union of Biochemistry and Molecular Biology, which is based on the reactions that proteases catalyze (Enzyme Nomenclature 1992), and the MEROPS System of peptidase clans and families that uses a hierarchical, structure-based classification of proteases (Rawlings and Barrett 1993).

11.2.1.1 EC System

The EC (Enzyme Committee) classification system is based on the type of catalyzed reaction and the nature of the protease active site. Proteases are located within the hydrolase class (EC 3.), acting on peptide bonds (peptidase subclass EC 3.4). All peptidases catalyze the same reaction, the hydrolysis of the peptide bond; however, they can be classified according to the position of the peptide bond to be hydrolyzed within the polypeptide chain. On this basis, they are divided into "exopeptidases," which act only near a terminus of a polypeptide chain, and "endopeptidases," that act internally in polypeptide chains (Enzyme Nomenclature 1992).

The exopeptidases act only near the ends of polypeptide chains and can liberate a single amino acid residue, dipeptides or tripeptides. Based on their site of action at the N- or C-terminus, they are classified as amino- and carboxypeptidases, respectively. The endopeptidases are characterized by their preferential action at the peptide bonds in the inner regions of the polypeptide chain away from the N- and C-termini. They are divided into five subgroups based on their catalytic mechanisms, which involve the presence of certain amino acids in the enzyme active site. Thus, the serine endopeptidases (EC 3.4.21) possess a serine (Ser) residue in their active site; the cysteine endopeptidases (EC 3.4.22) have a cysteine (Cys) group; aspartic endopeptidases (EC 3.4.23) often have two aspartic (Asp) residues, which are responsible for the catalytic activity; the activity of metallo-endopeptidases (EC 3.4.24) depends on the presence of a divalent metal ion in the active site, usually zinc (Zn^{2+}); and the threonine (Thr) endopeptidases (EC 3.4.25) employ a threonine residue instead of a serine one. Endopeptidases that could not be assigned to any of these groups are listed in sub-subclass EC 3.4.99.

11.2.1.2 MEROPS System

The MEROPS classification system is the newest of the systems and considers primary and tertiary structures to group sets of homologous peptidases into *Families* and sets of related families into *Clans* (Rawlings and Barrett 1993). The MEROPS system uses its own identification code. Each family is identified by a letter representing the catalytic type of the proteolytic enzymes it contains, that is, A (Aspartic), C (Cysteine), G (Glutamic), M (Metallo), N (Asparagine), P (Mixed), S (Serine), T (Threonine), and U (Unknown), followed by an arbitrarily assigned number. The name of each clan is identified with two letters: the first represents the catalytic type of the families included in the clan and the second is a capital letter of the alphabet assigned sequentially. The letter "P" is used for a clan containing families with more than one catalytic type (e.g., serine, threonine, and cysteine). This classification is under constant review (https://merops.sanger.ac.uk/index.shtml).

11.2.2 Protease Catalytic Types and Their Action Mechanism

The catalytic type of a protease relates to the chemical groups responsible for peptide bond hydrolysis. All proteases polarize the carbonyl group of the substrate peptide bond by stabilizing the oxygen in an oxyanion hole, which makes the carbon atom more vulnerable for attack by an activated nucleophile. The major differences between the catalytic types are the nature of the nucleophile and oxyanion stabilizer. In cysteine, threonine, and serine peptidases the nucleophile of the catalytic site is part of an amino acid, either a hydroxyl group (serine and threonine

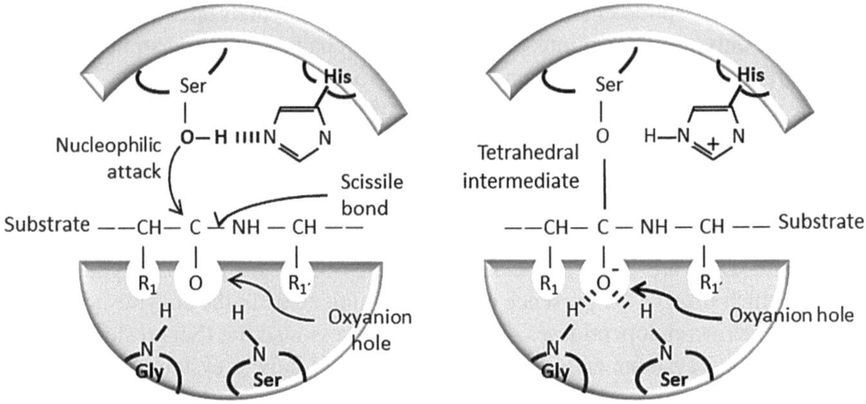

Fig. 11.1 Catalytic mechanism of serine peptidases

peptidases) or a sulfhydryl group (cysteine peptidases). The oxyanion hole is usually stabilized by two residues in the backbone of the protease (Barrett et al. 2004). On the other hand, metalloproteases, and glutamic and aspartic proteases, use water as nucleophile, activated by electrostatic interactions with the metal ion (Me^{2+}), or glutamate (Glu) or aspartate (Asp) residues, respectively. The oxyanion of these proteases is stabilized by Me^{2+}, Gln, and Asp, respectively (Fujinaga et al. 2004; Rawlings et al. 2014).

We explain further the catalytic action of serine and metallopeptidase families as they include bacterial alkaline proteases (Ellaiah et al. 2002; Singhal et al. 2012). The hydrolysis mechanism of *serine peptidases* usually proceeds in two distinct steps. First, the nucleophilic attack by the serine hydroxyl group on the carbonyl carbon of the substrate forms a tetrahedral intermediate (Fig. 11.1). This intermediate is stabilized by two backbone NH groups from Gly and Ser, respectively, which provide hydrogen bonds from the oxyanion-binding site to the carbonyl oxygen (Fig. 11.1) (Polgár 2005). The intermediate breaks down by general acid catalysis to acyl enzyme and the 'amino' portion of the substrate as the first product of reaction. As a result of this acylation step, the proton of the serine hydroxyl has been transferred by the imidazole group of a histidine residue to the amine leaving group (Fig. 11.1). In a second step, the acyl-enzyme intermediate is hydrolyzed through the reverse pathway of acylation, but in this addition–elimination reaction a water molecule instead of the serine residue is the attacking nucleophile (Barrett et al. 2004; Polgár 2005). Thus, it results in the hydrolysis of the peptide bond.

The *metallopeptidases*, as aspartic peptidases, use a water molecule to carry out the nucleophilic attack on a peptide bond (Fig. 11.2). A divalent metal cation, usually zinc, activates the water molecule. The metal ion is held in place by amino acid ligands, usually three in number (Fig. 11.2). Most of the metalloproteases are enzymes containing the His-Glu-Xaa-Xaa-His (HEXXH) motif, which forms a part of the site for binding the metal. The water molecule is also hydrogen bonded to a glutamic acid (Fig. 11.2). That carboxyl group serves as a general base to remove a proton and assists the attack of the same water molecule on the peptide carbonyl.

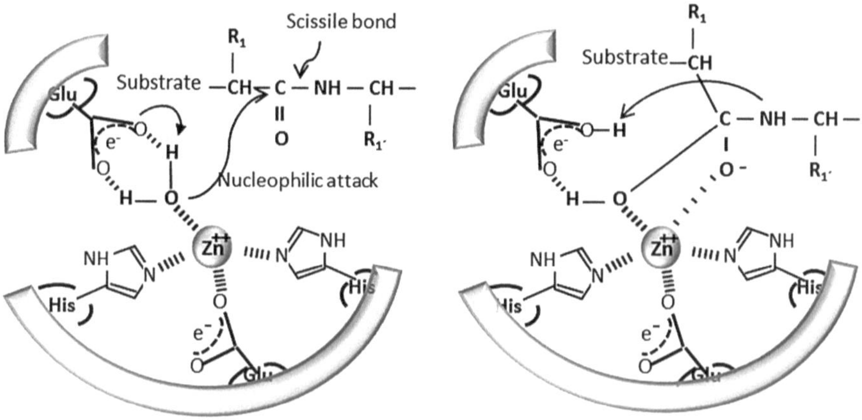

Fig. 11.2 Catalytic mechanism of metallopeptidases

Again, a proton must be transferred to the leaving nitrogen atom, and this could be derived from the glutamic acid. Thus, the glutamic acid would be acting in analogy to one of the aspartic acid groups in the aspartic peptidases and to the histidine in the serine and cysteine peptidases (Dunn 2001; Barrett et al. 2004).

11.3 Bacterial Alkaline Proteases

Proteases are referred to as acidic, neutral, or alkaline enzymes based on the optimum pH for their hydrolytic activity (Rao et al. 1998). Alkaline proteases mostly include enzymes with an optimum pH for activity between 9 and 11, an optimal temperature between 50 °C and 70 °C, and a molecular mass in the range of 30 to 45 kDa (Singhal et al. 2012; Jisha et al. 2013). Particularly, serine alkaline proteases comprise important technical enzymes used in the detergent, leather, and meat industries (Bhunia et al. 2012). Among alkaline protease-producing bacteria, there are neutralophilic, alkalotolerant, and alkaliphilic strains (Ellaiah et al. 2002; Fujinami and Fujisawa 2010). Extremophilic bacteria adapted to more than one environmental stressor, such as haloalkaliphiles, which survive under saline and alkaline conditions, also produce alkaline proteases (Raval et al. 2014). Thermophilic alkaliphiles, thermophilic haloalkaliphiles, and psychrophilic and psychrotrophic bacteria have also been reported to produce valuable alkaline proteases (Kasana 2010; Shrinivas and Naik 2011).

Taxonomically, one of the most studied genera of alkaline protease producers is *Bacillus* (Ito et al. 1998; Rao et al. 1998). Neutralophilic strains of this genus produce important alkaline serine proteases (Table 11.1). An example is Subtilisin Carlsberg, produced from fermentation of *Bacillus licheniformis* (Jacobs et al. 1985). Moreover, alkaliphilic *Bacillus* spp. also constitute an important source of alkaline proteases (Ito et al. 1998; Horikoshi 2004). Since the alkaliphilic *Bacillus* group was classified into nine species (Nielsen et al. 1995), numerous novel strains have been described and many of them are alkaline protease producers (Saeki et al. 2002;

Table 11.1 Bacterial alkaline proteases

Alkaline protease	Catalytic type	Optimum pH	p*I*	Molecular mass (kDa)	Bacterium	Reference
Savinase	Serine protease	10–12	>9.5	27	*Bacillus lentus*	Betzel et al. (1992), Margesin et al. (1992)
Subtilisin Carlsberg	Serine protease	10–12	8.6	29	*Bacillus licheniformis*	Jacobs et al. (1985), Margesin et al. (1992)
SAP	Serine protease	11.5	–	19	*Streptomyces* sp. YSA-130	Yum et al. (1994)
M	Serine protease	12.3	>10.6	28	*Bacillus* sp. KSM-K16	Kobayashi et al. (1995)
KP-43	Serine protease	11–12	9.7–9.9	43	*Bacillus* sp. KSM-KP43	Saeki et al. (2002)
S7 protease	Serine metalloprotease	11	4.5	44	*Streptomyces* sp. S7	Tatineni et al. (2008)
AprB	Serine protease	10	–	28	*Bacillus* sp. B001	Deng et al. (2010)
htrA-like	Serine protease	8	6.6	37	*Bacillus subtilis* DR8806	Farhadian et al. (2015)
SAPS-P1 and SAPS-P2	Serine protease	12 and 10	–	36 and 21	*Streptomyces* sp. strain AH4	Touioui et al. (2015)
S, N, and B	Metalloprotease	8.5, 7.5, and 7.0	–	36, 53, and 71	*Bacillus stearothermophilus* strain TLS3	Sookkheo et al. (2000)
PseA protease	Metalloprotease	8	–	35	*Pseudomonas aeruginosa* PseA	Gupta et al. (2005)
NprC1	Metalloprotease	8	–	35.5	*Bacillus cereus* SV1	Manni et al. (2008)

Kazan et al. 2005; Siegert et al. 2009; Deng et al. 2010; Denizci et al. 2010; Prakash et al. 2010). To this group belong the commercial high-alkaline proteases Savinase® and Esperase® (Novozymes Corp.) produced by alkalophilic *Bacillus* spp. (Table 11.1). These proteases are characterized by high activity and stability at elevated temperature and pH.

Other important alkaline protease producers are species of *Pseudomonas* and *Streptomyces* (Table 11.1). A solvent and thermoalkali stable protease is produced by *Pseudomonas putida* SKG-1; the protease was most active at pH 9.5 and 40 °C, and in the presence of organic solvent exhibited 72–191 % of relative activity (Singh et al. 2013). *Streptomyces* sp. strain AH4 produces two extracellular alkaline proteases, both enzymes stable within a wide range of temperature (45–75 °C) and pH (8.0–11.5), (Touioui et al. 2015). *Streptomyces* sp. S7 produces an alkaline keratinase with an optimal activity at 45 °C and pH 11 (Tatineni et al. 2008). Alkaline proteases with interesting biotechnological properties are also produced by *Stenotrophomonas maltophilia* (Waghmare et al. 2015), *Serratia* sp. (Bhargavi and Prakasham 2013), and *Vibrio* sp. (Manjusha et al. 2013), among many other microbial strains.

11.4 Bioprospection of Alkaline Protease-Producing Bacteria from Patagonia

11.4.1 The Patagonian Monte

Microorganisms are responsible for the turnover of soil organic carbon, energy flow, and nutrient transformations and mineralization (Hättenschwiler et al. 2011). Bioprospection of soil microorganisms is particularly interesting because soil is heterogeneous and its properties vary not only across the landscape and with depth, but also at a microenvironment scale (Burns et al. 2013). Vegetation, climate, natural events, and anthropic activities also influence the soil environment and its microbiota. Particularly, in arid ecosystems water availability controls soil biological processes, and resources are spatially and temporally discontinuously distributed (Noy-Meir 1973; Austin et al. 2004).

Soil proteases derive from different sources, including microorganisms, plants, and animals (Vranova et al. 2013). Protease production can help microorganisms to cope with nitrogen-limited environments such as arid soils where nitrogen is, after water, the most important nutrient controlling productivity (Belnap 1995; Theron and Divol 2014). The particular characteristics of arid ecosystems in addition to the inherent soil heterogeneity lead to a diversity of microorganisms adapted to the local prevailing conditions, showing properties such as optimal growth or extracellular enzyme activity at different pHs, salt concentrations, or temperatures. This diversity may also impact the community of protease-producing bacteria that release different amounts or types of proteases affecting the overall soil protease activity. Accordingly, a better understanding of the response of overall soil protease activity to spatial and temporal soil variations could contribute to delineating strategies for the screening of specific protease-producing microorganisms.

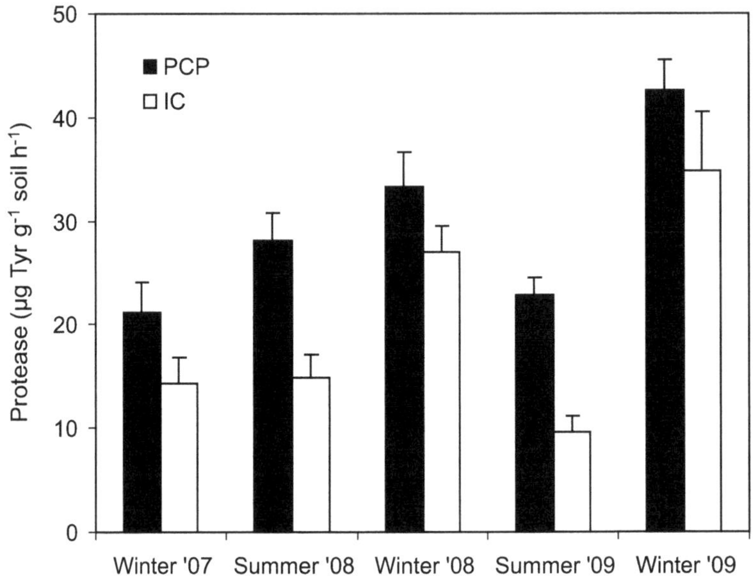

Fig. 11.3 Alkaline protease activity in plant-covered patch (PCP) and intercanopy area (IC) soils from the Patagonian Monte (Argentina), at different sampling dates (winter 2007, summers and winters 2008 and 2009)

The arid Patagonian Monte occupies the northeastern portion of the Chubut Province (between 42°–44°S and 64°–68°W), Patagonia, Argentina (León et al. 1998). The plant canopy cover is low (<60 % of the soil surface) and distributed in a patchy structure formed by clumps of shrubs and perennial grasses surrounded by intercanopy areas with sparse or no vegetation (Bisigato and Bertiller 1997; Mazzarino et al. 1998). Soils are exposed to wind and water erosion, as well as salinization and alkalinization processes associated with nonirrigated lands (Olivera et al. 2005). In alkaline soils of the Patagonian Monte (pH 8.23–8.61), the overall alkaline protease activity was higher in soils under plant-covered patches than in the intercanopy areas (Fig. 11.3; Olivera et al. 2014), which was possibly associated with the concentration of soil resources and microorganisms underneath plant patches (Mazzarino et al. 1996; Prieto et al. 2011). Moreover, a 3-year monitoring showed that the highest protease activity levels occurred in the winters with higher precipitations and soil moisture (Fig. 11.3; winters 2008 and 2009). Congruently, the best predictors of alkaline protease activity were soil moisture in combination with microbial biomass-C and total soil-N, suggesting that dry periods with low humidity and low N-input could affect microbial protease-producing communities and restrict soil proteolysis and N-availability (Olivera et al. 2014).

The selective conditions of the soils from the Patagonian Monte encouraged screening programs to isolate extracellular protease-producing bacteria with high activity at moderate temperature and high pH values. Sampling efforts especially concentrated in the shrub rhizosphere of plant-covered patches, taking into account that the desert conditions contribute to a low microbial activity, and that microbial development is more significant in the rhizosphere, as explained previously.

From the rhizosphere of *Atriplex lampa*, a perennial shrub able to colonize alkaline and saline soil microsites in the Patagonian Monte shrubland, was isolated the protease-producing strain PAT 05. After a polyphasic taxonomic study PAT 05 was classified in a novel species, *Bacillus patagoniensis*, together with strains belonging to the phenon 4a of the alkaliphilic *Bacillus* group (Nielsen et al. 1995). *B. patagoniensis* PAT 05^T grew between pH 7 and 10, with an optimum at about pH 8, between 5 ° and 40 °C, and with 15 % (w/v) NaCl (Olivera et al. 2005). This finding indicated that PAT 05^T was a moderately alkaliphilic bacteria, and it was also psychrotolerant and halotolerant (Olivera et al. 2005). The alkaliphilic *Bacillus* group constitutes an important source of extracellular enzymes for numerous industrial processes (Ito et al. 1998; Horikoshi 2004). In PAT 05^T culture supernatant, three proteolytic bands were detected: the most intense one had a high p*I* value (>10.3) and the two slighter ones had p*I* values of about 5.9 and 4.6 (Olivera et al. 2006). The most intense band corresponded to an alkaline serine protease with a molecular mass of 29.4 kDa (Olivera et al. 2006). Its optimal pH and temperature for activity were 11 ° and 60 °C, respectively. Interestingly, PAT 05^T proteases displayed higher activity at moderate temperatures (22 % and 38 % residual activity at 20 °C and 30 °C, respectively) than other alkalophilic *Bacillus* species (e.g., *Bacillus clausii* GMBAE 42, 28 % relative activity at 30 °C, and *B. clausii* I-52, 20 % at 40 °C) (Joo et al. 2003; Denizci et al. 2004; Olivera et al. 2006). Moreover, PAT 05^T protease activity was stable or conserved a significant residual activity in various surfactants (SDS, 1 % w/v; Triton X-100 and Tween 20, both 1 % v/v), chelators (1,10-phenanthroline and EDTA, both 10 mM), and oxidizing-reducing agents (H_2O_2, 10 % v/v; DTT 1 mM), and with high NaCl concentrations (63 % residual activity in 2 M NaCl) (Olivera et al. 2006). Altogether, PAT 05^T protease activity in high alkaline/saline conditions, its significant residual activity at moderate temperatures, and its keratinolytic activity suggest that this strain is a potential source of alkaline protease suitable for the detergent and leather industries.

11.4.2 The Southern Patagonian Coast

In marine environments, there are a high proportion of proteolytic bacteria (Atlas and Bartha 1981). As seawater is alkaline, proteases secreted by marine bacteria are stable and active in an alkaline pH range (Wu and Chen 2011). Hence, marine environments are promising for the isolation of alkaline protease-producing bacteria. Moreover, marine microorganisms from cold environments carry out their metabolic processes at low temperature by means of adaptive changes in proteins (particularly enzymes), translation systems, and in cellular lipids to maintain membrane fluidity and permeability (Russell 1990). Thus, their enzymes characteristically show high catalytic efficiency at low temperatures, which associates with an increase in the flexibility of the active site and, therefore, a lower thermal stability than their mesophilic counterparts (Feller and Gerday 2003; Gerday 2013). These unique properties of cold-active enzymes have gained much attention regarding their potential for biotechnological applications. For example, high enzyme activity at low/

moderate temperatures is an important feature for energy saving in processes that would be performed at room or tapwater temperature. The use of thermosensible enzymes also allows their selective inactivation in complex mixtures (Gerday et al. 2000; Moran et al. 2001).

Even though most of the microbial communities inhabiting marine Argentinean environments remain unexplored and unexploited (Dionisi et al. 2012), several studies focused on the isolation and characterization of marine protease-producing bacteria, particularly studies concentrated in the southern Patagonian coast where low temperatures select a cold-adapted microbiota that could produce cold-active enzymes. Some studies were conducted in the coast of Tierra del Fuego Province, located in the southernmost part of Patagonia, where the climate is moderately constantly cold and the surface seawater temperature ranged from a medium value of 9.7 °C in summer to 4.5 °C in July (Olivera et al. 2007). There, Cristóbal et al. (2011) isolated 230 cold-adapted bacterial strains from seawater and the gut of benthonic organisms from the Beagle Channel; 33 % of them showed protease activity under alkaline conditions at 4 °C and 47 % at 15 °C. Olivera et al. (2007) isolated 19 alkaline protease-producing strains from intertidal marine sediments of the Isla de Los Estados Reserve (Argentina) that belonged to the genera *Pseudoalteromonas* (8), *Shewanella* (6), *Colwellia* (3), *Planococccus* (1), and (1 strain) to the family *Flavobacteriaceae*. Interestingly, Tropeano et al. (2013) found that the same bacterial genera dominate the protease-producing bacteria from the Potter Cove marine environment in Antarctica. From Isla de Los Estados isolates, the highest thermosensitivity was shown by the alkaline extracellular proteases produced by *Colwellia* sp. IE1-3 (Fig. 11.4), which residual activity after 50 min at 40 °C was only 15 % (Olivera et al. 2007). The optimum pH for IE1-3 protease activity was approximately 9 and the optimum temperature was in the range 30–40 °C (Fig. 11.5). Moreover, the thermostability of the extracellular proteases produced by the isolates of the different genera varied considerably, showing an increasing trend in thermal stability from *Colwellia* (ΔG^*_{in} 96.6–99.0 kJ mol^{-1}) < *Pseudoalteromonas* (ΔG^*_{in} 100.6–104.8 kJ mol^{-1}) < *Shewanella*/*Planococcus* (ΔG^*_{in} 109.8–121.9 kJ mol^{-1})

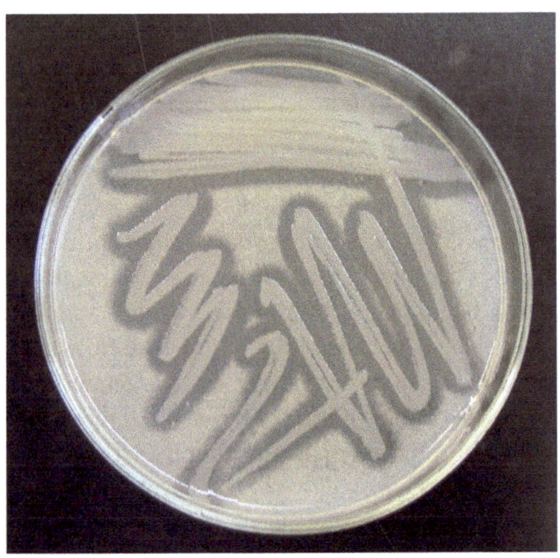

Fig. 11.4 *Colwellia* sp. IE1-3 grown on skim milk agar at 12 °C; clearing of the agar around the colonies indicated proteolytic activity

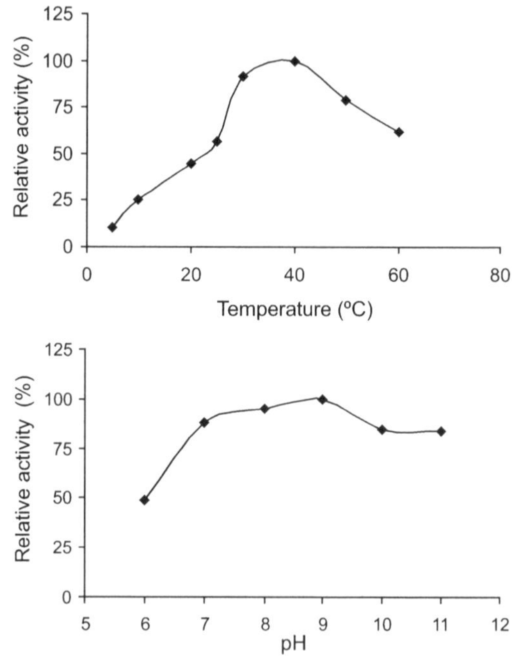

Fig. 11.5 Optimal temperature (**a**) and pH (**b**) of alkaline protease activity of *Colwellia* sp. IE1–3

(Olivera et al. 2007). The estimation of ΔG^*_{in}, which is the energy barrier that must be reached by an enzyme to become thermally irreversible inactivated, resulted in a useful approach for a comparative screening of cold-active proteases. Such results reflected the different thermal properties of the proteases that could coexist in the intertidal area of sandy beaches, which are exposed to cycles of changing environmental conditions that probably are more complex and harsher than other marine habitats (Brown and McLachlan 1990).

In another coastal area of Patagonia, the San Jorge Gulf (Chubut, Argentina), Estevao Belchior and Vacca (2006) studied the role of autochthonous bacteria of the intestinal tract of hake (*Merluccius hubbsi*) in fish spoilage. They isolated a protease-producing marine *Pseudoalteromonas* sp. (CR41) which was a psychrotrophic, moderately alkaliphilic and halophilic strain. It produced proteases active at alkaline pH that hydrolyzed hake muscle proteins, but such proteolytic activity was lower at 7 °C than at 22 °C (Estevao Belchior and Vacca 2006).

11.5 Biotechnological Applications of Alkaline Proteases: Focus on Patagonia

Biotechnological applications of alkaline proteases have been the topic of several reviews (Ito et al. 1998; Ellaiah et al. 2002; Fujinami and Fujisawa 2010; Singhal et al. 2012). In this section, we particularly focus on the potential usefulness of alkaline proteases for Patagonian industries, mainly in wool and fish processing activities.

In Argentina, sheep raising for wool and meat production is the predominant land use on the native Patagonian shrubland. Wool fiber is demanded by a segment of the market that appreciates its attributes and properties. Nevertheless, current differentiation strategies for such a market segment require industrial processes with low environmental impacts, in addition to the improvement of wool fiber attributes to promote wool valorization against synthetic fibers. Thus, one of the main needs is the development of wool antifelting/shrinking treatments that permit the washing of wool in domestic machines (Buschle-Diller 2003). Alkaline proteases with keratinolytic activity (keratinases) may attenuate wool fiber scales, imparting shrink resistance, softness, and whiteness to the wool (Buschle-Diller 2003; Shen 2010; Ibrahim et al. 2012). In contrast to the chlorine-Hercosett method, the currently used wool shrink-proofing process, enzymatic treatments do not produce hazardous pollutants and favor saving water, chemicals, and energy (Kotlińska and Lipp-Symonowicz 2011). Alkaline keratinases, such as those produced by *B. patagoniensis* PAT O5^{T} (Olivera et al. 2006) and other microbial proteases from sheep wool authochtonous bacteria (Iglesias et al. 2014), showed the capability to hydrolyze wool keratin, which has high recalcitrance to proteolytic degradation from the tight packing of its protein chains, making them interesting candidates for the development of shrink-proofing treatments for the wool industry.

Proteases may also be applied in the treatment of high-protein wastes (Kasana 2010), such as those produced by fish- and crustacean-processing plants located near the Patagonian ports. This kind of industry generates large amounts of solid and liquid wastes with high organic matter content that must be managed properly to avoid environmental pollution. By-products generated by the fish-processing sector may be as much as 75 % of the original processed material (Lopes et al. 2015). Such wastes constitute an important source of proteins, polysaccharides, polyunsaturated fatty acids, minerals and vitamins, and bioactive peptides (Blanco et al. 2007). Moreover, processing of the Patagonian shrimp *Pleoticus muelleri*, which is generally commercialized without head and frequently without shell, produces large amounts of shrimp head, shell, and tail wastes containing chitin, proteins, lipids, and pigments. These by-products could be used as substrates for the microbial production of valuable bioproducts. Particularly regarding protease production, cuttlefish by-products were used for the growth of protease-producing strains in Tunisia (Souissi et al. 2008). In addition, shrimp shell from Taiwan was used as the sole carbon/nitrogen source for protease production by the alkalophilic bacteria *Bacillus cereus* strain TKU006 (Wang et al. 2009). On the other hand, alkaline proteases could be applied in hydrolysates of fish by-products to produce, for example, bioactive peptides (Lassoued et al. 2015).

11.6 Conclusions

The increasing range of technological applications of alkaline proteases requires constant research to provide enzymes with properties fitting the new industrial requirements and demands of the market. Such research includes the bioprospecting of novel alkaline proteases. The microbial diversity that inhabits the different environments from the Argentine Patagonian region represents a valuable source of biotechnological products that remains mostly unexplored. Even though there have been as yet few studies about alkaline protease-producing bacteria from Patagonia, the promising characteristics of these enzymes encourage continuing the bio-prospection efforts. Further studies considering both novel cultivation strategies and new approaches such as metagenomics and high-throughput screening methods will be important to allow access to novel alkaline proteases from Patagonian bacterial communities.

Acknowledgments The authors gratefully acknowledge support from ANPCyT (PICT Start Up 2012–2004) and CONICET (PIP 11220120100050CO), Argentina.

References

Alkorta I, Aizpurua A, Riga P, Albizu I, Amezaga I, Garbisu C (2003) Soil enzyme activities as biological indicators of soil health. Rev Environ Health 18:65–73

Atlas RM, Bartha R (1981) Microbial ecology: fundamentals and applications. Addison-Wesley, Reading, Boston

Austin AT, Yahdjian L, Stark JM, Belnap J, Porporato A, Norton U, Ravetta DA, Schaeffer SM (2004) Water pulses and biogeochemical cycles in arid and semiarid ecosystems. Oecologia (Berl) 141:221–235

Barrett AJ, Rawlings ND, Woessner JF (eds) (2004) Handbook of proteolytic enzymes, 2nd edn. Elsevier-Academic Press, London

Belnap J (1995) Surface disturbances: their role in accelerating desertification. Environ Monit Assess 37:39–57

Betzel C, Klupsch S, Papendorf G, Hastrup S, Branner S, Wilson KS (1992) Crystal structure of the alkaline proteinase Savinase from *Bacillus lentus* at 1.4 Å resolution. J Mol Biol 223:427–445

Bhargavi PL, Prakasham RS (2013) A fibrinolytic, alkaline and thermostable metalloprotease from the newly isolated *Serratia* sp RSPB11. Int J Biol Macromol 61:479–486

Bhunia B, Basak B, Dey A (2012) A review on production of serine alkaline protease by *Bacillus* spp. J Biochem Tech 3:448–457

Bisigato AJ, Bertiller MB (1997) Grazing effects on patchy dryland vegetation in Northern Patagonia. J Arid Environ 36:639–653

Blanco M, Sotelo CG, Chapela MJ, Pérez-Martín RI (2007) Towards sustainable and efficient use of fishery resources: present and future trends. Trends Food Sci Technol 18:29–36

Brown AC, McLachlan A (1990) Ecology of sandy shores. Elsevier, Amsterdam

Burns RG, DeForest JL, Marxsen J, Sinsabaugh RL, Stromberger ME, Wallenstein MD, Weintraub MN, Zoppini A (2013) Soil enzymes in a changing environment: current knowledge and future directions. Soil Biol Biochem 58:216–234

Buschle-Diller G (2003) Substrates and their structure. In: Cavaco-Paulo A, Gübitz GM (eds) Textile processing with enzymes. Woodhead, Cambridge, pp 42–82

Cristóbal HA, López MA, Kothe E, Abate CM (2011) Diversity of protease-producing marine bacteria from sub-Antarctic environments. J Basic Microbiol 51:590–600

Deng A, Wua J, Zhang Y, Zhang G, Wen T (2010) Purification and characterization of a surfactant-stable high-alkaline protease from *Bacillus* sp. B001. Bioresour Technol 101:7100–7106

Denizci AA, Kazan D, Abeln ECA, Erarslan A (2004) Newly isolated *Bacillus clausii* GMBAE 42: an alkaline protease producer capable to grow under highly alkaline conditions. J Appl Microbiol 96:320–327

Denizci AA, Kazan D, Erarslan A (2010) *Bacillus marmarensis* sp. nov., an alkaliphilic, protease-producing bacterium isolated from mushroom compost. Int J Syst Evol Microbiol 60:1590–1594
Dionisi H, Lozada M, Olivera NL (2012) Bioprospection of marine microorganisms. Part II: Potential and challenges for Argentina. Rev Argent Microbiol 44:122–132
Dunn BM (2001) Determination of protease mechanism. In: Beynon R, Bond JS (eds) Proteolytic enzymes. A practical approach, 2nd edn. Oxford University Press, New York, pp 77–104
Ellaiah P, Srinivasulu B, Adinarayana K (2002) A review on microbial alkaline proteases. J Sci Ind Res 61:690–704
Estevao Belchior SG, Vacca G (2006) Fish protein hydrolysis by a psychrotropic marine bacterium isolated from the gut of hake (*Merluccius hubbsi*). Can J Microbiol 52:1266–1271
Farhadian S, Asoodeh A, Lagzian M (2015) Purification, biochemical characterization and structural modeling of a potential htrA-like serine protease from *Bacillus subtilis* DR8806. J Mol Catal B-Enzym 115:51–58
Feller G, Gerday C (2003) Psychrophilic enzymes: hot topics in cold adaptation. Nat Rev Microbiol 1:200–208
Folse HJ 3rd, Allison SD (2012) Cooperation, competition, and coalitions in enzyme-producing microbes: social evolution and nutrient depolymerization rates. Front Microbiol 3:338
Fujinaga M, Cherney MM, Oyama H, Oda K, James MN (2004) The molecular structure and catalytic mechanism of a novel carboxyl peptidase from *Scytalidium lignicolum*. Proc Natl Acad Sci USA 101:3364–3369
Fujinami S, Fujisawa M (2010) Industrial applications of alkaliphiles and their enzymes: past, present and future. Environ Technol 31:845–856
Gerday C (2013) Psychrophily and catalysis. Biology 2:719–741
Gerday C, Aittaleb M, Bentahir M, Chessa J, Claverie P, Collins T, D'Amico S, Dumont J, Garsoux G, Georlette D, Hoyoux A, Lonhienne T, Meuwis M, Feller G (2000) Cold-adapted enzymes: from fundamentals to biotechnology. TiBTech 18:103–107
Gupta A, Roy I, Khare SK, Gupta MN (2005) Purification and characterization of a solvent stable protease from *Pseudomonas aeruginosa* PseA. J Chromatogr A 1069:155–161
Hättenschwiler S, Fromin N, Barantal S (2011) Functional diversity of terrestrial microbial decomposers and their substrates. CR Biol 334:393–402
Horikoshi K (2004) Alkaliphiles. Proc Jpn Acad Ser B 80:166–178
Ibrahim NA, El-Shafei HA, Abdel-Aziz MS, Ghaly MF, Eid BM, Hamed AA (2012) The potential use of alkaline protease from *Streptomyces albidoflavus* as an ecofriendly wool modifier. J Textile Inst 103:490–498
Iglesias M, Sequeiros C, García S, Olivera NL (2014) Screening of keratinolytic bacteria from patagonian Merino wool with potential for textile processes. Biocell 38:87
Ito S, Kobayashi T, Ara K, Ozaki K, Kawai S, Hatada Y (1998) Alkaline detergent enzymes from alkaliphiles: enzymatic properties, genetics, and structures. Extremophiles 2:185–190
Jacobs M, Eliasson M, Uhlen M, Flock JI (1985) Cloning, sequencing and expression of subtilisin Carlsberg from *Bacillus licheniformis*. Nucleic Acids Res 13:8913–8926
Jisha NB, Smitha RB, Pradeep S, Sreedevi S, Unni KN, Sajith S, Priji P, Josh MS, Benjamin S (2013) Versatility of microbial proteases. Adv Enz Res 1:39–51
Joo HS, Kumar CG, Park GC, Paik SR, Chang CS (2003) Oxidant and SDS-stable alkaline protease from *Bacillus clausii* I-52: production and some properties. J Appl Microbiol 95:267–272
Kasana RC (2010) Proteases from psychrotrophs: an overview. Crit Rev Microbiol 36:134–145
Kazan D, Denizci AA, Kerimak Öner MN, Erarslan A (2005) Purification and characterization of a serine alkaline protease from *Bacillus clausii* GMBAE 42. J Ind Microbiol Biotechnol 32:335–344
Kobayashi T, Hakamada Y, Adachi S, Hitomi J, Yoshimatsu T, Koike K, Kawai S, Ito S (1995) Purification and properties of an alkaline protease from alkalophilic *Bacillus* sp. KSM-K16. Appl Microbiol Biotechnol 43:473–481
Kotlińska A, Lipp-Symonowicz B (2011) Research on the enzymatic treatment of wool fibres and changes in selected properties of wool. Fibres Text East Eur 19:88–93

Lassoued I, Mora L, Nasri R, Jridi M, Toldrá F, Aristoy MC, Barkia A, Nasri M (2015) Characterization and comparative assessment of antioxidant and ACE inhibitory activities of thornback ray gelatin hydrolysates. J Funct Foods 13:225–238
León RJC, Bran D, Collantes M, Paruelo JM, Soriano A (1998) Grandes unidades de la vegetación de la Patagonia extra andina. Ecol Aust 8:125–144
Lopes C, Antelo LT, Franco-Uría A, Alonso AA, Pérez-Martín R (2015) Valorisation of fish by-products against waste management treatments: comparison of environmental impacts. Waste Manag 46:103–112
Manjusha K, Jayesh P, Divya J, Sreelakshmi B, Priyaja P, Gopinath P, Saramma AV, Bright Singh IS (2013) Alkaline protease from a non-toxigenic mangrove isolate of *Vibrio* sp. V26 with potential application in animal cell culture. Cytotechnology 65:199–212
Manni L, Jellouli K, Agrebi R, Bayoudh A, Nasri M (2008) Biochemical and molecular characterization of a novel calcium-dependent metalloprotease from *Bacillus cereus* SV1. Proc Biochem 43:522–530
Margesin R, Palma N, Knauseder F, Schinner F (1992) Purification and characterization of an alkaline serine protease produced by a psychrotrophic *Bacillus* sp. J Biotechnol 24:203–206
Mazzarino MJ, Bertiller MB, Sain CL, Laos F, Coronato FR (1996) Spatial patterns of nitrogen availability, mineralization, and immobilization in northern Patagonia, Argentina. Arid Soil Res Rehabil 10:295–309
Mazzarino MJ, Bertiller MB, Sain C, Satti P, Coronato F (1998) Soil nitrogen dynamics in north-eastern Patagonia Steppe under different precipitation regimes. Plant Soil 202:125–131
Molière N, Turgay K (2013) General and regulatory proteolysis in *Bacillus subtilis*. In: Dougan DA (ed) Regulated proteolysis in microorganisms. Springer, Dordrecht, pp 73–103
Moran AJ, Hills M, Gunton J, Nano FE (2001) Heat-labile proteases in molecular biology applications. FEMS Microbiol Lett 197:59–63
Nielsen P, Fritze D, Priest FG (1995) Phenetic diversity of alkaliphilic *Bacillus* strains: proposal for nine new species. Microbiology 141:1745–1761
Enzyme Nomenclature (1992) Recommendations of the Nomenclature Committee of the International Union of Biochemistry and Molecular Biology on the Nomenclature and Classification of Enzymes. Academic Press, Orlando
Noy-Meir I (1973) Desert ecosystems: environment and producers. Annu Rev Ecol Syst 4:25–51
Olivera N, Siñeriz F, Breccia JD (2005) *Bacillus patagoniensis* sp. nov., isolated from the rhizosphere of *Atriplex lampa* in Patagonia, Argentina. Int J Syst Evol Microbiol 55:443–447
Olivera N, Sequeiros C, Siñeriz F, Breccia JD (2006) Characterization of alkaline proteases from a novel alkalitolerant bacterium *Bacillus patagoniensis*. World J Microbiol Biotechnol 22:737–743
Olivera N, Sequeiros C, Nievas ML (2007) Diversity and enzyme properties of protease-producing bacteria isolated from sub-Antarctic sediments of Isla de los Estados, Argentina. Extremophiles 11:517–526
Olivera NL, Prieto L, Carrera AL, Saraví Cisneros H, Bertiller MB (2014) Do soil enzymes respond to long-term grazing in an arid ecosystem? Plant Soil 378:35–48
Polgàr L (2005) The catalytic triad of serine peptidases. Cell Mol Life Sci 62:2161–2172
Prakash P, Jayalakshmi SK, Sreeramulu K (2010) Purification and characterization of extreme alkaline, thermostable keratinase, and keratin disulfide reductase produced by *Bacillus halodurans* PPKS-2. Appl Microbiol Biotechnol 87:625–633
Prieto LH, Bertiller MB, Carrera AL, Olivera NL (2011) Soil enzyme and microbial activities in a grazing ecosystem of Patagonian Monte, Argentina. Geoderma 162:281–287
Rao MB, Tanksale AM, Ghatge MS, Deshpande VV (1998) Molecular and biotechnological aspects of microbial proteases. Microbiol Mol Biol Rev 62:597–635
Raval VH, Pillai S, Rawal CM, Singh SP (2014) Biochemical and structural characterization of a detergent-stableserine alkaline protease from seawater haloalkaliphilic bacteria. Proc Biochem 49:955–962
Rawlings ND, Barrett AJ (1993) Evolutionary families of peptidases. Biochem J 290:205–218
Rawlings ND, Waller M, Barrett AJ, Bateman A (2014) MEROPS: the database of proteolytic enzymes, their substrates and inhibitors. Nucleic Acids Res 42:D503–D509

Russell RJ (1990) Cold adaptation of microorganisms. Philos Trans R Soc Lond B 326:595–611
Saeki K, Hitomi J, Okuda M, Hatada Y, Kageyama Y, Takaiwa M, Kubota H, Hagihara H, Kobayashi T, Kawai S, Ito S (2002) A novel species of alkaliphilic *Bacillus* that produces an oxidatively stable alkaline serine protease. Extremophiles 6:65–72
Shen J (2010) Enzymatic treatment of wool and silk fibres. In: Nierstrasz VA, Cavaco-Paulo A (eds) Advances in textile biotechnology. Woodhead, Cambridge, pp 171–192
Shrinivas D, Naik GR (2011) Characterization of alkaline thermostable keratinolytic protease from thermoalkalophilic *Bacillus halodurans* JB 99 exhibiting dehairing activity. Int Biodet Biodegrad 65:29–35
Siegert P, Wieland S, Engelskirchen J, Merkel M, Maurer K-H, Bessler C, Henkel AG, et al. (2009) Novel alkaline protease from *Bacillus gibsonii* and washing and cleaning agents containing said novel alkaline protease. Patent US 20090275493
Singh SK, Singh SK, Tripathi VR, Garg SK, Khare SK (2013) Downstream processing, characterization, and structure–function relationship of solvent-, detergent-, psychro-, thermo-, alkalistable metalloprotease from metal-, solvent-tolerant psychrotrophic *Pseudomonas putida* SKG-1 Isolate. Biotechnol Prog 29:99–108
Singhal P, Nigam VK, Vidyarthi AS (2012) Studies on production, characterization and applications of microbial alkaline proteases. Int J Adv Biotechnol Res 3:653–669
Sookkheo B, Sinchaikul S, Phutrakul S, Chen S-T (2000) Purification and characterization of the highly thermostable proteases from *Bacillus stearothermophilus* TLS33. Protein Express Purif 20:142–151
Souissi N, Ellouz-Triki Y, Bougatef A, Blibech M, Nasri M (2008) Preparation and use of media for protease producing bacterial strains based on by-products from cuttlefish (*Sepia officinalis*) and wastewaters from marine-product processing factories. Microbiol Res 163:473–480
Tatineni R, Doddapaneni KK, Potumarthi RC, Vellanki RN, Kandathil MT, Kolli N, Mangamoori LN (2008) Purification and characterization of an alkaline keratinase from *Streptomyces* sp. Bioresour Technol 99:1596–1602
Theron W, Divol B (2014) Microbial aspartic proteases: current and potential applications in industry. Appl Microbiol Biotechnol 98:8853–8868
Touioui SB, Jaouadi NZ, Boudjella H, Ferradji FZ, Belhoul M, Rekik H, Badis A, Bejar S, Jaouadi B (2015) Purification and biochemical characterization of two detergent-stable serine alkaline proteases from *Streptomyces* sp. strain AH4. World J Microbiol Biotechnol 31:1079–1092
Tropeano M, Vázquez S, Coria S, Turjanski A, Cicero D, Bercovich A, Mac Cormack W (2013) Extracellular hydrolytic enzyme production by proteolytic bacteria from the Antarctic. Pol Polar Res 34:253–267
Vranova V, Rejsek K, Formanek P (2013) Proteolytic activity in soil: a review. Appl Soil Ecol 70:23–32
Waghmare SR, Gurav AA, Mali AS, Nadaf NH, Jadhav DB, Sonawane KD (2015) Purification and characterization of novel organic solvent tolerant 98 kDa alkaline protease from isolated *Stenotrophomonas maltophilia* strain SK. Prot Expr Purif 107:1–6
Wang S-L, Chão C-H, Liang T-W, Chen C-C (2009) Purification and characterization of protease and chitinase from *Bacillus cereus* TKU006 and conversion of marine wastes by these enzymes. Mar Biotechnol 11:334–344
Wu J-W, Chen X-L (2011) Extracellular metalloproteases from bacteria. Appl Microbiol Biotechnol 92:253–262
Yum D-Y, Chung H-C, Bai D-H, Oh D-H, Yu J-H (1994) Purification and characterization of alkaline serine protease from an alkalophilic *Streptomyces* sp. Biosci Biotechnol Biochem 58:470–474

Chapter 12
Extremophilic Patagonian Microorganisms Working in Biomining

Patricia Chiacchiarini, Laura Lavalle, María Sofía Urbieta, Ricardo Ulloa, Edgardo Donati, and Alejandra Giaveno

Abstract The microorganisms known as extremophiles have become a powerful tool in the field of biotechnology. Among them, acidophilic and thermophilic microorganisms capable of oxidizing iron(II) or sulfur compounds are very important in ore-processing operations as they are able to enhance the dissolution of sulfide ores. The aim of this chapter is to describe the physiological and phylogenetic characteristics of the main acidophilic species and communities found in geothermal and mining environments in Neuquén Province, Patagonia Argentina, and the advances done by our research group in their application to biomining and bioremediation of heavy metals. Additionally, the chapter includes the description of a novel thermoacidophilic archaeon from the genus *Acidianus* (*Acidianus copahuensis*) autochthonous of the Copahue geothermal area isolated and characterized by our research group.

12.1 Introduction

The extremophiles are the microorganisms able to grow at one or more environmental conditions outside the range considered normal for human life, that is, the thermophiles, psychrophiles, halophiles, acidophiles, piezophiles, and alkaliphiles. Their capacity, not only to survive but also to function under harsh conditions, gives them vital importance in biotechnology, and they have forever changed our perceptions of the limits of living organisms. In this chapter we pay special attention to acidophiles because of their relevance in commercial-scale biomining and their potential applications in bioremediation (Du Plessis et al. 2007; Schippers 2007; Watling 2015).

P. Chiacchiarini • L. Lavalle • R. Ulloa • A. Giaveno (✉)
PROBIEN (CONICET, UNCo), Facultad de Ingeniería, Universidad Nacional del Comahue, Buenos Aires 1400, 8300 Neuquén, Argentina
e-mail: agiaveno@hotmail.com

M.S. Urbieta
CINDEFI (CONICET-UNLP), Fac. Cs. Exactas, UNLP, Calle 47 y 115, La Plata, Argentina

E. Donati
Industrial Fermentation Research and Development Center (CINDEFI), CONICET-UNLP, School of Science, National University of La Plata, La Plata, Argentina

N.L. Olivera et al. (eds.), *Biology and Biotechnology of Patagonian Microorganisms*, DOI 10.1007/978-3-319-42801-7_12

Acidophiles can be classified according to their optimum growth temperature: mesophiles up to ~40 °C, moderate thermophiles between ~40 °C and ~55 °C, and extreme thermophiles between ~55 °C and ~80 °C. The most relevant acidophiles to biomining applications are the iron- and sulfur-oxidizing bacteria and Archaea that grow autotrophically by fixing CO_2 from the atmosphere. The first used and better characterized leaching acidophilic bacteria belong to the genus *Acidithiobacillus* (formerly *Thiobacillus* (Kelly and Wood 2000) in the order *Gammaproteobacteria*. The species of this genus are the first isolates of extremely acidophilic sulfur- or iron(II)-oxidizing bacteria: the mesophilic *Acidithiobacillus (At.) thiooxidans* and *At. ferrooxidans* together with the moderately thermophilic *At. caldus*. Other important bacteria in bioleaching are the members of the genus *Leptospirillum* in the phylum Nitrospirae; they are also mesophilic, extreme acidophiles that derive energy only from the oxidation of iron(II) but not from sulfur compounds (Schippers 2007). All the extreme thermophilic metal sulfide-oxidizing microorganisms belong to the domain Archaea, specifically to the genera *Acidianus*, *Metallosphaera*, *Sulfolobus*, and *Sulfurococcus* in the order *Sulfolobales*; most of the mesophilic and moderately thermophilic Archaea used in biomining belong to the order *Thermoplasmatales* and the genus *Ferroplasma*.

Our research team has been working since 1997 in fields mainly related to isolation and characterization of acidophilic microorganisms from natural environments, especially from Neuquén Province, and their application in biotechnological processes such as metal ore bioleaching/biooxidation and heavy metals bioremediation at the laboratory scale. We have contributed to the study of the diversity of acidophilic microorganisms in Copahue-Caviahue geothermal region (CCG) using culture-dependent and culture-independent approaches (Chiacchiarini et al. 2009, 2010; Giaveno et al. 2013; Urbieta et al. 2012, 2014a). For some years now it has been clear that the high throughput of modern molecular ecology techniques to assess diversity and the need to obtain the microorganisms isolated for physiological and biotechnological studies oblige the use of a variety of techniques for the study of any habitat. The culture-dependent studies demonstrated that typically only 1–10 % of the total population from any biosphere is cultivable; in contrast, culture-independent approaches, mainly sequencing of the 16S rRNA gene, allow a more realistic picture of the ecosystems because almost the entire microbial community is considered. Nevertheless, traditional microbiological procedures remain essential because isolation and biochemical and physiological characterization of novel species in pure culture is fundamental to develop whole-cell applications (Lutton et al. 2013).

The aim of this chapter is to describe the physiological and molecular characterization of the acidophilic microorganisms isolated from different environments from Neuquén Province, Patagonian Argentina, and their applications in biomining and bioremediation of heavy metals. The main topics described are (1) a summary of the native acidophilic microorganisms isolated and characterized from the CCG system, from "La Silvita" and "Andacollo" mining areas in Neuquén, Argentina; (2) a description of a novel thermophilic archaeon isolated from the CCG system, its versatile metabolic characteristics and its phylogenetic characterization; and an outline of several examples of (3) biomining applications carried out at laboratory scale using ores from Argentinean mines and (4) heavy metals bioremediation assays.

12.2 The Copahue-Caviahue Geothermal System

12.2.1 Site Description

The Copahue-Caviahue geothermal (CCG) system is located on the Northwest corner of Neuquén Province, Patagonia Argentina. Its geology, geochemistry, volcanism, and thermalism have been studied during the past 25 years (Accorinti et al. 1991; Delpino and Bermúdez 1995; Mas et al. 1996; Vallés et al. 2004; Gammons et al. 2005; Caselli et al. 2006; Varekamp et al. 2006; Pedrozo et al. 2008). Nowadays, this area is still very attractive for diverse scientific fields (Vélez et al. 2011; Agusto et al. 2013; Temporetti et al. 2013; Monasterio et al. 2015). Interest in the biodiversity of microorganisms that inhabit this extremely low pH natural environment has increased significantly in recent years. Some of the studies focused on the ecology of the bacteria, yeast, plankton, and algae communities (Juárez and Vélez 1993; Pedrozo et al. 2001; Wendt-Potthoff and Koschorreck 2002; Lavalle et al. 2005; Russo et al. 2008; Chiacchiarini et al. 2009; Giaveno et al. 2009b; Chiacchiarini et al. 2010; Urbieta et al. 2012; 2014a; b).The CCG system has been the most important study area in our research projects and it deserves special attention. Table 12.1 shows the location, the physicochemical characteristics of the water, and the metal concentrations obtained by inductively coupled plasma mass spectrometry (ICP-MS) during a sampling campaign in the dry season.

The area includes the Copahue Volcano, an andesitic stratovolcano that presents a small acidic crater lake (pH 0.2–1.1) (37°51'S, 71°10.2'W), 2977 m a.s.l. On the east flank of the Volcano just below the crater lake an acid hot spring (NA) emerges and feeds the Rio Agrio, downstream there are two acidic hot springs, the South source (VA1) and the North source (VA2). Both hot springs feed the Upper Rio Agrio (URA, pH 0.5–2.5), the main water inputs of which are Rio Rojo, Rio Blanco, and Rio Jara (pH 6.0). After several waterfalls the URA discharges into the glacial Caviahue Lake (LC, pH 2.1–3.7), with a north and a south basin that overflows in one arm through the Lower Rio Agrio (LRA, pH 2.1–6.0) (Varekamp 2004; Pedrozo et al. 2008; Chiacchiarini et al. 2010). Approximately 10 km away LRA presents a great waterfall named "Salto del Agrio" (SA). Finally, the LRA mixes with the Rio Ñorquin and is further diluted (Varekamp et al. 2009). There is an anthropogenic influence on this area from the presence of two small villages, Caviahue and Copahue, that vary in population throughout the year, attracting many tourists to the therapeutic thermal complex in the summer and to a sky sports center in the winter. The water and sediment samples for isolation were collected during different field campaigns from URA and LRA, the crater lake of the volcano, Caviahue Lake (LC); and the hot springs of Copahue village: Las Maquinitas (LMi) and Las Máquinas (LMa). Pyrite, sulfur deposits, hematite, and jarosite were found on the banks of the streams and pools of the system. Four of the sites sampled, Laguna Verde Este (LVE), Baño 9 (B9), Agua del Limón (AL), and Laguna Sulfurosa (LS) are situated in the Copahue village. Three pools have been constructed to use the naturally acidic hydrothermal water and muds for therapeutic purposes. LVE received its name from the microalgae mats that cover its surface and borders

Table 12.1 Location and water chemistry of some sampling sites in the Caviahue geothermal region (CCG) system

Site	Location	pH	T (°C)	Conc. (mS cm^{-1})	SO_4^{2-}	Cl^-	Fe	K	Mg	Ca	Na	Mn
NA (Naciente del Agrio)	37°51′24″S, 71°09′14″W	0.8	70	65.1	5542.7	7408.5	816.1	485.0	552.2	13.2	867.4	18.8
VA1 (Vertiente Agrio Sur)	37°51′23″S, 71°09′04″W	0.8	45	63.1	8578.6	11428.6	834.6	504.9	696.2	14.8	911.7	21.2
VA2 (Vertiente Agrio Norte)	37°51′19″S, 71°09′09″W	1.0	42	42.0	8604.5	5053.8	537.8	151.6	341.7	17.1	364.9	10.2
CG (Cascada del Gigante)	37°53′11″S, 71°04′15″W	1.9	10	24.3	978.3	2239.8	510.0	147.5	281.2	13.0	326.0	11.3
CC (Cascada de la Culebra)	37°53′08″S, 71°04′14″W	1.9	10	24.5	932.9	2182.3	447.4	136.0	239.7	14.9	282.6	9.7
CV (Cascada de la Virgen)	37°52′59″S, 71°04′00″W	2.3	13	23.5	881.0	2124.9	450.6	134.5	266.3	15.0	282.6	10.1
LC (Lago Caviahue)	37°53′14″S, 71°02′46″W	3.1	16	1.17	280.9	48.2	21.3	4.4	18.1	3.2	14.0	0.7
PG (Puente Gendarmería)	37°49′42″S, 70°58′06″W	3.3	16	0.72	326.4	49.4	2.7	4.6	11.0	1.6	11.6	0.4
SA (Salto del Agrio)	37°48′43″S, 70°55′31″W	3.7	16	0.35	128.1	28.7	Nd	2.9	7.7	2.3	7.5	0.1
LS (Laguna Sulfurosa)	37°49′5.36″S, 71°5′54.38″W	3.0	55	1.12	611.7	28.7	3.2	12.6	4.8	5.8	25.2	0.1
LVE (Laguna Verde Este)	37°49′6.31″S, 71°5′49.01″W	2.4	28	2.60	410.8	3.4	6.6	8.7	1.8	3.2	17.3	0.1
B9 (Baño 9)	37°48′59.8″S, 71°5′48.52″W	2.0	50	3.38	196.6	13.8	7.7	5.4	2.2	0.8	12.6	0.1
AL (Agua del Limón)	37°49′0.13″S, 71°5′36.18″W	2.0	55	5.10	807.5	3.8	31.4	11.1	1.5	2.3	23.2	0.1
LMi (Las Maquinitas)	37°49′09″S, 71°05′12″W	2.5	85	8.60	2040.0	5.7	43.2	9.7	4.4	3.2	26.4	0.7
LMa (Las Máquinas)	37°50′2.61″S, 71°5′3.20″W	1.8	39	3.81	212.5	17.2	10.6	5.7	8.3	2.8	15	0.2

The metals and anions concentration, are expressed in mg l^{-1}

(Laguna Verde means green lagoon). The water of Laguna Sulfurosa (LS) is generally muddy from the presence of colloidal sulfur and gas that emerges in some places and becomes visible by bubbling at the surface. Las Maquinitas (LMi) site is located near the Copahue village and present the most extreme temperature found in the area. The sample site named Las Máquinas (LMa) is located in the homonymous geothermal manifestation area; there the water samples were taken from a moderate-temperature hydrothermal pool chosen because it is the least affected by human activity.

12.2.2 Isolation and Characterization of Native Acidophilic Microorganisms from the CCG System

In the CCG system, we have detected and isolated chemolitho-autotrophic and chemolitho-heterotrophic bacteria, archaea, yeasts, and filamentous fungi. This biodiversity is described in the following sections.

12.2.2.1 Mesophilic and Moderately Thermophilic Prokaryotes

In a first report, we described the isolation of iron-oxidizing strains of the genus *Acidithiobacillus* (on 9 K-agarose solid medium at pH 1.8), the physiology, and molecular characterization (Lavalle et al. 2005). The analysis of 16S rRNA genes by PCR/ARDRA using E*co*RI and E*co*RV enzymes was done to identify acidophilic mesophilic bacteria.

Figure 12.1 shows the influence of temperature on specific maximum growth rate (μ) at pH 1.8 on Fe(II) for isolated strains from the CCG system and a collection strain (DSM11477). At 30 °C almost all the isolates presented the highest μ

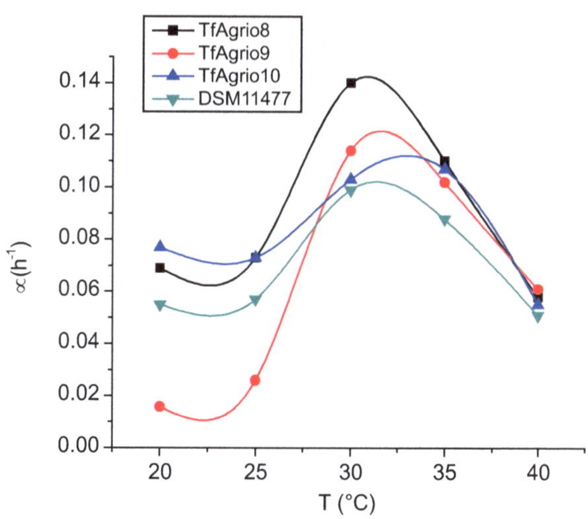

Fig. 12.1 Effect of temperature on specific maximum growth rate (μ) at pH 1.8 for iron(II)-oxidizing acidophilics isolated from the Caviahue geothermal region (CCG) system and a collection strain

values, except for TfAgrio10 that showed maximum μ values at 35 °C. All the native species of *Acidithiobacillus (At.) ferrooxidans* found in the CCG system proved to be less susceptible to 30 °C than the *At. ferrooxidans* collection strain (DSM11477), confirming the major capacity of indigenous strains for Fe(II) oxidation. One of the *At. ferrooxidans* species isolated (TfAgrio8) presented a highest specific growth rate on Fe(II) ($\mu = 0.14\ h^{-1}$) and a high productivity of iron(III) (5.6 mmol $Fe^{3+}\ l^{-1}\ h^{-1}$) at pH 1.8 and 30 °C. On the other hand, TfAgrio8 presented the highest sulfuric acid productivity (17.7 mmol l^{-1} day^{-1}) on sulfur growth. This better performance makes it suitable for possible use for acidification in the process of mineral leaching, solubilization of metals in groundwater, dissolution of phosphates, etc. (Lavalle et al. 2005).

To culture diverse acidophilic bacteria we used the overlay technique proposed by D.B. Johnson with the solid media FeTSBo and FeSo (Johnson 1995). Mesophilic iron-oxidizing bacteria grew at 30 °C as jarosite-encrusted colonies of varying size and morphology on FeTSBo plates from samples collected from different stations in Rio Agrio (CV, CG, LC, PG, SA, PT Puerta del Trolope, and AÑ on the confluence of the Rio Agrio with the nonacidic Rio Ñorquin) and the hot springs in Copahue (LVE, B9, LMi, and LMa). Additionally, moderately thermophilic iron-oxidizing bacteria developed on FeTSBo plates at 45 °C from samples collected from Upper Rio Agrio (NA, VA1, and VA2) and the hot springs in Copahue (LS, LVE, B9, AL, LMi, and LMa). Various species of *Leptospirillum*, also a chemolithotrophic iron(II)-oxidizing bacteria, were isolated and characterized from Rio Agrio (sampling stations CC, LC, PG, and SA) (Chiacchiarini et al. 2010).

On the other hand, several sulfur-oxidizing acidophiles were isolated in basal salt 0 K medium supplemented with 3 g l^{-1} sulfur at pH 3.0 from samples collected all through Upper Rio Agrio where pH was lower than 2.0. One of the isolates characterized as *At. thiooxidans* (TtAgrio6) presented the highest sulfuric acid productivity (26.5 mmol l^{-1} day^{-1}) compared to other native strains and to *At. thiooxidans* DSM11478 (24.3 mmol l^{-1} day^{-1}) (Fig. 12.2). TtAgrio6 was selected to study the feasibility of applying a bioleaching treatment to a contaminated municipal sludge, as we describe on Sect. 4.1.2.

Moderately thermophilic sulfur-oxidizing acidophiles grew at 45 °C on FeSo plates only from the sampling station NA from Upper Rio Agrio. *At. caldus*-like strains were isolated from B9, AL, and LMi.

Mesophilic and moderately thermophilic heterotrophic acidophiles grew on overlay plates inoculated with samples from VA1, VA2, LVE, B9, AL, LMi, and LMa. Heterotrophic colonies were gelatinous cream in color and presented various sizes and morphologies. Sulfate-reducing bacteria (SRB) were cultured at 30 °C and 45 °C from samples collected from the geothermal ponds LMa, LMi, and B9. Fluorescent in situ hybridization (FISH) analysis using specific probes on cultures from LMa samples showed that some of the microbial species were related to the genera *Leptospirillum*, *Sulfobacillus*, *Acidithiobacillus*, *Acidimicrobium*, and *Ferroplasma* (Giaveno et al. 2009a).

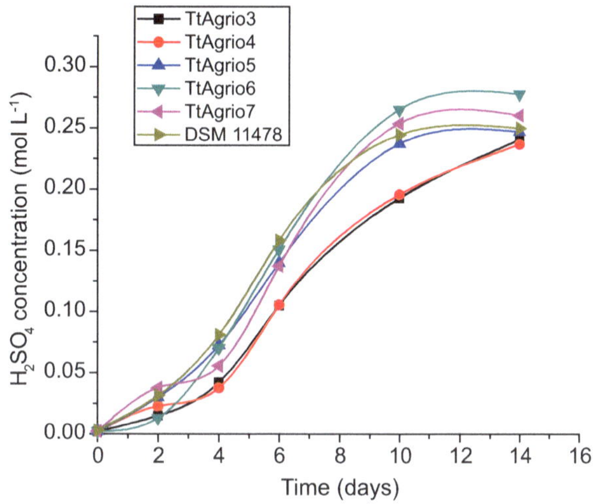

Fig. 12.2 Production of sulfuric acid of different sulfur-oxidizing species isolated from Upper Rio Agrio and the collection strain *Acidithiobacillus thiooxidans* DSM11478

12.2.2.2 Mesophilic Eukaryotes

In the past few years prokaryotic acidophilic microorganisms have been extensively studied in the CCG system as has been described in this chapter. However, the diversity of yeasts and filamentous fungi has received considerable less attention. The first study on yeast isolation from Rio Agrio, that also included the screening of metal-tolerant strains among the isolates, was reported by Lavalle et al. (2007). The concentration of yeast cells observed in the water samples was variable, between 45 and 149 cfu ml^{-1} in CC, CV, and SA samples. The genera *Cryptococcus* and *Rhodotorula* were the most abundant.

Isolated species such as *Rhodotorula mucilaginosa* and *Sporidiobolus salmonicolor* were identified by polymerase chain reaction-restriction fragment length (PCR-RFLP) analysis of the ITS1-5.8S-ITS2 region (Chiacchiarini et al. 2009). Russo et al. presented in 2008 the first complete assessment of yeast biodiversity in Rio Agrio.

The studies done by our research group exclusively focused on copper-, nickel-, cadmium-, and zinc-tolerant yeasts, with the aim of using them in metal uptake assays for bioremediation. A pink pigmented yeast isolate (Agrio-16) presented the highest tolerance to the mentioned metals. It was phenotypically characterized to species level according to the methods and keys proposed by Kurtzman and Fell (1998) and genotypically characterized by PCR-RFLP analysis with enzymes *Cfo*, *Hin*fI and *Hae*III. Both results confirmed that the isolate Agrio-16 belonged to the species *Rhodotorula mucilaginosa* (Esteve-Zarzoso et al. 1999; www.yeast-id.com database). Among filamentous fungi, species from the genera *Aspergillus* and *Penicillium* (at least three different species) were identified (Chiacchiarini et al. 2010).

12.2.2.3 Biodiversity in Geothermal Ponds and a Novel Thermoacidophilic Archaea: *Acidianus copahuensis*

The CCG system is the habitat of a wide variety of thermotolerant and thermophilic organisms, which are specially adapted to grow in this environment (Chiacchiarini et al. 2010; Urbieta et al. 2012; Giaveno et al. 2013). Urbieta et al. (2014a) recently reported the prokaryotic biodiversity of five representative ponds using two complementary molecular ecology techniques: phylogenetic analysis of 16S rRNA bacterial and archaeal genes and FISH (or CARD-FISH) for quantitative estimation of biodiversity. The results, supported by multivariate statistical analysis, showed that the biodiversity in Copahue ponds seemed to be determined by temperature. High-temperature ponds were dominated by Archaea, mainly apparently novel representatives from the orders *Sulfolobales* and *Thermoplasmatales* that had no close cultivated relatives. A deeper description of prokaryotic biodiversity on CCG system is given in the work published by Urbieta et al. (2012; 2014a; 2015a, b). These reports describe in detail the species that colonized Río Agrio and the several representative geothermal ponds in the area. The results presented allowed outlining preliminary geomicrobiological models in the CCG system. Detailed reading of these works is recommended to complete an overview of the ecological potential of the system.

Considering that most of the archaeal sequences detected in Copahue were distantly related to cultivated species and formed separate clades in the phylogenetic tree, it is possible to consider the CCG system as an excellent biological reservoir of potential novel species, mainly related to the sulfur cycle. In fact, we have reported a novel indigenous thermoacidophilic archaeon from the CCG system that we named *Acidianus copahuensis* ALE1 strain (DSM 29038).

We have done several studies to characterize this novel archaeon to understand its role in the iron and sulfur biogeochemical cycles in the CCG system and to evaluate its potential applications in biomining processes (Giaveno and Donati 2007; Giaveno 2010; Giaveno et al. 2011; Nodlovu et al. 2014). Some of these applications are described briefly later in this chapter.

Acidianus copahuensis has been isolated from three geothermal acidic hot springs (B9, LMi, and LMa) in Copahue. It is interesting to notice that although the three ponds presented high temperature and low pH values, the exact environmental conditions (especially temperature) differed were among the three (see Table 12.1), showing that *A. copahuensis* has a natural growth flexibility. The details of the isolation and characterization of *A. copahuensis* ALE1 strain were reported by Giaveno et al. (2013). The M88 culture medium, recommended by DSMZ (http:/dsmz.de/media/media88.htm) for the growth of thermoacidophilic archaea (Vartoukian et al. 2010; Stewart 2012), was supplemented with sulfur, tetrathionate, or yeast extract among other energy sources and cultivated at 65 °C. For solid-plate cultures Gellan Gum was used as the gelling agent because agar melts at temperatures above 50 °C (Lin and Casida 1984; Lindstrom and Sehlin 1989). The polar lipid analysis, carried out by thin layer chromatography, revealed the presence of diphosphatidylglycerol, phosphatidylinositol, phosphoglycolipids, and six different chromatographic mobility types of glycolipids. The cell morphology of *A. copahuensis* ALE1 strain was

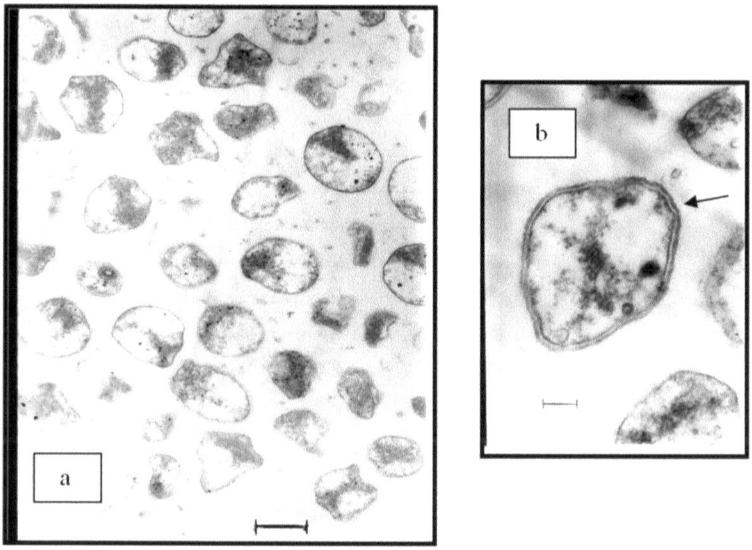

Fig. 12.3 Transmission electron micrographs (TEM) of *Acidianus copahuensis* cells growing in M88 medium. *Arrows* denote typical S-layer present in archaeal cells. *Bars* **a** 850 nm; **b** 210 nm

described by Giaveno et al. (2009b) using different microscopy techniques. Figure 12.3 shows the lobed cells of 0.5 to 1.0 μm with a typical archaeal envelope formed by a single membrane and covered by a paracrystalline glycoprotein layer known as the S layer that serves as protection against a hostile environment. These characteristics are similar to other members of the order *Sulfolobales*.

Phylogenetic analysis based on 16S rRNA gene sequence (Fig. 12.4) shows that strain ALE1 clusters together with members of the family *Sulfolobaceae* in the class *Thermoprotei*, within the phylum Crenarchaeota. However, the ALE1 strain appears in a separate branch from the other members of the genus *Acidianus* reported so far. This isolate has no close relatives according to the NCBI sequence database, not even considering uncultured species, which is even more surprising and reinforces the idea that *A. copahuensis* is autochthonous in the CCG system. Their closest relatives are members of the genus *Acidianus*, but they show low sequence similarity: *A. hospitalis* 91 %, *A. infernus* DSM 3191, *A. ambivalens* DSM 3772, *A. manzaensis* NA-1, and *A. sulfidivorans* DSM 18786 share 90 % of sequence similarity (Segerer et al. 1986; Fuchs et al. 1996; He et al. 2004; Yoshida et al. 2006; Plumb et al. 2007; Kletzin 2008; Liang et al. 2012).

The genome sequence of *Acidianus copahuensis* was obtained using a whole-genome shotgun (WGS) strategy with a 454-FLX Titanium pyrosequencer at INDEAR (Argentina), as recently reported by Urbieta et al. (2014b). The draft genome is 2,454,023 bases in length. The G-C content of the genomic DNA is 35.63 mol%. A total of 2,548 coding sequences (CDSs) and 52 structural RNAs (49 tRNAs and 3 rRNAs) were predicted. Forty-seven percent of the CDSs were classified as hypothetical proteins and 20 % as known enzymes; 34 % of the CDSs were

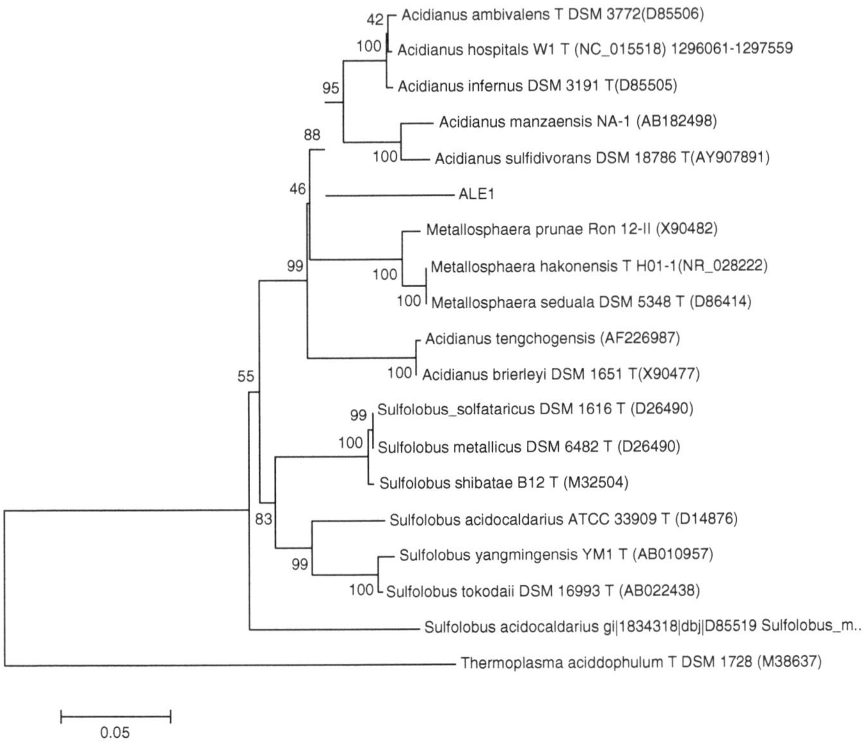

Fig. 12.4 Phylogenetic tree based on 16S rRNA gene sequences showing affiliation of strain ALE1 and selected members of the order *Sulfolobales*. Gene Bank accession numbers are given in *parentheses*. Tree was constructed using neighbour-joining and Kimura two-parameters methods. Bootstraps values are percentages based on 1000 resamplings. *T* type strain. *Bar* 0.05 sequence divergence

assigned to RAST subsystems. The genome of the *A. copahuensis* presents genes that might be related to the relevant metabolic features of the strain. Key enzymes for sulfur compound oxidation, such as sulfur oxygenase-reductase (SOR) and thiosulfatequinone oxidoreductase (TQO), were detected. Homologues of some of the Fox cluster enzymes, associated with iron oxidation, were also found. The genome presents proteins of the five major terminal oxidase complexes of *Sulfolobales*, so far reported in only two species (Auernik and Kelly 2008). Carbon fixation through the 3-hydroxypropionate-4-hydroxybutyrate cycle can be inferred by the presence of key enzymes of this pathway. An interesting discovery is the presence of aioAB genes encoding arsenite oxidase, which might be an indication of a bioenergetics use of arsenite. These genes were reported in *A. hospitalis* and *Sulfolobus tokodaii* genomes but not in other *Sulfolobales*. In an all-versus-all BLASTp comparison (e value, $1e^{-20}$) to *A. hospitalis* (the closest sequenced relative), *A. copahuensis* showed 789 unique proteins, and in a comparison with *Metallosphaera sedula* (another related archaeon with very similar metabolic features), 766 unique proteins were detected. This whole-genome shotgun (WGS) project has been deposited at DDBJ/EMBL/GenBank under the accession N°JFZT1160895 (http://www.uniprot.org/taxonomy/1160895) (Urbieta et al. 2014b).

It is already known that *Acidianus* species are thermoacidophilic Archaea (Segerer et al. 1986); nevertheless, the novel *A. copahuensis* ALE1 strain shows very interesting metabolic abilities even within the *Acidianus* genus. It was able to grow at all temperatures in a range between 55 °C and 80 °C in M88 medium supplemented with sucrose. Growth rate and cell numbers were influenced by temperature; the fastest growth was achieved at 75 °C with a generation time of 6.7 h. As pH, six different values from 1 to 5 were tested at 75 °C in M88 medium supplemented with sucrose. Growth was detected at all pH values, but was influenced by initial pH conditions. Optimal growth was detected at pH 3.0 with a generation time close to 7.0 h. Additionally, *A. copahuensis* was able to grow aerobically under heterotrophic and also autotrophic conditions at 75 °C when M88 medium was supplemented with sucrose, sulfur, or tetrathionate as energy source, respectively. Finally, *A copahuensis* is also able to grow anaerobically using iron (III) or sulfur as the electron acceptor and sulfur or hydrogen as the electron donor under autotrophic conditions. These results showed that *A. copahuensis* is capable of producing biosulfidogenesis under anaerobic conditions (oxidation of hydrogen and reduction of sulfur) (Giaveno et al. 2013).

12.3 Acidophilic Microorganisms from La Silvita and Andacollo Mining Areas

Besides microbiological studies on the CCG system, our group carried out the assessment of native microorganisms in different mineral areas of the Province of Neuquén. Metal sulfide ores are niches to a wide acidophilic microbiota that offer potential application in biotechnological processes.

La Silvita mine is an epithermal deposit with veins of lead in the andesitic tuffs of Eocene, located at the northeast of the town of Loncopué (38°01′22″S, 70°32′10″W). The average grade for the ore reported is 4.18% Pb, 3.44% Zn, 8.2 g t^{-1} Au, and 3.0 g t^{-1} Ag (Leanza et al. 2011). In the La Silvita polymetallic mine the samples were collected from an acid mine drainage with pH 1.5 and from an iron-rich runoff with pH 4.1. The search of microorganisms in these samples was only focused on the species *Leptospirillum ferrooxidans*. The influence of pH and temperature on the specific growth rate and ferrous iron oxidation rate of the isolated species was studied to assess their performance in bioleaching processes. All native *L. ferrooxidans* species showed slightly higher μ than the collection strain *L. ferrooxidans* DSMZ 2705 at pH 1.4 and 25 °C. For example, μ for isolate Lf-LS 04 and DSMZ 2705 at 25 °C and pH 1.4 were 0.050 h^{-1} and 0.020 h^{-1}, and the q_{Fe2+} were 0.19 and 0.17 g l^{-1} h^{-1}, respectively (Lavalle et al. 2008). Additionally, data obtained in growth kinetic assays carried out with native strains of *L. ferrooxidans* in batch culture at different temperatures and pH values were used to develop an artificial neural network (ANN) model. The ANN is a nonlinear estimation technique widely used in data processing that has recently gained much importance because of its range of applicability in assessing biological systems. The final ANN model proved to be an efficient and robust tool in predicting two variables simultaneously, the cell concentration and ferrous iron concentration, in the growth of strains of *L. ferrooxidans*, when the temperature, pH, and time were fed as inputs to the network (Lavalle et al. 2012).

Andacollo mining district, located between the towns of Andacollo and Huinganco, in the northwest of the Neuquén province, is another interesting location for the study of microbial biodiversity with biomining purposes. There gold mines are exploited by gallery excavation with the subsequent metal concentration carried out by froth flotation. Nine samples of water, silt, and rocks were collected from the mines Buena Vista and San Pedro, and their surrounding areas, to search for autochthonous bioleaching microorganisms. The sampling sites included some areas that are currently in operation and others that belong to previous waste stockpiles. Although the results shows that iron and sulfur oxidizers were present in all samples collected, a community obtained from a sample named Relave Viejo (an old stockpile) showed the best performance for reaching the highest iron and copper solubilization and acid production (Ulloa et al. 2012). The biodiversity of the environmental samples was studied by FISH and denaturing gradient gel electrophoresis (DGGE). The results show the presence of microorganisms belonging to the genera *Acidithiobacillus, Leptospirillum, Sulfobacillus, Desulfovibrio*, and *Acidovorax*. Most of the microorganisms of the genus *Acidithiobacillus* were related to *Acidithiobacillus (At.) ferrooxidans* but high intraspecific variability within this species was found. Archaea species were not detected by the methods used. At present, we are studying biodiversity on Relave Viejo sample by high-throughput sequencing of 16S rRNA genes and bioinformatics analysis tools. This work will allow us to increase knowledge of the biodiversity in the Andacollo mining habitat and to correlate it with other ecological parameters.

12.4 Biomining and Bioremediation Applications

Biomining is the generic term that describes the extraction of metal from ores and concentrates using microbiological technology. This area of biotechnology has seen considerable growth in scale and application since the 1960s following advances in our understanding of microorganisms suitable for the task and improvement in engineering designs for commercial biomining operations (Rawlings and Johnson 2007). Biomining includes two different processes: bioleaching and biooxidation. Bioleaching is an effective technology for metal extraction from low-grade ores and mineral concentrates because of its simplicity, low operating cost, and low environmental pollution. Metal recovery (copper, zinc, cobalt, nickel, etc.) from sulfide minerals is based on the activity of chemolithotrophic bacteria. In contrast to bioleaching wherein the metal of interest is solubilized, in the biooxidation process the valuable metal remains in the solid phase and the surrounding undesired mineral matrix is degraded by selective dissolution mediated by microorganisms. Mineral biooxidation is a viable technology specially used in the pretreatment of refractory sulfide gold ores and concentrates.

In the following sections we discuss the contributions of our research group to the use of native microorganisms from Patagonia in the bioprocessing of ores.

Industrial activities related to metals have increasing impact on the environment and ultimately on human life quality. To mitigate this impact, efforts must be made to enhance metal recovery from industrial waste products, sewage sludge, and soil contaminated with heavy metals. Bioremediation is an eco-friendly biotechnology that has shown successful results. Here, we also provide examples of bioremediation assays carried out with acidophilic bacteria and pigmented yeast isolated from the CCG system.

12.4.1 *Mesophilic and Moderately Thermophilic Microorganisms in Bioprocessing*

12.4.1.1 Bioleaching and Biooxidation

The characteristics of the most important microorganisms involved in these processes were mentioned in the introduction of this chapter. The main experiments carried out by our research group using indigenous bacteria from Patagonia, described in Sect. 3, are summarized in the following paragraphs. Bioleaching assays performed on La Silvita and Capillitas ores and the biooxidation of a gold concentrate from Andacollo are presented here.

La Silvita is a polymetallic sulfide mine where the main sulfides are sphalerite (ZnS), galena (PbS), pyrite (FeS_2), and chalcopyrite ($CuFeS_2$). The proportions of the valuable metals are 23.87 % Fe, 10.41 % Pb, 8.36 % Zn, and 0.06 % Cu. The capacity of two native *Leptospirillum ferrooxidans* species (Lf-LS02 and Lf-LS04) to catalyze the mineral leaching process was evaluated at laboratory scale (shake flask experiments). Lf-LS04 presented the best performance, achieving 100 % of zinc and copper recoveries in 20 and 12 days, respectively. Iron solubilization was 42 % after 33 days. These recoveries were greater than those obtained using the reference strain *L. ferrooxidans* DSMZ 2705 or in abiotic conditions (Lavalle et al. 2008). Isolate Lf-LS04 was selected to be used in the bioleaching of La Silvita ore in an airlift bioreactor (Giaveno et al. 2007a, b). The experiments were carried out in a 12-l reverse-flow airlift reactor using iron-free 9 K medium at pH 1.8, mineral particle size <74 μm, pulp density of 1 %, and superficial air velocity of 0.01 m s^{-1}. Bioleaching of copper was completed within 44 days; 98.0 % of zinc and 65.0 % of iron were leached after 65 days. The metal recovery from La Silvita sulfide ore with strain Lf-LS04 was higher than recoveries obtained previously with a collection strain of *Acidithiobacillus ferrooxidans*, showing the effectiveness of native species and Lf-LS04 potential in future commercial bioleaching operations.

Another sulfide ore from Capillitas (Catamarca, Argentina) containing mainly marcasite (FeS_2), sphalerite (ZnS), and chalcopyrite ($CuFeS_2$) was evaluated in a bioleaching experiment. In this case, a moderately thermophilic acidophilic consortium from the CCG system was used. The consortium included species related to *L. ferriphilum*, *At. caldus*, and an uncharacterized moderately thermophilic acidophilic heterotrophic bacterium. A shake flask experiment was performed at 42 °C, pH 2, pulp density of 2 % w/v, and particle size <74 μm. Zinc and copper solubilization was 80.0 % and 98.3 %, respectively, with daily maximum productivity of 4.8 ppm zinc and 63.8 ppm copper. This moderately thermophilic consortium proved to be 6 and 40 times more efficient than the traditional chemical solubilization for each metal, respectively, and also more efficient than another mesophilic consortium previously used (unpublished results).

Our research group also attempted the biooxidation of a gold concentrate from Andacollo area using an autochthonous consortium obtained from samples from Relave Viejo that showed the highest rates of iron and sulfur oxidation. The inoculated system reached 50.0 % iron solubilization whereas in abiotic systems iron was not released to the media. The presence of soluble iron indicated that the consortium used pyrite and probably arsenopytire as energy source. Gold recovery by cyanidation after biooxidation was 94.8 % in inoculated flasks and 67.0 % in abiotic systems (Ulloa et al. 2012). These results reinforce the advantage of using indigenous microorganisms in biomining processes.

12.4.1.2 Bioremediation

Our research group studies the capability of certain microorganisms to be used in the bioremediation of heavy metal-contaminated environments, mainly through different biotechnologies: bioleaching and biosorption. Here we describe two examples: the bioremediation of sludge by leaching using *At. thiooxidans* and a biosorption assay through yeast biomass.

A strain of *At. thiooxidans* (TtAgrio6) isolated from the CCG system was selected to evaluate the use of bioleaching to remediate a metal-contaminated municipal sludge coming from a secondary sedimentary tank from Parque Industrial of Neuquén wastewater treatment. Bioleaching experiments were carried out in shake flasks containing the sewage sludge and elemental sulfur. TtAgrio6 culture was able to solubilize 70 % of zinc and 25 % of chromium. At the end of the bioleaching process, the metal concentrations in the treated mud were below the legal permitted values (Chiacchiarini et al. 2010).

Biosorption has emerged as an important cost-effective alternative for the removal of toxic metal from industrial effluents because conventional techniques such as ion exchange, membrane filtration, oxidation–reduction, chemical precipitation, adsorption, reverse osmosis, and evaporative recovery are high cost and have negative impact on the environment (Wang and Chen 2009). Microorganisms such as yeasts are potential bioremediators, removing metals via active or passive uptake (Volesky 2003). Consequently, several yeast strains isolated from Río Agrio were tested for copper, nickel, cadmium, and zinc tolerance. The isolate Agrio-16, characterized as *Rhodotorula mucilaginosa*, was selected to study its ability to uptake copper and nickel from diluted solutions. The results showed that Agrio-16 was efficient in the removal of Cu(II) and Ni(II), especially at low pH values (2.5–4.5). The maximum sorption capacity for Ni(II) uptake was 62.4 mg g^{-1} and 68.9 mg g^{-1} at pH 2.5 and 4.5, respectively, and 160.5 mg g^{-1} for Cu(II) (Lavalle et al. 2007).

12.4.2 Thermophiles in Biomining

Although the first microorganisms used in biomining were essentially mesophiles or moderate thermophiles, biodiversity studies (especially sequencing of the whole microbial community by 16S rRNA genes) on samples from industrial applications, particularly tanks, began to reveal the presence of thermophilic bacteria and Archaea. Bioleaching and biooxidation are highly exothermic processes (more than 30 MJ kg^{-1} of oxidized sulfur). The use of mesophiles requires cooling systems that increase the cost of the processes and are energy consuming (Petersen 2010). Some of the thermophiles used in biomining are *At. caldus*, *Acidimicrobium*, *Sulfobacillus*, and *Ferroplasma* with an optimum temperature range between 45 °C and 55 °C and extreme thermophiles with optimum temperatures between 60 °C and 85 °C, as the Archaea from the genera *Sulfolobus*, *Acidianus*, and *Metallosphaera*. Also, thermophiles present a viable alternative to the bioleaching of refractory mineralogical species such as arsenopyrite (FeAsS), enargite (Cu_3AsS_4), and chalcopyrite ($CuFeS_2$) (Du Plessis et al. 2007).

12.4.2.1 Strategy to Optimize Gold Recovery

To evaluate the behavior of a thermophilic consortium, and more specifically the role of *Acidianus copahuensis*, in a biooxidation process we designed an assay using a polymetallic sulfide refractory gold ore concentrate processed in the Andacollo ore treatment plant and concentrated by flotation (Giaveno 2010). Metals with commercial interest, such as Au (160 ppm), Ag (288 ppm), Fe (24.1 %), Pb (2.8 %), Zn (1.2 %), and Cu (0.3 %) were detected in the sample. It is interesting to remark that arsenic concentration (1523.4 ppm) was high enough to have inhibitory effect. The mineralogical composition exhibited quartz (SiO_2), feldspar ($KAlSi_3O_8$), pyrite (FeS_2), arsenopyrite (AsFeS), sphalerite (ZnS), chalcopyrite ($CuFeS_2$), and traces of galena (PbS). The presence of pyrite and arsenopyrite in the ore indicated that much of the gold might be occluded in the mineral matrix, which is an obstacle to traditional commercial exploitation. The assay, that included an abiotic control, was performed in shake flasks at 150 rpm, 75 °C, with 1 % pulp density in M88 culture medium at pH 2. The abiotic control showed a rapid increase in pH, probably caused by the dissolution of basic species present in the ore. In inoculated systems pH decreased from the activity of sulfur-oxidizing microorganisms (Fig. 12.5). In spite of the high arsenic content in the ore, microbial activity was not completely inhibited.

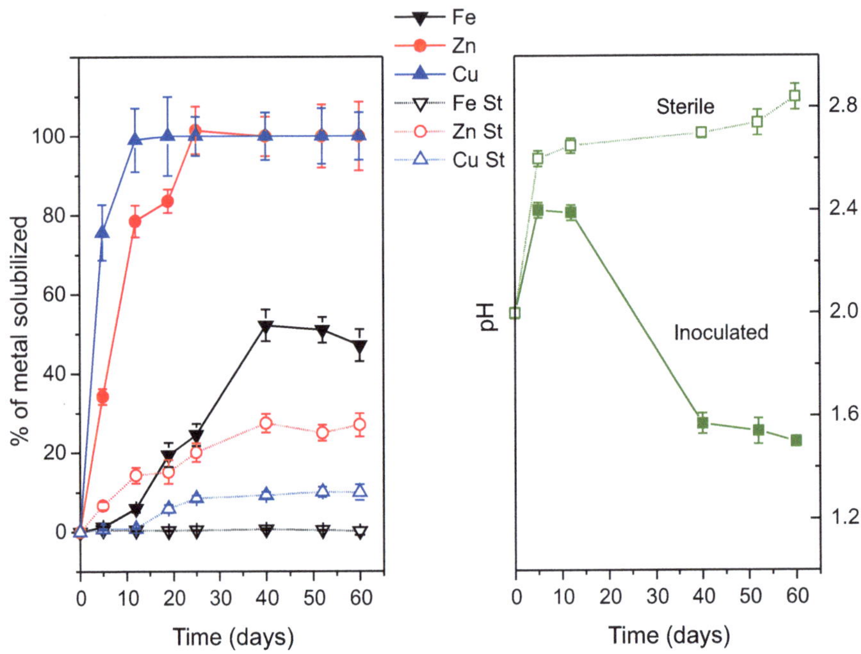

Fig. 12.5 Percentages of metals solubilized and pH evolution along the Andacollo ore biooxidation assay using a thermophilic consortium

The thermophilic consortium efficiently solubilized zinc and copper (Fig. 12.5). Iron solubilization is associated with pyrite dissolution, and it is a parameter that allows following the evolution of the biooxidation process. The abiotic control showed no appreciable iron solubilization because the increased pH favored precipitation of iron compounds and impeded pyrite dissolution. Gold recovery was similar to that obtained using a mesophilic native consortium from Andacollo mines; however, the thermophilic consortium achieved maximum gold recovery faster, increasing the process productivity (Giaveno 2010).

12.5 Conclusions

In our research work during the past years, we have detected and/or isolated quite interesting acidophilic species and consortia from regional mine ores and the Copahue Caviahue Geothermal system in Neuquén, Patagonia. Particularly, we have isolated, characterized, and reported a novel species of thermoacidophilic crenarchaeon, *Acidianus copahuensis*.

We have presented several reports concerning technical application in biomining bioremediation of heavy metals at the laboratory scale. Bioleaching using acidophilic iron- and sulfur-oxidizing microorganisms has been successfully applied to the recovery of metals from sulfide minerals and in the pretreatment (biooxidation) of refractory gold minerals.

On the other hand, techniques based on noncultured microorganisms have shown the great biodiversity present in mining environments such as Andacollo and in the fascinating CCG system. Our results show there is an extremophilic microbial richness waiting to be characterized and, most importantly, cultivated. To achieve these goals, it will be necessary to extend our understanding on various aspects of acidic, high-temperature ecosystems as well as the biochemistry and growth requirements of the extreme microbial life that inhabits them.

Acknowledgments Our research work has been supported by Universidad Nacional del Comahue and CONICET grants. The authors are very grateful to Dr. Ana Monasterio from EPROTEN (Ente Provincial de Termas del Neuquén) for getting the permission to work in the CCG area and the management and technical staff of CORMINE (Corporación Minera del Neuquén S.E.P.) for their invaluable help. The authors thank Professors and staff from Asentamiento Universitario Zapala (Fac. Ingeniería, UNComahue) for assistance during sampling and conditioning of the mineral samples.

References

Accorinti J, Squadrone M, Wenzel M, Perez A (1991) Valoración de las propiedades antimicrobianas del agua del volcán Copahue Neuquén Argentina. Arch Argent Dermatol 41:229–237

Agusto M, Tassi F, Caselli AT, Vaselli O, Rouwet D, Capaccioni B, Caliro S, Chiodini G, Darrah T (2013) Gas geochemistry of the magmatic-hydrothermal fluid reservoir in the Copahue–Caviahue Volcanic Complex (Argentina). J Volcanol Geoth Res 257:44–56

Auernik KS, Kelly RM (2008) Identification of components of electron transport chains in the extremely thermoacidophilic crenarchaeon *Metallosphaera sedula* through iron and sulfur compound oxidation transcriptomes. Appl Environ Microbiol 74:7723–7732

Caselli AT, Dapeña C, Agusto M, Delgado Huertas A (2006) Geothermal Copahue volcano system, Argentina. New stable isotope and geochemical data. In: Proceedings of the Vth South American Symposium on Isotope Geology, Uruguay, pp 332–336

Chiacchiarini P, Lavalle L, Giaveno A, Donati E (2009) Acidophilic microorganisms from geothermal Copahue Volcano system. Assessment of biotechnological applications. Adv Mater Res 71-73:87–91

Chiacchiarini P, Lavalle L, Giaveno A, Donati E (2010) First assessment of acidophilic microorganisms from geothermal Copahue-Caviahue system. Hydrometallurgy 104:334–341

Delpino D, Bermúdez A (1995) Eruptions of pyroclastic sulphur at crater lake of Copahue Volcano, Argentina. In: International Union of Geodesy and Geophysics. XXI General Assembly, Abstracts, p 128

Du Plessis CA, Batty JD, Dew DW (2007) Commercial applications of thermophile bioleaching. In: Rawlings DE, Johnson DB (eds) Biomining. Springer, Berlin, pp 57–80

Esteve-Zarzoso B, Belloch C, Uruburu F, Querol A (1999) Identification of yeast by RFLP analysis of the 5.8S rRNA gene and the two ribosomal internal transcribed spacer. Int J Syst Bacteriol 49:329–337

Fuchs T, Huber H, Burggraf S, Stetter KO (1996) 16S rDNA-based phylogeny of the archaeal order *Sulfolobales* and reclassification of *Desulfurolobus ambivalens* as *Acidianus ambivalens* comb. nov. Syst Appl Microbiol 1:1956–1960

Gammons CH, Wood SA, Pedrozo F, Varekamp JC, Nelson B, Shope CL, Baffico G (2005) Hydrogeochemistry and rare earth element behavior in a volcanically acidified watershed in Patagonia, Argentina. Chem Geol 222:249–267

Giaveno MA (2010) Biolixiación y biooxidación de minerales utilizando cepas nativas con capacidad termofílica Tesis de doctorado, Doctor en Cs. Exactas, área Química; Cs. Exactas UNLP. Available on line in http://hdl.handle.net/10915/2699

Giaveno MA, Donati E (2007) Bioleaching of a zinc sulfide ore by thermophilic consortia isolated from Copahue volcano. Adv Mater Res 20-21:79–82

Giaveno MA, Lavalle L, Chiacchiarini P, Donati E, Sand W (2007a). Airlift reactors: characterization and applications in biohydrometallurgy. In: Donati ER, Sand W (eds) Microbial processing of metal sulfides. Springer, Dordrecht, pp 169–191

Giaveno A, Lavalle L, Chiacchiarini P, Donati E (2007b) Bioleaching of zinc from low-grade complex sulfide ores in an airlift by isolated *Leptospirillum ferrooxidans*. Hydrometallurgy 89:117–126

Giaveno A, Chiacchiarini P, Cordero C, Lavalle L, Huergo J, Donati E (2009a) Oxidative capacity of native strains from Copahue geothermal system in the pretreatment of a gold sulfide ore. Adv Mater Res 71-73:473–476

Giaveno MA, Huergo J, Lavalle L, Sand W, Donati E (2009b) Molecular and morphological characterization of cultures from the extreme environmental area of Copahue Volcano-Argentina. Adv Mater Res 71-73:93–96

Giaveno MA, Pettinari G, González Toril E, Aguilera A, Urbieta MS, Donati E (2011) The influence of two thermophilic consortia on troilite (FeS) dissolution. Hydrometallurgy 106:19–25

Giaveno MA, Urbieta MS, Ulloa JR, Toril EG, Donati ER (2013) Physiologic versatility and growth flexibility as the main characteristics of a novel thermoacidophilic *Acidianus* strain isolated from Copahue geothermal area in Argentina. Microb Ecol 65:336–346

He Z, Zhong H, Li Y (2004) *Acidianus tengchongensis* sp. nov., a new species of acidothermophilic archaeon isolated from an acidothermal spring. Curr Microbiol 48:159–163

Johnson DB (1995) Selective solid media for isolation and enumerating acidophilic bacteria. J Microbiol Methods 23:205–218

Juárez AB, Vélez CG (1993) Sobre la presencia de *Chlorellakessleri* (Chlorococcales, Chlorophyceae) en aguas del Complejo Termal Copahue (prov. de Neuquen, Argentina). Bol Soc Argent Bot 29(1/2):105–107

Kelly DP, Wood AP (2000) Reclassification of some species of *Thiobacillus* to the newly designated genera *Acidithiobacillus* gen. nov., *Halothiobacillus* gen. nov. and *Thermithiobacillus* gen. nov. Int J Syst Evol Microbiol 50:511–516

Kletzin A (2008) Oxidation of sulphur and inorganic sulphur compounds in *Acidianus ambivalens*. In: Dahl C, Fredrich GC (eds) Microbial sulfur metabolism. Springer, Berlin, pp 184–200

Kurtzmann CP, Fell JW (1998) The yeasts: a taxonomic study. Elsevier, Amsterdam

Lavalle L, Chiacchiarini P, Pogliani C, Donati E (2005) Isolation and characterization of acidophilic bacteria from Patagonia Argentina. Process Biochem 40:1095–1099

Lavalle L, Portillo M, Chiacchiarini P, Donati E (2007) Heavy metal tolerance and copper uptake in yeast isolated from Patagonia Argentina. Adv Mater Res 20:639–642

Lavalle L, Giaveno A, Pogliani C, Donati E (2008) Bioleaching of a polymetallic sulfide mineral by native strains of *Leptospirillum ferrooxidans* from Patagonia Argentina. Process Biochem 43:445–450

Lavalle A, Curia L, Lavalle L, Giaveno A, Donati E (2012) Artificial neural network to predict the growth of *Leptospirillum ferrooxidans* in 9K defined medium. Int J Eng Res Appl 2:1406–1416

Leanza H, Arregui C, Carbone O, Danieli J, Vallés J Eds (2011) Relatorio del XVIII Congreso Geológico Argentino. Geología y Recursos Naturales de la Provincia del Neuquén. Buenos Aires. Asociación Geológica Argentina ISBN 978-987-22403-3-2

Liang CL, Xia JL, Nie ZY, Yang Y, Ma CY (2012) Effect of sodium chloride on sulfur speciation of chalcopyrite bioleached by the extreme thermophile *Acidianus manzaensis*. Bioresour Technol 110:462–467

Lin CC, Casida LE Jr (1984) Gerlerite as a gelling agent in media for the growth of thermophilic microorganisms. Appl Environ Microbiol 47:427–429

Lindstrom BE, Sehlin HM (1989) High efficiency of plating of the thermophilic sulfur-dependent archaebacterium *Sulfolobus acidocaldarius*. Appl Environ Microbiol 55:3020–3021

Lutton E, Schellevis R, Shanmuganathan A (2013) Culture-dependent methods increase observed soil bacterial diversity from Marcellus shale temperate forest in Pennsylvania. J Stud Res 2:9–16

Mas GR, Mas LC, Bengochea L (1996) Hydrothermal surface alteration in the Copahue Geothermal field (Argentina). Proceedings of the Twenty First Workshop Geothermal Reservoir Engineering, January 22-24, 1996. SGP-TR-151-34. Stanford University, California, 241–246

Monasterio A, Valles J, Pettinari G, Setti M, López-Galindo A, Baschini M (2015) Maduración de peloides en ambiente natural: experiencia en las Termas de Copahue. In Balnea10 IV CICAP BOI; Maraver F, Vela L, Ankli W, Eds. 255-264. ISBN 978-84-606-9368-0

Nodlovu S, Simate G, Mchibwa K, Giaveno MA (2014) Characterization of nanoprecipitates formed from the forced hydrolysis of bioleach liquors under different pH conditions. J Ind Eng Chem 20:3578–3583

Pedrozo F, Kelly L, Diaz M, Temporetti P, Baffico G, Kringel R, Friese K, Mages M, Geller W, Woelfl S (2001) First results on the water chemistry, algae and trophic status of an Andean acidic lake system of volcanic origin in Patagonia (Lake Caviahue). Hydrobiologia 452:129–137

Pedrozo FL, Temporetti PF, Beamud G, Diaz MM (2008) Volcanic nutrient inputs and trophic state of Lake Caviahue, Patagonia, Argentina. J Volcanol Geoth Res 178:205–212

Petersen J (2010) Modelling of bioleach processes: connection between science and engineering. Hydrometallurgy 104:404–409

Plumb J, Haddad CM, Gibson JA, Franzmann PD (2007) *Acidianus sulfidivorans* sp. nov., an extremely acidophilic, thermophilic archaeon isolated from a solfatara on Lihir Island, Papua New Guinea, and emendation of the genus description. Int J Syst Evol Microbiol 57:1418–1423

Rawlings DE, Johnson BD (eds) (2007) Biomining. Springer, Berlin

Russo G, Libkind D, Sampaio J, van Broock M (2008) Yeast diversity in the acidic Rio Agrio-Lake Caviahue volcanic environment (Patagonia, Argentina). FEMS Microbiol Ecol 65:415–424

Schippers A (2007) Microorganisms involved in bioleaching and nucleic acid-based molecular methods for their identification and quantification. In: Microbial Processing of Metal Sulfides. E. Donati and W. Sand Eds. Springer.The Netherlands. 3–33

Segerer A, Neuner AM, Kristjansson JK, Stetter K (1986) *Acidianus infernus* gen. nov. sp. nov. and *Acidianus brierleyi* comb. nov.: facultatively aerobic, extremely acidophilic thermophilic sulfur-metabolizing archaebacteria. Int J Syst Bacteriol 36:559–564

Stewart EJ (2012) Growing unculturable bacteria. J Bacteriol 194:4151–4160

Temporetti P, Snodgrass K, Pedrozo F (2013) Dynamics of phosphorus in sediments of a naturally acidic lake. Int J Sediment Res Issue 1:90–102

Ulloa R, Giaveno A, Donati E (2012) Adaptation of acidophilic microbial community from Andacollo to be used in biomining processes. XI Jornadas Argentinas de Tratamiento de Minerales, Neuquén, Argentina 301-306. ISBN978-987-604-311-3

Urbieta MS, González Toril E, Aguilera A, Giaveno MA, Donati E (2012) First prokaryotic biodiversity assessment using molecular techniques of an acidic river in Neuquén, Argentina. Microb Ecol 64:91–104

Urbieta MS, Toril EG, Giaveno MA, Bazan AA, Donati ER (2014a) Archaeal and bacterial diversity in five different hydrothermal ponds in the Copahue region in Argentina. Syst Appl Microbiol 37:429–441

Urbieta MS, Rascovan N, Castro C (2014b) Draft genome sequence of the novel thermoacidophilic archaeon *acidianus copahuensis* strain ALE1, isolated from the copahue volcanic area in Neuquén, Argentina. Genome Announc 2(3):e00259–14. doi:10.1128/genomeA.00259-14

Urbieta MS, González Toril E, Bazán ÁA, Giaveno MA, Donati E (2015a) Comparison of the microbial communities of hot springs waters and the microbial biofilms in the acidic geothermal area of Copahue (Neuquén, Argentina). Extremophiles 19:437–450

Urbieta MS, Porati GW, Segretín AB, González Toril E, Giaveno MA, Donati ER (2015b) Copahue geothermal system: a volcanic environment with rich extreme prokaryotic biodiversity. Microorganisms 3:344–363

Vallés JM, Baschini MT, Pettinari GR, García N (2004) Characterization of Muds and Waters of the Copahue Geothermal Field, Neuquen Province, Patagonia Argentina. 8th International Congress on Applied Mineralogy 1, 507-510. ISBN 85-98656-01-1, www.icam2004.org

Varekamp JC (2004) Copahue volcano: a modern terrestrial analog for the opportunity landing site? Eos 85(41):401–407

Varekamp J, Maarten deMoor J, Merrill M, Colvin A, Goss A, Vroon P, Hilton D (2006) Geochemistry and isotopic characteristics of the Caviahue-Copahue volcanic complex, Province of Neuquén, Argentina. Geol Soc Am Spec Pap 407:317–342

Varekamp JC, Ouimette AP, Herman SW, Flynn KS, Bermudez A, Delpino D (2009) Naturally acid waters from Copahue volcano, Argentina. Appl Geochem 24:208–220

Vartoukian SR, Palmer RM, Wade WG (2010) Strategies for culture of 'unculturable' bacteria. FEMS Microbiol Lett 309:1–7

Vélez ML, Euillades P, Caselli A, Blanco M, Martínez Díaz J (2011) Deformation of copahue volcano: inversion of InSAR data using a genetic algorithm. J Volcanol Geoth Res 202:117–126

Volesky B (2003) Sorption and biosorption. Sorbex, Montreal

Wang J, Chen C (2009) Biosorbents for heavy metals removal and their future. Biotechnol Adv 27:195–226

Watling HR (2015) Review of biohydrometallurgical metals extraction from polymetallic mineral resources. Minerals 5:1–60

Wendt-Potthoff K, Koschorreck M (2002) Functional groups and activities of bacteria in a highly acidic volcanic mountain stream and lake in Patagonia, Argentina. Microb Ecol 43:92–106

Yoshida N, Nakasato M, Ohmura N, Ando A, Saiki H, Ishii M, Igarashi Y (2006) *Acidianus manzaensis* sp.nov., a novel thermoacidophilic archaeon growing autotrophically by the oxidation of H_2 with the reduction of Fe^{3+}. Curr Microbiol 53:406–411

Chapter 13
Microorganisms from Patagonian Aquatic Environments for Use in Aquaculture

Cynthia Sequeiros, Marisa E. Garcés, Melania Fernández, Sergio F. Martínez Díaz, Diego Libkind, and Nelda Lila Olivera

Abstract Argentinean Patagonia is an extensive area scarcely explored from the microbiological point of view. Pristine and cold aquatic Patagonian environments as well as their fauna harbor a microbial diversity with interesting characteristics for application in aquaculture. This chapter presents the importance of aquaculture in recent decades and the problems associated with its fast and intensive development. We focus on Patagonian microorganisms and their products with potential application in aquaculture either as probiotics or as a source of specific nutritional components (i.e., astaxanthin). Thus, we analyze the probiotic characteristics of different bacteria from marine and freshwater environments. We also pay particular attention to antimicrobial agent production (i.e., bacteriocins) and adhesion properties of such strains. Moreover, we introduce Patagonian native yeasts as a promising natural source of astaxanthin and other valuable compounds for fish and crustacean aquaculture.

C. Sequeiros (✉)
Centro para el Estudio de Sistemas Marinos (CESIMAR-CCT CENPAT-CONICET), Blvd. Brown 2915, Puerto Madryn, Chubut, Argentina
e-mail: sequeiro@cenpat-conicet.gob.ar

M.E. Garcés • N.L. Olivera
Patagonian Institute for the Study of Continental Ecosystems (IPEEC-CENPAT), National Scientific and Technical Research Council (CONICET), Argentina

M. Fernández
Centro para el Estudio de Sistemas Marinos (CESIMAR - CCT CENPAT - CONICET), Blvd. Brown 2915, Puerto Madryn, Chubut, Argentina

Secretaría de Ciencia, Tecnología e Innovación Productiva de la Provincia de Chubut, Belgrano 914, Rawson, Chubut, Argentina

S.F. Martínez Díaz
Microbiology and Molecular Biology Laboratory, Interdisciplinary Center of Marine Science, CICIMAR-IPN, Playa el Conchalito sn, La Paz, Baja California Sur CP. 23090, Mexico

D. Libkind
Andean-Patagonian Institute for Biological and Geoenvironmental Technologies (IPATEC) CONICET, National University of Comahue, Argentina

N.L. Olivera et al. (eds.), *Biology and Biotechnology of Patagonian Microorganisms*, DOI 10.1007/978-3-319-42801-7_13

13.1 Introduction

Aquaculture is one of the fastest growing food-producing sectors and today provides almost half of all fish for human food (FAO 2014). With worldwide aquaculture expanding, the risks associated with the emergence and transmission of pathogenic bacteria have increased. The frequent use of antibiotics in aquaculture farms has led to the emergence of antibiotic multiresistant strains (Heuer et al. 2009). In addition, consumer awareness for safe food has also increased, highlighting the importance of searching for new strategies in aquaculture feeding and health management (Balcázar et al. 2006; Lara-Flores 2011).

Probiotics or beneficial microorganisms have emerged as remarkable feed additives in aquaculture, and now they are routinely incorporated into commercial fish feed formulations (Merrifield and Zhou 2011). During the past decade, much research has highlighted the potential benefits of probiotics for aquatic animal health and performance (Verschuere et al. 2000; Nayak 2010; Merrifield et al. 2010; Pérez-Sánchez et al. 2014). The use of microorganisms and their products as feed additives for aquaculture is also having a strong global growth (Hasan 2001). Aquaculture probiotics and natural feed additives are used for several purposes such as to enhance the immune system, to promote growth, to achieve desired meat and skin pigmentation, and to improve the organoleptic properties of farmed organisms (Hasan 2001; Lovatelli and Chen 2009).

Patagonia is one of the largest natural regions in southern South America, where aquatic environments include the glaciers and lakes of the Andean Patagonia, shallow lakes and rivers of the Patagonian Plateau, and extensive marine coasts (Quirós and Drago 1999). These diverse aquatic environments and their microbial communities still remain substantially understudied. Such mostly pristine environments constitute a source of microorganisms adapted to extreme conditions (low temperature, high ultraviolet radiation, etc.) with potential new biotechnological properties.

This chapter focuses on Patagonian microorganisms and their products with potential application in aquaculture either as probiotics or as a source of specific nutritional components. Particularly, we discuss the probiotic characteristics of bacterial strains from marine and freshwater environments, and the production of astaxanthin (used to provide a pinkish-red color to salmon, trout, and crustaceans) and other valuable compounds produced by native yeasts.

13.2 The Importance of Microorganisms in Aquaculture

13.2.1 Probiotic Bacteria as an Environmentally Friendly Alternative

In aquaculture, farmed animals are in constant interaction with the aquatic environment. Opportunistic pathogens in high densities can be ingested either with the feed or through the water (Verschuere et al. 2000). Thus, the adjacent ambient environment has a much larger influence on the health status of farmed aquatic animals

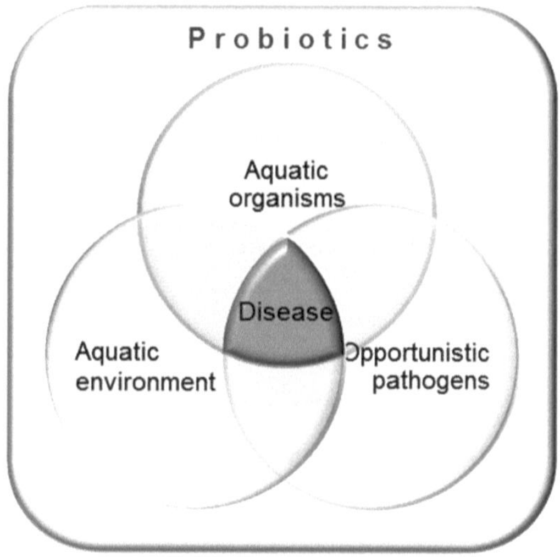

Fig. 13.1 Schematic representation of the relationship between probiotics and aquaculture components

than on terrestrial animals. Aquaculture probiotics are defined as live microorganisms that when provided via the diet or rearing water produce benefits on the host by modifying the host-associated or ambient microbial community, by ensuring improved digestion and feed utilization, by stimulating the immune system and reproductive efficiency, or by improving the water quality and controlling pathogens (Verschuere et al. 2000; Merrifield et al. 2010). Consequently, probiotic bacteria could benefit not only the host but also the surrounding environment.

Disease in aquatic organisms can result from imbalance among the physicochemical properties of the aquatic environment, the cultured organisms, and the presence of opportunistic pathogens within farm systems. Probiotics have the capability to act on these three components, preventing disease (Fig. 13.1).

Multiple mechanisms have been proposed to explain the beneficial effects of probiotics. The main actions are the production of antimicrobial compounds (i.e., organic acids, hydrogen peroxide, bacteriocins), the competition for adhesion sites and nutrients, the probiotic enzymatic contribution to host digestion, the enhancement of host immune responses, the improvement of water quality, and the disruption of quorum sensing (Verschuere et al. 2000; Balcázar et al. 2006; Merrifield et al. 2010). Figure 13.2 shows possible probiotic modes of action on aquaculture components.

In the assessment of the probiotic potential of a microorganism, an essential attribute is its ability to inhibit fish and other aquatic organism pathogens, which is evaluated through in vitro tests or in vivo challenge tests (Verschuere et al. 2000; Balcázar et al. 2007; Kesarcodi-Watson et al. 2008). Probiotic bacteria producing inhibitory compounds in the intestine of the host, on its surface, or in the water environment are thought to constitute a barrier against opportunistic pathogen proliferation, reducing disease problems and economic losses in aquaculture (Verschuere et al. 2000). As a result of their metabolism, some bacteria have the capability to produce antagonistic compounds such as hydrogen peroxide, diacetyl, siderophores, organic acids, and bacteriocins, which makes them good candidates as prophylactic and therapeutic agents (Verschuere et al. 2000; Gatesoupe 2008).

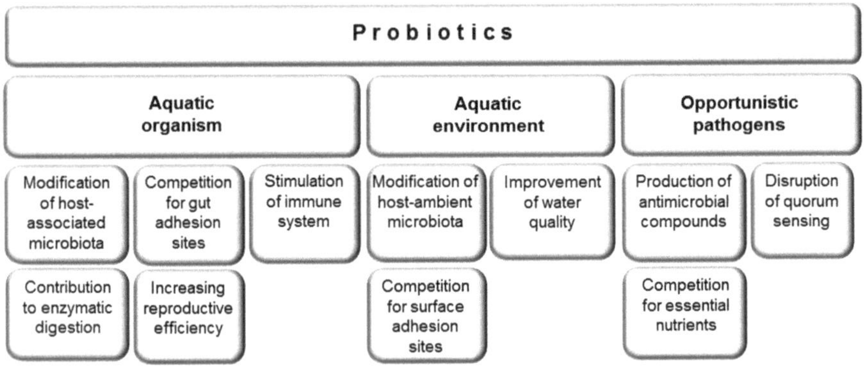

Fig. 13.2 Possible probiotic modes of action on aquaculture components

Another attribute considered as a good criterion for aquaculture probiotic preselection is the ability of a microorganism to compete for adhesion sites (Verschuere et al. 2000; Balcázar et al. 2006). This attribute is related with the microbial capability to adhere to epithelial cells and mucosal surfaces, which may contribute to protect skin surfaces, gills, and the gastrointestinal tract against pathogen invasion (Verschuere et al. 2000; Vine et al. 2004).

The sources and the criteria used for the isolation and the selection of microorganisms are critical to achieve efficient probiotics. This process should include rigorous microorganism assessments under in vitro and in vivo conditions, pilot testing, and scaling. Probiotic candidates should not be pathogenic, not only for the host species but also for aquatic animals in general and for human consumers (Merrifield et al. 2010). It would be desirable that new probiotics will not present a risk to the host or have neither virulence nor antibiotic resistance genes (Verschuere et al. 2000).

Frequently, novel probiotic bacteria are isolated from wild or cultivated organism microbiota and different environments. Bacteria with probiotic potential for aquaculture include gram-negative strains from the genera *Vibrio* (Austin et al. 1995; Riquelme et al. 2001), *Pseudomonas* (Spanggaard et al. 2001), *Shewanella* (Varela et al. 2010; Tapia-Paniagua et al. 2012), *Aeromonas* (Brunt and Austin 2005), and *Roseobacter* (Ruiz-Ponte et al. 1999). Gram-positive strains such as *Bacillus* spp. (Moriarty 1998; Rengpipat et al. 1998; Newaj-Fyzul et al. 2007; Ghosh et al. 2007; Zhou et al. 2009; Lin et al. 2012; Moreira de Souza et al. 2012) and lactic acid bacteria (LAB) (Robertson et al. 2000; Panigrahi et al. 2005; Kim and Austin 2006; Balcázar et al. 2007; Vendrell et al. 2008; Gatesoupe 2008; Merrifield et al. 2011; Pérez-Sánchez et al. 2011a) have also been applied in aquaculture. Many of the probiotic strains used in vivo belongs to LAB, mainly to the genera *Lactobacillus*, *Carnobacterium*, *Lactococcus*, *Leuconostoc*, *Enterococcus*, and *Pediococcus* (Pérez-Sánchez et al. 2014). *Bacillus* strains have also been extensively used in aquaculture (Pérez-Sánchez et al. 2014). Both gram-positive groups are Generally Recognized as Safe (GRAS) by the Food and Drug Administration (FDA), meaning that these bacteria are not harmful to animals or humans.

A large number of bacteria investigated for aquaculture applications were isolated from farmed fish or from different sources within hatcheries (Table 13.1). On the other hand, fewer strains come from pristine environments (Table 13.1). The latter are generally more sensitive to antibiotics than are hatchery isolates (Pérez-Sánchez et al. 2011b; Sequeiros et al. 2015). However, they are possibly not as adapted to hatchery conditions as those isolates from farms.

13.2.2 Yeasts as a Natural Source of Feed Additives

Yeasts are a large and heterogeneous group of eukaryotic microorganisms sorted within the phyla Ascomycota and Basidiomycota. They are widely distributed in nature and have been even found as part of the gut microbiota of both wild and farmed fish (for a review, see Gatesoupe 2007; Navarrete and Tovar-Ramírez 2014). Yeast are used in biotechnology for different purposes, such as production of fermented foods, alcoholic fermentation, recombinant protein and vitamin synthesis, and biological control (for a review, see Johnson 2013a, b). Because of its excellent nutritional content and availability (given their wide industrial applications), yeast and its components are extensively used in feed for terrestrial and aquatic animals. Currently, additional applications as a functional feed additive (probiotic live yeast), their cellular components (cell walls, yeast extracts), or as a source of products of higher purity such as β-glucans and nucleotides are being developed (Encarnacao 2016). Industrial yeasts are commonly used in aquaculture, either as live food organisms, or after processing as a feed ingredient (Stones and Mills 2004). It has been shown that yeasts or their derivates can enhance growth, survival, and gut maturation, and improve the immune and antioxidant systems in fish and shrimps (Burgents et al. 2004; Ozorio et al. 2010; Andrews et al. 2011). Yeasts are well known in animal nutrition because they can act as a producer of polyamines, which enhance intestinal maturation (Peulen et al. 2000); furthermore, they are good sources of vitamins B, E, and D, as well as single cell proteins, essential amino acids, lipids, and in some cases even carotenoids that can provide color and antioxidant activity. Comparable to bacteria, beneficial effects of some yeast species when used as probiotics are the results of simultaneous action of various mechanisms such as modulation of some aspects of local and systemic immune responses, trapping of bacterial toxin or pathogenic bacterial cells on the yeast surface, and maintenance of intestinal epithelium integrity (dos Santos Martins and Nicoli 2015). The larger cell size of yeasts (can be a hundred times larger than bacterial cells) may explain the fact that the introduction of a low yeast population (10^4 CFU g^{-1}) through the feed can induce beneficial effects in the host (Navarrete and Tovar-Ramírez 2014). The mechanisms responsible for probiotic yeast survival in the digestive system are still scarcely known. A cellular protective role of lipid droplets against stress conditions has been recently suggested (Zamith-Miranda et al. 2016). Several studies have shown that yeasts can successfully replace part of dietary fish meal in different fish species (Mahnken et al. 1980; Rumsey et al. 1991; Oliva-Teles and Goncalves 2001; Fournier et al. 2002; Olvera-Novoa et al. 2002; Sealey et al. 2007).

Table 13.1 Aquaculture probiotics from different sources tested in vivo

Probiotic strains	Source	Tested species	References
Vibrio sp. C33 and *Pseudomonas* sp. 11	Hatchery seawater (fish-rearing water), Chile	Scallop larvae (*Argopecten purpuratus*)	Riquelme et al. (2001)
Aeromonas media strain A199	American Type Culture Collection, ATCC 33907. Fish farm effluent, England	Silver perch (*Bidyanus bidyanus*)	Lategan et al. (2004)
Lactococcus lactis CLFP 100, *Leuconostoc mesenteroides* CLFP 196 and *Lactobacillus sakei* CLFP 202	Salmonids of fish farms, Spain	Rainbow trout (*Oncorhynchus mykiss*)	Balcázar et al. (2007)
Carnobacterium maltaromaticus and *C. divergens*	Rainbow trout from fish farms, Scotland	Rainbow trout (*Oncorhynchus mykiss*)	Kim and Austin (2006)
Leuc. mesenteroides CLFP 196 and *Lb. plantarum* CLFP 238	Salmonids of fish farms, Spain	Rainbow trout (*Oncorhynchus mykiss*)	Vendrell et al. (2008)
Lb. rhamnosus (GG)	American Type Culture Collection, ATCC 53103. Human feces, United States	Rainbow trout (*Oncorhynchus mykiss*)	Panigrahi et al. (2007))
Lb. plantarum CLFP 3, *L. lactis* CLFP 25 and *Leuc. mesenteroides* CLFP 68	Rainbow trout from fish farms, Spain	Rainbow trout (*Oncorhynchus mykiss*)	Pérez-Sánchez et al. (2011a)
Pseudomonas	Fish farms, Denmark	Rainbow trout (*Oncorhynchus mykiss*)	Spanggaard et al. (2001)
L. lactis AR21	Rotifer mass culture, Belgium	Rotifer (*Brachionus plicatilis*)	Shiri Harzevili et al. (1998)
L. lactis D1813	Wild shrimps from Tachibana Bay, Japan	Kuruma shrimps (*Marsupenaeus japonicus*)	Maeda et al. (2014)
Lb. paracasei subsp. *tolerans* F2	*Ramnogaster arcuata* from Bahıa Blanca Estuary, Argentina	Rainbow trout (*Oncorhynchus mykiss*)	López Cazorla et al. (2015)
Lb. pentosus H16	*Merluccius hubbsi* from coast of the Chubut Province, Argentina	*Artemia franciscana*	Garcés et al. (2015)
Carnobacterium sp. strain K1	Lønningdahl and Hardanger culture stations, Norway	Atlantic salmon (*Salmo salar*)	Robertson et al. (2000)

Despite all of this, only a few species of yeast are used commercially in the aquaculture industry (Table 13.2), the most widely commercialized species being *Saccharomyces cerevisiae*, the same species (although different strains) with which bread, some beer styles, some wine varieties, spirits, and bioethanol are produced (Coutteau et al. 1990). Therefore, many of these strains are better known as baker's

Table 13.2 List of relevant yeast species for aquaculture and examples of their use and benefits

Yeast	Taxonomy	Primary habitat	Aquaculture use	Known benefit	References
Saccharomyces cerevisiae	Ascomycota, Saccharomycetales, Saccharomycetaceae	Soil and bark	Commercial	Improved immune response, growth, feed efficiency, blood biochemistry, survival rate	Siwicki et al. (1994), Mohanty et al. (1996), Oliva-Teles and Goncalves (2001), Ortuño et al. (2002), Li and Gatlin (2004), Hisano et al. (2008)
S. cerevisiae var. *boulardii*	Ascomycota, Saccharomycetales, Saccharomycetaceae	Uncertain	Commercial	Faster digestive system maturation, protection against pathogens, enzymatic stimulation	Quentel et al. (2005), Wache et al. (2006)
Cyberlindnera jadinii (*Candida utilis*)	Ascomycota, Saccharomycetales, Phaffomycetaceae	Uncertain	Commercial, source of SCP	Immune response stimulation	Siwicki et al. (1994)
Debaryomyces hansenii	Ascomycota, Saccharomycetales, Debaryomycetaceae	Ubiquitous, halotolerant	Experimental	Faster digestive system maturation	Tovar-Ramirez et al. (2002, 2004)
Phaffia rhodozyma	Basidiomycota, Agaricomycotina, Cystofilobasidiales	Leaves and exudates	Commercial, Source of astaxanthin	Improved pigmentation, larval survival, resistance to pathogens	Scholz et al. (1999), Newaj-Fyzul et al. (2014)

SCP single cell protein

yeast or brewer's yeast. The former comes as a pure and primary culture, grown under strict conditions on a sugar substrate such as molasses, making a very consistent base for the production of yeast extracts, autolyzed yeast, yeast cell walls, and their derivate nucleotides and β-glucans. In contrast, brewer's yeast is normally generated as a by-product of beer production in brewery industries, and after alcoholic fermentation it eventually becomes a residue that can be exploited for aquaculture feed. The nutritional content is similar but less consistent to that in baker's yeast, and contains more trace minerals such as selenium and chromium (Tacon 2012). Another source of *S. cerevisiae* yeasts is from bioethanol production industries where they are harvested after having performed alcoholic fermentation of sugarcane, beet sugar, or grain syrup. Selling prices are normally low for this type of yeast; however the quality, and the protein content, are less consistent (Tacon 2012). Within *Saccharomyces cerevisiae*, the probiotic attributes of the *boulardii* subgroup of strains have been attributed to effects on enteric pathogens, intestinal barrier function integrity, antiinflammatory effects, immune stimulation, and trophic effects on the intestinal mucosa (Jacobson 2015). Current research is identifying new yeast probiotics outside the *S. cerevisiae* ssp. *boulardii* cluster. For example, *Candida utilis* (current valid name is *Cyberlindnera jadinii*), also known as torula yeast, is being used for some time in aquaculture given it has been tested as a protein source in diets for aquatic animals, and it has been found that feed acceptance and survival are not affected when used as partial fish meal replacement (Martin et al. 1993). Experiments with this yeast have been carried out using tilapia (Olvera-Novoa et al. 2002), abalone (Britz 1996), and shrimp (Gamboa-Delgado et al. 2016), among others. Nowadays, this yeast has been commercially available for decades given that several companies have achieved massive production of this yeast for commercial purposes. In addition, the halotolerant yeast *Debaryomyces hansenii* has been considered an excellent probiotic candidate in fish aquaculture as in several experiments it showed capability to enhance growth, survival, and gut maturation and to improve the immune and antioxidant systems in fish larvae and juveniles (Navarrete and Tovar-Ramírez 2014).

Another important yeast for aquaculture is *Phaffia rhodozyma* (synonym *Xanthophyllomyces dendrorhous*). In contrast to previous cases, this yeast belongs to the basidiomyectous fungi and possesses several unique characteristics, being the most important in terms of this chapter for its unique ability to accumulate the valuable carotenoid pigment astaxanthin (Andrewes and Starr 1976). Astaxanthin is fundamental for fish and crustacean coloration given animals cannot synthesize carotenoids, which must be provided in their feed for deposition into the flesh, carapace, or plumage (fish, crustaceans, or birds, respectively). Astaxanthin is valuable commercially mainly for pigmentation of salmonids raised in aquaculture. It is the most expensive ingredient in salmonid feeds, and it has an estimated market more than US $200 million per year (Schmidt et al. 2011). This market is supplied primarily by chemically synthesized astaxanthin and, based upon not just its applications in coloring but also its antioxidative and health-promoting properties (Guerin et al. 2003), the global astaxanthin commercial market has grown continuously in the last decade (Schmidt et al. 2011). However, because the synthesis is relatively difficult and expensive compared to simpler carotenoids such as β-carotene, several companies and academic laboratories have investigated biological sources of astaxanthin. Among these sources, *P. rhodozyma* and the microalga *Haematococcus pluvialis* have attracted much interest because of their ability to synthesize high levels of astaxanthin and because *P. rhodozyma* can reach high cell densities in fermentor cultures (>50 g dry cell weight per liter) (Echevarri-Erasun and Johnson 2002). Since its discovery in 1977 it was suggested that *P. rhodozyma* could be a source of astaxanthin for salmonids raised in aquaculture (Johnson et al. 1977, 1980a) and as a carotenoid source for other animals (Johnson et al. 1980b). One of the main drawbacks of utilizing *P. rhodozyma* as a commercial astaxanthin source is the low levels of astaxanthin found in wild isolates and the thick cell wall and capsule of the yeast, which apparently hinders astaxanthin uptake (Echevarri-Erasun and Johnson 2002). However, through mutation strategies (chemical mutagenesis and random screening) several astaxanthin-hyperproducing strains have been developed that accumulate more than 10,000 $\mu g\ g^{-1}$ dry yeast. Companies have adopted some of these strains for industrial production and commercialization of astaxanthin. These hyperproducing mutants have been valuable for the elucidation of genes involved in astaxanthin biosynthesis, which allowed genetic engineering as an alternative way of generating improved strains. For this, genetic tools were developed that allow directed transformation of *P. rhodozyma* strains (Adrio and Veiga 1995).

Regarding the limited absorption of astaxanthin because of the yeast's thick cell wall, interesting processes have been developed to circumvent this problem. For example, the co-inoculation with *Bacillus circulans* into fermentors of growing *P. rhodozyma* facilitated uptake of the pigment by rainbow trout by the secreted lytic enzymes that partially hydrolyzed the cell wall (Johnson 2003; Okagbue and Lewis 2008). Mechanical breakage also facilitated uptake of the astaxanthin in the flesh of rainbow trout. Other means of pigment liberation by the yeast, including optimization of drying conditions, mechanical breakage, and enzyme treatment, have been developed by the biotechnology industries (Schmidt et al. 2011).

The accumulating knowledge of the biology, genomics, and metabolomics of *P. rhodozyma* will have a marked impact on the development of improved strains and industrial production processes.

13.3 Patagonian Microorganisms and Their Products with Potential Application in Aquaculture

13.3.1 Probiotic Potential of Bacteria from Marine and Freshwater Environments

To obtain new and more effective probiotics, it is desirable to explore environments with microorganisms adapted to extreme environmental conditions, which may be useful in varying conditions that take place in aquaculture production. The Patagonian region is known for possessing unique features, including lower temperatures than in similar areas of the Northern Hemisphere during winter periods and particularly high temperatures in summer (Paruelo et al. 1998). It is expected that under these conditions highly tolerant bacteria may be isolated to be used in fluctuating environmental conditions.

In Argentina, the screening of bacteria with probiotic properties for use in aquaculture started a few years ago (Pasteris et al. 2009, 2011; Sequeiros et al. 2010, 2015; Sica et al. 2012; Fernández 2013; Garcés et al. 2015; López Cazorla et al. 2015). These studies characterized the strains through in vitro and in vivo experiments. Table 13.3 shows some bacteria isolated from organisms (algae, invertebrates, and fish) and from marine sediments (northeast coast of Chubut Province) and freshwater sediments (Rio Santa Cruz, Rio Chubut) in Patagonia, with antimicrobial activity against fish pathogens.

Lactococcus lactis TW34 showed a strong inhibitory activity against *Lactococcus garvieae* and other gram-positive fish pathogens (Table 13.3). *L. garvieae* is the etiological agent of lactococcosis, an emerging pathology affecting different wild and farmed fish species all over the world (e.g., rainbow trout, yellowtail, tilapia, and catfish) in both marine and freshwater aquaculture (Vendrell et al. 2006; Aguado-Urda et al. 2011; Reimundo et al. 2011). TW34 antimicrobial activity is the result of the production of a bacteriocin that was identified as Nisin Z (Sequeiros et al. 2015). This bacteriocin was highly thermostable, retaining antimicrobial activity even after autoclaving (121 °C for 15 min), (Sequeiros et al. 2010). It was also very stable in a wide pH range (3–7), showing a slight activity decreasing from pH 9 up to 11 (Sequeiros et al. 2010). Moreover, its activity remained constant up to 6 months when it was stored at −20 °C (Sequeiros et al. 2010). Figure 13.3 shows Nisin Z structure, a lantibiotic peptide containing thioether-bridged amino acids (lanthionines and methyllanthionines) (Chatterjee et al. 2006). These thioether rings are responsible for the high thermostability, acid tolerance, and high antimicrobial activity of this class of bacteriocin (de Vuyst and Vandamme 1994; Kuipers 2010). To date, only two strains from marine origin have been described as Nisin Z producers: *L. lactis* subsp. *lactis* isolated from the intestine of olive flounder from Busan, Korea (Heo et al. 2012) and *L. lactis* TW34 (Table 13.3). Interestingly, TW34 bacteriocin activity was stable at a wider pH range than bacteriocin from *L. lactis* from olive flounder. The knowledge about bacteriocins from marine microbial sources is at an early stage. Nisin is the only bacteriocin approved for food applications, and the European Union regulates the authorization, marketing, and use of probiotics as feed additives. *L. lactis* TW34 as a Nisin-producing strain would be a promising alternative for application as a feed additive in the prevention of lactococcosis in aquaculture systems (Sequeiros et al. 2015). Moreover, *L. lactis* TW34 showed high sensitivity to a wide variety of antibiotics, which is highly valuable for the safe use of the probiotic (Sequeiros et al. 2010). As the optimal temperature for TW34 antimicrobial activity production was as low as 15 °C (four times higher than at 30 °C), this strain is particularly attractive for salmoniculture (Sequeiros et al. 2010). For example, rainbow trout is sensitive to lactococcosis and its optimum water temperature for culture is below 21 °C (FAO 2005–2015).

Another interesting marine bacterium isolated from the Patagonian coast was *Lactobacillus pentosus* H16 (Table 13.3). Its antimicrobial activity against *Vibrio*

Table 13.3 Bacteria with antimicrobial activity against fish pathogens isolated from aquatic environments and organisms of northeastern Patagonia

Strain	Place	Source	Inhibited fish pathogens	Reference
Lactococcus lactis TW34	Coast of Chubut Province	*Odontesthes platensis*	*L. garvieae; L. piscium; C. piscicola; Sp. Iniae*	Sequeiros et al. (2010)
Carnobacterium sp. T4	Santa Cruz River	*Oncorhynchus mykiss*	*C. piscicola*	Garcés et al. (2014)
Carnobacterium sp. T15	Santa Cruz River	*Oncorhynchus mykiss*	*C. piscicola; L. garvieae*	Garcés et al. (2014)
Carnobacterium sp. M15	Santa Cruz River	*Oncorhynchus mykiss*	*C. piscicola*	Garcés et al. (2014)
Lactobacillus pentosus H16	Coast of Chubut Province	*Merluccius hubbsi*	*A. salmonicida; V. alginolyticus*	Garcés et al. (2015)
Bacillus sp. T39	Coast of Chubut Province	*Pinguipes brasilianus*	*C. piscicola; Ls. anguillarum*	Fernández et al. (2015)
Carnobacterium sp. PM22	Coast of Chubut Province	*Enteroctopus megalocyathus*	*A. salmonicida*	Fernández et al. (2015)
Lactobacillus sp. M26	Coast of Chubut Province	*Acanthistius patachonicus*	*L. garvieae; C. piscicola; A. salmonicida; Y. ruckeri; V. alginolyticus; Ls. anguillarum*	Fernández et al. (2015)

A. Aeromonas, C. Carnobacterium, Ls. Listonella, L. Lactococcus, Sp. Streptococcus, V. Vibrio, Y. Yersinia

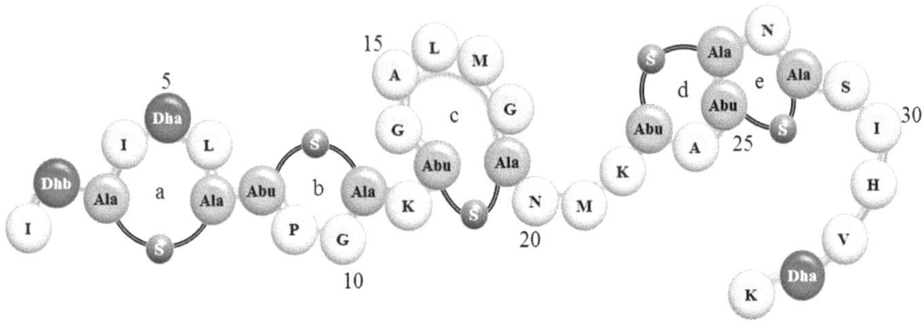

Fig. 13.3 Schematic representation of the primary structure of posttranslationally modified Nisin Z. Lanthionine and methyllanthionine characteristic structures forming the thioether rings are labeled *a–e*. Dehydrated residues are *dark shaded* and those forming lanthionine and methyllanthionine bridges are *clear shaded*

alginolyticus and *Aeromonas salmonicida* was associated with a pH decrease, possibly caused by the production of organic acids characteristics of LAB metabolism, which suppresses the growth of many other bacteria (Garcés et al. 2015). In challenge in vivo experiments, H16 prevented *Artemia franciscana* mortality when it was administered previously to *V. alginolyticus* inoculation (Garcés et al. 2015). These in vivo results were attributed to several H16 protective mechanisms against fish pathogens, including its antimicrobial activity and its capability to produce cell-bound biosurfactants and selectively adhere to mucosal surfaces. All this conferred H16 competitive advantages for a rapid colonization of the intestinal tract and the displacement of pathogenic strains.

13.3.2 Native Yeasts as Source of Valuable Compounds

Patagonian natural habitats have proven to be an incredible resource of biotechnologically relevant yeasts, as is depicted in Chap. 18 of this book. Among these, a few species have potentiality for aquaculture and are reviewed in this section.

As previously mentioned, *Saccharomyces*, in particular *S. cerevisiae*, is by far the most used and studied yeast, and thus the yeast is subjected to larger research and experimental trials in aquaculture. However, several other yeasts are becoming gradually more important, such as *Cyberlindnera jadinii* (*Candida utilis*), *Debaryomyces hansenii*, and *Phaffia rhodozyma* (Table 13.2).

Availability of yeast biomass in Patagonia is mainly related to the wine industries and most recently to the high number of craft breweries that emerged in the Andean Patagonia. Although brewer's yeast is being currently used for animal feed, to the best of our knowledge none of this yeast biomass has been yet tested for the use in local aquaculture. Autochthonous strains of *Saccharomyces cerevisiae* in Patagonia are mostly related to wineries (reviewed in Chap. 15 of this book); however, at least two isolates were obtained from natural environments (Libkind et al. 2011). However, the dominant *Saccharomyces* species in the region are the psychrotolerant *S. uvarum* and *S. eubayanus*, both of which are considered natives of the Andean Patagonian forests (Libkind et al. 2011). The latter was originally described in Patagonia and is considered one of the ancestors of the lager yeast, with which lager beer is brewed. The application of *S. eubayanus* for the brewing industry is a matter of current study, which will possibly result in innovative beers but also in a good source of new yeast biomass for its assessment as a feed additive for aquaculture. We believe both species have potential as a source of SCP (single cell protein) and vitamins, but whether they have any probiotic activity as their sister species *S. cerevisiae* remains to be tested. The fact that these species are cold adapted might ensure better fitness of survival in the aquaculture systems of Patagonia when compared to mesophilic yeasts such as *Cyberlindnera jadinii* and *S. cerevisiae*. Further studies are required to assess the potentiality of these Patagonian autochthonous yeasts for aquaculture.

Several studies have accumulated relative to the biodiversity and biotechnology of carotenoid-synthesizing yeasts native from Patagonian natural environments, in particular, Andean water bodies of glacial origin. Interesting yeasts belonging to the basidiomycetous genera *Rhodotorula*, *Rhodosporidium*, *Sporobolomyces*, *Sporidiobolus*, *Cystofilobasidium*, *Dioszegia*, and *Phaffia* were isolated and identified from Patagonian natural habitats (Libkind et al. 2003, 2009a; de García et al. 2007; Russo et al. 2008; Brandão et al. 2011). Novel species from Patagonia for all these genera, except *Phaffia*, were obtained and many have already been formally described (Libkind et al. 2005, 2009b, 2010). The ability to produce cell biomass and carotenoid pigments for specific isolates and species has been tested using semi-synthetic medium (MMS), and agro-industrial byproducts (cane molasses, corn syrup, raw malt extract) as carbon sources. Maximum pigment production (300 $\mu g\ g^{-1}$) was achieved by the strain CRUB 1046 (Libkind and van Broock 2006). This strain was later recognized as a novel yeast species and was described as *Cystofilobasidium lacus-mascardii* (Libkind et al. 2009b). We also studied fatty acids (FA) profiles in six carotenoid-producing yeast species isolated from temperate aquatic environments in Patagonia. Total FAs ranged from 2 % to 15 % of dry biomass. Linoleic, oleic, palmitic, and α-linolenic acids were the major FA constituents, which accounted for as much as 40 %, 34 %, 13 %, and 9 % of total FAs, respectively. The high percentage of polyunsaturated fatty acids (PUFAs) found in Patagonian yeasts, in comparison to other yeasts, is indicative of their cold-adapted metabolism. Our results suggested that Patagonian yeasts may be considered an interesting source of essential PUFAs. *C. lacus-mascardii* produced about 13 % of its dry weight in lipids constituted mainly by linoleic acid (Libkind et al. 2008a).

As a result of their multiple interesting traits, the production of biomass and carotenoid pigments by the Patagonian native species *Cystofilobasidium lacus-mascardii* was optimized using factorial design and surface response techniques and then scaled to 10-l fermentors. After the optimization, the yeast produced 33 % more carotenoid pigments and the composition detected was torularodin (16.9 %), torulene (25 %), γ-carotene (20.3 %), β-carotene (6.7 %), and several unknown pigments (31.1 %) (Libkind 2006). Live yeast biomass of this species was later applied during 2 months as feed additive for reared rainbow trout, showing good acceptance and no evidence of negative effects when compared to controls (Libkind and Baez unpublished).

Probably the most interesting case is that of the native Patagonian populations of the astaxanthin-producing yeast *Phaffia rhodozyma*, which were obtained from fungal stromata samples in Andean Patagonia and in fewer occasions also from freshwater samples (Libkind et al. 2007, 2008b; Brandão et al. 2011). Patagonian populations were found to be genetically different from those found in the Northern Hemisphere and those later obtained in natural environments in New Zealand and Australia (Libkind et al. 2007; David-Palma et al. 2014). The production of astaxanthin in these unique Patagonian populations was confirmed as well as the accumulation of unusually high proportions of PUFAs (70.5 %). The production level of the pigment in these native wild isolates was low for biotechnological purposes but provides a new genomic background for genetic improvements. The complete genome sequence of a representative strain of the Patagonian population was recently obtained (Bellora et al., submitted).

13.4 Conclusions

A wide variety of compounds with potential application in aquaculture such as bacteriocins, biosurfactants, carotenoid pigments, organic acids, and polyunsaturated fatty acids are synthesized by different Patagonian strains. These microorganisms could become a potential source of compounds of biotechnological importance. It is therefore necessary to continue with screening programs to improve our limited understanding about the Patagonian microbial world, including new metagenomic techniques in combination with proteomics.

Acknowledgments The authors gratefully acknowledge support from ANPCyT (PICT 2012-1004) and CONICET (PIP 11220120100050CO), Argentina. Also to Universidad Nacional del Comahue (project B171) and ANPCyT (PICT 2014-2542) granted to DL.

References

Adrio JL, Veiga M (1995) Transformation of the astaxanthin-producing yeast *Phaffia rhodozyma*. Biotechnol Lett 9:509–512

Aguado-Urda M, López-Campos GH, Gibello A, Cutuli MT, López-Alonso V, Fernández-Garayzábal JF, Blanco MM (2011) Genome sequence of *Lactococcus garvieae* 8831, isolated from rainbow trout lactococcosis outbreaks in Spain. J Bacteriol 193:4263–4264

Andrewes AG, Starr MP (1976) (3R, 3′R)-Astaxanthin from the yeast *Phaffia rhodozyma*. Phytochemistry 15:1009–1011

Andrews SR, Sahu NP, Pal AK, Mukherjee SC, Kumar S (2011) Yeast extract, brewer's yeast and spirulina in diets for *Labeo rohita* fingerlings affect haemato-immunological responses and survival following *Aeromonas hydrophila* challenge. Res Vet Sci 91:103–109

Austin B, Stuckey LF, Robertson PAW, Effendi I, Griffith DRW (1995) A probiotic strain of *Vibrio alginolyticus* effective in reducing diseases caused by *Aeromonas salmonicida*, *Vibrio anguillarum* and *Vibrio ordalii*. J Fish Dis 18:93–96
Balcázar JL, de Blas I, Ruiz-Zarzuela I, Cunningham D, Vendrell D, Múzquiz JL (2006) The role of probiotics in aquaculture. Vet Microbiol 114:173–186
Balcázar JL, de Blas I, Ruiz-Zazuela I, Vandrell D, Gironés O, Muzquiz JL (2007) Enhancement of the immune response and protection induced by probiotic lactic acid bacteria against furunculosis in rainbow trout (*Oncorhynchus mykiss*). FEMS Immunol Med Microbiol 51:185–193
Brandão LR, Libkind D, Vaz ABM, Espírito Santo L, Moliné M, de García V, van Broock M, Rosa CA (2011) Yeasts from an oligotrophic lake in Patagonia (Argentina): diversity, distribution and synthesis of photo-protective compounds and extracellular enzymes. FEMS Microbiol Ecol 76:1–13
Britz PJ (1996) The suitability of selected protein sources for inclusion in formulated diets for the South African abalone, *Haliotis midae*. Aquaculture 140:63–73
Brunt J, Austin B (2005) Use of a probiotic to control lactococcosis and streptococcosis in rainbow trout, *Oncorhynchus mykiss* (Walbaum). J Fish Dis 28:693–701
Burgents JE, Burnett KG, Burnett LE (2004) Disease resistance of Pacific white shrimp, *Litopenaeus vannamei*, following the dietary administration of a yeast culture food supplement. Aquaculture 231:1–8
Chatterjee C, Patton GC, Cooper L, Paul M, van der Donk WA (2006) Engineering dehydro amino acids and thioethers into peptides using lacticin 481 synthetase. Chem Biol 13:1109–1117
Coutteau P, Lavens P, Sorgeloos P (1990) Baker's yeast as a potential substitute for live algae in aquaculture diets: artemia as a case study. J World Aquacult Soc 21:1–9
David-Palma M, Libkind D, Sampaio JP (2014) Global distribution, diversity hotspots and niche transitions of an astaxanthin-producing eukaryotic microbe. Mol Ecol 23:921–932
de García V, Brizzio S, Libkind D, Buzzini P, van Broock M (2007) Biodiversity of cold-adapted yeasts from runoff glacial rivers in Patagonia, Argentina. FEMS Microbiol Ecol 59:331–341
de Vuyst L, Vandamme E (1994) Nisin, a lantibiotic produced by *Lactococcus lactis* subsp. *lactis*: properties, biosynthesis, fermentation and applications. In: de Vuyst L, Vandamme E (eds) Bacteriocins of lactic acid bacteria. Microbiology, genetics and applications. Chapman & Hall, London, pp 151–221
Dos Santos Martins F, Nicoli JR (2015) Mechanisms of action of probiotic yeasts. In: Venema K, do Carmo AP (eds) Probiotics and prebiotics: current research and future trends. Caister Academic Press, Norfolk, UK, pp 105–114
Echevarri-Erasun C, Johnson EA (2002) Fungal carotenoids. In: Khachtourians GG, Arora DK (eds) Applied mycology and biotechnology, vol 2, Agriculture and food production. Elsevier, Amsterdam, pp 45–85
Encarnacao P (2016) Functional feed additives in aquaculture feeds. In: Nates SF (ed) Aquafeed formulation. Elsevier, Amsterdam, pp 217–236
FAO (2005–2015) Cultured Aquatic Species Information Programme. *Oncorhynchus mykiss*. Text by Cowx IG. In: FAO Fisheries and Aquaculture Department [online]. Rome. Updated 15 June 2005. http://www.fao.org/fishery/culturedspecies/Oncorhynchus_mykiss/en. Accessed 18 Nov 2015
FAO (2014) The State of World Fisheries and Aquaculture 2014. FAO, Rome
Fernández M (2013) Bacterias marinas con propiedades antagónicas hacia ictiopatógenos del género *Vibrio*. Thesis, Facultad de Ciencias Naturales, Universidad Nacional de La Patagonia San Juan Bosco, Puerto Madryn, Argentina. INV: 00220T
Fernández M, Garcés M, Sequeiros C, Olivera NL (2015) Diversidad de bacterias marinas aisladas de la costa Patagónica argentina con propiedades probióticas relevantes para acuicultura. XVI Congreso Nacional de Biotecnología y Bioingeniería. Del 21–26 junio 2015. Guadalajara. México. http://smbb.com.mx/congresos%20smbb/guadalajara15/PDF/XVI/trabajos/VII/VIIC-08.pdf
Fournier V, Gouillou-Coustans MF, Metailler R, Vachot C, Moriceau J, Le Delliou H, Huelvan C, Desbruyeres E, Kaushik SJ (2002) Nitrogen utilisation and ureogenesis as affected by dietary nucleic acid in rainbow trout (*Oncorhynchus mykiss*) and turbot (*Psetta maxima*). Fish Physiol Biochem 26:177–188

Gamboa-Delgado J, Fernández-Díaz B, Nieto-López M, Cruz-Suárez LE (2016) Nutritional contribution of torula yeast and fish meal to the growth of shrimp *Litopenaeus vannamei* as indicated by natural nitrogen stable isotopes. Aquaculture 453:116–121

Garcés M, Olivera NL, Sequeiros C (2014) Bacteria with probiotic characteristics isolated from Patagonian fish. Simposio de la Red Latinoamericana de Investigaciones en el Sistema Modelo de Pez Cebra, LAZEN 2014. Valparaiso, Chile

Garcés ME, Sequeiros C, Olivera NL (2015) Marine *Lactobacillus pentosus* H16 protects *Artemia franciscana* from *Vibrio alginolyticus* pathogenic effects. Dis Aquat Org 113:41–50

Gatesoupe FJ (2007) Live yeasts in the gut: natural occurrence, dietary introduction, and their effects on fish health and development. Aquaculture 267:20–30

Gatesoupe FJ (2008) Updating the importance of lactic acid bacteria in fish farming: natural occurrence and probiotic treatments. J Mol Microbiol Biotechnol 14:107–114

Ghosh S, Sinha A, Sahu C (2007) Effect of probiotic on reproductive performance in female livebearing ornamental fish. Aquacult Res 38:518–526

Guerin M, Huntley ME, Olaizola M (2003) *Haematococcus* astaxanthin: applications for human health and nutrition. Trends Biotechnol 21:210–216

Hasan MR (2001) Nutrition and feeding for sustainable aquaculture development in the third millennium. In: Subasinghe RP, Bueno P, Phillips MJ, Hough C, McGladdery SE, Arthur JR (eds) Aquaculture in the Third Millennium. Technical Proceedings of the Conference on Aquaculture in the Third Millennium, Bangkok, Thailand, 20–25 Feb 2000. pp. 193–219. NACA, Bangkok and FAO, Rome

Heo WS, Kim EY, Kim YR, Hossain MT, Kong IS (2012) Salt effect of nisin Z isolated from a marine fish on the growth inhibition of *Streptococcus iniae*, a pathogen of streptococcosis. Biotechnol Lett 34:315–320

Heuer OE, Kruse H, Grave K, Collignon P, Karunasagar I, Angulo FJ (2009) Human health consequences of use of antimicrobial agents in aquaculture. Clin Infect Dis 49:1248–1253

Hisano H, Falcon DR, Barros MM, Pezzato LE (2008) Influence of yeast derivates on growth performance and survival of juvenile prawn *Macrobrachium amazonicum*. Ciênc Anim Bras 9(3):657–662

Jacobson JK (2015) Yeasts as probiotics: established in animals, but what about man? In: Venema K, do Carmo AP (eds) Probiotics and prebiotics: current research and future trends. Caister Academic Press, Norfolk, UK, pp 115–134

Johnson EA (2003) *Phaffia rhodozyma*: colorful odyssey. Int Microbiol 6:169–174

Johnson EA (2013a) Biotechnology of non-*Saccharomyces* yeasts: the ascomycetes. Appl Microbiol Biotechnol 97:503–517

Johnson EA (2013b) Biotechnology of non-*Saccharomyces* yeasts: the basidiomycetes. Appl Microbiol Biotechnol 97:7563–7577

Johnson EA, Conklin DE, Lewis MJ (1977) The yeast *Phaffia rhodozyma* as a dietary pigment source for salmonids and crustaceans. J Fish Res Board Can 34:2417–2421

Johnson EA, Villa TG, Lewis MJ (1980a) *Phaffia rhodozyma* as an astaxanthin source in salmonid diets. Aquaculture 20:123–134

Johnson EA, Lewis MJ, Grau CR (1980b) Pigmentation of egg yolks with astaxanthin from the yeast *Phaffia rhodozyma*. Poultry Sci 59:1777–1782

Kesarcodi-Watson A, Kaspar H, Lategan MJ, Gibson L (2008) Probiotics in aquaculture: the need, principles and mechanisms of action and screening processes. Aquaculture 274:1–14

Kim DH, Austin B (2006) Innate immune responses in rainbow trout (*Oncorhynchus mykiss*, Walbaum) induced by probiotics. Fish Shellfish Immunol 21:513–524

Kuipers A (2010) Microbial production of thioether-stabilized peptides. Dissertation, University of Groningen, NL. http://www.rug.nl/research/portal/files/2608474/13complete.pdf. Accessed 15 Nov 2015

Lara-Flores M (2011) The use of probiotic in aquaculture: an overview. Int Res J Microbiol 2:471–478

Lategan MJ, Torpy FR, Gibson LF (2004) Biocontrol of saprolegniosis in silver perch *Bidyanus bidyanus* (Mitchell) by *Aeromonas media* strain A199. Aquaculture 235:77–88

Li P, Gatlin DM (2004) Dietary brewers yeast and the prebiotic GroBiotic AE influence growth performance, immune responses and resistance of hybrid striped bass (*Morone chrysops* × *M. saxatilis*) to *Streptococcus iniae* infection. Aquaculture 231:445–456

Libkind D (2006) Carotenogenic yeasts from aquatic environments in northwestern Patagonia. Biotechnological applications. PhD Thesis, Tucumán National University, Argentina

Libkind D, van Broock MR (2006) Biomass and carotenoid pigments production by Patagonian native yeasts. World J Microbiol Biotechnol 22:687–692

Libkind D, Brizzio S, Ruffini A, Gadanho M, van Broock MR, Sampaio JP (2003) Molecular characterization of carotenogenic yeasts from aquatic environments in Patagonia, Argentina. Antonie van Leeuwenhoek J Microbiol 84:313–322

Libkind D, Gadanho M, van Broock MR, Sampaio JP (2005) *Sporidiobolus longiusculus* sp. nov. and *Sporobolomyces patagonicus* sp. nov., novel yeasts of the Sporidiobolales isolated from aquatic environments in Patagonia, Argentina. Int J Syst Evol Microbiol 55:503–509

Libkind D, Ruffini A, van Broock M, Alves L, Sampaio JP (2007) Biogeography, host-specificity, and molecular phylogeny of *Phaffia rhodozyma* and its sexual form, *Xanthophyllomyces dendrorhous*. Appl Environ Microbiol 73:1120–1125

Libkind D, Arts M, van Broock M (2008a) Fatty acid composition of cold-adapted carotenogenic basidiomycetous yeasts. Rev Argent Microbiol 40:193–197

Libkind D, Moliné M, de García V, Fontenla S, van Broock M (2008b) Characterization of a novel South American population of the astaxanthin-producing yeast *Xanthophyllomyces dendrorhous* (*Phaffia rhodozyma*). J Ind Microbiol Biotechnol 35:151–158

Libkind D, Moline M, Sampaio J, van Broock M (2009a) Yeasts from high altitude lakes: influence of UV radiation. FEMS Microbiol Ecol 69:353–362

Libkind D, Gadanho M, van Broock M, Sampaio JP (2009b) *Cystofilobasidium lacus-mascardii* sp. nov., a new basidiomycetous yeast species isolated from aquatic environments of the Patagonian Andes and *Cystofilobasidium macerans* sp. nov., the sexual stage of *Cryptococcus macerans*. Int J Syst Evol Microbiol 59:622–630

Libkind D, Sampaio JP, van Broock M (2010) Cystobasidiomycetes yeasts from Patagonia (Argentina): description of *Rhodotorula meli* sp. nov. from glacial meltwater. Int J Syst Evol Microbiol 60:2251–2256

Libkind D, Hittinger CT, Valério E, Gonçalves C, Dover J, Johnston M, Gonçalves P, Sampaio JP (2011) Microbe domestication and the identification of the wild genetic stock of lager-brewing yeast. Proç Natl Acad Sci USA 35:14539–14544

Lin S, Mao S, Guan Y, Luo L, Luo L, Pan Y (2012) Effects of dietary chitosan oligosaccharides and *Bacillus coagulans* on the growth, innate immunity and resistance of koi (*Cyprinus carpio koi*). Aquaculture 342:36–41

López Cazorla A, Sica MG, Brugnoni LI, Marucci PL, Cubitto MA (2015) Evaluation of *Lactobacillus paracasei* subsp. *tolerans* isolated from Jenyn's sprat (*Ramnogaster arcuata*) as probiotic for juvenile rainbow trout *Oncorhynchus mykiss* (Walbaum, 1792). J Appl Ichthyol 31:88–94

Lovatelli A, Chen JX (2009) Use of environmental friendly feed additives and probiotics in Chinese aquaculture. FAN-FAO Aquacult Newsl 42:32–35

Maeda M, Shibata A, Biswas G, Korenaga H, Kono T, Itami T, Sakai M (2014) Isolation of lactic acid bacteria from kuruma shrimp (*Marsupenaeus japonicus*) intestine and assessment of immunomodulatory role of a selected strain as probiotic. Mar Biotechnol 16:181–192

Mahnken CVW, Spinelli J, Waknitz FW (1980) Evaluation of an alkane yeast (*Candida* sp.) as a substitute for fish meal in Oregon moist pellet: feeding trials with coho salmon (*Oncorhynchs kisutch*) and rainbow trout (*Salmo gairdneri*). Aquaculture 20:41–56

Martin AM, Goddard S, Bemistera P (1993) Production of *Candida utilis* biomass as aquaculture feed. J Sci Food Agric 61:363–370

Merrifield D, Zhou Z (2011) Probiotic and prebiotic applications in aquaculture. J Aquacult Res Dev S1:e001. doi:10.4172/2155-9546.S1-e001

Merrifield DL, Dimitroglou A, Foey A, Davies SJ, Baker RTM, Bøgwald J, Castex M, Ringø E (2010) The current status and future focus of probiotic and prebiotic applications for salmonids. Aquaculture 302:1–18

Merrifield DL, Bradley G, Harper GM, Baker RTM, Munn CB, Davies SJ (2011) Assessment of the effects of vegetative and lyophilized *Pediococcus acidilactici* on growth, feed utilization, intestinal colonization and health parameters of rainbow trout (*Oncorhynchus mykiss* Walbaum). Aquacult Nutr 17:73–79

Mohanty SN, Swain SK, Tripathi SD (1996) Rearing of catla (*Catla catla* Ham.) spawn on formulated diets. J Aquacult Tropics 11:253–258

Moreira de Souza D, Medeiros Suita S, Pereira Leivas Leite FPL, Romano LA, Wasielesky W, Ballester ELC (2012) The use of probiotics during the nursery rearing of the pink shrimp *Farfantepenaeus brasiliensis* (Latreille, 1817) in a zero exchange system. Aquacult Res 43:1828–1837

Moriarty DJW (1998) Control of luminous *Vibrio* species in penaeid aquaculture ponds. Aquaculture 164:351–358

Navarrete P, Tovar-Ramírez D (2014) Use of yeasts as probiotics in fish aquaculture. In: Hernandez-Vergara MP, Perez-Rostro CI (eds) Sustainable aquaculture techniques. INTECH, pp 135–171

Nayak SK (2010) Probiotics and immunity: a fish perspective. Fish Shellfish Immunol 29:2–14

Newaj-Fyzul AH, Al-Harbi BA, Austin B (2014) Review: developments in the use of probiotics for disease control in aquaculture. Aquaculture 431:1–11

Newaj-Fyzul A, Adesiyun AA, Mutani A, Ramsubhag A, Brunt J, Austin B (2007) *Bacillus subtilis* AB1 controls *Aeromonas* infection in rainbow trout (*Oncorhynchus mykiss*, Walbaum). J Appl Microbiol 103:1699–1706

Okagbue R, Lewis MJ (2008) Influence of mixed culture conditions on yeast-wall hydrolytic activity of *Bacillus circulans* WL-12 and on extractability of astaxanthin from the yeast *Phaffia rhodozyma*. J Appl Bacteriol 59:243–255

Oliva-Teles A, Goncalves P (2001) Partial replacement of fishmeal by brewer's yeast *Saccharomyces cerevisae* in diets for sea bass *Dicentrarchus labrax* juveniles. Aquaculture 202:269–278

Olvera-Novoa MA, Martínez-Palacios CA, Olivera-Castillo L (2002) Utilization of torula yeast (*Candida utilis*) as a protein source in diets for tilapia (*Oreochromis mossambicus* Peters) fry. Aquacult Nutr 8:257–264

Ortuño J, Cuesta A, Rodríguez A, Esteban MA, Meseguer J (2002) Oral administration of yeast, *Saccharomyces cerevisiae*, enhances the cellular innate immune response of gilthead seabream (*Sparus aurata* L.). Vet Immunol Immunopathol 85:41–50

Ozorio ROA, Turini BGS, Môro GV, Oliveira LST, Portz L, Cyrino JEP (2010) Growth, nitrogen gain and indispensable amino acid retention of pacu (*Piaractus mesopotamicus*, Holmberg 1887) fed different brewers yeast (*Saccharomyces cerevisiae*) levels. Aquacult Nutr 16:276–283

Panigrahi A, Kiron V, Puangkaew J, Kobayashi T, Satoh S, Sugita H (2005) The viability of probiotic bacteria as a factor influencing the immune response in rainbow trout *Oncorhynchus mykiss*. Aquaculture 243:241–254

Panigrahi A, Kiron V, Satoh S, Hiron I, Kobayashi T, Sugita H, Puangkaew J, Aoki T (2007) Immune modulation and expression of cytokine genes in rainbow trout *Oncorhynchus mykiss* upon probiotic feeding. Dev Comp Immunol 31:372–382

Paruelo JM, Beltran A, Jobbagy E, Sala OE, Golluscio RA (1998) The climate of Patagonia: general patterns and controls on biotic. Ecol Austral 8:85–101

Pasteris SE, Roig Babot G, Otero MC, Bühler MI, Nader-Macías ME (2009) Beneficial properties of lactic acid bacteria isolated from a *Rana catesbeiana* hatchery. Aquacult Res 40:1605–1615

Pasteris SE, Guidoli MG, Otero MC, Bühler MI, Nader-Macías ME (2011) *In vitro* inhibition of *Citrobacter freundii*, a red-leg syndrome associated pathogen in raniculture, by indigenous *Lactococcus lactis* CRL 1584. Vet Microbiol 151:336–344

Pérez-Sánchez T, Balcázar JL, Merrifield DL, Carnevali O, Gioacchino G, de Blas I et al (2011a) Expression of immune related genes in rainbow trout (*Oncorhynchus mykiss*) induced by probiotic bacteria during *Lactococcus garvieae* infection. Fish Shellfish Immunol 31:196–201

Pérez-Sánchez T, Balcázar JL, García Y, Halaihel N, Vendrell D, De Blas I, Merrifield DL, Ruiz-Zarzuela I (2011b) Identification and characterization of lactic acid bacteria isolated from rainbow trout, *Oncorhynchus mykiss* (Walbaum), with inhibitory activity against *Lactococcus garvieae*. J Fish Dis 34:499–507

Pérez-Sánchez T, Ruiz-Zarzuela I, Blas I, Balcázar JL (2014) Probiotics in aquaculture: a current assessment. Rev Aquacult 6:133–146

Peulen O, Deloyer P, Grandfils C, Loret S, Dandrifosse G (2000) Intestinal maturation induced by spermine in young animals. Livestock Prod Sci 66:109–120

Quentel C, Gatesoupe FJ, Aubin J, Lamour F, Abiven A, Baud M, Labbé L, Forraz M (2005) Ofimer probiotic study on rainbow trout. I. Resistance against *Yersinia ruckeri* and humoral immune response of rainbow trout (*Oncorhynchus mykiss*) submitted to probiotic treatment with *Saccharomyces cerevisiae* var. *boulardii*. In: Howell B, Flos R (eds) Lessons from the Past to Optimise the Future. Aquaculture Europe 2005, Trondheim, Norway, 5–9 Aug 2005. EAS Special Publication, vol 35. European Aquaculture Society, Oostende, Belgium, pp 380–381

Quirós R, Drago E (1999) The environmental state of Argentinean lakes: an overview. Lakes Reserv Res Manag 4:55–64

Reimundo P, Pignatelli M, Alcaraz LD, D'Auria G, Moya A, Guijarro JA (2011) Genome sequence of *Lactococcus garvieae* UNIUD074, isolated in Italy from a lactococcosis outbreak. J Bacteriol 193:3684–3685

Rengpipat S, Phianphak W, Piyatiratitivorakul S, Menasveta P (1998) Effects of a probiotic bacterium on black tiger shrimp *Penaeus monodon* survival and growth. Aquaculture 167:301–313

Riquelme CE, Jorquera MA, Rojas AI, Avendaño RE, Reyes N (2001) Addition of inhibitor-producing bacteria to mass cultures of *Argopecten purpuratus* larvae (Lamarck, 1819). Aquaculture 192:111–119

Robertson PAW, O'Dowd C, Burrells C, Williams P, Austin B (2000) Use of *Carnobacterium* sp. as a probiotic for Atlantic salmon (*Salmo salar* L.) and rainbow trout (*Oncorhynchus mykiss* Walbaum). Aquaculture 185:235–243

Ruiz-Ponte C, Samain JF, Sanchez JL, Nicolas JL (1999) The benefit of a *Roseobacter* species on the survival of scallop larvae. Mar Biotechnol 1:52–59

Rumsey GL, Hughes SG, Smith RR, Kinsella JE, Shetty KJ (1991) Digestibility and energy values of intact, disrupted and extracts from dried yeast fed to rainbow trout (*Oncorhynchus mykiss*). Animal Feed Sci Technol 33:185–193

Russo G, Libkind D, Sampaio JP, van Broock M (2008) Yeast diversity at the volcanic acidic environment of the Lake Caviahue and Rio Agrio (Patagonia, Argentina). FEMS Microbiol Ecol 65:415–424

Schmidt I, Schewe H, Gassel S, Jin C, Buckingham J, Sandmann MHG, Schrader J (2011) Biotechnological production of astaxanthin with *Phaffia rhodozyma/Xanthophyllomyces dendrorhous*. Appl Microbiol Biotechnol 89:555–571

Scholz U, Garcia Diaz G, Ricque D, Cruz Suarez LE, Vargas Albores F, Latchford J (1999) Enhancement of vibriosis resistance in juvenile *Penaeus vannamei* by supplementation of diets with different yeast products. Aquaculture 176:271–283

Sealey WM, Barrows FT, Johansen KA, Overturf K, LaPatra SE, Hardy RW (2007) Evaluation of the ability of partially autolyzed yeast and Grobiotic-A to improve disease resistance in rainbow trout. N Am J Aquacult 69:400–406

Sequeiros C, Vallejo M, Marguet ER, Olivera NL (2010) Inhibitory activity against the fish pathogen *Lactococcus garvieae* produced by *Lactococcus lactis* TW34, a lactic acid bacterium isolated from the intestinal tract of a Patagonian fish. Arch Microbiol 192:237–245

Sequeiros C, Garcés ME, Vallejo M, Marguet ER, Olivera NL (2015) Potential aquaculture probiont *Lactococcus lactis* TW34 produces nisin Z and inhibits the fish pathogen *Lactococcus garvieae*. Arch Microbiol 197:449–458

Shiri Harzevili AR, Van Duffel H, Dhert P, Swings J, Sorgeloos P (1998) Use of a potential probiotic *Lactococcus lactis* AR21 strain for the enhancement of growth in the rotifer *Brachionus plicatilis* (Müller). Aquacult Res 29:411–417

Sica MG, Brugnoni LI, Marucci PL, Cubitto MA (2012) Characterization of probiotic properties of lactic acid bacteria isolated from an estuarine environment for application in rainbow trout (*Oncorhynchus mykiss* Walbaum) farming. Antonie van Leeuwenhoek J Microbiol 101:869–879

Siwicki AK, Anderson DP, Rumsey GL (1994) Dietary intake of immunostimulants by rainbow trout affects non-specific immunity and protection against furunculosis. Vet Immunol Immunopathol 41:125–139

Spanggaard B, Huber I, Nielsen J, Sick EB, Pipper CB, Martinussen T, Slierendrecht WJ, Gram L (2001) The probiotic potential against vibriosis of the indigenous microflora of rainbow trout. Environ Microbiol 3:755–765

Stones CS, Mills DV (2004) The use of live yeast and yeast culture products in aquaculture. Int Aquafeed 7:28–34

Tacon P (2012) Yeast in Aquaculture. International AquaFeed. November–December. 14–18

Tapia-Paniagua ST, Díaz-Rosales P, León-Rubio JM, de La Banda IG, Lobo C, Alarcón FJ et al (2012) Use of the probiotic *Shewanella putrefaciens* Pdp11 on the culture of Senegalese sole (*Solea senegalensis* Kaup 1858) and gilthead seabream (*Sparus aurata* L.). Aquacult Int 20:1025–1039

Tovar-Ramirez D, Zambonino J, Cahu C, Gatesoupe FJ, Vázquez-Juárez R, Lésel R (2002) Effect of live yeast incorporation in compound diet on digestive enzyme activity in sea bass (*Dicentrarchus labrax*) larvae. Aquaculture 204:113–123

Tovar-Ramirez D, Zambonino-Infante JL, Cahu C, Gatesoupe FJ, Vazquez-Juarez R (2004) Influence of dietary live yeast on European sea bass (*Dicentrarchus labrax*) larval development. Aquaculture 234:415–427

Varela JL, Ruiz-Jarabo I, Vargas-Chacoff L, Arijo S, León-Rubio JM, García-Millán I, Martín del Río MP, Moriñigo MA, Mancera JM (2010) Dietary administration of probiotic Pdp11 promotes growth and improves stress tolerance to high stocking density in gilthead seabream *Sparus auratus*. Aquaculture 309:265–271

Vendrell D, Balcázar JL, Ruiz-Zarzuela I, de Blas I, Gironés O, Múzquiz JL (2006) *Lactococcus garvieae* in fish: a review. Comp Immunol Microbiol 29:177–198

Vendrell D, Balcázar JL, de Blas I, Ruiz-Zarzuela I, Gironés O, Múzquiz JL (2008) Protection of rainbow trout (*Oncorhynchus mykiss*) from lactococcosis by probiotic bacteria. Comp Immunol Microbiol 31:337–345

Verschuere L, Rombaut G, Sorgeloos P, Veratraete W (2000) Probiotic bacteria as biological control agents in aquaculture. Microbiol Mol Biol Rev 64:655–671

Vine NG, Leukes WD, Kaiser H, Daya S, Baxter J, Hecht T (2004) Competition for attachment of aquaculture candidate probiotic and pathogenic bacteria on fish intestinal mucus. J Fish Dis 27:319–326

Wache Y, Auffray F, Gatesoupe FJ, Zambonino J, Gayet V, Labbe L (2006) Cross effects of the strain of dietary *Saccharomyces cerevisiae* and rearing conditions on the onset of intestinal microbiota and digestive enzymes in rainbow trout, *Onchorhynchus mykiss,* fry. Aquaculture 258:470–478

Zamith-Miranda D, Palma ML, Matos GM, Schiebel JG, Maya-Monteiro CM, Aronovich M, Bozza PT, Bozza FA, Nimrichter L, Montero-Lomeli M, Marques ETA Jr, Martins FS, Douradinha B (2016) Lipid droplet levels vary heterogeneously in response to simulated gastrointestinal stresses in different probiotic *Saccharomyces cerevisiae* strains. J Funct Foods 21:193–200

Zhou XX, Wang YB, Li WF (2009) Effect of probiotic on larvae shrimp (*Penaeus vannamei*) based on water quality, survival rate and digestive enzyme activities. Aquaculture 287:349–353

Chapter 14
Indigenous Lactic Acid Bacteria Communities Associated with Spontaneous Malolactic Fermentations in Patagonian Wines: Basic and Applied Aspects

Danay Valdés La Hens, Bárbara M. Bravo-Ferrada, Natalia S. Brizuela, Elizabeth E. Tymczyszyn, Axel Hollmann, Lucrecia Delfederico, and Liliana Semorile

Abstract During winemaking, complex microbial interactions take place and microorganisms showing a selective advantage emerge in a given period as the dominant populations. At the beginning, yeasts, responsible for the alcoholic fermentation (AF), consume sugars present in grapes to yield ethanol and carbon dioxide, leading the transformation of must into wine. The tolerance of lactic acid bacteria (LAB) to low pH and high ethanol are the main factors that select their occurrence in winery ecosystems. LAB guide a secondary biological process, the malolactic fermentation (MLF), which produces deacidification of wine, enhancing its microbial stability and modifying the wine aroma profile. When MLF takes place spontaneously, it is carried out by one or more species of indigenous LAB present in grapes and cellars, naturally adapted to the regional peculiarities of wine. Thus, it is highly advisable to study the indigenous microbiota, best adapted to the agro-ecological conditions of a specific wine-producing area, to select the most representative strains with *terroir* characteristics for their use as starter cultures. In this chapter, we summarize the studies conducted so far on LAB population diversity in some Patagonian wines, as well as the methods and criteria to select potential indigenous malolactic cultures for these wines, including the adaptation to processing conditions.

D. Valdés La Hens (✉) • B.M. Bravo-Ferrada • N.S. Brizuela • E.E. Tymczyszyn
L. Delfederico • L. Semorile
Laboratorio de Microbiología Molecular, Instituto de Microbiología Básica y Aplicada, Departamento de Ciencia y Tecnología, Universidad Nacional de Quilmes, Roque Sáenz Peña N° 352, (B1876BXD) Bernal, Buenos Aires, Argentina
e-mail: dvhens@unq.edu.ar

A. Hollmann
Laboratorio de Microbiología Molecular, Instituto de Microbiología Básica y Aplicada, Departamento de Ciencia y Tecnología, Universidad Nacional de Quilmes, Roque Sáenz Peña N° 352, (B1876BXD) Bernal, Buenos Aires, Argentina

Laboratorio de Biointerfaces y Sistemas Biomiméticos - Universidad Nacional de Santiago del Estero (UNSE) y Centro de Investigaciones y Transferencia de Santiago del Estero (CITSE) - CONICET-UNSE, Ruta Nacional N° Km 1125 S/N, (4206) Santiago del Estero, Argentina

N.L. Olivera et al. (eds.), *Biology and Biotechnology of Patagonian Microorganisms*, DOI 10.1007/978-3-319-42801-7_14

14.1 Introduction

14.1.1 Generalities of Lactic Acid Bacteria (LAB)

The lactic acid bacteria (LAB) are a diverse group of related microorganisms that have been used for millennia, on both an industrial and artisanal scale, in food and beverage fermentation processes, extending the shelf life of the product and improving the organoleptic qualities. Besides their long history of safe use for human consumption, LAB are also used in producing silage and agricultural animal feeds (Reid et al. 2003; Klaenhammer et al. 2005). The genetic diversity of the bacteria, evidenced by the expression of a large number of metabolic activities, allows them to inhabit a variety of ecological niches where carbohydrate-rich substances are available, ranging from plants and material of plants origin, food matrices such as dairy products, meats, vegetables, sourdough bread, and wine, to human or animal mucosal surfaces such as the oral cavity, vagina, and gastrointestinal tract (Pfeiler and Klaenhammer 2007). The bacteria can improve the digestive health of young animals, and also have human medical applications, as in the case of probiotic strains, which have a part in maintaining health, regulating the immune system, and managing disease, that it is just starting to be understood (Selle and Klaenhammer 2013; Belizário and Napolitano 2015).

From a functional classification point of view, the LAB are a group of chemoorganotrophic rods or cocci, gram-positive, non-spore-forming, and catalase-negative bacterial species, able to produce lactic acid as the main end product from carbohydrate fermentation. Usually, these organisms are anaerobes or aerotolerants, and exclusively fermentative, because of their lack of cytochromes and their inability to synthesize porphyrins. Nevertheless, recent evidence has shown, at least in *Lactococcus lactis*, *Lactobacillus plantarum*, *Streptococcus agalactiae*, and *Enterococcus faecalis*, that electron transport chain function can be reconstituted and so the possibility of aerobic respiration, if heme precursors or menaquinone are added to the culture medium (Brooijmans et al. 2009).

The main discrepancy in the LAB taxonomy is the lack of correlation between phylogenetic placement and metabolic properties. Besides their morphology, the homofermentative or heterofermentative character is a deciding factor in their classification. Among the cocci, the bacteria from the genus *Pediococcus* are homofermenters (produce more than 85 % lactic acid from glucose) and those from the genera *Leuconostoc* and *Oenococcus* are heterofermentative (produce CO_2, ethanol, and acetic acid in addition to lactic acid). The lactobacilli are divided into three groups: (I) strict homofermenters (yet not identified in wine), (II) facultative heterofermenters, and (III) strict heterofermenters. The strictly homofermentative lactobacilli do not ferment pentose and form two molecules of lactic acid from one molecule of glucose by the Embden–Meyerhoff pathway. In facultative heterofermenters (group II), one glucose molecule leads to two molecules of lactic acid, as in group I bacteria, but the pentoses are fermented into lactic and acetic acid by the heterofermentative pentose phosphate pathway. The bacteria in group III do not possess the fructose-1,6-diphosphate aldolase and ferment glucose into CO_2, lactic and acetic acid, and ethanol by the pentose phosphate pathway, and pentose into lactic and acetic acid as bacteria from group II (Ribereau-Gayon et al. 2006).

Phylogenetically related to *Bacilli*, LAB belongs to one of the six families of the order *Lacobacillales*. The most diverse family is *Lactobacillaceae*, which includes the genus *Lactobacillus*, with more than 100 species recognized. Other industrially important genera are *Lactococcus*, *Enterococcus*, *Oenococcus*, *Pediococcus*, *Streptococcus*, and *Leuconostoc*. Although the genus *Bifidobacterium* is traditionally included among LAB, and has been considered relatives to lactobacilli based on the analysis of the murein structure, nowadays it is known that it belongs to the phylum Actinobacteria, and is poorly phylogenetically related to genuine LAB. Besides, all the 29 species belonging to the genus *Bifidobacterium* degrade hexoses through a peculiar metabolic pathway, the so-called *bifid shunt*, that is, the fructose-6-phosphate pathway, where the key enzyme is fructose-6-phosphoketolase (Felis and Dellaglio 2007).

The analysis of complete genomes has allowed a better understanding of the evolutionary divergence of LAB and reveals a trend of relatively recent and ongoing reduction in genome size (van de Gutche et al. 2006), suggesting that the bulk of the genes were lost in the adaptation to nutrient-rich environments. The last common ancestor of *Lactobacilalles* appears to have lost between 600 and 1200 genes and gained about 100 during its divergence from the *Bacilli* ancestor (Makarova and Koonin 2007). The extent of genome reduction varies among LAB, with such members as *Oenococcus oeni* having only 1700 predicted ORFs (open reading frames) compared with the 3000 of *Lactobacillus (Lb.) plantarum* (Pfeiler and Klaenhammer 2007). Detailed comparative analysis of genomic data emphasized the remarkable diversity within the LAB group at numerous taxonomic levels, that is, order, family, group, genus, and even species (Douillard and de Vos 2014). This diversity results from the interactions between genome and environment. Many LAB species have reduced biosynthetic properties consequent to the abundance and variety of nutrients in their habitat. The loss of metabolic genes is compensated by genome enrichment in genes encoding for transporters (ABC or PTS systems), allowing LAB to use nutrients and by-products from their niche. Other LAB species or strains maintain a broad ecological flexibility, which may result in a high adaptation capacity to adapt to drastic environmental changes.

14.1.2 LAB in Winemaking

Winemaking is a complex microbial process in which yeasts, mainly *Saccharomyces cerevisiae*, produce the alcoholic fermentation (AF)-consuming sugars present in grapes to yield ethanol, leading the transformation of must into wine. Also LAB have a significant part, guiding a secondary biological process, the malolactic fermentation (MLF), which takes place during or after AF. MLF is responsible for the conversion of L-malic acid to L-lactic acid and CO_2 and is carried out by one or more

LAB species, mainly *O. oeni* and *Lb. plantarum* (López et al. 2008; Miller et al. 2011). The MLF produces deacidification of wine with a concomitant increase in the pH, which is a desirable effect in wines with high acidity. Another contribution of MLF refers to the microbial stability by the removal of malic acid as a possible carbon substrate, and the modification of the wine aroma profile, linked to the enzymatic activity of LAB (Henick-Kling 1993; Maicas et al. 1999; Ugliano et al. 2003; Lerm et al. 2010). When MLF takes place spontaneously, it is carried out by one or more species of indigenous LAB present in grapes and cellars (Lonvaud-Funel 1999; Rodas et al. 2003; Bae et al. 2006). However, the inhibiting agents to which LAB are exposed in musts and wines may cause MLF failure. Also, spontaneous MLF could have unpredictable results, such as a considerable increase in volatile acidity, the consumption of residual sugars, and the formation of undesirable metabolites such as biogenic amines (López et al. 2008).

Conversion of L-malic acid to L-lactic acid is carried out by a single enzyme called malolactic enzyme (MLE), in the presence of Mn^{2+} and NAD^{+}, using a mechanism that is still unclear. The description of the complete *mle* operon from *O. oeni* (Labarre et al. 1996) has allowed heterologous expression of MLE in several organisms including *Escherichia coli*, *Lb. plantarum*, *S. cerevisiae*, and *Schizosaccharomyces pombe*, capable of performing MLF in an optimized way (Schümann et al. 2012). Nevertheless, the use of genetically modified organisms in food production has some concerns that remain unsolved. In this context, the use of MLF starter cultures, with selected but naturally obtained LAB strains, is becoming a usual practice in wine industry, because it can improve the process and enhance the quality and safety of wines. LAB strains isolated from a specific geographic region take advantage of their natural adaptation to the wine characteristics while maintaining the regional peculiarities. Thus, it is highly advisable to study the indigenous microbiota, best adapted to the agro-ecological conditions of a specific wine-producing area, as starter cultures (Ruiz et al. 2010; González-Arenzana et al. 2012; Bokulich et al. 2014).

O. oeni is the major LAB used as commercial starter cultures for MLF. However, different studies have shown that *Lb. plantarum* strains can grow in wines (du Plessis et al. 2004; G-Alegría et al. 2004; López et al. 2008; Cho et al. 2011; Miller et al. 2011) and possess resistance mechanisms to tolerate high ethanol concentration and low pH (G-Alegría et al. 2004; Lee et al. 2012), and the first commercial culture of this LAB species has recently been released as malolactic starter (Lerm et al. 2011). On the other hand, it has been demonstrated that some strains of *Lb. plantarum* possess interesting antimicrobial activities that inhibit other oenological LAB strains (Rojo-Bezares et al. 2007; Knoll et al. 2008). *Lb. plantarum* also shows a more diverse enzymatic profile than *O. oeni*, which could be important in the modification of a wine aroma profile (Lerm et al. 2011). *Lb. plantarum* is encountered in a variety of environmental niches and meats, vegetables, and fermented beverages. This species has one of the largest genomes known among LAB, which is thought to be related to the ecological flexibility and the diversity of environmental niches where *Lb. plantarum* is encountered.

Selection of LAB strains for wine inoculation should essentially be based on a high malolactic activity under harsh environment conditions encountered (low pH, high levels of ethanol and sulfite) (Guzzo et al. 2000; Spano and Massa 2006; Fiocco et al. 2007). Further features to be taken into account correspond to interesting oenological properties such as the presence of enzymatic activities involved in malic acid degradation and the production of desirable wine aromas, and the absence of biogenic amine synthesis (Grimaldi et al. 2000; Liu 2002; Moreno-Arribas et al. 2000, 2003; Bartowsky and Henschke 2004).

14.1.3 Patagonian Wines

Located at 37°5′ and 42°5′ southern latitude, Argentinean North Patagonia is one of the southernmost winegrowing regions of the world that has optimal agro-ecological conditions for high-quality viticulture and a long winemaking tradition. The presence of strong winds dries the environment, which helps to prevent the spread of diseases, making possible the development of organic wines. Moreover, the wide temperature range, particularly during the ripening of fruits, enables the grapes to ripen slowly and to develop a better accumulation of flavors. Then, the wine is the result of a set of environmental factors that affect the quality of the grapes as well as the diversity of microorganisms present on grape surfaces and in musts during the winemaking process. Some of the varieties of red grape cultivated are Cabernet Sauvignon, Malbec, Merlot, and Pinot Noir. Young dry red (85 %) and white wines (12 %) from neutral *Vitis vinifera* varieties are those mostly produced (Lopes et al. 2007). In musts of Patagonian red grapes the L-malic acid content represents, on average, 53 % of the titratable acidity (36 % of total acidity), reaching values of 66 % (45 % of total acidity) in Pinot Noir (Crisóstomo 2007). The MLF of these wines takes place spontaneously, and it is often unpredictable. For this reason, it would be interesting to dispose of indigenous malolactic cultures to ensure that the process is successfully performed, keeping the characteristics of the wine *terroir*.

In this chapter, we summarize the studies on LAB population diversity in some Patagonian wines, as well as the methods and criteria to select potential indigenous malolactic cultures for these wines.

14.2 Diversity of Indigenous LAB During Spontaneous MLF

Microbial succession is a universal phenomenon observed in spontaneous fermentation processes, which is a reflection of microbial interactions, competition for intrinsic growth factors such as nutrients, and resistance to inhibitory environmental conditions such as high acidity and ethanol (Ultee et al. 2013). Consequently, microorganisms showing a selective advantage emerge in a given period as the dominant populations during fermentation (Sánchez et al. 2010). Although the *V. vinifera* phyllosphere is colonized by a complex microbial community that substantially modulates wine qualities (Barata et al. 2012), only a few grape-surface microbes can actually develop and survive under the harsh environment conditions (Bokulich et al. 2012). During AF the growth of yeasts is stimulated by sugars. Later, the large amount of yeast biomass produced will die and autolyze, releasing amino acids and vitamins that may encourage the growth of LAB species along the MLF process (Fleet 2003). At the end of AF, the tolerance of LAB to low pH and high ethanol are the main factors that select their occurrence in winery ecosystems. If yeast growth is delayed, various species of LAB and acetic acid bacteria may grow and inhibit the AF, affecting the winemaking process (Fleet 2003). The LAB involved in this process (and commonly found on grape surfaces and in the winery environment) are composed of acid- and ethanol-tolerant strains, primarily from the four genera *Lactobacillus*, *Pediococcus*, *Leuconostoc*, and *Oenococcus* (Mills et al. 2008). Grape juices produced from healthy and mature grapes have low LAB populations, usually less than 10^3 cfu ml^{-1}.

14.2.1 Culture-Independent Methods to Analyze the Diversity of LAB Communities from Patagonian Wines

The description of the diversity of LAB species present in wine is highly dependent on the methodology used, environmental conditions, and grape cultivar. Traditional microbiological methods, such as isolation on selective growth media, microscopic observation, and enumeration of microbial populations show biases, increasing the detection of those species better adapted. However, the detection of minor species requires the use of more efficient growth media that mimic conditions under which LAB species grow in their natural habitat. For this reason, culture-dependent techniques fail to reveal the true diversity of wine LAB communities, because the populations numerically less important, under stress, or in a viable but non-cultivable (VBNC) state, are hardly recovered, leading to errors in assessing the microbial ecology of complex ecosystems (Cocolin et al. 2011). Thus, culture-independent methods are essential to fully understand the composition and dynamics of microbial communities throughout the winemaking process. For this purpose total bacterial DNA is extracted directly from wine samples and used in PCR assays to amplify housekeeping genes (typically constitutive genes required for the maintenance of basic cellular function), resulting in PCR amplicons different from the microorganisms present in the sample. The combination of PCR with denaturing gradient gel electrophoresis (PCR-DGGE) is a very sensitive tool to resolve complex microbial communities and for species identification. The PCR-DGGE technique has been applied using different regions of housekeeping genes for studying the microbial ecology of wine fermentations. The bacterial chromosomal regions typically targeted are the hypervariable domains V1, V3, V6, and V8 of the *16S rRNA* gene.

However, in some species, the *16S rRNA* gene is present in multiple copies within the same bacteria, which can have a different sequence, especially in V1 and V6 domains (Coenye and Vandamme 2003). The encoding gene for the β-subunit of RNA polymerase (*rpo*B gene) was proposed as an alternative gene for PCR-DGGE analysis because multiple copies of *rpo*B have never been reported for bacteria (Dahllof et al. 2000; Renouf et al. 2006). Depending on the housekeeping gene selected, the detection and discrimination of bacterial species varies. The discriminatory power of the *16S rRNA* gene is low among *Pediococcus* species. In contrast, differences could be seen with the *rpo*B gene, notably between *P. parvulus* and *P. damnosus*, allowing the identification of LAB cocci species in mixed samples and their distinction from other LAB species, mainly of *Lactobacillus* genera (Spano et al. 2006; Renouf et al. 2006).

To describe LAB species diversity in Patagonian wines, we have applied the culture-independent PCR-DGGE technique, using two genome regions, a fragment of the housekeeping *rpoB* gene, and the V3 variable region of the *16S rRNA* gene, to achieve a more complete description of LAB diversity through the combined use of both gene regions (Valdés La Hens et al. 2015). Samples from Merlot wines, 2008 and 2012 vintages, as well as from Pinot Noir wines, 2010 vintage, were collected according to the following schedule: at the end of AF (day 0, MLF1), at day 14 (MLF2), and at day 35 (MLF3). Also, a final MLF stage of a Pinot Noir wine, 2012 vintage, was also analyzed. Only the AF of the Merlot wines 2008 vintage was promoted by must inoculation with the commercial culture *S. cerevisiae* F10 Laffort (Bordeaux, France) [guided AF (GAF)]. In the remaining vinifications, the AF was spontaneous [natural AF (NAF)]. MLF was spontaneous in all wines analyzed; it is important to note that the Patagonian cellar sampled had never used commercial malolactic starters.

In the Patagonian Merlot wines, 14 LAB species were identified by PCR-*rpoB*/DGGE and PCR-*16S rRNA*-V3/DGGE, with the constant presence of the species *O. oeni* and *Lb. plantarum* at all MLF stages analyzed (Fig. 14.1). The LAB species found in the Merlot NAF wines (2008 vintage) were *O. oeni*, *Lb. buchneri*, *Lb. casei*, *Lb. paracasei*, *Lb. collinoides*, *Lb. fermentum*, *Lb. plantarum*, *Lb. reuteri*, *Lb. rhamnosus*, *Lb. ghizouensis*, *P. pentosaceus*, and *P. ethanolidurans* (Fig. 14.1a, b). The *Lactobacillus* species identified from the Merlot 2012 vintage were *Lb. plantarum*, *Lb. rhamnosus*, *Lb. casei*, and *Lb. kunkeei* (Fig. 14.1c, d). Two cocci species were founded as well, *O. oeni* and *P. ethanolidurans*. On the other hand, in the Merlot GAF wines, *O. oeni*, *Lb. plantarum*, *Lb. guizhouensis*, *Lb. rhamnosus*, *Lb. reuteri*, *Lb. collinoides*, *Lb. casei*, *L. mesenteroides*, *Pediococcus pentosaceus*, and *P. ethanolidurans* were the LAB species identified with MLF (Fig. 14.1a, b). All the detected species belonging to the *Lactobacillus* genus, have been previously described as oenological species with favorable sensory properties, except *Lb. kunkeei*, detected in the Merlot wine 2012 vintage, which has been reported in association to damaged grapes (Bae et al. 2006) and with sluggish/stuck fermentations (Edwards et al. 1998a, b). This study reported, for the first time, the presence of *Lb. guizhouensis* in wine samples.

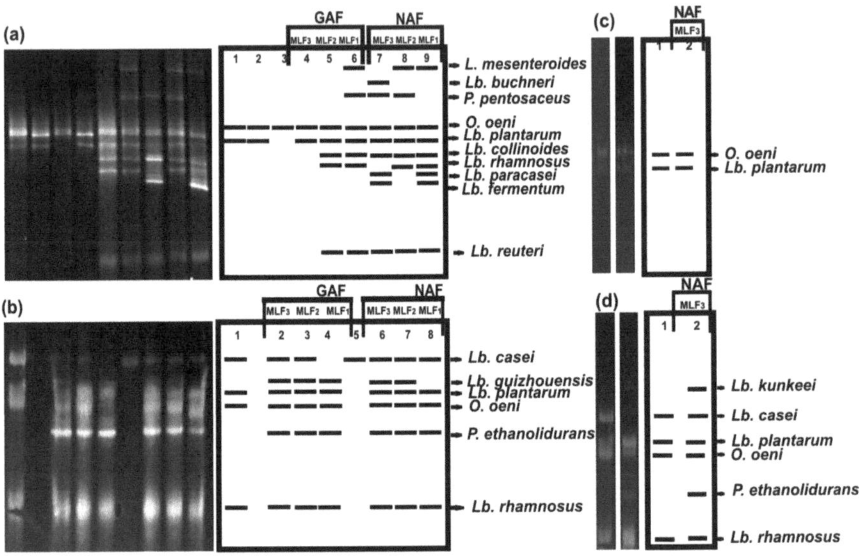

Fig. 14.1 Polymerase chain reaction–denaturing gradient gel electrophoresis (PCR-DGGE) profiles of lactic acid bacteria (LAB) species associated with Merlot wine samples from 2008 and 2012 vintages, at different stages of natural alcoholic fermentation (AF) and malolactic fermentation (MLF) (NAF-MLF) and guided AF (GAF-MLF). **a** PCR-*rpo*B/DGGE of Merlot 2008: *1*, *Lactobacillus (Lb.) plantarum* and *Oenococcus (O.) oeni* control strains; *2*, mix of DNA amplification products from *O. oeni* and *Lb. plantarum*; *3*, *O. oeni* control strain; *4–6*, three MLF stages of Merlot with GAF; *7–9*, three MLF stages of Merlot with NAF. **b** PCR-*16S rRNA* V3/DGGE of the Merlot 2008: *1*, *Lb. plantarum*, *Lb. casei*, *Lb. rhamnosus*, and *O. oeni* control strains; *2–4*, three MLF stages of Merlot with GAF; *5*, *Lb. casei* control strain; *6–8*, three MLF stages of Merlot with NAF. **c** PCR-*rpo*B/DGGE of Merlot 2012: *1*, *Lb. plantarum* and *O. oeni* control strains; 2, MLF_3. **d** PCR-*16S rRNA* V3/DGGE of Merlot 2012: *1*, *Lb. plantarum*, *Lb. casei*, *Lb. rhamnosus*, and *O. oeni* control strains; *2*, MLF_3

The combined use of two targeted gene regions in the Patagonian Pinot Noir wines led to identification of 11 species of *Lactobacillus*, 3 species of *Pediococcus*, 1 *Lactococcus*, and *O. oeni*, a total of 16 LAB species (Fig. 14.2). Also, 2 species belonging to *Bacillus* could be observed at the beginning of MLF: *B. circulans* and *B. subtilis*, which could be the result of a soil contamination. The species *O. oeni*, *Lb. plantarum*, *Lb. guizhouensis*, *Lc. lactis*, *P. pentosaceus*, and *P. ethanolidurans* were detected at all MLF stages (Fig. 14.2a, b). Interestingly, *L. lactis*, which was observed only in the Patagonian Pinot Noir wines, has seldom been detected in wines and also on grape surfaces (Bae et al. 2006; Mesas et al. 2011; González-Arenzana et al. 2013).

Comparing the results obtained by PCR-*rpo*B/DGGE and PCR-*16S rRNA*-V3/DGGE, it appears that the *rpo*B gene shows a greater ability for discriminating among the main LAB cocci species found in wines. These results were somewhat expected, because the primers chosen were based on the *rpo*B gene sequence that is usually selected to discriminate cocci species (Renouf et al. 2006). However, *P. ethanolidurans*

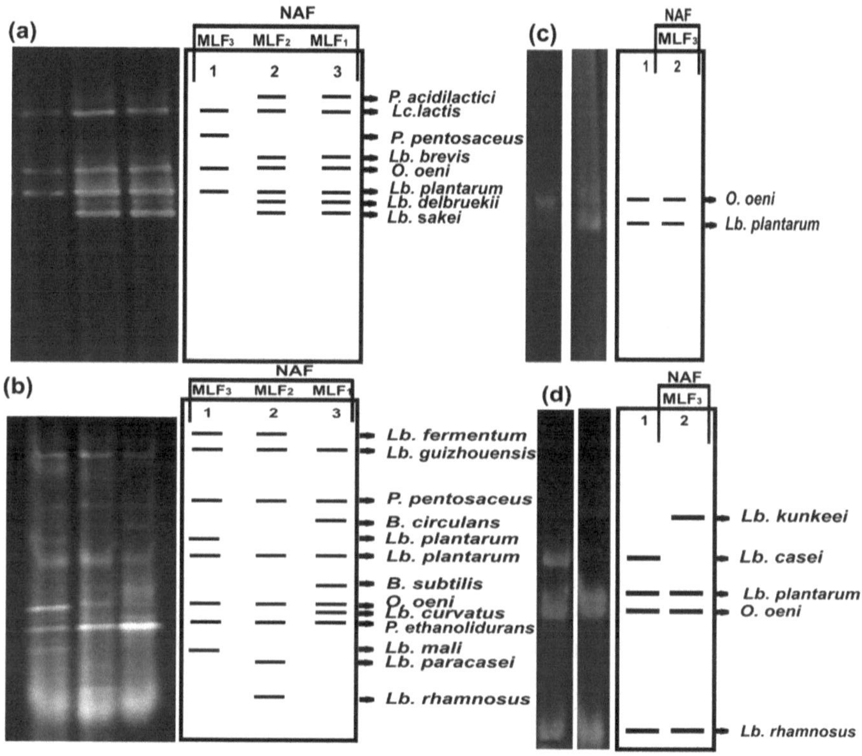

Fig. 14.2 PCR-DGGE profiles of LAB species associated to Pinot Noir wine samples from 2010 and 2012 vintages, at different stages of natural AF and MLF (NAF-MLF). **a** PCR-*rpo*B/DGGE of Pinot noir 2012: *1–3*, three MLF stages. **b** PCR-*16S rRNA* V3/DGGE of Pinot Noir: *1–3*, three MLF stages. **c** PCR-*rpo*B/DGGE of Pinot Noir 2012: *1*, *Lb. plantarum* and *O. oeni* control strains; *2*, final MLF stage. **d** PCR-*16S rRNA* V3/DGGE of Pinot Noir 2012: *1*, *Lb. plantarum*, *Lb. casei*, *Lb. rhamnosus*, and *O. oeni* control strains; *2*, final MLF stage

was not detected by the PCR-*rpo*B/DGGE gene, suggesting that the primers selected have some limitations that were solved by using the V3 region of the *16S rRNA* gene. The presence of *O. oeni* and *Lb. plantarum* was easily detected by both gene regions from Merlot as well as from Pinot Noir wines. In agreement with our results, González-Arenzana et al. (2013) detected 13 LAB species (9 by PCR-*16S rRNA*/DGGE of V4–V5 region and 4 by PCR-*rpo*B/DGGE), many of which were also detected in the Patagonian wines. Therefore, our results reinforce the idea that the use of the *16S rRNA* V3 gene region in combination with the *rpo*B gene enhances the knowledge of diversity from the LAB community, given that different species that are present in varying proportion could be detected. By applying the PCR-DGGE technique, our research showed the greatest amount of LAB described so far through spontaneous MLF in red wines. The prevalence of *O. oeni* and *Lb. plantarum* in all wine samples suggests that both species are involved in leading the spontaneous MLF in the Patagonian Pinot Noir and Merlot wines tested.

14.2.2 Culture-Dependent Strategies Applied to Isolation and Identification of LAB from Patagonian Wines

The community analysis of LAB present in wine by culture-dependent methods requires the isolation of the bacteria before their identification. The culture media used as support of LAB growth are generally rich in nutrients and growth factors, because lactic bacteria are unable to synthesize all growth requirements. Different selective growth media may be employed such as MRS or MRS plus ethanol or tomato juice (Bae et al. 2006), and MLO for *O. oeni* (formerly *Leuconostoc oenos*) (Caspritz and Radler 1983). In our study, species of *Lactobacillus*, *Pediococcus*, and *Leuconostoc* were isolated on MRS, and *O. oeni* on MLO, both supplemented with cycloheximide and/or pimaricin to prevent growth of yeasts and filamentous fungi. The cultures were incubated at 25–28 °C, under anaerobic conditions to prevent the growth of acetic acid bacteria (Bae et al. 2006), and the colonies were microscopically examined. Gram-positive bacteria with a negative catalase reaction were considered to belong to the LAB group. The presumptive identification of isolates must be completed by additional tests, such as the determination of the lactic acid production pathway (homo- or heterofermentative) and the analysis of ability to ferment carbohydrates. In the latter case, different species may have the same fermentation profile, which may lead to misidentification, so the results need to be carefully interpreted (Ribereau-Gayon et al. 2006). Physiological and biochemical criteria involved in the identification of LAB can be confusing, as most species of this group have similar nutritional and growth requirements (Vandamme et al. 1996).

Numerous methods for molecular identification of LAB of oenological interest allow achieving results quickly and reliably and with implementation of low cost. The methods are essentially based on direct analysis of the DNA obtained from pure cultures derived from a single colony, either through the polymorphism present in the restriction patterns of specific gene sequences (PCR-RFLP), the species-specific PCR reactions, or the complete or partial sequencing of housekeeping genes such as *rpo*B and *16S rRNA*, where the sequence obtained is compared with gene sequences deposited in databases.

In studies carried out on Patagonian wines (Bravo-Ferrada et al. 2013, 2016; Valdés La Hens et al. 2015), the LAB isolates obtained were identified by sequence analysis of a fragment of the *rpoB* gene. A 294-bp region was amplified using the primers rpoB1o, rpoB1, and rpoB2 (Renouf et al. 2006), and the amplicons were subjected to restriction analysis with *Aci*I, *Hinf*I, and *Mse*I enzymes. Restriction products were detected by agarose gel electrophoresis, and the profiles obtained were compared with those belonging to type strains. This method has some limitations to differentiate the species *Lb. sakei*, *Lb. sanfrancisensis*, and *P. damnosus* because of the similarity of their restriction profiles. Furthermore, restriction fragments less than 25 bp are difficult to visualize in the agarose gel, preventing their analysis (Fig. 14.3). An alternative methodology that we have used was PCR-ARDRA (restriction analysis of amplified *16S rRNA* gene), consisting in the amplification of about 1500 bp of this gene, using the primers pH and pA described by Ulrike et al. (1989), and the subsequent amplicons digestion with the enzymes *Mse*I and *Bfa*I.

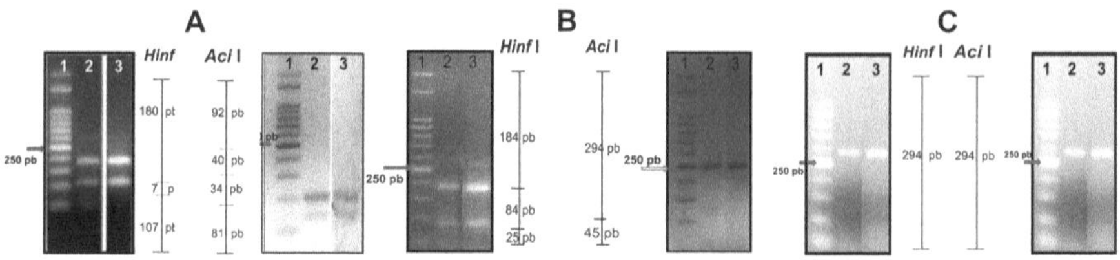

Fig. 14.3 Polymerase chain reaction–restricted fragment length polymorphism (PCR-RFLP) profiles of *rpo*B gene from *Lb. plantarum* (**a**), *O. oeni* (**b**), and *Lb. rhamnosus* (**c**). **a** *Line 1*, MW marker; *line 2*, *Lb. plantarum* ATCC 14917; *line 3*, *Lb. plantarum* UNQLp 155. **b** *Line 1*, MW marker; *line 2*, *O. oeni* ATCC 27310; *line 3*, *O. oeni* UNQOe 71.1. **c** *Line 1*, MW marker; *line 2*, *Lb. rhamnosus* ATCC 8530; *line 3*, *Lb. rhamnosus* UNQLrh 144

By applying the methodologies described here, LAB species isolated and identified from Patagonian red wines, varieties Merlot, Pinot Noir, and Malbec (vintages 2008, 2010, 2011, and 2013) were *Lb. plantarum* and *O. oeni*, as the main species, and *Lb. brevis*, *Lb. rhamnosus*, *L. mesenteroides*, *P. parvulus*, *P. pentosaceus*, *Lb. paracasei*, *Lb. fermentum*, and *P. acidilactici* in minor proportion (Bravo-Ferrada et al. 2013, 2016; Valdés La Hens et al. 2015).

14.2.3 Genetic Heterogeneity of Patagonian Oenococcus oeni *and* Lactobacillu. plantarum *Strains*

Because *Lb. plantarum* and *O. oeni* were the most prevalent species recovered from Patagonian wine samples, the isolates belonging to these species were subjected to a genetic heterogeneity study that was performed by RAPD-PCR using the Coc primer (Cocconcelli et al. 1995). The UPGMA dendrogram generated from the electrophoretic profiles illustrates the genomic variability and the clonal relationships between each single genomic fingerprinting (Fig. 14.4a, b) (Valdés La Hens et al. 2015). The number of different genotypes was 19 of 76 isolates for *Lb. plantarum* and 20 of 68 isolates for *O. oeni*. Clusters 1 and 19 from *Lb. plantarum* included the largest number of members (14 and 9 isolates, respectively), with cluster 1 being the most abundant for the 2012 Pinot Noir and cluster 19 the predominant one for the 2010 Pinot Noir. Among the *O. oeni* isolates, cluster 1 was the predominant one for the Merlot wine, and clusters 7 and 12 were the main genotypes for the 2012 Pinot Noir. *Lb. plantarum* strains isolated from different Patagonian wines were grouped into different clusters. All clusters from both *Lb. plantarum* and *O. oeni* included members that showed a 100 % similarity among them, meaning that members of the same strain were repeatedly isolated. These results were in agreement with others previously reported by our laboratory (Bravo-Ferrada et al. 2013).

There were some interesting differences between the Patagonian wines analyzed here: the 2008 Merlot showed the most complex polymorphism for *Lb. plantarum* species, but the least complex for *O. oeni* species. For the latter, the highest strain

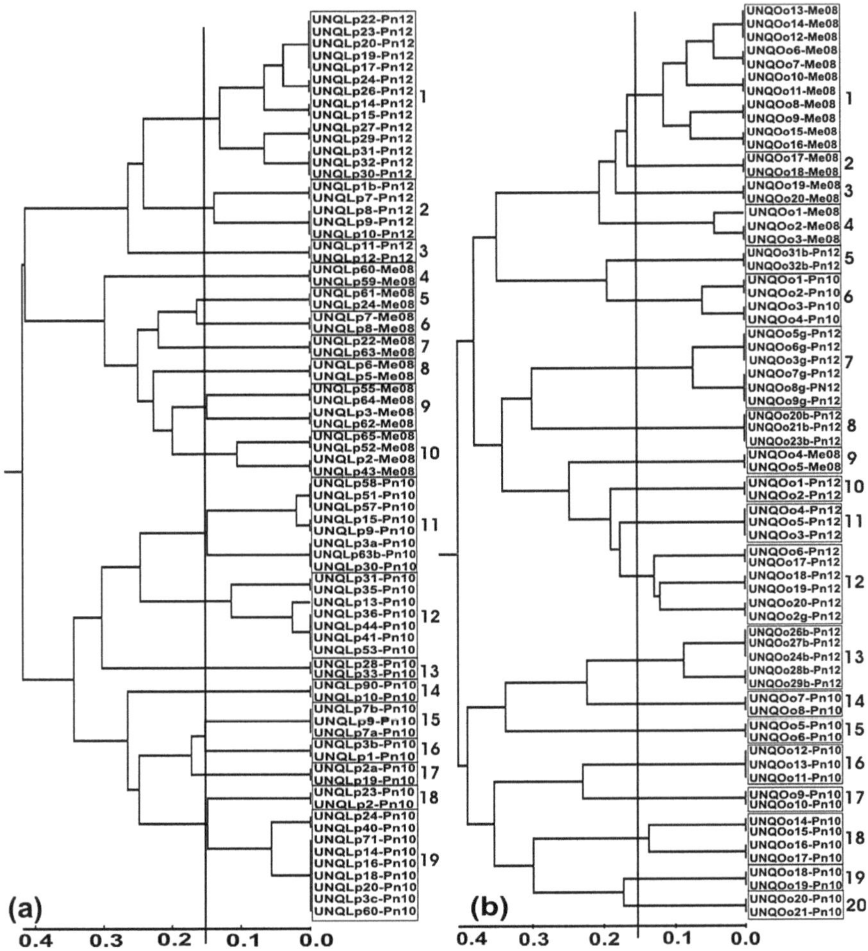

Fig. 14.4 Clustering of LAB isolates from wine samples (Merlot and Pinot Noir). Dendrograms were obtained by the unweighted pair group method using arithmetic averages analysis of Coc random amplification of polymorphic DNA (RAPD)-PCR. **a** *Lb. plantarum* isolates. **b** *O. oeni* isolates

diversity was observed in the 2010 Pinot Noir. We found a good correlation between RAPD profiles and wine samples, with only two cases where isolates coming from different samples showed similar RAPD patterns. The genetic diversity data from both *Lb. plantarum* and *O. oeni* isolates demonstrate a considerable genotypic heterogeneity through the MLF in all vinifications studied. This finding suggests that there are several strains, none of them implanted in the cellar, that come from grapes and lead the spontaneous MLF, highlighting the need to use an indigenous MLF starter to exert better process control.

14.3 Framework for Selection of New Starter Cultures: Stress Tolerance and Enzymatic Activities

After isolation of LAB strains to be employed as starter cultures, they should be subjected to various selection tests before use at pilot or industrial scale. Some of the criteria considered in that selection include tolerance of LAB strains to low pH and high ethanol and molecular SO_2 concentrations, good growth characteristics under the winemaking conditions, compatibility with the *S. cerevisiae* yeast used in the AF, ability to survive in the production process, inability to produce biogenic amines, and production of aromatic compounds that may potentially contribute to a favorable wine aroma profile (Lerm et al. 2010). In the case of the red wines produced in the Patagonian region, it will be necessary to verify that LAB strains can develop the MLF in the harsh environment of these wines, including the adaptation to processing conditions (low winemaking temperature, pH 3.3–3.8, molecular SO_2 0.25–2.50 mg l^{-1}, lysozyme 100–500 ppm, ethanol 13–14 %) (Lopes et al. 2007; Bravo-Ferrada et al. 2013).

Ethanol is the major metabolite produced by yeasts during the AF, which produce adverse effects on the growth and metabolic activities of LAB, and has an integral role in the ability of these bacteria to survive in the wine environment and perform the MLF. The analysis of growth of Patagonian *Lb. plantarum* and *O. oeni* strains, in presence of two ethanol concentrations (10 % and 14 %), allowed the observation that all the strains showed maximum cell populations in 10 % ethanol compared to 14 % ethanol. A concentration of 14 % ethanol was strongly inhibitory to most of *Lb. plantarum* and *O. oeni* strains, only a few strains being able to reach population values around 10^9 cfu ml^{-1} (Bravo-Ferrada et al. 2013, 2016).

The pH of wine is another determinant factor for the success of MLF. The amount of LAB species that can survive and proliferate is dependent on the wine pH (Kunkee 1967; Rosi et al. 2003; G-Alegría et al. 2004). Although the optimal growth pH for *O. oeni* is 4.8, and 6.5 for *Lb. plantarum*, G-Alegría et al. (2004) reported that some strains of *O. oeni* and *Lb. plantarum* were able to grow at pH 3.2. The growth behavior of Patagonian *Lb. plantarum* and *O. oeni* strains was tested in MRS and MLO broth, respectively, at different pH (6.5, 4.8, 3.5, 3.6, 3.7, and 3.8). Although all strains were able to grow at different pH conditions during 15 days, when the pH was reduced from the optimum 6.5 (for *Lb. plantarum*) and 4.8 (for *O. oeni*), to 3.8, 3.7, 3.6, and 3.5, the maximum cell populations were drastically reduced (Bravo-Ferrada et al. 2013, 2016).

As was already mentioned, another factor to consider in wine fermentations is the addition of SO_2 to grape must, because it is essential in the growth of LAB strains and in the development of MLF. Sulfite is an antimicrobial compound widely used in winemaking as a chemical preservative, because improves the fermentation, inhibiting the growth of undesirable bacteria and yeast, further having antioxidant activity (Romano et al. 1993). The allowed SO_2 concentrations in wine vary between 20 and 400 mg l^{-1}, depending on the legislation, the type of wine, and the country of manufacture (Ribereau-Gayon et al. 2006). The SO_2 is added to the grape must in general at the beginning of AF, after completion of this stage, and to the finished product, for microbiological protection. The percentage of distribution of the different forms depends on the pH (Bauer and Dicks 2004). At the pH values of the wine, the fraction of active or molecular SO_2 remaining free can range from 1 % to 10 %, which explains the greater biological stability present in more acidic wines. When the effect of molecular SO_2 (concentrations of 0.25, 1.25, and 2.5 mg l^{-1}) on the growth capacity of Patagonian *Lb. plantarum* and *O. oeni* strains was assessed, in MRS and MLO broth at pH 3.5, respectively, it was observed that a concentration of 2.5 mg l^{-1} strongly inhibited the growth of the strains, probing their sensitivity to the bactericidal activity of SO_2 (Bravo-Ferrada et al. 2013, 2016).

Lysozyme is an enzyme that has been proposed as an alternative to SO_2 to control the growth of LAB and to promote the delay of MLF. This enzyme is very effective against gram-positive bacteria and acts by breaking the glycosidic bond β-(1,4) between *N*-acetylmuramic acid and *N*-acetylglucosamine, components of cell-wall peptidoglycan (Bartowsky et al. 2005). Both the sensitivity of LAB to lysozyme and the dosage of the enzyme are parameters to be considered in determining the effectiveness of lysozyme to inhibit LAB and MLF (Bartowsky 2003). By analysis of the growth of Patagonian strains of *Lb. plantarum* and *O. oeni*, in MRS or MLO added to different lysozyme concentrations (50, 250, 500 ppm), a greater resistance to bacteriolytic activity was observed in the *Lb. plantarum* strains, whereas the *O. oeni* strains were notably sensitive to this activity (Bravo-Ferrada et al. 2013, 2016).

14.3.1 Effect of LAB on Flavor Development

Deacidification and greater microbiological stability of wine are some of the effects of the MLF. Also, this fermentation has a positive influence on the development of the organoleptic properties of wine, modifying flavors derived from fruit and producing aromatic compounds (Matthews et al. 2004; Bartowsky 2005). In winemaking technology, microbial enzymes provide many benefits including the formation of the wine aroma, which has become the main focus of interest. Enzymes are also involved in the improving of color in red wines, and they can solve problems associated to wine filtration. Therefore, it is crucial to analyze, in oenological LAB strains, the potential to influence wine composition, and therefore the processing, organoleptic properties, and quality of wine.

The sensory properties of wine are the result of a large number of individual compounds (Matthews et al. 2004). Four groups of these compounds, the monoterpenes, C_{13}-isoprenoids, benzene derivatives, and aliphatic compounds, can occur linked to sugars to form glycosides. Monoterpenes contribute to the varietal character that typifies wine and are important aromatics compounds. These compounds exhibit an immobilized reservation of aroma, and breaking the bond between sugar and terpene contributes to increase their intensity and therefore the wine flavor (Ribereau-Gayon et al. 2006). In our laboratory, the glycosidase activity was analyzed in eight selected

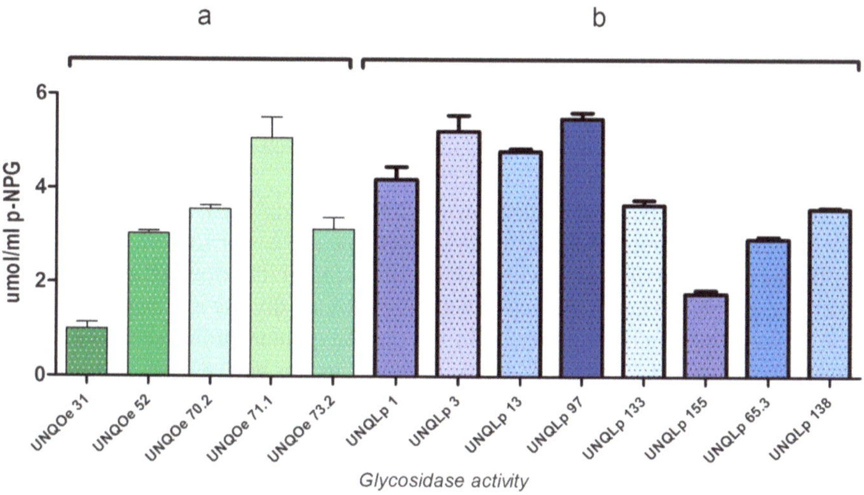

Fig. 14.5 Glycosidase activity on the substrate *p*-nitrophenyl-β-D-glucopyranoside (p-NPG) from five selected *O. oeni* (**a**) and eight *Lb. plantarum* (**b**) strains. Enzyme activity is expressed as μmol of *p*-nitrophenol released from a volume of 1 ml of cell culture containing 1 g of dry weight cells. Data are shown as mean ± SD of two independent measurements

Patagonian *Lb. plantarum* strains (Fig. 14.5), finding that all of them were positive for this activity, although with quantitative differences among strains (Bravo-Ferrada et al. 2013). On the other hand, when oenological Patagonian *O. oeni* strains were studied, some of them exhibited glycosidase activity (Fig. 14.5), and others did not show such enzyme activity (Brizuela 2013). However, it should be pointed out that glycosidase activity is influenced by some environmental factors such as pH, temperature, presence of sugars, and ethanol (Grimaldi et al. 2000, 2005).

On the other hand, tannins are polyphenolic, water-soluble compounds present in plants, the function of which is unknown, but it is believed to be related to respiration processes and defense mechanisms of plants against insect attack. Tannin acyl hydrolase is an inducible enzyme known as tannase. The presence of tannase activity in LAB may confer advantages in winemaking processes, by reducing the astringency of wine. Studies on Patagonian LAB strains (Bravo-Ferrada et al. 2013, 2016) showed that the *Lb. plantarum* strains studied showed tannase activity, whereas this activity was not detected in any of the *O. oeni* strains tested.

One of the most important aromatic compounds produced by LAB during MLF is 2,3-butanedione, which at low concentrations (about 1.4 mg ml^{-1}) contributes positively to the wine aroma, supplying buttery notes and adding complexity to the wine (Martineau and Henick-Kling 1995; Bartowsky and Henschke 2004; Swiegers et al. 2005), although at high concentrations it depreciates the quality. This compound is formed as an intermediate product of citric acid metabolism (Bartowsky et al. 2002). The metabolism of citric acid begins at the end of MLF, and the maximum concentration of 2,3-butanedione is reached when malic acid is exhausted (Bartowsky and Henschke 2004). The degradation of citric acid is dependent of the

LAB strains and other factors such as pH, contact of wine with the lees, and SO_2 level (Martineau and Henick-Kling 1995; Nielsen and Richelieu 1999). The ability to consume citrate, determined by a qualitative method, was analyzed in the Patagonian LAB strains (Bravo-Ferrada et al. 2013, 2016), and all the *Lb. plantarum* strains showed the ability to consume citrate, whereas in the *O. oeni* strains the activity was dependent on the strain analyzed.

Other criteria used for the selection of LAB to be employed as starter cultures is the inability to produce biogenic amines (BA), low molecular substances generated by decarboxylation of amino acids that are often found in fermented products. The chemical properties and biological functions of BA are diverse and can have beneficial or harmful effects on humans. So far, the Patagonian strains screened for the presence of *hdc*, *tdc*, and *ptc*A genes, involved in the biosynthesis of histamine, tyramine, and putrescine production, respectively, were negative (Bravo-Ferrada et al. 2013, 2016; Brizuela 2013).

14.3.2 Acclimation Treatment of Starter Cultures and Malolactic Activity

To survive "re-introduction" into the hostile wine environment, without a decrease of viable cell numbers and a subsequent loss of malolactic activity, the starter cultures should be subjected to a treatment of acclimation. This adaptation essentially consists of the acquisition of resistance mechanisms that enable bacteria to regulate the intracellular pH to maintain active the metabolic machinery. Adaptation also implies a modification of the membrane fluidity and structure, and the synthesis of the so-called stress proteins or HSP (heat shock proteins). The pre-adaptation of starter cultures is slow (24–48 h) and requires microbiological expertise. Winemakers can choose to produce their own acclimation medium for starter cultures, which usually contain grape juice or apple juice supplemented with other nutrients (Pilone and Kunkee 1972; Kunkee 1974; Costello et al. 1983; Henick-Kling 1993). An example of such medium may include grape juice diluted in water 1:1 and 0.5 % v/v yeast extract, pH 4.5. Once the cell culture reaches a density of 10^7 at 10^9 cfu ml^{-1}, it is directly inoculated in wine. Other winemakers prepare the starter cultures with diluted wine to reduce the risk of microbial contamination and to increase the tolerance of bacteria to ethanol (Hayman and Monk 1982; Nault et al. 1995).

The preexposure of bacteria to stress conditions, such as ethanol and low pH, has a positive effect on survival (Bourdineaud et al. 2003). Recommended treatments are the incubation in acclimation media with sublethal ethanol concentrations of 4–10 % v/v (Cecconi et al. 2009; Solieri et al. 2010; Lerm et al. 2011) and pHs that range between 4.6 and 3.5. For these reasons, acclimation media used for Patagonian *Lb. plantarum* and *O. oeni* strains were formulated with ethanol concentrations of 6 % or 10 %. The growth behavior and the ability to consume malic acid were tested in a synthetic wine, at laboratory scale, inoculating bacteria directly into the wine or subjecting them previously to an acclimation treatment. The synthetic wine was designed according the characteristics of the Patagonian Pinot Noir wines

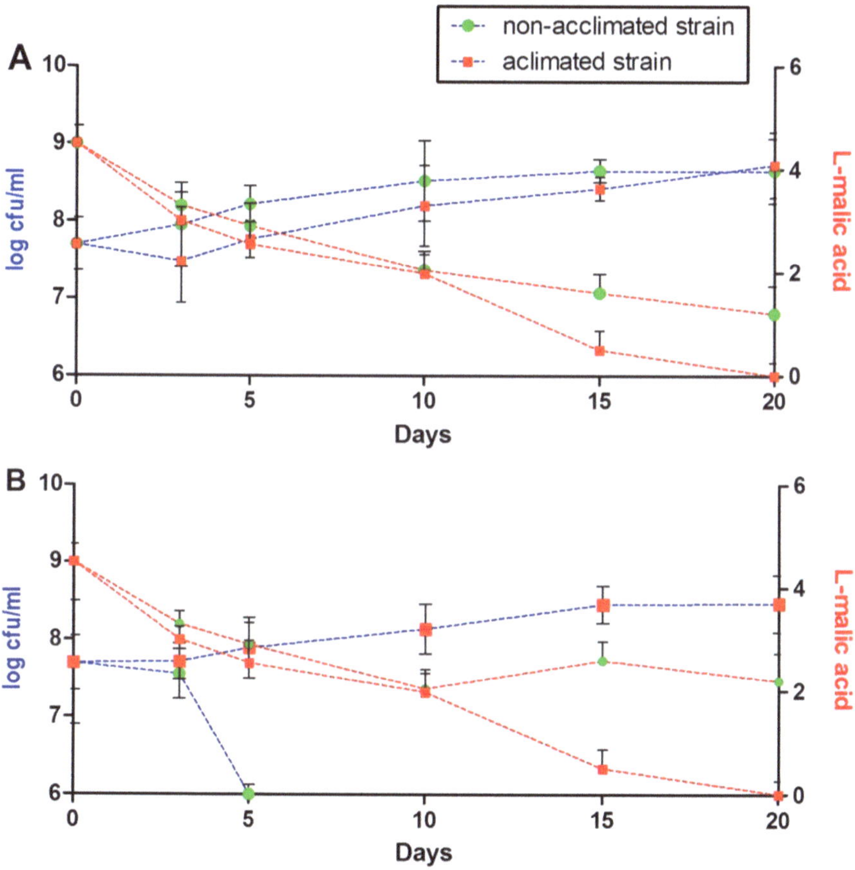

Fig. 14.6 Evolution of bacterial population (*blue line*) (log cfu ml^{-1}) and the L-malic acid concentration (*red line*) (g l^{-1}) during MLF in a synthetic wine: *Lb. plantarum* UNQLp 155 (**a**) and *O. oeni* UNQOe 73.2 (**b**)

(14 % ethanol, pH 3.5) (Lopes et al. 2007; Bravo-Ferrada et al. 2013). Figure 14.6a shows that the *Lb. plantarum* strain UNQLp 155 was able to remain viable and grow after direct inoculation in the synthetic wine. However, when this strain was previously acclimated, the number of viable cells and the consumption of malic acid were higher. The *O. oeni* strain UNQOe 73.2 was unable to remain viable after its direct inoculation in the synthetic wine. However, once acclimated in media containing 6 % or 10 % ethanol, this strain was able to remain viable and grow in the synthetic wine during 15 days (Fig. 14.6b) (Bravo-Ferrada et al. 2013, 2014, 2016).

Capozzi et al. (2010) found similar results working with *O. oeni* strains, and found that these bacteria were unable to grow in synthetic wine with 13 % ethanol, despite having been isolated from wines with 12.5 % and 13.9 % ethanol. da Silveira et al. (2002) suggest that the failure of the MLF, after direct inoculation of *O. oeni* strains in wine, can be explained by the deleterious effects of ethanol on the bacteria, in combination with the low pH in the wine.

14.3.3 Acclimation Effect on Cell Surface and Membrane Properties

As previously mentioned, the main wine stress factor is the high ethanol concentration whose target is the cytoplasm membrane. Ethanol is a low weight molecule without charge that can disrupt the membrane, producing proton efflux and affecting ATP biosynthesis, L-malate uptake (Leão and van Uden 1984; da Silveira et al. 2002), and loss of intracellular compounds such as NAD^+, NADH, and AMP (cofactors of malolactic activity) (Osman and Ingram 1985) with the consequent loss of viability. In *O. oeni*, ethanol concentrations higher than 8 % induce extensive membrane disorganization. In addition, a deleterious effect during MLF has been reported from the synergistic action of ethanol and low pH (da Silveira et al. 2002). For example, MLF was successful at 20 % v/v ethanol when pH was 5.0, but at pH 3.0 it requires that ethanol concentration be lower than 12 % (Capucho and San Romao 1994). As the tolerance of malolactic starter cultures to wine conditions is crucial for a successful of MLF, acclimation with sublethal ethanol concentrations becomes necessary. Previous work reported that acclimation in media containing ethanol within 4–10 % v/v (Cecconi et al. 2009; Solieri et al. 2010; Lerm et al. 2011) and low pH (3.5–4.6) can increase bacterial tolerance during winemaking. One of the main resistance mechanisms reported is the change in the fatty acid composition of the membrane that produces a decrease in the fluidity (Bastianini et al. 2000; Chu-Ky et al. 2005; Grandvalet et al. 2008), the induction of the small heat shock protein Lo18 (Guzzo et al. 1997), and modification of cytoplasm and membrane protein profiles (da Silveira et al. 2004; Teixeira et al. 2002).

The inoculation of Patagonian *Lb. plantarum* strains in wine-like medium (Bravo-Ferrada et al. 2013) induced a rapid disruption of membrane integrity, measured by multiparametric flow cytometry. This membrane cell damage affected both cultivability and ability to consume malic acid (Bravo-Ferrada et al. 2014). However, when cells were previously acclimated in 6 % or 10 % v/v ethanol, the membrane damage was lower after wine inoculation (14 % v/v ethanol, pH 3.5). As a consequence, this treatment improves the bacterial cultivability and the L-malic acid consumption in wine-like medium, in agreement with data reported for *O. oeni* strains (da Silveira et al. 2002). The acclimation media, besides ethanol and acids, must be rich in nutrients and growth factors to promote the expression of resistance mechanisms (Lerm et al. 2011). In addition, high concentrations of glucose and fructose have been reported to favor the development of LAB biomass in adverse conditions (Maicas et al. 2000).

The effect of ethanol on the membrane lipid composition of *O. oeni* has been widely studied. In some cases, a decrease in bilayer fluidity, with an increase in unsaturated lipid ratio and in membrane proteins, was observed (da Silveira et al. 2002), whereas others reported an increase in the saturated fatty acids composition with an increase in lactobacillic acid levels (Cyc19:0) in relationship with higher resistance to ethanol stress (Grandvalet et al. 2008; Teixeira et al. 2002). The fatty acid composition of three Patagonian *Lb. plantarum* strains, grown in MRS broth, showed that 16:0 and 18:1 were the most abundant and that cycC19:0 was approximately 10–20 % (Bravo-Ferrada et al. 2015a). Upon acclimation, important changes were observed in the fatty acid composition. The two main effects were a reduction of the unsaturation degree, by a drastic decrease of 18:1 fatty acid, and a decrease in the hydrocarbon chain lengths. These changes were more drastic when cultures were previously acclimated at higher ethanol concentrations (Bravo-Ferrada et al. 2015a). These results are in concordance with data reported by van Bokhorst-van de Veen et al. (2011). These authors also reported an activation of the novo biosynthesis of fatty acids. The regulation of the degree of saturation/unsaturation makes the cytoplasm membrane less permeable to ethanol and short-chain fatty acids could counteract the interdigitations of the lipid bilayer by action of ethanol, maintaining the membrane properties unaltered (Weber and de Bont 1996).

The analysis of correlation between fatty acids composition and malic acid consumption (MAC) reveals that, although MAC is strain specific, the increase of saturated and short-chain fatty acids favored malolactic activity (Bravo-Ferrada et al. 2015a). Acclimation treatment not only produces changes at the level of the lipid bilayer: changes of the Z-potential values of acclimated cells reveal modifications on the macromolecules exposed on the surface, such as the cell wall (Bravo-Ferrada et al. 2015b). Similar changes in the cell wall were observed by transmission electron microscopy (TEM) in other *Lb. plantarum* strains exposed to ethanol (van Bokhorst-van de Veen et al. 2011). This change in the cell-surface properties have been correlated with a higher damage after dehydration processes, visualized as an increase in the roughness on the surface by atomic force microscopy (AFM) (Bravo-Ferrada et al. 2015b). However, for other strains of *Lb. plantarum*, isolated from Patagonian wines, acclimation increases their recovery after freezing and freeze-drying, maintaining a higher resistance to inoculation in wine-like medium after preservation processes (Bravo-Ferrada et al. 2015c).

Acknowledgments This work was funded by grants from Universidad Nacional de Quilmes (Programa Microbiología Molecular Básica y Aplicaciones Biotecnológicas), Comisión de Investigaciones Científicas de la Provincia de Buenos Aires (CIC-PBA), and ANPCyT (PICTO UNQ 2006 N° 36474, PICT SU 2012 N° 2804, PICT 2013 N° 1481). BMBF and NB are fellows of Consejo Nacional de Investigaciones Científicas y Técnicas (CONICET). LS is a member of the Research Career of CIC-BA; AH and ET are members of the Research Career of CONICET.

References

Bae S, Fleet GH, Heards GM (2006) Lactic acid bacteria associated with wine grapes from several Australian vineyards. J Appl Microbiol 100:712–727

Barata A, Malfeito-Ferreira M, Loureiro V (2012) The microbial ecology of wine grape berries. Int J Food Microbiol 153:243–259

Bartowsky EJ (2003) Lysozyme and winemaking. Aust NZ Grapegrow Winemak 473a:101–104

Bartowsky E (2005) *Oenococcus oeni* and malolactic fermentation: moving into the molecular arena. Aust J Grape Wine Res 11:174–187

Bartowsky EJ, Henschke PA (2004) The 'buttery' attribute of wine diacetyl desirability, spoilage and beyond. Int J Food Microbiol 96:235–252

Bartowsky EJ, Francis IL, Bellon JR, Henschke PA (2002) Is buttery aroma perception in wines predictable from the diacetyl concentration? Aust J Grape Wine Res 8:180–185
Bastianini A, Granchi L, Guerrini S, Vincenzini M (2000) Fatty acid composition of malolactic *Oenococcus oeni* strains exposed to pH and ethanol stress. Ital J Food Sci 12:333–342
Bauer R, Dicks LMT (2004) Control of malolactic fermentation in wine. A review. S Afr J Enol Vitic 25:74–88
Belizário JE, Napolitano M (2015) Human microbiomes and their roles in dysbiosis, common diseases, and novel therapeutic approaches. Front Microbiol 6:1050
Bokulich NA, Joseph CML, Allen G, Benson AK, Mills DA (2012) Next-generation sequencing reveals significant bacterial diversity of botrytized wine. PLoS One 7:e36537
Bokulich NA, Thorngate JH, Richardson PM, Mills DA (2014) Microbial biogeography of wine grapes is conditioned by cultivar, vintage, and climate. Proc Natl Acad Sci USA 111:E139–E148
Bourdineaud JP, Nehmé B, Tesse S, Lonvaud-Funel A (2003) The *fis*H gene of the wine bacterium *Oenococcus oeni* is involved in protection against environmental stress. Appl Environ Microbiol 69:2511–2520
Bravo-Ferrada BM, Hollmann A, Delfederico L, Valdés La Hens D, Caballero A, Semorile L (2013) Patagonian red wines: selection of *Lactobacillus plantarum* isolates as potential starter cultures for malolactic fermentation. World J Microbiol Biotechnol 29:1537–1549
Bravo-Ferrada BM, Tymczyszyn EE, Gómez-Zavaglia A, Semorile L (2014) Effect of acclimation medium on cell viability, membrane integrity and ability to consume malic acid in synthetic wine by oenological *Lactobacillus plantarum* strains. J Appl Microbiol 116:360–367
Bravo-Ferrada BM, Gómez-Zavaglia A, Semorile L, Tymczyszyn E (2015a) Effect of the fatty acid composition of acclimated oenological *Lactobacillus plantarum* on the resistance to ethanol. Lett Appl Microbiol 60:155–161
Bravo-Ferrada BM, Gonçalves S, Semorile L, Santos NC, Tymczyszyn EE, Hollmann A (2015b) Study of surface damage on cell envelope assessed by AFM and flow cytometry of *Lactobacillus plantarum* exposed to ethanol and dehydration. J Appl Microbiol 118:1409–1417
Bravo-Ferrada BM, Brizuela N, Gerbino E, Gómez-Zavaglia A, Semorile L, Tymczyszyn EE (2015c) Effect of protective agents and previous acclimation on ethanol resistance of frozen and freeze-dried *Lactobacillus plantarum* strains. Cryobiology 71:522–528
Bravo-Ferrada BM, Hollmann A, Brizuela N, Valdés La Hens D, Tymczyszyn EE, Semorile L (2016) Growth and consumption of L-malic acid in wine-like medium by acclimated and non-acclimated cultures of Patagonian *Oenococcus oeni* strains. Folia Microbiol. doi:10.1007/s12223-016-0446-y
Brizuela N (2013) Selection of *Lactobacillus plantarum* and *Oenococcu soeni* isolates to be used as indigenous starter cultures for malolactic fermentation of Patagonian wines. Graduate Thesis, UNQ
Brooijmans R, Smit B, Santos F, van Riel J, de Vos WM, Hugenholtz J (2009) Heme and menaquinone induced electron transport in lactic acid bacteria. Microb Cell Fact 8:28
Capozzi V, Russo P, Beneduce L, Weidmann S, Grieco F, Guzzo J, Spano G (2010) Technological properties of *Oenococcus oeni* strains isolated from typical southern Italian wines. Lett Appl Microbiol 50:327–334
Capucho I, San Romao MV (1994) Effect of ethanol and fatty acids on malolactic activity of *Oenococcus oeni*. J Appl Microbiol Biotechnol 42:391–395
Caspritz G, Radler F (1983) Malolactic enzyme of *Lactobacillus plantarum*. Purification, properties and distribution among bacteria. J Biol Chem 258:4907–4910
Cecconi D, Milli A, Rinalducci S, Zolla L, Zapparoli G (2009) Proteomic analysis of *Oenococcus oeni* freeze-dried culture to assess the importance of cell acclimation to conduct malolactic fermentation in wine. Electrophoresis 30:2988–2995
Cho GS, Krauss S, Huch M, du Toit M, Franz CM (2011) Development of a quantitative PCR for detection of *Lactobacillus plantarum* starters during wine malolactic fermentation. J Microbiol Biotechnol 21:1260–1266

Chu-Ky S, Tourdot-Marechal R, Marechal PA, Guzzo J (2005) Combined cold, acid, ethanol shocks in *Oenococcus oeni*: effects on membrane fluidity and cell viability. Biochim Biophys Acta 1717:118–124
Cocconcelli PS, Porro D, Galandini S, Senini L (1995) Development of RAPD protocol for typing of strains of lactic acid bacteria and enterococci. Lett Appl Microbiol 1:376–379
Cocolin L, Campolongo S, Alessandria V, Dolci P, Rantsiou K (2011) Culture-independent analysis and wine fermentation: an overview of achievements 10 years after first application. Ann Microbiol 61:17–23
Coenye T, Vandamme P (2003) Extracting phylogenetic information from whole-genome sequencing projects: the lactic acid bacteria as a test case. Microbiology 149:3507–3517
Costello MD, Morrison GJ, Lee TH, Flee GH (1983) Numbers and species of lactic acid bacteria in wines during vinification. Food Tech Assoc Aust 35:14–18
Crisóstomo B (2007) Caracterización fisicoquímica de mostos de uva de la Región Sur destinados a vinificación. Graduate Thesis, UNCO
da Silveira MG, San Romão V, Loureiro-Dias MC, Rombouts FM, Abee T (2002) Flow cytometry assessment of membrane integrity of ethanol-stressed *Oenococcus oeni* cells. Appl Environ Microbiol 68:6087–6093
da Silveira MG, Baumgartner M, Rombouts FM, Abee T (2004) Effect of adaptation to ethanol on cytoplasmic and membrane protein profiles of *Oenococcus oeni*. Appl Environ Microbiol 70:2748–2755
Dahllof I, Baillie H, Kjelleberg S (2000) *rpo*B-Based microbial community analysis avoids limitations inherent in *16S rRNA* gene intraspecies heterogeneity. Appl Environ Microbiol 66:3376–3380
Douillard FP, de Vos WM (2014) Functional genomics of lactic acid bacteria: from food to health. Microb Cell Fact 13(suppl 1):S8
du Plessis HW, Dicks LM, Pretorius IS, Lambrechts MG, du Toit M (2004) Identification of lactic acid bacteria isolated from South African brandy base wines. Int J Food Microbiol 91:19–29
Edwards CG, Haag KM, Collins MD, Hutson R, Huang YC (1998a) *Lactobacillus kunkeei* sp nv.: a spoilage organism associated with grape juice fermentations. J Appl Microbiol 84:698–702
Edwards CG, Haag KM, Collins MD (1998b) Identification of some lactic acid bacteria associated with sluggish/stuck fermentations. Am J Enol Vitic 49:445–448
Felis GE, Dellaglio F (2007) Taxonomy of *Lactobacilli* and *Bifidobacteria*. Curr Issues Intest Microbiol 8:44–61
Fiocco D, Capozzi V, Goffin P, Hols P, Spano G (2007) Improved adaptation to heat, cold, and solvent tolerance in *Lactobacillus plantarum*. Appl Microbiol Biotechnol 77:909–915
Fleet GH (2003) Yeast interactions and wine flavor. Int J Food Microbiol 86:11–22
G-Alegría E, López I, Ignacio Ruiz J, Sáenz J, Fernández E, Zarazaga M, Dizy M, Torres C, Ruiz-Larrea F (2004) High tolerance of wild *Lactobacillus plantarum* and *Oenococcus oeni* strains to lyophilization and stress environmental conditions of acid pH and ethanol. FEMS Microbiol Lett 230:53–61
González-Arenzana L, Santamaría P, López R, Tenorio C, López-Alfaro I (2012) Dynamics of indigenous lactic acid bacteria in wine fermentation from La Rioja (Spain) during three vintages. Environ Microbiol 63:12–19
González-Arenzana L, López R, Santamaría P, López-Alfaro I (2013) Dynamics of lactic acid bacteria populations in Rioja wines by PCR-DGGE. Comparison with culture-dependent methods. Appl Microbiol Biotechnol 97:6931–6941
Grandvalet C, Assad-García JS, Chu-Ky S, Tollot M, Guzzo J, Gresti J, Tourdot-Maréchal R (2008) Changes in membrane lipid composition in ethanol- and acid-adapted *Oenococcus oeni* cells: characterization of the *cfa* gene by heterologous complementation. Microbiology 154:2611–2619
Grimaldi A, McLean H, Jiranek V (2000) Identification and partial characterization of glycosidic activities of commercial strains of the lactic acid bacteria *Oenococcus oeni*. Am J Enol Vitic 51:362–369
Grimaldi A, Bartowsky E, Jiranek V (2005) Screening of *Lactobacillus* spp. and *Pediococcus* spp. for glycosidase activities that are important in oenology. J Appl Microbiol 99:1061–1069

Guzzo J, Delmas F, Pierre F, Jobin MP, Samyn B, van Beeumen J, Cavin JF, Divies C (1997) A small heat shock protein from *Leuconostoc oenos* induced by multiple stresses and during stationary growth phase. Lett Appl Microbiol 24:393–396

Guzzo J, Jobin MP, Delmas F, Fortier LC, Garmyn D, Tourdot-Maréchal R, Lee B, Diviès C (2000) Regulation of stress response in *Oenococcus oeni* as a function of environmental changes and growth phase. Int J Food Microbiol 55:27–31

Hayman DC, Monk PR (1982) Starter culture preparation for the induction of malolactic fermentation in wine. Food Tech Assoc Aust 34:14–18

Henick-Kling T (1993) Malolactic fermentation. In: Fleet GH (ed) Wine microbiology and biotechnology, 1st edn. Harwood Academic, Chur, Switzerland

Klaenhammer TR, Barrangou R, Buck BL, Azcárate-Peril MA, Altermann E (2005) Genomic features of lactic acid bacteria effecting bioprocessing and health. FEMS Microbiol Rev 29:393–409

Knoll C, Divol B, du Toit M (2008) Genetic screening of lactic acid bacteria of oenological origin for bacteriocin-encoding genes. Food Microbiol 25:983–991

Kunkee RE (1967) Control of malolactic fermentation induced by *Leuconostoc citrovorum*. Am J Enol Vitic 18:71–77

Kunkee RE (1974) Malolactic fermentation and winemaking. Adv Chem Res 137:151–170

Labarre C, Diviès C, Guzzo J (1996) Genetic organization of the *mle* locus and identification of a *mle*R-like gene from *Leuconostoc oenos*. Appl Environ Microbiol 62:4493–4498

Leão C, van Uden N (1984) Effects of ethanol and other alkanols on passive proton influx in the yeast *Saccharomyces cerevisiae*. Biochim Biophys Acta 774:43–48

Lee SG, Lee KW, Park TH, Park JY, Han NS, Kim JH (2012) Proteomic analysis of proteins increased or reduced by ethanol *of Lactobacillus plantarum* ST4 isolated from Makgeoli, traditional Korean rice wine. J Microbiol Biotechnol 22:516–525

Lerm E, Engelbrecht L, du Toit M (2010) Malolactic fermentation: the ABC's of MLF. S Afr J Enol Vitic 31:186–192

Lerm E, Engelbrecht L, du Toit M (2011) Selection and characterisation of *Oenococcus oeni* and *Lactobacillus plantarum* South African wine isolates for use as malolactic fermentation starter cultures. S Afr J Enol Vitic 32:280–295

Liu SQ (2002) Malolactic fermentation in wine: beyond deacidification. J Appl Microbiol 92:598–601

Lonvaud-Funel A (1999) Lactic acid bacteria in the quality improvement and depreciation of wine. Antonie Van Leeuwenhoek 76:317–331

Lopes CA, Rodríguez ME, Sangorrín M, Querol A, Caballero AC (2007) Patagonian wines: the selection of an indigenous yeast starter. J Ind Microbiol Biotechnol 34:539–546

López I, López R, Santamaría P, Torres C, Ruiz-Larrea F (2008) Performance of malolactic fermentation by inoculation of selected *Lactobacillus plantarum* and *Oenococcus oeni* strains isolated from Rioja red wine. Vitis 47:123–129

Maicas S, Gil JV, Pardo I, Ferrer S (1999) Improvement of volatile composition of wines by controlled inoculation of malolactic bacteria. Food Res Int 32:491–496

Maicas S, Pardo I, Ferrer S (2000) The effects of freezing and freeze-drying of *Oenococcus oeni* upon induction of malolactic fermentation in red wine. Int J Food Sci Technol 35:75–79

Makarova KS, Koonin EV (2007) Evolutionary genomics of lactic acid bacteria. J Bacteriol 189:1199–1208

Martineau B, Henick-Kling T (1995) Formation and degradation of diacetyle in wine during alcoholic fermentation with *Saccharomyces cerevisiae* strain EC 1118 and malolactic fermentation with *Leuconostoc oenos* strain MCW. Am J Enol Vitic 46:442–448

Matthews A, Grimaldi A, Walker M, Bartowsky E, Grbin P, Jiranek V (2004) Lactic acid bacteria as a potential source of enzymes for use in vinification. Appl Environ Microbiol 70:5715–5731

Mesas JM, Rodríguez MC, Alegre MT (2011) Characterization of lactic acid bacteria from musts and wines of three consecutive vintages of Riveira Sacra. Lett Appl Microbiol 52:258–268

Miller BJ, Franz CM, Cho GS, du Toit M (2011) Expression of the malolactic enzyme gene (*mle*) from *Lactobacillus plantarum* under winemaking conditions. Curr Microbiol 62:1682–1688
Mills DA, Phister T, Neely E, Johannsen E (2008) Wine fermentation. In: Cocolin L, Ercolini D (eds) Molecular techniques in the microbial ecology of fermented foods. Springer, Berlin, pp 162–192
Moreno-Arribas MV, Torlois S, Joyeux A, Bertrand A, Lonvaud-Funel A (2000) Isolation, properties and behavior of tyramine-producing lactic acid bacteria from wine. J Appl Microbiol 88:584–593
Moreno-Arribas MV, Polo CM, Jorganes F, Muñoz R (2003) Screening of biogenic amine production by lactic acid bacteria isolated from grape must and wine. Int J Food Microbiol 84:117–123
Nault I, Gerbaux V, Larpent JP, Vayssier Y (1995) Influence of pre-culture conditions on the ability of *Leuconostoc oenos* to conduct malolactic fermentation in wine. Am J Enol Vitic 46:357–362
Nielsen JC, Richelieu M (1999) Control of flavour development in wine during and after malolactic fermentation by *Oenococcus oeni*. Appl Environ Microbiol 65:740–745
Osman Y, Ingram LO (1985) Mechanism of ethanol inhibition of fermentation in *Zymomonas mobilis* CP4. J Bacteriol 164:173–180
Pfeiler EA, Klaenhammer TR (2007) The genomics of lactic acid bacteria. Trends Microbiol 15:546–553
Pilone GJ, Kunkee RE (1972) Characterization and energetics of *Leuconostoc oenos* ML-34. Am J Enol Vitic 23:61–70
Reid G, Jass J, Sebulsky MT, McCormick JK (2003) Potential uses of probiotics in clinical practice. Clin Microbiol Rev 16:658–672
Renouf V, Claisse O, Lonvaud-Funel A (2006) Inventory and monitoring of wine microbial consortia. Appl Microbiol Biotechnol 75:149–164
Ribereau-Gayon P, Glories Y, Maujean A, Dubourdieu D (2006) Handbook of enology, vol 2, 2nd edn, The chemistry of wine stabilization and treatments. Wiley, New York
Rodas AM, Ferrer S, Pardo I (2003) 16S-ARDRA, a tool for identification of lactic acid bacteria isolated from grape must and wine. Syst Appl Microbiol 26:412–422
Rojo-Bezares B, Sáenz Y, Navarro I, Zarazaga M, Ruiz-Larrea F, Torres C (2007) Co-culture inducible bacteriocin activity of *Lactobacillus plantarum* strain J23 isolated from grape must. Food Microbiol 24:482–491
Romano P, Suzzi G, Zironi R, Comi G (1993) Biometric study of acetoin production in *Hanseniaspora guilliermondii* and *Kloekera apiculata*. Appl Environ Microbiol 59:1838–1841
Rosi I, Fia G, Canuti V (2003) Influence of different pH values and inoculation time on the growth and malolactic activity of a strain of *Oenococcus oeni*. Aust J Grape Wine Res 9:194–199
Ruiz P, Seseña S, Izquierdo PM, Llanos Palop M (2010) Bacterial biodiversity and dynamics during malolactic fermentation of Tempranillo wines as determined by a culture-independent method (PCR-DGGE). Appl Microbiol Biotechnol 86:1555–1562
Sánchez A, Rodríguez R, Coton M, Coton E, Herrero M, García LA et al (2010) Population dynamics of lactic acid bacteria during spontaneous malolactic fermentation in industrial cider. Food Res Int 43:2101–2107
Schümann C, Michmayr H, Eder R, del Hierro AM, Kulbe K, Mathiensen G, Nguyen TH (2012) Heterologous expression of *Oenococcus oeni* malolactic enzyme in *Lactobacillus plantarum* for improved malolactic fermentation. AMB Express 2:19
Selle K, Klaenhammer TR (2013) Genomic and phenotypic evidence for probiotic influences of *Lactobacillus gasseri* on human health. FEMS Microbiol Rev 37:915–935
Solieri L, Genova F, De Paola M, Giudici P (2010) Characterization and technological properties of *Oenococcus oeni* strains from wine spontaneous malolactic fermentations: a framework for selection of new starter cultures. J Appl Microbiol 108:285–298
Spano G, Massa S (2006) Environmental stress response in wine lactic acid bacteria: beyond *Bacillus subtilis*. Crit Rev Microbiol 32:77–86

Spano G, Beneduce L, de Palma L, Quinto M, Vernile A, Massa S (2006) Characterization of wine *Lactobacillus plantarum* by PCR-DGGE and RAPD-PCR analysis and identification of *Lactobacillus plantarum* strains able to degrade arginine. World J Microbiol Biotechnol 22:769–773

Swiegers JH, Bartowsky EJ, Henschke PA, Pretorius IS (2005) Yeast and bacteria modulation of wine aroma and flavor. Aust J Grape Wine Res 11:139–173

Teixeira H, Gonçalves MG, Rozés N, Ramos A, San Romão MV (2002) Lactobacillic acid accumulation in the plasma membrane of *Oenococcus oeni*: a response to ethanol stress? Microb Ecol 43:146–153

Ugliano M, Genovese A, Moio L (2003) Hydrolysis of wine aroma precursors during malolactic fermentation with four commercial starters of *Oenococcus oeni*. J Agric Food Chem 51:5075–5078

Ulrike E, Rogall T, Blocker H, Emde M, Bottger EC (1989) Isolation and direct complete nucleotide determination of entire genes. Characterization of a gene encoding for 16S ribosomal RNA. Nucleic Acids Res 17:7843–7853

Ultee A, Wacker A, Kunz D, Löwestein R, König H (2013) Microbial succession in spontaneously fermented grape must before, during and after stuck fermentation. S Afr J Enol Vitic 34:68–78

Valdés La Hens D, Bravo-Ferrada BM, Delfederico L, Caballero A, Semorile L (2015) Prevalence of *Lactobacillus plantarum* and *Oenococcus oeni* during spontaneous malolactic fermentation in Patagonian red wines revealed by polymerase chain reaction-denaturing gradient gel electrophoresis with two targeted genes. Aust J Grape Wine Res 21:49–56

van Bokhorst-van de Veen H, Abee T, Tempelaars M, Bron PA, Kleerebezem M, Marco ML (2011) Short- and long-term adaptation to ethanol stress and its cross-protective consequences in *Lactobacillus plantarum*. Appl Environ Microbiol 77:5247–5256

van de Gutche M, Penaud S, Grimaldi C, Barbe V, Bryson K, Nicolas P, Robert C et al (2006) The complete genome sequence of *Lactobacillus bulgaricus* reveals extensive and ongoing reductive evolution. Proc Natl Acad Sci USA 103:9274–9279

Vandamme P, Pot B, Gillis M, de Vos P, Kersters K, Swings J (1996) Polyphasic taxonomy, a consensus approach to bacterial systematic. Microbiol Rev 60:407–438

Weber FJ, de Bont JAM (1996) Adaptation mechanisms of microorganisms to the toxic effects of organic solvents on membranes. Biochim Biophys Acta 1286:225–245

Part III
Yeast Biotechnology

Chapter 15
Saccharomyces in Traditional and Industrial Fermentations from Patagonia

María Eugenia Rodríguez, Andrea Cecilia Origone, Melisa González Flores, and Christian Ariel Lopes

Abstract A large variety of fermented foods and beverages with traditional and cultural value have been described in the world, including industrial products such as wine, cider, and beer as well as traditional ones. In contrast with the massive scientific information available about the microbiota responsible for winemaking, yeasts responsible for most traditional fermented beverages around the world remain undiscovered. Both industrial and traditional fermentation processes coexist in Patagonia, making this region an ideal scenario for fermentative yeast diversity studies. The most relevant feature of this area is the fact that most traditional processes are produced at low temperature (below 20 °C), which directly affects the microbial diversity. We identified and characterized fermentative yeasts present during industrial fermentations of wine and cider and traditional fermentations (chichas) obtained from wild apples and *Araucaria araucana* seeds, substrates typically

M.E. Rodríguez
Instituto de Investigación y Desarrollo en Ingeniería de Procesos, Biotecnología y Energías Alternativas (PROBIEN, Consejo Nacional de Investigaciones Científicas y Técnicas de la República Argentina–Universidad Nacional del Comahue),
Buenos Aires 1400 (8300), Neuquén, Argentina

Facultad de Ciencias Médicas, Universidad Nacional del Comahue,
Buenos Aires 1400 (8300), Neuquén, Argentina

A.C. Origone • M. González Flores
Instituto de Investigación y Desarrollo en Ingeniería de Procesos, Biotecnología y Energías Alternativas (PROBIEN, Consejo Nacional de Investigaciones Científicas y Técnicas de la República Argentina–Universidad Nacional del Comahue),
Buenos Aires 1400 (8300), Neuquén, Argentina

C.A. Lopes (✉)
Instituto de Investigación y Desarrollo en Ingeniería de Procesos, Biotecnología y Energías Alternativas (PROBIEN, Consejo Nacional de Investigaciones Científicas y Técnicas de la República Argentina–Universidad Nacional del Comahue),
Buenos Aires 1400 (8300), Neuquén, Argentina

Facultad de Ciencias Agrarias, Universidad Nacional del Comahue, Buenos Aires 1400 (8300), Neuquén, Argentina
e-mail: christian.lopes@probien.gob.ar

N.L. Olivera et al. (eds.), *Biology and Biotechnology of Patagonian Microorganisms*, DOI 10.1007/978-3-319-42801-7_15

used by aboriginal communities to prepare soft alcoholic beverages. As a general rule, only *Saccharomyces cerevisiae* strains were obtained from wines and ciders, and they showed a close genetic relationship with European strains of this species. In traditional fermentations, commercial bakery and European wine strains of *S. cerevisiae* were detected as pure or mixed cultures with *Saccharomyces uvarum*, a cryotolerant species. This last species was also isolated from *A. araucana* seeds in Patagonian forests together with *Saccharomyces eubayanus*, another cryotolerant species of the genus. Genetic information obtained from the analysis of *S. uvarum* from apple chichas evidenced a closer relationship to industrial (European) strains than to natural (Patagonian) strains of this species. North Patagonia is an interesting scenario to study cryotolerant (*S. uvarum* and *S. eubayanus*) yeast diversity studies, and a source of new strains with potential biotechnological interest.

15.1 *Saccharomyces* in Fermented Beverages Around the World

Fermentation is one of the oldest and most economical methods of producing and preserving food. All over the world, humans since the first Neolithic civilizations have produced fermented food and beverages. Besides being a way to preserve the quality and safety of raw materials, fermentation is also a biological process employed to obtain products used as medicines, analgesics, disinfectants, etc. Four main fermentation processes are recognized: alcoholic, lactic acid, acetic acid, and alkali fermentations (Blandino et al. 2003). From them, alcohol fermentation that results in the production of ethanol as the main chemical product obtained from yeast metabolic activity is the focus of this chapter.

The earliest archaeological evidence for the production of an alcoholic fermented beverage made of rice, honey, and fruits dated to 7000 BC in China (McGovern et al. 2004). Since then, different fermented products of great nutritional and cultural value have been elaborated from diverse raw materials (sources of sugars for yeast growth), in different climates and with different levels of technological complexity (Marshall and Mejía 2011; Nout 2003). In Patagonia, particularly, fermented food and beverage elaboration processes coexist with different degrees of industrialization and tradition, including wine, cider, and different regionally relevant beverages associated with aboriginal communities, in general called *chichas*. The preparation of many indigenous or traditional fermented foods and beverages remains today as a house art. In most cases, beverages are produced in homes, villages, and small-scale industries. In contrast, the preparation of other beverages, such as wines, beers, some ciders, sake, and some distillates, has evolved to a high-volume production system (on a commercial or industrial scale).

In this chapter, we focus on fermented beverages produced under traditional and industrial conditions in Patagonia, with particular emphasis on the yeasts involved in these processes. Yeasts are the common factor to most of these fermented beverages, because many species are able to transform the sugars into ethanol through a metabolic process known as alcoholic fermentation. Although alcoholic fermentation has been empirically used by humanity for centuries, the direct implication of yeasts in these processes was not discovered until 1860 through the experiments of Louis Pasteur. Previously, the preparation of fermented foods and beverages had been carried out only in an artisan way and without any knowledge of the role of the microorganisms involved. After the discovery of Pasteur, studies on yeasts and their influence on fermented beverages flavour (taste and aroma) grew systematically. This unicellular fungi converts carbohydrates into ethanol, carbon dioxide. and a multitude of esters, higher alcohols, organic acids, and carbonyl compounds that influence the organoleptic properties of the fermented products and constitute their secondary or fermentative aroma (de la Roza et al. 2003; Fleet 2003).

From an ecological point of view, fermentations are fluctuating environments that expose yeasts to a variety of stresses including high osmolarity (caused by the presence of high sugar concentrations), anaerobiosis, acid stress, nutrient depletion (particularly nitrogen, lipids, and vitamins), and ethanol toxicity (Marsit et al. 2015). Even in this apparently adverse habitat, a myriad of yeast species have been described in fermented beverages, mostly in early stages of fermentation where ethanol concentrations are still low, with little effect on yeast viability. Although yeast diversity is quite variable according to each particular fermented beverage, in most cases it is composed by yeasts belonging to the ascomycetous genera *Saccharomyces*, *Hanseniaspora*, *Pichia*, *Debaryomyces*, *Metschnicowia*, *Kluyveromyces*, and *Torulaspora*, among others (Jespersen 2003; Fleet 2003; Greppi et al. 2013). From all these genera, *Saccharomyces* species dominate the alcoholic fermentations in most cases. Table 15.1 summarizes different fermented alcoholic beverages traditionally or industrially elaborated around the world, in which *Saccharomyces* yeast species were identified as part of their microbiota. From the total of seven natural species recognized nowadays in the *Saccharomyces* genus, that is, *S. arboricolus*, *S. cerevisiae*, *S. eubayanus*, *S. kudriavzevii*, *S. mikatae*, *S. paradoxus*, and *S. uvarum* (Almeida et al. 2014; Boynton and Greig 2014), only *S. cerevisiae* and *S. uvarum* have been repeatedly isolated from different fermented beverages. Among them, *S. cerevisiae* is by far the most commonly detected *Saccharomyces* species in alcoholic fermentations around the world (Table 15.1). This dominance results from the ability of this species to produce and accumulate ethanol, to tolerate different stress conditions typically present in fermentations, and to compete with other yeasts at the temperate (25–35 °C) temperatures typical in most fermentative processes around the world. Also, *Saccharomyces uvarum* was detected in some fermentations carried out at low temperatures, as the result of its cryophilic nature, in specific regions, mainly in Europe (Table 15.1).

From a physiological aspect, *S. uvarum* produces less acetic acid and ethanol but more glycerol than *S. cerevisiae*, as well as particular volatile compounds such as phenylethanol and phenylethyl acetate (Masneuf-Pomarède et al. 2010).

A particular mention must be given of the species *S. eubayanus*. This species is the last taxon described within the *Saccharomyces* genus, from isolates obtained from *Nothofagus* trees, including *N. pumilio* and *N. antarctica*, as well as from the stromata of their parasitic fungi, *Cyttaria* spp., in Northern Patagonia forests

Table 15.1 Representative list of *Saccharomyces* species identified in different traditional and industrial alcoholic fermentations around the world, according to available public literature

Continent	Beverage	Major ingredients	*Saccharomyces*	Country	References
Africa	Kachasu	Masau (*Ziziphus mauritana*)	*S. cerevisiae*	Zymbabwe	Nyanga et al. (2013)
	Chikokivana	Maize, finger millet	*S. cerevisiae*	Zymbabwe	Jespersen (2003)
	Mukumbi	Marula (*Sclerocarya birrea*) fruits	*S. cerevisiae*	Zymbabwe	Mpofu et al. (2008)
	Munkoyo	Maize	*S. cerevisiae*	Zambia	Jespersen (2003)
	Bussa	Maize	*S. cerevisiae*	Kenia	Jespersen (2003)
	Agadagigi	Plantain	*S. cerevisiae; S. uvarum*	Nigeria	Jespersen (2003)
	Burukutu	Sorghum, maize	*S. cerevisiae*	Nigeria	Jespersen (2003)
	Palm wine	Palm (*Elaeis* spp.; *Raphia* spp.; *Borassus* spp.; *Cocos* spp.; *Phoenix* spp.; *Nypa* spp.) sap	*S. cerevisiae*	Burkina Faso/Cameroon/Ghana/Nigeria	Jespersen (2003), Chandrasekhar et al. (2012), Santiago-Urbina and Ruiz-Terán (2014)
			S. bayanus; S. uvarum	Nigeria	Santiago-Urbina and Ruiz-Terán (2014)
	Pito beer	Sorghum	*S. cerevisiae*	Ghana/Nigeria	Jespersen (2003)
	Dolo	Sorghum	*S. cerevisiae*	Burkina Faso/Togo	Jespersen (2003)
	Tej	Honey	*S. cerevisiae*	Ethiopia	Bahiru et al. (2006)
	Ogol	Honey, mangle (*Blighia unijungata*) bark	*S. cerevisiae*	Ethiopia	Teramoto et al. (2005)
	Bouza	Coarsely ground grains	*S. cerevisiae; S. uvarum*	Egypt/Turkey	Hancioglu and Karapinar (1997)
	Tchapalo	Sorghum	*S. cerevisiae*	Cote d'Ivoire	N'guessan et al. (2011)
	Attieké	Cassava	*S. cerevisiae*		Coulin et al. (2006)
	Wine	Grape	*S. cerevisiae*	South Africa	Fay and Benavides (2005)

America	Mezcal	*Agave tequilana* var. *azul*/*A. angustifolia*	*S. cerevisiae*	Mexico	Lappe-Oliveras et al. (2008)
		Agave salmiana	*S. cerevisiae*	Mexico	Verdugo Valdez et al. (2011)
	Raicilla	*A. angustifolia*/*A. inaequidens*/*A. maximiliana*	*S. cerevisiae*	Mexico	Lappe-Oliveras et al. (2008)
	Prickly pear wine	*Opuntia streptocantha*	*S. cerevisiae*	Mexico	Rodríguez-Lerma et al. (2011)
	Pulque	*Agave atrovirens*/ *A. americana*/ *A. mapisaga*/*A. salmiana*	*S. cerevisiae*	Mexico	Estrada-Godina et al. (2001), Nout (2003), Lappe-Oliveras et al. (2008), Marcial-Quino et al. (2015)
			S. bayanus; S. paradoxus	Mexico	Lappe-Oliveras et al. (2008)
	Bocanora	*Agave angustifolia*	*S. cerevisiae*	Mexico	Alvarez-Ainza et al. (2015)
	Tequila	*Agave tequilana* var. *Azul*	*S. cerevisiae*	Mexico	Lachance (1995)
	Cachaça	Sugarcane	*S. cerevisiae*	Brazil	Vila Nova et al. (2009), Gomes et al. (2010), Badotti et al. (2014)
	Cauim	Peanuts, rice	*S. cerevisiae*	Brazil	Lacerda Ramos et al. (2010)
	Chicha	Maize, corn, fruits (pineapple, apple), roots (cassava, *Aracacia xanthorrhiza*)	*S. cerevisiae*	Argentina/Bolivia/Chile/Colombia/ Ecuador/Perú	Blandino et al. (2003), Chaves-Lopez et al. (2014), Rodríguez et al. (2011)
			S. uvarum	Argentina	Rodríguez et al. (2011)
	Champús	Maize, rice, wheat	*S. cerevisiae*	Bolivia/Chile/Colombia/Ecuador/Perú	Chaves-Lopez et al. (2014)
	Guarapo	Sugarcane (*Saccharum officinarum*)	*S. cerevisiae; S. paradoxus*	Bolivia/Chile/Colombia/Ecuador/Perú	Pataro et al. (2000)
	Mudai	Pehuen (*Araucaria araucana*) seeds	*S. cerevisiae*	Argentina	Rodríguez et al. (2014)

(continued)

Table 15.1 (continued)

Continent	Beverage	Major ingredients	*Saccharomyces*	Country	References
	Cider	Apple (*Malus domestica*; *Malus pumila*)	*S. cerevisiae*	Argentina	Barbagelata (2010), González Flores et al. (2015)
			S. uvarum	Argentina	González Flores et al. (2015)
	Wine	Grape	*S. cerevisiae*	Argentina/Brazil/Chile/Perú/United States	Lopes et al. (2002), Baffi et al. (2011), Schuller et al. (2012)
Asia	Kodo ko jaanr	Finger millet (*Eleusine coracana*)	*S. cerevisiae*	India	Tamang et al. (2012)
	Bhaati jaanr	Rice	*S. cerevisiae*	India	Tamang et al. (2012)
	Zutho	Rice	*S. cerevisiae*	India	Tamang et al. (2012)
	Huangjiu	Rice	*S. cerevisiae*	China	Chen and Xu (2010)
	Sake	Rice	*S. cerevisiae*	China/Japan	Aidoo et al. (2006), Blandino et al. (2003)
	Takju beer	Rice, wheat, barley, maize, millet	*S. cerevisiae*	Korea	Aidoo et al. (2006)
	Ruou nep than	Rice	*S. cerevisiae*	Vietnam	Aidoo et al. (2006)
	Jnard/Jaanrl/Thumba	Finger millet, rice, maize, wheat	*S. cerevisiae*	Buthan/Nepal/India	Aidoo et al. (2006)
	Palm wine	Sap of coconut, date, or palmyra palm	*S. cerevisiae*	Bangladesh/India/Indonesia/Malaysia/Sri Lanka/Philipines/Thailand	Aidoo et al. (2006)
	Wine	Grape (*Vitis vinifera*)	*S. cerevisiae*	China/India/Israel/Turkey	Fay and Benavides (2005), Liti et al. (2006), Chavan et al. (2009), Schuller et al. (2012), Liu et al. (2015)
Europe	Beer	Barley	*S. cerevisiae; S. pastorianus*	Various countries	Various reviews
	Slivovitz	Plum	*S. cerevisiae*	Poland	Satora and Tuszynski (2005)
	Boza	Wheat, rye, millet, maize	*S. cerevisiae*	Albania/Bulgaria/Romania	Hancioglu and Karapinar (1997), Blandino et al. (2003), Botes et al. (2007)

	Cider	Apple	*S. cerevisiae*	France/Ireland/Spain	Cabranes et al. (1990), Dueñas et al. (1994), Naumov et al. (2001), Morrisey et al. (2004), Coton et al. (2006), Suárez Valles et al. (2007), Pérez-Través et al. (2014)
			S. uvarum	France/Germany/Ireland/Switzerland	Naumov et al. (2001), Almeida et al. (2014), Pérez-Través et al. (2014)
			S. bayanus	France/Spain	Coton et al. (2006), Suárez Valles et al. (2007)
			S. kluyveri	Spain	Cabranes et al. (1990)
			S. cerevisiae × *S. uvarum* × *S. kudriavzevii*	France	Masneuf et al. (1998), Pérez-Través et al. (2014)
	Perry	Pear (*Pyrus* spp.)	*S. uvarum*	Germany	Pérez-Través et al. (2014)
	Wine	Grape	*S. cerevisiae*	Austria/France/Germany/Hungary/Italy/Slovakia/Spain/Switzerland	Naumov et al. (2002), Dellaglio et al. (2003), Ciani et al. (2004), Demuyter et al. (2004), Fay and Benavides (2005), González et al. (2006), Lopandic et al. (2008), Díaz et al. (2013), Almeida et al. (2014)
			S. uvarum	Austria/Czech Republic/France/Hungary/Italy/Slovakia/Spain	Naumov et al. (2002), Dellaglio et al. (2003), Lopandic et al. (2008), Almeida et al. (2014), Pérez-Través et al. (2014)
			S. bayanus	Spain/Switzerland	González et al. (2006), Díaz et al. (2013)
			S. pastorianus	Slovakia	Naumov et al. (2002)
			S. cerevisiae × *S. kudriavzevii*	Austria/Croatia/Germany/Spain/Switzerland	González et al. (2006), Peris et al. (2012)
			S. cerevisiae × *S. bayanus*	Spain	González et al. (2006)
			S. cerevisiae × *S. kudriavzevii* × *S. uvarum*	Switzerland	González et al. (2006), Peris et al. (2012), Almeida et al. (2014)
Oceania	Wine	Grape	*S. cerevisiae*	New Zealand	Fay and Benavides (2005)

(Libkind et al. 2011). A similar interrogate was established during some years with the origin of numerous independently formed hybrids between *S. cerevisiae* and *S. kudriavzevii* detected in wines, beers, and ciders, particularly in Europe (Table 15.1). In some cases these hybrids corresponded to commercial starters used for many years in these beverages for their capacity to ferment at low temperatures without knowing their hybrid nature (Bradbury et al. 2006; Lopandic et al. 2007; González et al. 2008, Table 15.1). Until 2008, *S. kudriavzevii* species had been only represented by a few strains isolated from decayed leaves of Japan and deposited at the culture collection of the Institute for Fermentation from Osaka (IFO, actually NBRC) (Naumov et al. 2000). However, the potential explanation for the European hybridization between *S. cerevisiae* and *S. kudriavzevii* in natural or artificial habitats required the existence of a European population of *S. kudriavzevii*. In 2008, Sampaio and Gonçalves isolated *S. kudriavzevii* from oak bark samples in Portugal, and 2 years later, Lopes et al. (2010) confirmed the presence of this species in oaks from Spain. Genetic analysis of these new *S. kudriavzevii* strains from Europe demonstrated a closer relationship with the *S. kudriavzevii* portion of hybrid genomes than Asian strains (Lopes et al. 2010; Peris et al. 2014). Again, where and under what conditions hybrids between these two species were formed are still unknown. Future surveys of *Saccharomyces* species diversity in natural and fermentative environments are necessary to gain a much better understanding of the ecological interactions between these species, including the hybridization phenomenon, and their impacts in the colonization of man-manipulated environments, which may be of interest for the fermentation industry.

The discovery of *S. eubayanus* also enabled scientists to clearly separate the species *S. uvarum* from the hybrid strains grouped as *S. bayanus*. Both are associated with such fermented beverages as wine, cider, and beer, but only *S. uvarum* was isolated from natural habitats (Sampaio and Gonçalves 2008; Libkind et al. 2011; Almeida et al. 2014; Rodríguez et al. 2014). Some years ago, taxonomists considered *S. uvarum* and *S. bayanus* to be varieties of the species *S. bayanus* because of their physiological and genetic similarity (Vaughan-Martini and Martini 2011). For that reason, some reports in public databases mentioned *S. bayanus* indistinctly to refer to both *S. bayanus* hybrids and *S. uvarum*. Recent genomic studies confirmed finally that *S. uvarum* is the sister species of *S. eubayanus* and that *S. bayanus* is a group of hybrids between *S. uvarum* and *S. eubayanus*, as previously mentioned (Pérez-Través et al. 2014; Peris et al. 2014).

Summarizing, several hybridization events seemed to have occurred between the excellent performance fermentative yeast species *S. cerevisiae* and other cryotolerant species such as *S. kudriavzevii*, *S. eubayanus*, and even *S. uvarum* (Morales and Dujon 2012; Boynton and Greig 2014). These events were associated with the ability to ferment at low temperatures, with the hybrids possessing selectable phenotypic characteristics in this context (hybrid vigour) (Belloch et al. 2008; Morales and Dujon 2012). The near absence of hybrid strains among natural isolates, even in habitats where two different species coexist (Sampaio and Gonçalves 2008; Lopes et al. 2010; Libkind et al. 2011; Rodríguez et al. 2014), reinforces the fact that adaptation to a human-related habitat is the selective pressure involved in hybrid formation.

As can be seen in Table 15.1, most microbiological studies about the diversity of yeasts in alcoholic beverages are associated with wines and, to a lesser extent, with cider and beer. Yeast diversity studies in traditional fermented beverages around the world are still infrequent, which is probably why hybrid strains among different *Saccharomyces* species or yeast species different from the well-known *S. cerevisiae* have not been discovered in traditional beverages around the world, in comparison with industrial processes.

Patagonia presents an important peculiarity that could influence the diversity of *Saccharomyces* species isolated from this region. This particularity is the low annual average temperature (below 20 °C). It could be expected that the strong influence of temperature would be particularly observed in traditional artisanal beverages elaboration that lack a proper temperature control, and it is less evident in industrial processes as wine and cider elaboration where temperature is a technologically controlled parameter. Low ambient temperature favours the development of cryotolerant fermentative yeast species that, at higher temperatures, are less competitive than other fast-growing species as such as *S. cerevisiae*.

Among the *Saccharomyces* genus, only *S. kudriavzevii*, *S. eubayanus*, and *S. uvarum* (as well as most hybrids involving portions of the genomes of these three species) have been associated with growth at low temperatures (Lopes et al. 2010; Peris et al. 2012; Libkind et al. 2011). Among them, only *S. uvarum* and hybrids including *S. bayanus*, *S. pastorianus*, and some *S. cerevisiae* × *S. kudriavzevii* strains have been obtained from fermented beverages elaborated at low temperatures (Table 15.1). In contrast, *S. eubayanus* and *S. kudriavzevii* have been, to date, only found in natural habitats (Sampaio and Gonçalves 2008; Lopes et al. 2010; Libkind et al. 2011; Bing et al. 2014; Rodríguez et al. 2014).

It is important to note that from their origins, different yeast strains of *Saccharomyces* and particularly *S. cerevisiae* have been directly or indirectly selected by humans to ferment sugar-rich substrates, in an evolutionary process called domestication. Human intervention has evolved wild *S. cerevisiae* strains into distinct domesticated variants (Fay and Benavides 2005; Diezmann and Dietrich 2009; Steensels and Verstrepen 2014). Some examples of traits acquired during this domestication process in *S. cerevisiae* wine strains include increased copper tolerance (Warringer et al. 2011), flavour production (Hyma et al. 2011), and fructose fermentation capacity (Novo et al. 2009). In brewing strains of this species, physiological features such as flocculation behaviour (Verstrepen et al. 2003) and improved maltose utilization (Voordeckers et al. 2012) were also selected. Because of this phenomenon, there are currently several nondomesticated ("wild") and domesticated ("industrial") genetic lineages of *S. cerevisiae* (Liti et al. 2009). Recent studies have confirmed this situation in *S. cerevisiae* ecology: diverged populations of wild *S. cerevisiae* exist independently of domesticated isolates (Fay and Benavides 2005; Liti et al. 2009; Sicard and Legras 2011; Wang et al. 2012). Genetic analysis of these feral or wild and industrial strains revealed that the genetic diversity within industrial strains is rather limited compared to the full spectrum of natural biodiversity.

As it was previously mentioned, *S. uvarum* is the other yeast species that can be isolated from both natural habitats and some fermented beverages; however, only a few reports have focused on the distribution of this species, and just one report mentioned its possible domestication (Almeida et al. 2014), probably because the number of strains available in comparison with *S. cerevisiae* is low. In this sense, Patagonia is an interesting place to analyse the diversity of this species because it is recovered from different sources including natural habitats and traditional and industrial beverages.

15.2 *Saccharomyces* in Fermented Beverages from Patagonia

15.2.1 Industrial Beverages

Different domestication environments as grape wine, rice wine, cider, and beer elaboration processes have been responsible for the selection of different yeast strains showing particular physiological and genetic traits. Yeasts in these beverages have suffered a strong selective pressure provoked during centuries of human manipulations and dispersion, which reduces the geographic structure typical of natural populations (Boynton and Greig 2014). The following paragraphs refer to *Saccharomyces* populations in wines and ciders, two different industrial beverages elaborated in Patagonia.

15.2.1.1 Wines

As previously mentioned, wine is one of the most studied fermented beverages around the world, especially from a microbiological point of view. The first historical evidence of grape wine is the presence of calcium salt from tartaric acid and a specific resin in a pottery jar found in Iran and dated to 5400–500 BC, from the Neolithic period (McGovern et al. 1996). On the other hand, the first evidence of the presence of *Saccharomyces cerevisiae* as the responsible agent of wine fermentation was obtained from residues of one of the earliest known wine jars from Egypt (3150 BC) by means of molecular procedures (Cavalieri et al. 2003). Later colonization by the Romans spread winemaking all around the Mediterranean; by 500 BC, wine was being produced in Sicily, Italy, France, Spain, Portugal, and northern Africa. It was only in the sixteenth century that European explorers introduced the vine, and wine, into the New World, including Mexico, Argentina, Peru, and Chile (Pretorius 2000). Even with this barely recent history in South America, local vineyards reflect their own *terroir*, the history, culture, and traditions of each particular vinicultural region.

In Argentina, various wine *Saccharomyces* diversity studies have been carried out in recent years, involving particularly the winegrowing areas of Mendoza Province (Mercado et al. 2007), North Patagonia (Lopes et al. 2002, 2006), and San Juan Province (Toro and Vazquez 2002). Local studies about *Saccharomyces* biota present in wine fermentations in Argentina have demonstrated that the exclusive *Saccharomyces* yeast species responsible for natural procedures is *S. cerevisiae* (Lopes et al. 2002; Mercado et al. 2007).

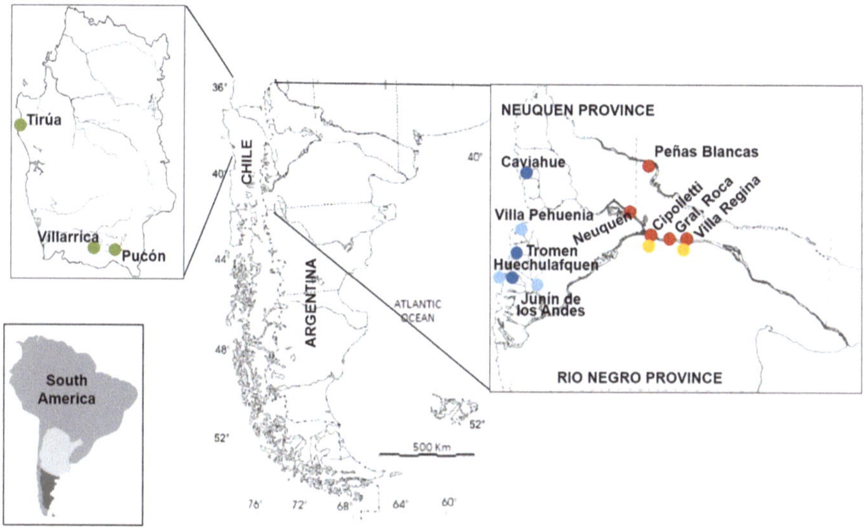

Fig. 15.1 Location of sampling areas in Northwestern Patagonia of Argentina (Neuquén and Río Negro provinces) and Chile. *Circles* indicate the sampled sites and *color* thereof indicates the sampled substrate (*red circle* wine, *yellow circle* cider, *light blue circle* Mudai, *dark blue circle A. araucana*, *green circle* apple chicha). *Left low corner*, from *light* to *dark gray*: South America, Argentina and Patagonia

North Patagonia in particular is the southernmost wine-producing region of Argentina and one of the most southern regions in the world. Most cellars are located in the Upper Valley of the Negro River, the Upper Valley of the Colorado River, and Low Valley of Neuquén River in a region called "Comahue" in Northern Patagonia (Fig. 15.1). A particularly interesting characteristic of this region is that both traditional and modern cellars are simultaneously established. In traditional cellars, wine production is essentially based on the natural fermentation of grape musts; however, in modern cellars, foreign commercial *S. cerevisiae* yeast starters are employed to conduct wine fermentations. The same cellar can also ferment their musts with or without yeast starters to obtain different final products. Additionally, the great expanse of the North Patagonian region (200 km among the most distant cellars) make difficult to find common climatological and edaphic features and, hence, similar wine characteristics. To overcome the wine quality variation in traditional cellars and the loss of wine regional identity in modern cellars provoked by the use of foreign starters, a strain of *S. cerevisiae* isolated from this region has been selected in our laboratory (Lopes et al. 2007a). The great extent of this region and different winemaking practices adopted by the diverse cellars means that the implantation capacity of this or other starters is not always optimal (Lopes et al. 2007b). This observation was evidenced using mtDNA-RFLP patterns or other alternative and useful molecular markers (Lopes et al. 2007b) (Fig. 15.2). For that reason, the most useful strategy is the selection of specific strains isolated and selected to be used in each different cellar or subregion.

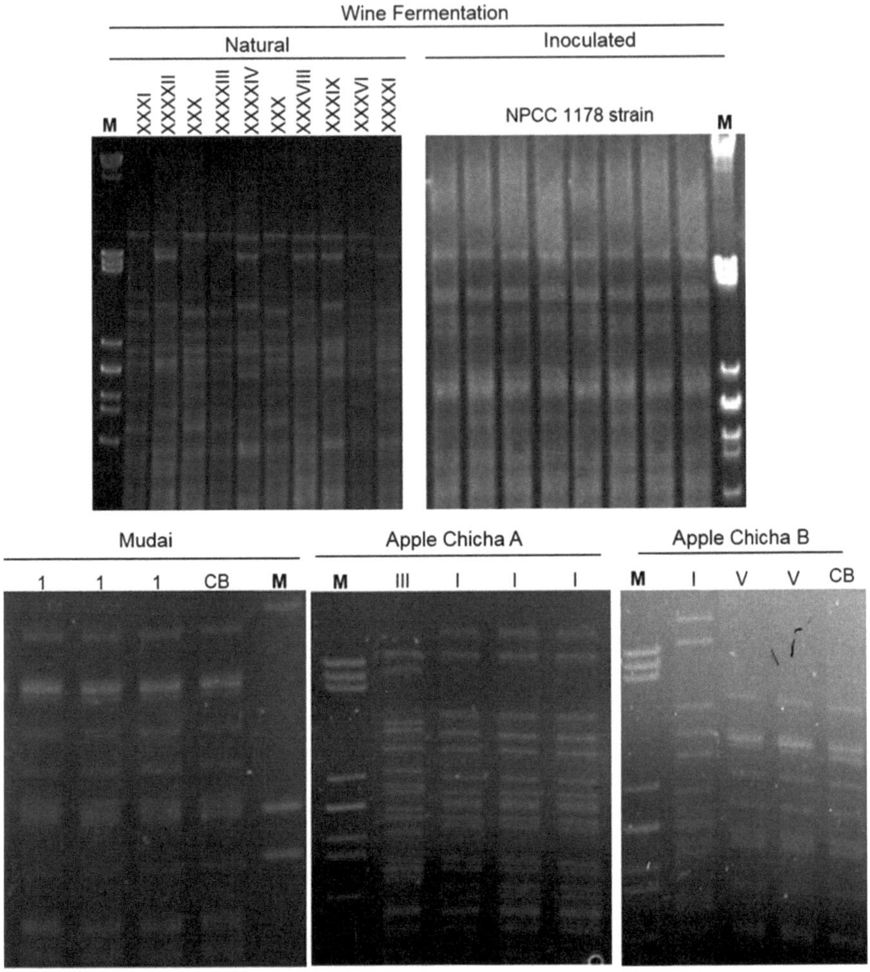

Fig. 15.2 mtDNA-RFLP patterns of some *Saccharomyces cerevisiae* isolates from natural and inoculated wine fermentation, from Mudai beverage and two apple chicha sampled (A and B). *Numbers* at the *top* of each gel indicate de mtDNA profile found in each beverage. *M* DNA size marker corresponding to lambda DNA digested with HindIII. *CB* commercial bakery yeast

Genetic studies based in multigene sequence analysis evidenced that all *S. cerevisiae* Patagonian strains studied showed a European origin (Peris et al. 2012); that is, they showed alleles similar to those detected in wine strains from Europe (Fig. 15.3). Moreover, the most frequent alleles in North Patagonian wine strains are also the most frequent in wine strains from Europe, South Africa, and Chile, evidencing a common origin of *S. cerevisiae* wine strains (Peris et al. 2012). The lack of *Saccharomyces* species different from *S. cerevisiae* could result from either the total absence of other species or the isolation methodology. Most wine yeast diversity studies, such as those carried out in our laboratory, employed isolation temperatures around 25–30 °C that select *S. cerevisiae* over other cryotolerant *Saccharomyces* species such as *S. uvarum* or even some hybrids adapted to grow at low temperatures.

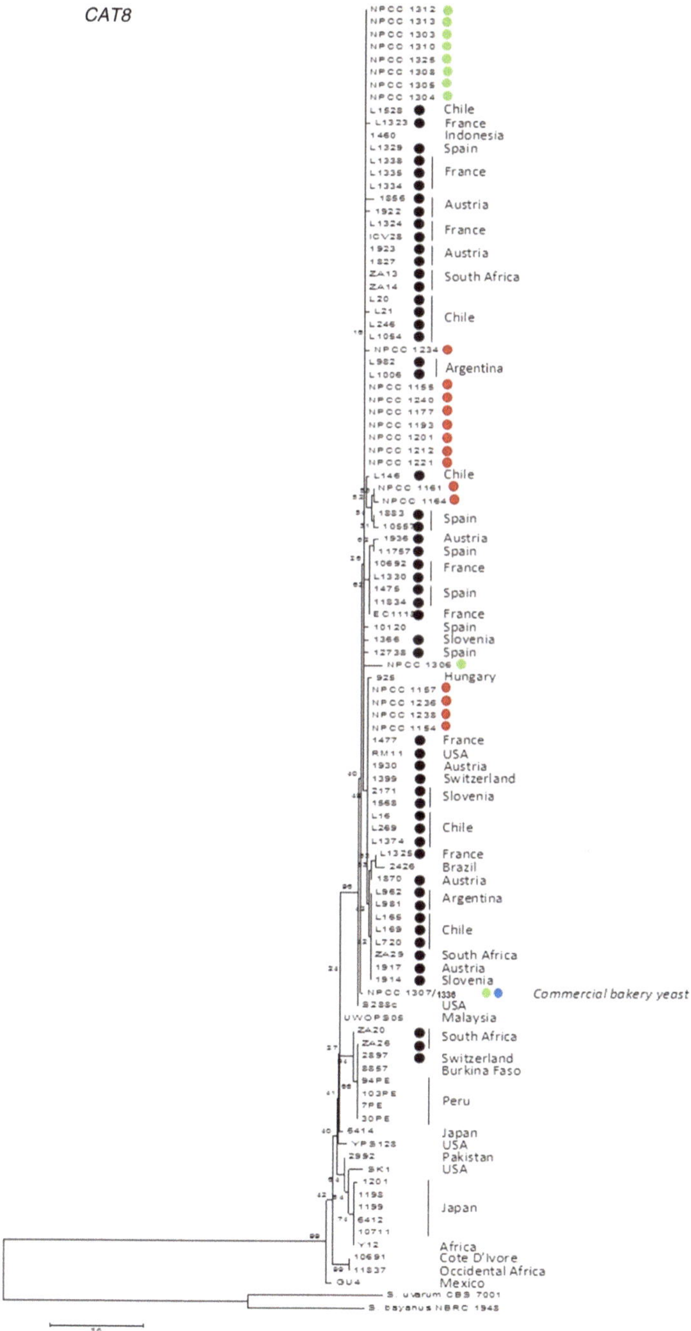

Fig. 15.3 Phylogenetic tree obtained with partial sequences genes *CAT8* and *GAL4* from *S. cerevisiae* yeast strains of different origins including reference strains of *Saccharomyces* genus. *Numbers* at the nodes correspond to bootstrap values based on 1000 pseudoreplicates. The scale is given in nucleotide substitutions per site. Patagonian wine (*red circle*) wine cider (*green circle*) origin. Wine (*black circle*) origin from around the world

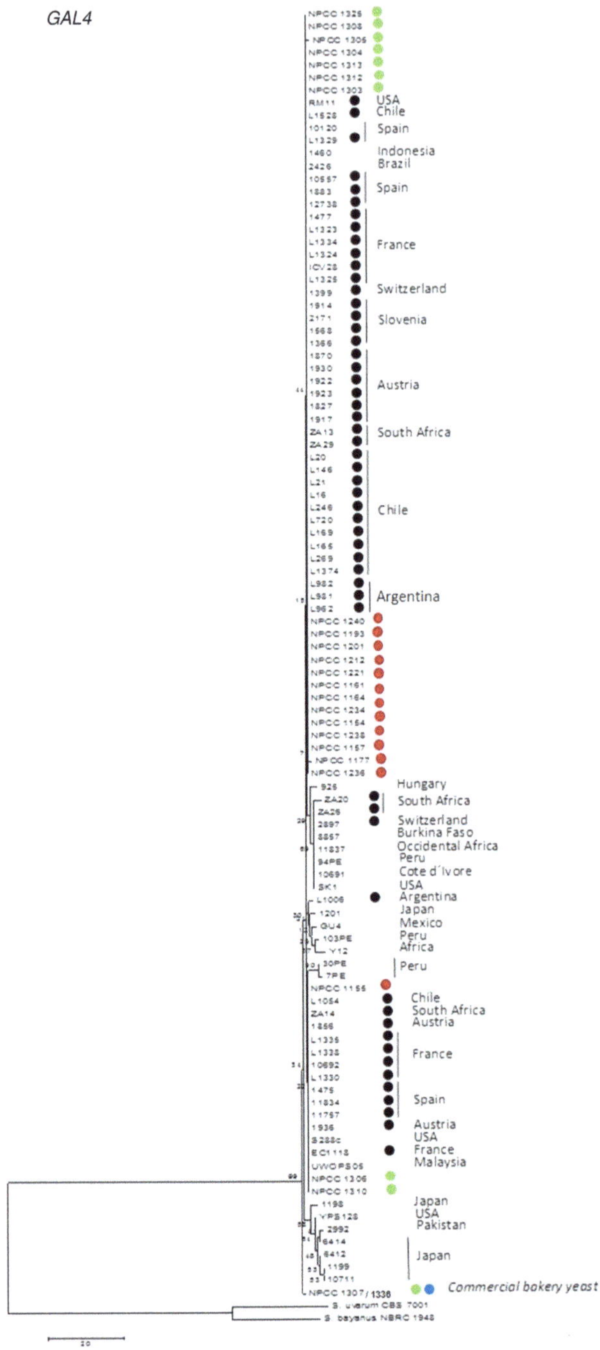

Fig. 15.3 (continued)

15.2.1.2 Ciders

The cultivated apple tree (*Malus domestica*) is one of the most important fruit crops in temperate regions. Only some specific varieties, showing small and bitter fruits, are used to produce ciders; others are consumed as a dessert. The history of apple domestication has just begun to be unravelled; genetic studies have demonstrated a Central Asian origin for the wild species *Malus sieverii* (Cornille et al. 2012). During late Neolithic times or the early Bronze Age, travellers introduced Asian wild apples through the Danube to Europe, where it evidenced a large secondary genetic contribution of the European wild species *Malus sylvestris* (Cornille et al. 2012). Romans perfected orchard economies based on the use of apples, and, during the last 2000 years, domesticated apples have diversified worldwide. However, ciders were produced in Western Europe by the Celts, even before the invasions of the Romans, using native apples. Apples were introduced to America by colonists in the sixteenth to seventeenth centuries. Cider is a refreshing soft alcoholic beverage popular worldwide. In Europe, ciders are produced and consumed in different countries such as Germany, France, Spain, Italy, Portugal, Belgium, The Netherlands, Ireland, Scotland, and England. In America, it is mainly produced in the USA, Mexico, Chile, and Argentina. Mendoza, Río Negro, and San Juan are the principal cider-producing provinces in Argentina.

Only a few studies about the genetic and physiological diversity of *Saccharomyces* yeasts associated with the elaboration of ciders are available in the literature. However, an interesting aspect to be considered is that both *S. cerevisiae* and *S. uvarum* have been recovered from this source, according to some reports (Table 15.1). In Argentina, only a few reports exist about yeasts in ciders and they come from investigations carried out in our laboratory. In a first study, we evaluated the total yeast biota present during the whole process of natural cider fermentations in two different cellars located in different sites in North Patagonia (Fig. 15.1). As was also observed in natural or spontaneous wine fermentations, different *S. cerevisiae* strains, evidenced by mtDNA-RFLP patterns and killer biotype, were detected in these cider fermentations, particularly in middle and end stages (Barbagelata 2010).

In a second and more recent study, an alternative protocol was employed for yeast recovery. In this case, different apple musts obtained from local cellars were naturally fermented at 13 °C and 26 °C and yeast isolation was carried out at the same temperatures. Again, yeasts belonging to the *Saccharomyces* genus dominated middle and end stages of fermentations carried out at different scales (1 and 5000 l) and at both 13 °C and 26 °C. Among *Saccharomyces* isolates, the species *S. cerevisiae* was in the majority in most conditions, with the exception of a must fermented at 13 °C in which *Saccharomyces uvarum* predominated (75 % and 62 % of total biota at middle and end stages, respectively). The intraspecific variability of *S. uvarum* isolates (56 isolates in total) was then evaluated by mtDNA-RFLP and RAPD-PCR methods, evidencing the presence of different strains (González Flores et al. 2015). However, the variability percentages in the *S. uvarum* cider population were lower than the same in populations from both natural habitats and apple chicha for the same species (see next section in this chapter).

Summarizing, a strong effect of fermentation temperature on yeast diversity was demonstrated in this assay, allowing us to obtain naturally present *S. uvarum* strains for the first time in ciders from South America.

15.2.2 Traditional Beverages

Traditional fermented beverages are generally associated with ancient cultures, although in some cases families that inherited the habit from their ancestors still regionally elaborate them. The elaboration of these beverages can be even part of the economic family support. Elaboration procedures and technologies of most of these beverages are orally transmitted from generation to generation, as in the case of the Mapuche people. The Mapuche community was the most important aboriginal group inhabiting the temperate forests in Andean (Argentina and Chile) Patagonia (de Mosbach 1992; Donoso and Lara 1996). This typically gatherer pre-Hispanic community used, and to a lesser extent still does, various wild and domesticated available fruits such as as beach strawberries (frutilla, *Fragaria chiloensis*), "Maqui" or Chilean wineberry (*Aristotelia chilensis*), "Calafate" or Magellan barberries (*Berberis* spp.), and feral apples and pears (*Malus* spp., *Pyrus* spp.), among others, to produce fermented beverages generally called "chichas" (Pardo and Pizarro 2005).

15.2.2.1 Mudai (or Muday)

One of the most interesting cases of study is a traditional fermented beverage, called Mudai, generally used in religious ceremonies by Mapuche communities. This soft beverage is made from the seeds, *ngulliw* in the Mapuche language, of the *Araucaria araucana* tree. This tree, commonly called Pehuen, is a gymnosperm endemic of the lower slopes of the Chilean and Argentinian southcentral Andes, typically above 1000 m altitude. In Argentina, it occupies a narrow strip on the Patagonian Andes ranging from 37°50′ to 39°20′ latitude in Neuquén Province. Pehuen seeds have also constituted an important source of carbohydrates for the Mapuche peoples from this area, who are in fact also called Araucanians (from *Araucaria araucana*) or Pehuenches (Pehuen people) because of their strong relationship with this tree. Pehuen seeds are eaten raw, boiled, or roasted and are often ground into flour to be used as an ingredient in soups and to make bread and beverages as Mudai (Herrmann 2005). The name Mudai is also used for chichas elaborated from wheat cultivated by Mapuche people.

In a recent study our laboratory, the yeasts associated with Mudai have been analyzed (Rodríguez et al. 2014). Fermentations were carried out in three different areas in Northwestern Patagonia (Neuquén Province): Villa Pehuenia (latitude 38°54′00″, longitude 71°19′58″, altitude 1200 m), Junín de los Andes (latitude 39°57′03″, longitude 71°04′15″, altitude 902 m), and Huechulafquen (latitude 39°79′90″, longitude 71°22′57″, altitude 875 m), during the Southern Hemisphere autumn (Fig. 15.1).

Only two yeast species, *Hanseniaspora uvarum* and *S. cerevisiae*, were recovered from different fermentation stages of this beverage.

The molecular characterization of the *S. cerevisiae* isolates obtained in our work evidenced a genetic homogeneity frequently observed in inoculated fermentations, in which the selected yeast starter dominates the fermentation process but is not expected in natural, non-inoculated processes as Mudai fermentations (Lopes et al. 2007b). MtDNA-RFLP analysis of a typical commercial bakery strain showed the same molecular pattern detected in our Mudai fermentations (Fig. 15.2), clearly evidencing a cross-contamination in this traditional fermented product with yeast strains commercially available to produce bread.

This foreign nature of the *S. cerevisiae* isolates obtained from Mudai was also evidenced by gene sequencing (Fig. 15.3). The use of commercial bakery yeasts in bread making by people from the Mapuche communities has previously been reported (Pardo and Pizarro 2005) and, for that reason, it is probably an unintentional contamination. The most relevant aspect to be considered from our results is the fact that the obtained products could exhibit worse properties and less desirable sensory attributes than those obtained with the natural original biota (Marini et al. 2009). It is well known that the introduction of foreign cultures cause a strong modification of the yeast microbiota by means of a replacement of the native *Saccharomyces* strains or the formation of intraspecific hybrids between native and introduced yeasts. This phenomenon was recently observed for *S. cerevisiae* strains isolated from Brazilian *cachaça* fermentations (Badotti et al. 2014). Our work is the first evidence from an ecological-molecular point of view of the impact of commercial yeast in traditional fermentations resulting in a clear and radical substitution of the natural yeast diversity. In fact, the analysis of fermentative yeast diversity in the surface of the raw material used for Mudai elaboration (*A. araucana* seeds) as well as in bark samples of this tree in different regions of Andean North Patagonia, evidenced the existence of native *Saccharomyces* species, not detected in any Mudai fermentation (Rodríguez et al. 2014). To our surprise, all *Saccharomyces* isolates recovered from *A. araucana* seed and bark samples belonged to the cryotolerant species *S. eubayanus* and *S. uvarum*. The absence of native *S. cerevisiae* strains in *A. araucana* may indicate a recent substitution of native *S. uvarum* or *S. eubayanus* strains as responsible for *Mudai* fermentations by commercial *S. cerevisiae* strains. In this sense, the best candidate is the species *S. uvarum* because it is frequently found in low-temperature fermentation processes from European regions of oceanic or continental climates, in which it coexists with and even replaces *S. cerevisiae* as the main yeast responsible for wine and cider fermentations (Table 15.1). Moreover, *S. uvarum* was also detected in regional apple chichas, as is mentioned later in this chapter.

S. eubayanus, the other candidate responsible for ancient Mudai fermentations, was only associated with natural habitats (mostly bark samples) in Argentinean Patagonia (Libkind et al. 2011), Wisconsin (USA) (Peris et al. 2014), as well as from Far East Asia (Bing et al. 2014).

To elucidate the participation of one or another yeast species in ancient Mapuche fermentations, samples of ceramic jars used by the Mapuche people to ferment or store fermented beverages and recovered from Andean North Patagonia by local anthropologists are being analyzed in our laboratory. These analyses would permit us to obtain information about the *Saccharomyces* strains potentially associated to fermented beverages before the introduction of commercial *S. cerevisiae* bakery strains to North Patagonian region. However, the results, although promising, are being still analyzed. Both the introduction of commercial *S. cerevisiae* bakery cultures by Mapuche people for bread elaboration and the higher temperatures in these homes as a product of human facilities can be associated with the replacement of original *S. uvarum/S. eubayanus* cryotolerant native strains by *S. cerevisiae* in Mudai fermentations.

15.2.2.2 Apple Chicha

Another traditional fermented beverage elaborated by Mapuche communities and studied by our research group is the apple chicha. In contrast to *Mudai* fermentations elaborated from seeds of a native plant, *Araucaria araucana*, this beverage is elaborated by using feral apple trees (*Malus* sp.) in a similar but more artisanal way than ciders. As previously mentioned in Sect. 13.2.1.2, apples, including those found in the Chilean region south of the Bío-Bío River, where chichas were analyzed, were introduced to America by Spanish conquistadors. This region was later abandoned by the Spaniards during the Araucanian wars and kept under Mapuche domination until the nineteenth century. Mapuche people exploited the fruits of apple trees (*manshanás-aliwen*, in the Mapuche language) as food but also to produce the apple chicha, a fresh soft alcoholic beverage. In Patagonia, on both sides of the Andes mountains, the extent of land occupied by apple trees was so great that this region was even called “el país de las manzanas” (the country of apples). In the Chilean side of the Andes, these apples were planted for Spanish citizens from Valdivia or Villarrica cities; however, in the Argentinean side they seemed to have been introduced and propagated by the Mapuches and remained in the region since then (Bandieri 2005). A genetic study carried out on some of these feral apples in Argentina evidenced the existence of different feral populations and a strong human influence in their geographic distribution (Calvo et al. 2010).

In our laboratory, apple chichas elaborated in three different areas of Chile: Tirúa (38°20′14.40″S, 73°29′46.66″O), Villarrica (39°16′47.44″S, 72°13′50.81″O), and Pucón (39°16′05.33″S, 71°58′42.94″O) were analyzed in a first experimental approximation (Fig. 15.1). As observed in Mudai fermentations, a low diversity of yeast species was found in the traditional apple fermentations evaluated (Rodríguez et al., 2011). *H. uvarum* and *Metschnikowia pulcherrima* were the dominant species during initial stages of fermentation, and both *S. cerevisiae* and *S. uvarum* were detected exclusively in middle and end stages. Some fermentations were absolutely dominated by *S. cerevisiae*, and others evidenced the presence of *S. uvarum* in different proportions. This last cryotolerant species was even dominant in some fermentations, with different strains evidenced by both mtDNA-RFLP patterns and multilocus sequence analysis (Figs. 15.2–15.4). Regarding the pres-

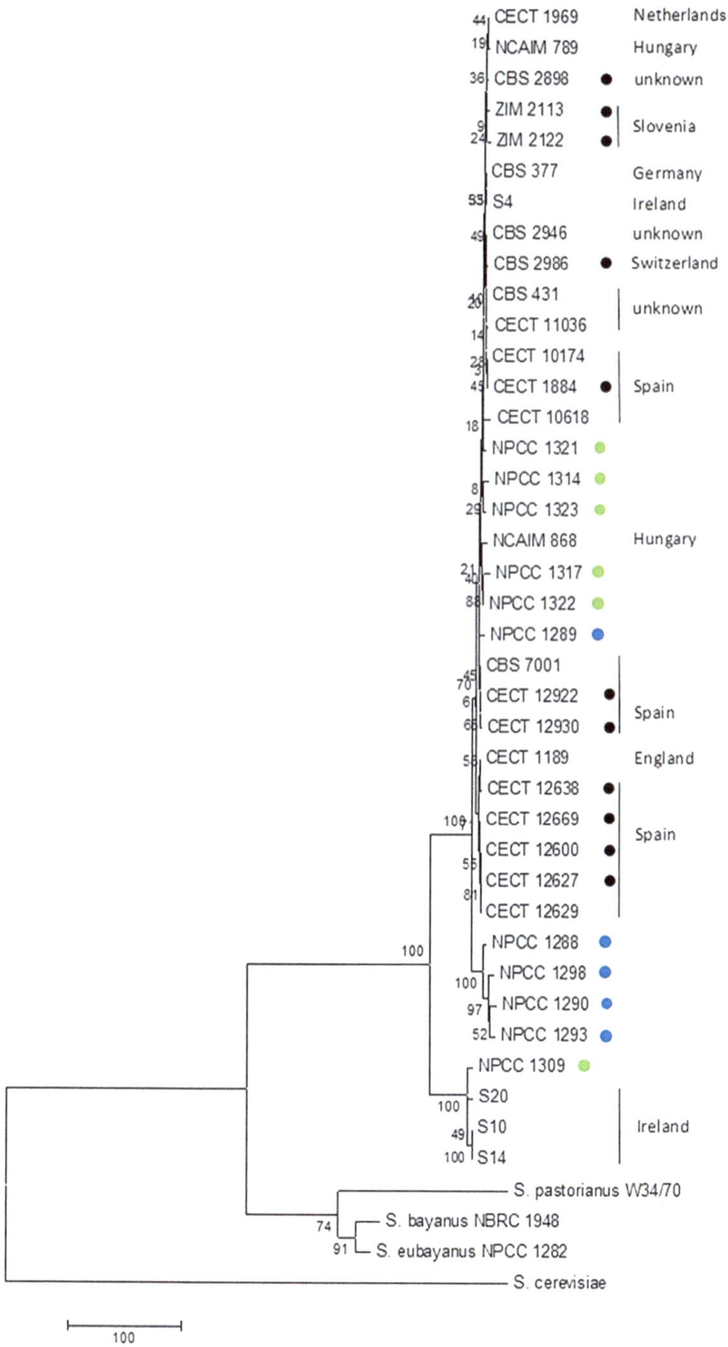

Fig. 15.4 Phylogenetic tree obtained with partial sequences of eight concatenated nuclear genes (*BRE5*, *CAT8*, *CYC3*, *CYR1*, *EGT2*, *GAL4*, *MET6* and *MNL1*) from *S. uvarum* yeast strains of different origins and reference strains of *Saccharomyces* genus. *Numbers* at the nodes correspond to bootstrap values based on 1000 pseudoreplicates. The scale is given in nucleotide substitutions per site. *A. araucaria* (*light blue circle*), cider (*green circle*) and (*black circle*) wine origin

ence of *S. cerevisiae* in the apple chichas analyzed in this study, the multilocus sequence analysis demonstrated that all strains obtained belonged to the large group of wine/European strains (Fig. 15.3). This group is mainly conformed of representative strains from Europe and America, including wine strains from Argentinean and Chilean Patagonia, mentioned previously in this chapter. As was previously observed for *Mudai*, the presence of a commercial baker yeast strain among the *S. cerevisiae* populations was also confirmed in one apple chicha (Fig. 15.2). In contrast to that observed in *Mudai*, where a bakery yeast strain completely dominated the fermentations, the presence of this strain in apple chichas was limited to only one of them, evidencing a higher diversity of naturally occurring, although not native, *S. cerevisiae* strains. This observation is consistent with the elaboration procedure of the two beverages: *Mudai* is performed using seed must that is boiled before fermentation; this procedure eliminates all natural microbiota present on the seed surfaces. On the other hand, apple chicha is made by natural fermentation of apple juice containing the complete yeast communities present on fruit surfaces.

Although the origin of *S. cerevisiae* isolates obtained from apple chichas seemed to be wine or European (Fig. 15.3), a question to answer is the origin of the *S. uvarum* found in these beverages. Our approximation to this answer based in multilocus sequences and the information obtained from previous works about domestication of *S. uvarum* (Almeida et al. 2014) could give together some information to solve this interrogative. Although multilocus sequencing may overestimate divergence regarding completely genome-based analysis (Almeida et al. 2014), some aspects can be compared because of the number of gene regions evaluated in our work. The possibilities to count a regional natural population of *S. uvarum* from Patagonian *A. araucana* trees (see previous section) (Rodríguez et al. 2014), as well as with strains isolated from industrialized products such as ciders or wines, allowed us to evaluate this relationship (Fig. 15.4).

The percentages of nucleotide divergence between *S. uvarum* strains from *A. araucana* and the remaining strains, isolated from fermentative or "domesticated" habitats such as cider, wine, and chicha, is about 1 %. This value is similar to the divergence values detected between populations A and B in *S. eubayanus* (Peris et al. 2014) and A and B in *S. uvarum* (Almeida et al. 2014). According to our results, *S. uvarum* strains from apple chichas were genetically closer to those from European ciders and wines, as well as our cider strains. Nevertheless, a certain degree of genetic flux with North Patagonian isolates from natural habitats could be also suggested based in our data (Rodríguez et al., 2011). A more complete genetic study including the total genome sequence of these *S. uvarum* strains is presently being carried out to evaluate their genetic relationships with strains from other origins, including strains from apple chichas, ciders, and natural habitats from Patagonia.

Altogether, our results indicate that North Patagonia results an interesting scenario for *S. uvarum* and *S. eubayanus* diversity studies, and a source of new strains with potential biotechnological interest.

References

Aidoo KE, Nout MJR, Sarkar PK (2006) Occurrence and function of yeasts in Asian indigenous fermented foods. FEMS Yeast Res 6:30–39

Almeida P, Gonçalves C, Teixeira S, Libkind D, Bontrager M, Masneuf-Pomarède I, Albertin W, Durrens P, Sherman DJ, Marullio P, Hittinger CT, Gonçalves P, Sampaio JP (2014) A Gondwanan imprint on global diversity and domestication of wine and cider yeast *Saccharomyces uvarum*. Nat Commun 5:4044

Alvarez-Ainza ML, Zamora-Quiñonez KA, Moreno-Ibarra GM, Acedo-Féliz E (2015) Genomic diversity of *Saccharomyces cerevisiae* yeast associated with alcoholic fermentation of bacanora produced by artisanal methods. Appl Biochem Biotechnol 175:2668–2676

Badotti F, Villaça ST, Arias A, Rosa CA, Barrio E (2014) Two interbreeding populations of *Saccharomyces cerevisiae* strains coexist in cachaça fermentations from Brazil. FEMS Yeast Res 14:289–301

Baffi MA, dos Santos Bezerra C, Arévalo-Villena M, Briones-Perez AI, Gomes E, Da Silva R (2011) Isolation and molecular identification of wine yeasts from a Brazilian vineyard. Ann Microbiol 61:75–78

Bahiru B, Mehari T, Ashenafi M (2006) Yeast and lactic acid flora of tej, an indigenous Ethiopian honey wine: variations within and between production units. Food Microbiol 23:277–282

Bandieri S (2005) Historia de la Patagonia. Ed Sudamericana, Buenos Aires

Barbagelata R (2010) Ecología de levaduras asociadas a la elaboración de sidras patagónicas. MSc Thesis, Universidad Nacional del Comahue

Belloch C, Orilc S, Barrio E, Qurol A (2008) Fermentative stress adaptation of hybrids within the *Saccharomyces* sensu stricto complex. Int J Food Microbiol 122:188–195

Bing J, Han P, Liu W, Wang Q, Bai F (2014) Evidence for a far East Asian origin of lager beer yeast. Curr Biol 24:R380–R381

Blandino A, Al-Aseeri ME, Pandiella SS, Cantero D, Webb C (2003) Cereal-based fermented foods and beverages. Food Res Int 36:527–543

Botes A, Todorov SD, von Mollendorff JW, Botha A, Deicks LMT (2007) Identification of lactic acid bacteria and yeast from boza. Process Biochem 42:267–270

Boynton PJ, Greig D (2014) The ecology and evolution of non-domesticated *Saccharomyces* species. Yeast 31:449–462

Bradbury JE, Richards KD, Niederer HA, Lee SA, Dunbar PR, Gardner RC (2006) A homozygous diploid subset of commercial wine yeast strains. Antonie van Leeuwenhoek 89:27–37

Cabranes C, Moreno J, Mangas JJ (1990) Dynamics of yeast populations during cider fermentation in the Asturian region of Spain. Appl Environ Microbiol 56:3881–3884

Calvo P, Carrera A, Sánchez E, Poverene M (2010) Genetic diversity in wild apple (*Malus* sp.) populations in Argentina. Am J Plant Sci Biotechnol 3:99–105

Cavalieri D, McGovern PE, Hartl DL, Mortimer R, Polsinelli M (2003) Evidence for *S. cerevisiae* fermentation in ancient wine. J Mol Evol 57:S226–S232

Chandrasekhar K, Sreevani S, Seshapani P, Pramodhakumari J (2012) A review on palm wine. Int J Res Biol Sci 2:33–38

Chavan P, Mane S, Kulkarni G, Shaikh S, Ghormade V, Nerkar DP, Shouche Y, Deshpande MV (2009) Natural yeast flora of different varieties of grapes used for wine making in India. Food Microbiol 26:801–808

Chaves-Lopez C, Serio A, Grande-Tovar CD, Cuervo-Mulet R, Delgaso-Ospina J, Paparella A (2014) Traditional fermented foods and beverages from a microbiological and nutritional perspective: the Colombian heritage. Comp Rev Food Sci Food Saf 13:1031–1048

Chen S, Xu Y (2010) The influence of yeast strains on the volatile flavour compounds of Chinese rice wine. J Inst Brew 116:190–196

Ciani M, Mannazzu I, Marinangeli P, Climenti F, Martini A (2004) Contribution of winery-resident *Saccharomyces cerevisiae* strains to spontaneous grape must fermentation. Antonie van Leeuwenhoek 85:159–164

Cornille A, Gladieux P, Smulders MJM, Roldán-Ruiz I, Laurens F, Le Cam B, Nersesyan A, Clavel J, Olonova M, Feugey L, Gabrielyan I, Zhang XG, Tenaillon MI, Giraud T (2012) New insight into the history of domesticated apple: secondary contribution of the European wild apple to the genome of cultivated varieties. PLoS Genet 8:e1002703

Coton E, Coton M, Levert D, Casaregola S, Sohier D (2006) Yeast ecology in French cider and black olive natural fermentations. Int J Food Microbiol 108:130–135

Coulin P, Farah Z, Assanvo J, Spillmann H, Puhan Z (2006) Characterisation of the microflora of attiéké, a fermented cassava product, during traditional small-scale preparation. Int J Food Microbiol 106:131–136

de la Roza C, Laca A, García L, Díaz M (2003) Ethanol and ethyl acetate production during the cider fermentation from laboratory to industrial scale. Process Biochem 38:1451–1456

de Mosbach EW (1992) Botanica indígena de Chile. Museo Chileno de Arte Precolombino Andrés Bello, Santiago, Chile

Dellaglio F, Zapparilo G, Malacrinó P, Suzzi G, Torriani S (2003) *Saccharomyces bayanus* var. *uvarum* and *Saccharomyces cerevisiae* succession during spontaneous fermentations of Recioto and Amarone wines. Ann Microbiol 53:411–425

Demuyter C, Lollier M, Legras JL, Le Jeune C (2004) Predominance of *Saccharomyces uvarum* during spontaneous alcoholic fermentation, for three consecutive years, in an Alsatian winery. J Appl Microbiol 97:1140–1148

Díaz C, Molina AM, Nähring J, Fischer R (2013) Characterization and dynamic behavior of wild yeast during spontaneous wine fermentation in steel tanks and amphorae. Biomed Res Int 2013:1–13

Diezmann S, Dietrich FS (2009) *Saccharomyces cerevisiae*: population divergence and resistance to oxidative stress in clinical, domesticated and wild isolates. PLoS One 4:e5317

Donoso C, Lara A (1996) Utilización de los bosques nativos en Chile: Pasado, Presente y Futuro. In: Armesto J, Villagrán C, Arroyo MK (eds) Ecología de los bosques nativos de Chile. Editorial Universitaria, Santiago, pp 363–384

Dueñas M, Irastorza A, Fernández K, Bilbao A, Huerta A (1994) Microbial populations and malolactic fermentation of apple cider using traditional and modified methods. J Food Sci 59:1060–1064

Estrada-Godina AR, Cruz-Guerrero AE, Lappe P, Ulloa M, Garcia-Garibay M, Gómez-Ruiz L (2001) Isolation and identification of killer yeasts from *Agave* sap (aguamiel) and pulque. World J Microbiol Biotechnol 17:557–560

Fay JC, Benavides JA (2005) Evidence for domesticated and wild populations of *Saccharomyces cerevisiae*. PLoS Genet 1:e5

Fleet GH (2003) Yeast interactions and wine flavour. Int J Food Microbiol 86:11–22

Gomes FCO, Lacerda ICA, Libkind D, Lopes CA, Carvajal J, Rosa CA (2010) Traditional foods and beverages from South America: microbial communities and production strategies. In: Krause J, Fleischer O (eds) Industrial fermentation: food processes, nutrient sources and production strategies. Nova Science, New York, pp 79–114

González Flores M, Di Niccolo R, Vera Macaya DL, Rodriguez ME, Lopes CA (2015) Yeast biodiversity in patagonian ciders: effect of temperature, apple variety and fermentation scale. In: Abstracts of the V Jornadas Sudamericanas de Biología y Biotecnología de Levaduras, Recife, Brazil

González S, Barrio E, Gafner J, Querol A (2006) Natural hybrids from *Saccharomyces cerevisiae*, *Saccharomyces bayanus* and *Saccharomyces kudriavzevii* in wine fermentations. FEMS Yeast Res 6:1221–1234

González SS, Barrio E, Querol A (2008) Molecular characterization of new natural hybrids between *Saccharomyces cerevisiae* and *Saccharomyces kudriavzevii* from brewing. Appl Environ Microbiol 74:2314–2320

Greppi A, Rantsiou K, Padonou W, Hounhouigan J, Jespersen L, Jakobsen M, Cocolin L (2013) Determination of yeast diversity in ogi, mawè, gowé and tchoukoutou by using culture-dependent and -independent methods. Int J Food Microbiol 165:84–88

Hancioglu O, Karapinar M (1997) Microflora of Boza, a traditional fermented Turkish beverage. Int J Food Microbiol 35:271–274

Herrmann TM (2005) Knowledge, values, uses and management of the Araucaria araucana forest by the indigenous Mapuche Pewenche people: a basis for collaborative natural resource management in southern Chile. Nat Resour Forum 29:120–134

Hyma KE, Saerens SM, Verstrepen KJ, Fay JC (2011) Divergence in wine characteristics produced by wild and domesticated strains of *Saccharomyces cerevisiae*. FEMS Yeast Res 11:540–551

Jespersen L (2003) Occurrence and taxonomic characteristics of strains of *Saccharomyces cerevisiae* predominant in African indigenous fermented foods and beverages. FEMS Yeast Res 3:191–200

Lacerda Ramos C, Gonzaga de Almeida E, de Melo Pereira GV, Gomes Cardoso P, Souza Días E, Freitas Schwan R (2010) Determination of dynamic characteristics of microbiota in a fermented beverage produced by Brazilian Amerindians using culture-dependent and culture-independent methods. Int J Food Microbiol 140:225–231

Lachance MA (1995) Yeast communities in natural tequila fermentation. Antonie van Leeuwenhoek 68:151–160

Lappe-Oliveras P, Moreno-Terrazas R, Arrizón-Gaviño J, Herrera-Suarez T, García-Mendoza A, Gschaedler-Mathis A (2008) Yeasts associated with the production of Mexican alcoholic non-distilled and distilled Agave beverages. FEMS Yeast Res 8:1037–1052

Libkind D, Hittinger CT, Valério E, Gonçalves S, Dover J, Johnston M, Gonçalves P, Sampaio JP (2011) Microbe domestication and the identification of the wild genetic stock of lager-brewing yeast. Proc Natl Acad Sci USA 108:14539–14544

Liti G, Barton DBH, Louis EJ (2006) Sequence diversity, reproductive isolation and species concepts in *Saccharomyces*. Genetics 174:839–850

Liti G, Carter DM, Moses AM, Warringer J, Parts L, Hames SA, Davey RP, Roberts IN, Burt A, Koufopanou V, Tsai IJ, Bergman CM, Bensasson D, O'Kelly MJT, Van Oudenaarden A, Barton DBH, Bailes E, Nguyen Ba AN, Jones M, Kuail MA, Gudhead I, Sims S, Smith F, BlonBlerg A, Durbin R, Louis EJ (2009) Population genomics of domestic and wild yeasts. Nature (Lond) 458:337–341

Liu N, Qin Y, Song Y, Ye D, Yuan W, Pei Y, Xue B, Liu Y (2015) Selection of indigenous *Saccharomyces cerevisiae* strains in Shanshan County (Xinjiang, China) for winemaking and their aroma-producing characteristics. World J Microbiol Biotechnol 31:1781–1792

Lopandic K, Gangle H, Wallner E, Tscheik G, Leitner G, Querol A, Borth N, Breitenbach M, Prillinger H, Tiefenbrunner W (2007) Genetically different wine yeasts isolated from Austrian vine-growing regions influence wine aroma differently and contain putative hybrids between *Saccharomyces cerevisiae* and *Saccharomyces kudriavzevii*. FEMS Yeast Res 7:953–965

Lopandic K, Tiefenbrunner W, Gangl H, Mandl K, Berger S, Leitner G, Abd-Ellah GA, Querol A, Gardner RC, Sterflinger K, Prillinger H (2008) Molecular profiling of yeasts isolated during spontaneous fermentations of Austrian wines. FEMS Microbiol 8:1063–1075

Lopes CA, van Broock M, Querol A, Caballero A (2002) *Saccharomyces cerevisiae* wine yeast populations in a cold region in Argentinean Patagonia. A study at different fermentation scales. J Appl Microbiol 93:608–615

Lopes CA, Lavalle TL, Querol A, Caballero A (2006) Combined use of killer biotype and mtDNA-RFLP patterns in a patagonian wine *Saccharomyces cerevisiae* diversity study. Antonie van Leeuwenhoek 89:147–156

Lopes CA, Rodríguez ME, Sangorrín M, Querol A, Caballero A (2007a) Patagonian wines: the selection of an indigenous yeast starter. J Ind Microbiol Biotechnol 34:539–546

Lopes CA, Rodríguez ME, Sangorrín M, Querol A, Caballero A (2007b) Patagonian wines: implantation of an indigenous strain of *Saccharomyces cerevisiae* in fermentations conducted in traditional and modern cellars. J Ind Microbiol Biotechnol 34:139–149

Lopes CA, Barrio E, Querol A (2010) Natural hybrids of *S. cerevisiae* × *S. kudriavzevii* share alleles with European wild populations of *Saccharomyces kudriavzevii*. FEMS Yeast Res 10:412–421

Marcial-Quino J, Garcia-Ocón B, Mendoza-Espinoza JA, Gómez-Manzo S, Enríquez-Flores S, Sierra-Palacios E (2015) Molecular identification of yeast of the pulque by PCR-DGGE, a traditional Mexican beverage. Int J Res 3:1–15

Marini MM, Gomes FCO, Silva CLC, Cadete RM, Badotti F, Oliveira ES, Cardoso CR, Rosa CA (2009) The use of selected starter *Saccharomyces cerevisiae* strains to produce traditional and industrial cachaça: a comparative study. World J Microbiol Biotechnol 25:235–242

Marshall E, Mejía D (2011) Traditional fermented food and beverages for improved livelihoods. FAO, Rome

Marsit S, Mena A, Bigey F, Sauvage FX, Couloux A, Guy J, Legras JL, Barrio E, Dequin S, Galeote V (2015) Evolutionary advantage conferred by an eukaryote-to-eukaryote gene transfer event in wine yeasts. Mol Biol Evol 32:1695–1707

Masneuf I, Hansen J, Groth C, Piskur J, Dubourdieu D (1998) New hybrids between *Saccharomyces* sensu stricto yeast species found among wine and cider production strains. Appl Environ Microbiol 64:3887–3892

Masneuf-Pomarède I, Bely M, Marullo P, Lonvaud-Funel A, Dubourdieu D (2010) Reassessment of phenotypic traits for *Saccharomyces bayanus* var. *uvarum* wine yeast strains. Int J Food Microbiol 139:79–86

McGovern PE, Glusker DL, Exner LJ, Voigt MM (1996) Neolithic resinated wine. Nature (Lond) 381:480–481

McGovern PE, Zhang J, Tang J, Zhang Z, Hall GR, Moreau RA, Nuñez A, Butrym ED, Richards MP, Wang C, Cheng G, Zhao Z, Wang C (2004) Fermented beverages of pre- and proto-historic China. Proc Natl Acad Sci USA 101:17593–17598

Mercado L, Dalcero A, Masuelli R, Combina M (2007) Diversity of *Saccharomyces* strains on grapes and winery surfaces: analysis of their contribution to fermentative flora of Malbec wine from Mendoza (Argentina) during two consecutive years. Food Microbiol 24:403–412

Morales L, Dujon B (2012) Evolutionary role of interspecies hybridization and genetic exchanges in yeasts. Microbiol Mol Biol Rev 75:721–739

Morrisey WF, Davenport B, Querol A, Donson ADW (2004) The role of indigenous yeasts in traditional Irish cider fermentations. J Appl Microbiol 97:647–655

Mpofu A, Kock JLF, Pretorious EE, Pohl CH, Zvauya R (2008) Identification of yeasts isolated from mukumbi, a Zimbabwean traditional wine. J Sust Dev Africa 10:88–102

N'guessan KF, Brou K, Jacques N, Casaregola S, Dje KM (2011) Identification of yeasts during alcoholic fermentation of tchapalo, a traditional sorghum beer from Cote d'Ivoire. Antonie van Leeuwenhoek 99:855–864

Naumov GI, James SA, Naumova ES, Louis EJ, Roberts IN (2000) Three new species in the *Saccharomyces* sensu stricto complex: *Saccharomyces cariocanus*, *Saccharomyces kudriavzevii* and *Saccharomyces mikatae*. Int J Syst Evol Microbiol 50:1931–1942

Naumov GI, Nguyen HV, Naumova ES, Michel A, Aigle M, Gaillardin C (2001) Genetic identification of *Saccharomyces bayanus* var. *uvarum*, a cider-fermenting yeast. Int J Food Microbiol 65:163–171

Naumov GI, Naumova ES, Antunovics Z, Sipiczki M (2002) *Saccharomyces bayanus* var. *uvarum* in Tokaj wine-making of Slovakia and Hungary. Appl Microbiol Biotechnol 59:727–730

Nout MJR (2003) Traditional fermented products from Africa, Latin America and Asia. In: Boekhout T, Robert V (eds) Yeasts in food: beneficial and detrimental aspects. Behrs, Hamburg, pp 451–473

Novo M, Bigey F, Beyne E, Galeote V, Gavory F, Mallet S, Cambon B, Legras JL, Wincker P, Casaregola S, Dequin S (2009) Eukaryote-to-eukaryote gene transfer events revealed by the genome sequence of the wine yeast *Saccharomyces cerevisiae* EC1118. Proc Natl Acad Sci USA 106:16333–16338

Nyanga LK, Nout MJR, Smid EJ, Boekhout T, Zwietering MH (2013) Fermentation characteristics of yeasts isolated from traditionally fermented masau (*Ziziphus mauritiana*) fruits. Int J Food Microbiol 166:426–432

Pardo O, Pizarro JL (2005) La chicha en el Chile precolombino. Mare Nostrum, Santiago

Pataro C, Guerra JB, Petrillo-Peixoto ML, Mendonça-Hagler LC, Linardi VR, Rosa CA (2000) Yeast communities and genetic polymorphism of *Saccharomyces cerevisiae* strains associated with artisanal fermentation in Brazil. J Appl Microbiol 89:24–31

Pérez-Través L, Lopes C, Querol A, Barrio E (2014) On the complexity of the *Saccharomyces bayanus* taxon: hybridization and potential hybrid speciation. PLoS One 9:e93729

Peris D, Lopes CA, Arias A, Barrio E (2012) Reconstruction of the evolutionary history of *Saccharomyces cerevisiae* × *S. kudriavzevii* hybrids based on multilocus sequence analysis. PLoS One 7:e45527

Peris D, Sylvester K, Libkind D, Gonçalves P, Sampaio JP, Alexande WG, Hittinger CT (2014) Population structure and reticulate evolution of *Saccharomyces eubayanus* and its lager-brewing hybrids. Mol Ecol 23:2031–2045

Pretorius IS (2000) Tailoring wine yeast for the new millennium: novel approaches to the ancient art of winemaking. Yeast 16:675–729

Rodríguez ME, Rebozzzio R, Barbagelata R, Sangorrín MP, Lopes CA (2011) Saccharomyces in traditional fermentations from Patagonia. 29th International Specialised Symposium on Yeast, Guadalajara, Jalisco, México

Rodríguez ME, Pérez-Través L, Sangorrín MP, Barrio E, Lopes CA (2014) *Saccharomyces eubayanus* and *Saccharomyces uvarum* associated with the fermentation of *Araucaria araucana* seeds in Patagonia. FEMS Yeast Res 14:948–965

Rodríguez-Lerma GK, Gutiérrez-Moreno K, Cárdenas-Manríquez M, Botello-Alvarez E, Jiménez-Islas H, Rico-Martínez R, Navarrete-Bolaños JL (2011) Microbial ecology studies of spontaneous fermentation: starter culture selection for prickly pear wine production. J Food Sci 76:M346–M352

Sampaio JP, Gonçalves P (2008) Natural populations of *Saccharomyces kudriavzevii* in Portugal are associated with oak bark and are sympatric with *S. cerevisiae* and *S. paradoxus*. Appl Environ Microbiol 74:2144–2152

Santiago-Urbina JA, Ruiz-Terán F (2014) Microbiology and biochemistry of traditional palm wine produced around the world. Int Food Res J 21:1261–1269

Satora P, Tuszynski T (2005) Biodiversity of yeasts during plum Wegierka Zwykla spontaneous fermentation. Food Technol Biotechnol 43:277–282

Schuller D, Cardoso F, Sousa S, Gomes P, Gomes AC, Santos MAS, Casal M (2012) Genetic diversity and population structure of *Saccharomyces cerevisiae* strains isolates from different grape varieties and winemaking regions. PLoS One 7(2):e32507

Sicard D, Legras JL (2011) Bread, beer and wine: yeast domestication in the *Saccharomyces* sensu stricto complex. C R Biol 334:229–236

Steensels J, Verstrepen KJ (2014) Taming wild yeast: potential of conventional and nonconventional yeasts in industrial fermentations. Annu Rev Microbiol 68:61–80

Suárez Valles B, Pando Bedriñana R, González García A, Querol Simón A, Rodríguez Madrera R (2007) A molecular genetic study of natural strains of *Saccharomyces* isolated from Asturian cider fermentations. J Appl Microbiol 103:778–786

Tamang JP, Tamang N, Thapa S, Dewan S, Tamang B, Yonzan H, Rai AK, Cheltri R, Chakrabarty J, Kharel N (2012) Microorganisms and nutritional value of ethnic fermented foods and alcoholic beverages of North East India. Indian J Trad Knowl 11:7–25

Teramoto Y, Sato R, Ueda S (2005) Characteristics of fermentation yeast isolated from traditional Ethiopian honey wine, ogol. Afr J Biotechnol 4:160–163

Toro ME, Vazquez F (2002) Fermentation behaviour of controlled mixed and sequential cultures of *Candida cantarellii* and *Saccharomyces cerevisiae* wine yeasts. World J Microbiol Biotechnol 18:347–354

Vaughan-Martini A, Martini A (2011) *Saccharomyces* Meyen ex Reess (1870). In: Kurtzman CP, Fell JW, Boekhout T (eds) The yeasts: a taxonomic study, vol 2. Elsevier, London, pp 733–746

Verdugo Valdez A, Segura García L, Kirchmayer M, Ramírez Rodríguez P, González Esquinca A, Coria R, Gschaedler Mathis A (2011) Yeast communities associated with artisanal mescal fermentations from *Agave salmiana*. Antonie van Leeuwenhoek 100:497–506

Verstrepen KJ, Derdelinckx G, Verachtert H, Delvaux FR (2003) Yeast flocculation: what brewers should know. Appl Microbiol Biotechnol 61:197–205

Voordeckers K, Brown CA, Vanneste K, van der Zande E, Voet A, Maere S, Verstrepen KJ (2012) Reconstruction of ancestral metabolic enzymes reveals molecular mechanisms underlying evolutionary innovation through gene duplication. PLoS Biol 10:e1001446

Wang Q, Liu W, Liti G, Wang SA, Bai FY (2012) Surprisingly diverged populations of *Saccharomyces cerevisiae* in natural environments remote from human activity. Mol Ecol 22:5404–5417

Warringer J, Zörgö E, Cubillos FA, Zia A, Gjuvsland A, Simpson JT, Forsmark A, Durbin R, Omholt SW, Louis EJ, Liti G, Moses A, Blomberg A (2011) Trait variation in yeast is defined by population history. PLoS Genet 7:e1002111

Chapter 16
Wild Yeasts Selection for High-Quality Patagonian Wines

Silvana María del Mónaco, Yolanda Leticia Curilén, Sebastián Mario Ezequiel Bravo, Adriana Beatriz Simes, Viviana Andrea Carreño, Ramona del Carmen Maturano, and Adriana Catalina Caballero

Abstract The use of yeast starters indigenous to each winegrowing area is today a worldwide oenological practice aiming to imprint a differential character to the wine. This chapter summarizes the research carried out towards the design and development of indigenous wine yeast starters from Argentinean Patagonia that are currently nonexistent in the market. The results shown include diversity and criteria

---

S.M. del Mónaco • R.C. Maturano
Instituto de Investigación y Desarrollo en Ingeniería de Procesos, Biotecnología y Energías Alternativas (PROBIEN), Consejo Nacional de Investigaciones Científicas y Técnicas de la República Argentina (CONICET)- Universidad Nacional del Comahue, Buenos Aires 1400, 8300 Neuquén, Provincia del Neuquén, Argentina

Y.L. Curilén • A.B. Simes • V.A. Carreño
Facultad de Ciencias y Tecnología de los Alimentos, Universidad Nacional del Comahue, 25 de Mayo y Reconquista, 8336 Villa Regina, Provincia de Río Negro, Argentina

S.M.E. Bravo
Instituto de Investigación y Desarrollo en Ingeniería de Procesos, Biotecnología y Energías Alternativas (PROBIEN), Consejo Nacional de Investigaciones Científicas y Técnicas de la República Argentina (CONICET)- Universidad Nacional del Comahue, Buenos Aires 1400, 8300 Neuquén, Provincia del Neuquén, Argentina

Facultad de Ciencias y Tecnología de los Alimentos, Universidad Nacional del Comahue, 25 de Mayo y Reconquista, 8336 Villa Regina, Provincia de Río Negro, Argentina

A.C. Caballero (✉)
Instituto de Investigación y Desarrollo en Ingeniería de Procesos, Biotecnología y Energías Alternativas (PROBIEN), Consejo Nacional de Investigaciones Científicas y Técnicas de la República Argentina (CONICET)- Universidad Nacional del Comahue, Buenos Aires 1400, 8300 Neuquén, Provincia del Neuquén, Argentina

Facultad de Ciencias y Tecnología de los Alimentos, Universidad Nacional del Comahue, 25 de Mayo y Reconquista, 8336 Villa Regina, Provincia de Río Negro, Argentina

Facultad de Ingeniería, Universidad Nacional del Comahue, Buenos Aires 1400, 8300 Neuquén, Provincia del Neuquén, Argentina
e-mail: adrianacaballero1@gmail.com; adriana.caballero@facta.uncoma.edu.ar; aradriana.caballero@probien.gob.ar

N.L. Olivera et al. (eds.), *Biology and Biotechnology of Patagonian Microorganisms*, DOI 10.1007/978-3-319-42801-7_16

for *Saccharomyces* and non-*Saccharomyces* wild yeast strain selection as well as their contribution to the physicochemical and sensorial qualities of wines when they are used either singly or in combined forms. The advantages of these indigenous starters over the foreign commercial ones to imprint a differential character on regional wines are also discussed.

16.1 Introduction

Fermented beverages and foods have had a significant role in most societies worldwide for millennia. Among these, wine has been present since the dawn of civilization and has followed humans along their paths of migration. Archaeologists have found evidence for wine production in Iran and Egypt at 6000 BC and 3000 BC, respectively (McGovern et al. 1997; Cavalieri et al. 2003). From there, this biotechnology expanded into the Old World (Europe, 2000 BC to 100 AD) and Northern Africa (1000 BC) through the Mediterranean Sea and through the Spicy Way towards Northern India and China (100 BC). Finally, from the sixteenth century, vines and winemaking were introduced into the New World: South America (1500 AC), South Africa (1600 AC), North America (1700), and Australia and New Zeland (1800) (Pretorius 2000). According to the last official report of the Office International de la Vigne et du Vin (OIV 2016), there are presently 7.5 million ha of vineyards across the world distributed into different winegrowing regions and *terroirs* (geography, history, and culture and traditions of each region) where 274.4 million hectoliters (hl) of wine are produced annually.

From early times, fermentation was presented as a valuable tool to preserve fresh food products through extending their shelf life, freshness, and wholesomeness, accounting for a rapid increase of wine fermentation technology. However, many centuries of accumulating practical knowledge had to pass before Pasteur, in the second half of the nineteenth century, demonstrated the key role of yeast strains in winemaking (Chambers and Pretorious 2010). From the past two centuries, yeasts have been an integral part of pioneering work in the field of fermentations, showing evidence that the impact of yeasts on wine quality is intimately linked to their ecology and biological activities (Fleet 2007). The most important practical consequence of this knowledge has been the development of selected yeast starters for microbiological control of the fermentative process (Fleet and Heard 1993), an innovation into winemaking practices that revolutionized the wine industry (Pretorius 2000; Fleet 2008). Today, the use of starters composed of indigenous wine yeasts selected from each winegrowing area is a widespread oenological practice aiming to impart a differential character to the wine as well as to obtain quality certifications (Fleet 2008; Tufariello et al. 2014; Alves et al. 2015). Commercial starters for alcoholic fermentation actually found in the market are composed of yeast strains isolated from world winegrowing regions other that those in Argentina, even Patagonia. Research carried out to develop locally this biotechnology is summarized in this chapter.

16.2 Argentine Vitiviniculture: the Patagonian Winegrowing Region

Vitiviniculture has historically occupied a place of importance in the agricultural industries from Argentina. The country is currently the fifth largest producer (annual production of 15 million hl), the seventh largest consumer (10.3 of 238.7 million hl consumed globally), and the tenth largest exporter (annual exportations of 2.7 million hl) of wines in the world (OIV 2016). Despite this, the need to satisfy the demands of a globalized and increasingly competitive international market, where consumer preference is the key factor that defines the success of a wine, impose upon the productive sectors new challenges that require technological innovation.

In Argentine, vine cultivation is carried out in a long area of 1700 km, extended along the Andes, in which at least six winegrowing regions can be distinguished (Catania and Avagnina 2010); one of them, the Patagonia region, located below 39°SL, is the southernmost region from Argentina and one of the most southern regions in the world (Fig. 16.1). It has optimal agro-ecological conditions for high-quality viticulture

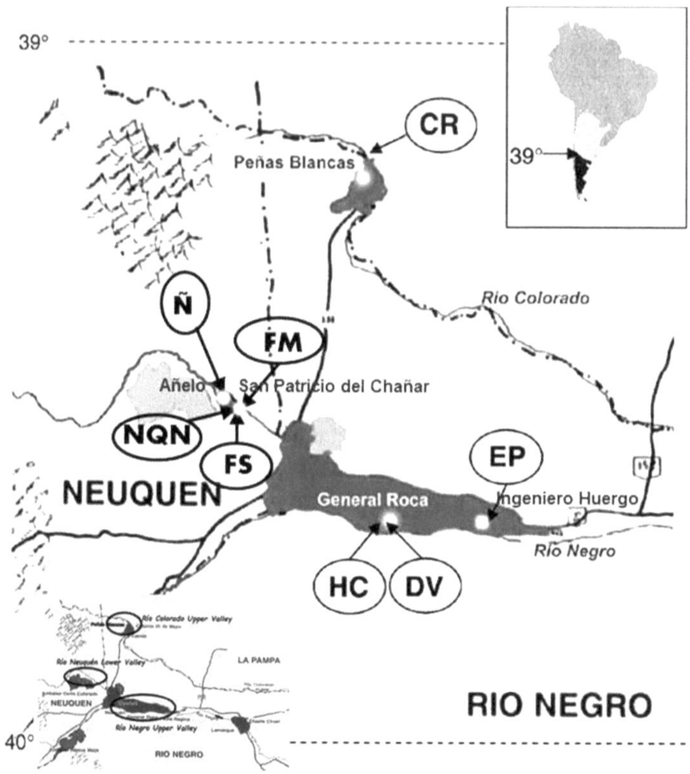

Fig. 16.1 Patagonian winegrowing region. Location of sampled Patagonian cellars and vineyards (EP, DV, HC, and CR in Río Negro Province and FS, FM, NQN, and Ñ in Neuquén Province). *Dark grey*: cultured vine areas. *Light spots*: municipalities (towns). *Right top corner*: South America (*dark grey*)–Argentina (*light grey*) and Argentinean Patagonia (*black*). *Left bottom corner*: *circles* indicate viticultural areas within Patagonia region. (*Source:* INV and modified)

and a long winemaking tradition (Le Guillou 2000). For these reasons, although the wine industry is still a secondary economic activity, it represents a very interesting alternative to diversify local production.

Regional wine production is mostly oriented to the elaboration of young and dry wines from red *Vitis vinifera* L. varieties (80 %) such as Malbec (the "brand image" of Argentine red wines) as well as Merlot and Pinot noir, two varieties that found in Patagonia the optimal ecological conditions to express all their oenological potential (Weizman 2009). Currently, regional wine production is based on both spontaneous alcoholic fermentations of the grape musts or conducted fermentations using commercial yeast starters (guided fermentations) (Lopes et al. 2007a).

Spontaneous fermentations can give high-quality wines with a unique regional character (Rainieri and Pretorius 2000; Vilanova and Sieiro 2006; Fleet 2008). However, inconsistencies in wine quality and diminished predictability of the process are frequently observed in these fermentation types (Fleet 2008). Guided fermentations, in contrast, show more predictable processes giving standardized, ordinary character wines, with greater consistency in quality but lacking in flavor complexity and without terroir notes (Vilanova and Sieiro 2006; Fleet 2008; Terrell et al. 2015; Alves et al. 2015). Some studies have reported that the use of locally selected yeast strains can positively affect the final quality of the wine (Vilanova and Massneuf-Pomarede 2005; Carrascosa et al. 2012; Tufariello et al. 2014; Alves et al. 2015). These data support the relevance of the diversity surveys to select wine yeasts better adapted to the ecological and technological features of each particular winegrowing area, preserving its own natural biodiversity (Bokulich et al. 2014; Gilbert et al. 2014).

16.3 Yeast Diversity Associated with Patagonian Red Winemaking

The chemical composition of wine is the foundation of its sensorial quality (Swiegers et al. 2005). In young wines, as those mainly produced in the Comahue region, chemical composition is determined by two main factors, grape quality and microbial ecology associated with winemaking (Cole and Noble 1997).

Winemaking is a complex microbial ecosystem that involves interactions between filamentous fungi, yeasts, and bacteria with different physiological and metabolic characteristics (Pretorius 2000; Fleet 2003; Barata et al. 2012). Among these, yeasts retain a predominant role, because they promote alcoholic fermentation, the biochemical process whereby grape must is transformed into wine (Fleet and Heard 1993). A succession of yeast species and strains associated with grape and winery surfaces develop during spontaneous grape juice fermentation (Fleet and Heard 1993; Pretorius 2000; Jolly et al. 2006; Fleet 2008). Ecological studies have shown that this yeast diversity is significantly influenced by the geographical, cultural, and technological features of each particular winegrowing region, which defines the *terroir* (Capozzi et al. 2015). The knowledge and the understanding of how this microbial terroir contributes to the natural environment of vineyard and to the identity of wine are processes that start at the harvest of the grapes, and then evolve along the different stages of fermentation (Van Leeuwen and Seguin 2006; Bokulich et al. 2013; Pinto et al. 2015).

Microbiological studies initiated several years ago have allowed characterizing the diversity of yeast biota associated with red winemaking in the Patagonian region. The results for yeast diversity obtained from ecological studies are highly dependent on the methodology used in its experimental approach. Unsuitable sampling schemes or unsuitable analytical strategies can bias the results of studies on diversity of yeasts associated with winemaking or other complex ecosystems (Fleet and Heard 1993; Lachance 2003; Barata et al. 2012), and several reviews on the analytical approaches to the study of overall yeast ecology have been published (Boundy-Mills 2006; Ciani et al. 2002; Kurtzman et al. 2011). The main aim of the microbiological studies on wine yeast diversity initiated several years ago in the Patagonian region was to create a Patagonian indigenous yeast bank to provide data on oenological properties for potential applications in regional wine production. However, and to obtain ecologically meaningful conclusions, experimental approaches were carried out taking into consideration the advice of Fleet and Heard (1993) and later Lachance (2003), in particular those related to sample size and sample replication, sampling strategies, and the nature and quality of substrates.

Eight regional representative cellars located in Río Negro (HC, DV, and EP) and Río Colorado (CR) Upper Valleys (Río Negro Province), and in Río Neuquén Lower Valley (FS, FM, NQN, and Ñ) (Neuquén Province), and their respective vineyards were selected for the studies (Fig. 16.1). The diversity of yeast biota associated with grape surfaces and grape musts belonging to red vine cultivars Malbec, Merlot, and Pinot noir were analyzed using culture-dependent methods and direct plating on unselective (GPY agar) and selective (L-Lys agar) media for the yeast isolation (van Broock et al. 1996; Lopes et al. 2007b; Caballero et al. 2004; 2008; Curilén et al. 2009; del Mónaco et al. 2014a). Yeast diversity associated with cellars was also analyzed, but, in this case, only four wineries were sampled (HC, EP, DV, CR) and before isolation, yeast samples were enriched (Sangorrín et al. 2007). Ripe, whole, and healthy grapes from Merlot, Malbec, and Pinot noir varieties were picked in the vineyards during 1993–1998 and 2006–2009 vintages at harvest time. In each vineyard, berries were picked from 20 % of total plants previously selected by random, in a relation of one berry for each plant. Yeast samples were obtained by agitation followed sonication of each berry in pure and sterile water as described by van Broock et al. 1996. Industrial grape must samples (1 l) from the same cellars and grape varieties were collected before (initial) and during spontaneous alcoholic fermentation (middle and end stages). Grape and must yeasts were isolated on GPY and L-Lys agar. To analyze yeast diversity associated with cellar surfaces, four empty vats from each winery and three different internal areas from each vat were sampled approximately 4 weeks before harvest. All yeasts were identified according to the methods and keys proposed by Kurtzman and Fell (1998), and their identity was confirmed by polymerase chain reaction-restriction fragment length polymorphism (PCR-RFLP) analysis of the ITS1-5, 8S-ITS2 region from the nuclear rDNA

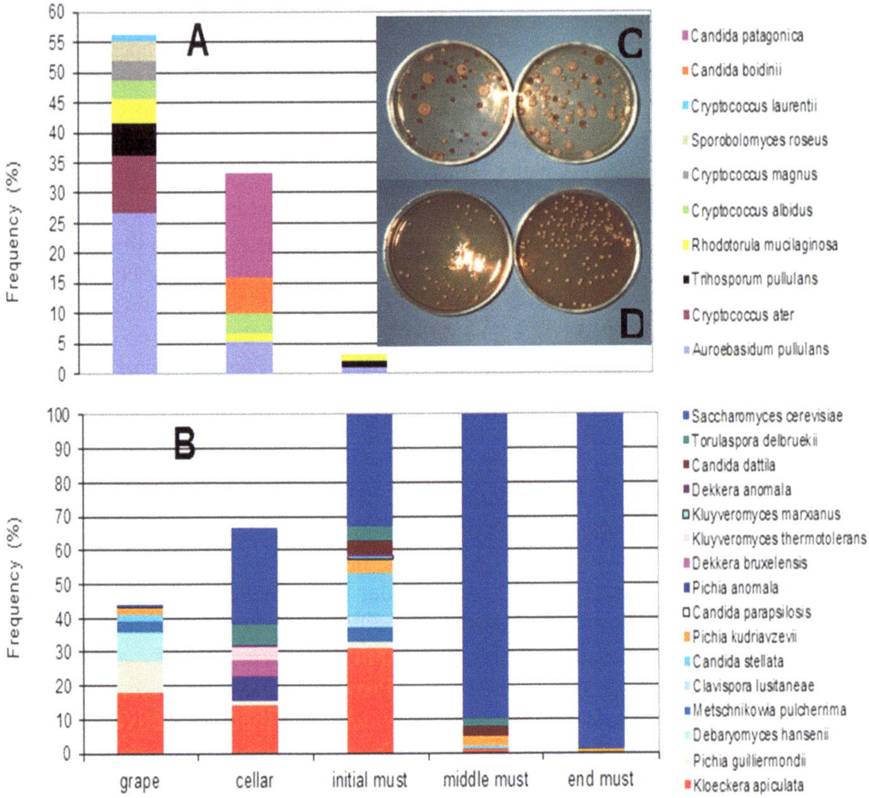

Fig. 16.2 Diversity of yeasts associated with Patagonian winegrowing region. **a** Aerobic yeasts. **b** Facultative yeasts. **c** Isolation plates of grape yeasts dislodged by shaking (*left*) and sonication (*right*). **d** Isolation plates of end must yeasts at two different dilution samples. Sample size: grape $n=536$, cellar $n=150$, initial musts (14°Bmé) $n=322$), middle must (6°Bmé) $n=320$), and end must (<0°Bmé) $n=397$. 1°Bmé ≈ 17 g total reducing sugar

gene complex (Esteve Zarzoso et al. 1999) or D1/D2 26S rDNA gene sequencing (Kurtzman and Robnett 1998).

The results obtained from these studies are shown in Fig. 16.2. Leaving out the yeast-like fungus *Aureobasidium pullulans*, the major species on the surface of the grapes (27 %), at harvest time the grape yeast biota arer divided between aerobic yeasts (30 %), dominated by basidiomycetous species (Fig. 16.2a) and facultative, weakly fermentative ascomycetous (43 %) with *Kloeckera apiculata* as the predominant species (18 %) (Fig. 16.2b). Cellar yeast biota is mostly facultative, strongly fermentative ascomycetous (67 %) with *Saccharomyces cerevisiae* as the major species (30 %) (Fig. 16.2b). This result and the extremely low occurrence (<0.2 %) of *S. cerevisiae* in grapes observed in this study (only one of grape yeast isolates belonged to *S. cerevisiae*) are in agreement with the postulate by Martini School on the "winery origin" of this species (Rosini 1984; Martini 1993; Vaughan-Martini and Martini 1995; Martini 2003; Schuller et al. 2005; Santamaría et al. 2005; Mercado et al. 2007).

Regarding the fermentative process, the composition of yeast species shown in the initial must evidences the contribution that both grape and cellar yeast biota make to the fermentation process as well as the selective pressure of the must in favor of fermentative species, *S. cerevisiae* in particular.

Additionally, and in agreement with that reported by other authors (Pretorius 2000; Costantini et al. 2009; Jolly et al. 2014; Capozzi et al. 2015), even though *Saccharomyces cerevisiae* is the most important yeast in spontaneous fermentations, other yeasts belonging to non-*Saccharomyces* species such as *Kloeckera apiculata*/*Hanseniaspora uvarum*, *Candida dattila*, *Pichia kudriavzevii*, and *Torulaspora delbruekii* are also able to remain in Patagonian musts and during fermentation periods in appropriate concentrations to significantly contribute to the sensorial quality of wine (Fig. 16.2b).

Although yeasts have been studied and used for decades, their unequivocal characterization at strain level has been made possible only more recently with the development of molecular techniques such as comparative genomic hybridization array (CGH array), genome sequence and functional annotation, restriction fragment length polymorphism of mitochondrial DNA (mtDNA-RFLP), pulsed-field gel electrophoresis (PFGE), random amplification of polymorphic DNA (RAPD), PCR amplification with intron splice site primers, PCR amplification of delta elements (δ sequences) microsatellite analyses (SSRs or microsatellite typing), and strain typing multilocus sequence typing (MLST) (Querol and Ramón 1996; Baleiras Couto et al. 1996; Pérez et al. 2001a; Schuller et al. 2004; Tofalo et al. 2013). Among these, the mtDNA-RFLP analysis is a molecular technique largely utilized in biodiversity studies for its ease of use and reliability. On the other hand, the usefulness of the killer biotype in the fingerprinting of wine *S. cerevisiae* strains as well as of other yeast species has been emphasized by some authors (Vaughan-Martini et al. 1996; Buzzini and Martini 2000; Sangorrín et al. 2002). Moreover, the natural distribution of yeasts producing killer toxins and sensitive to those toxins were demonstrated to be related to phylogeny as well as the ecological circumstances of the strains (Ganter and Starmer 1992).

The indigenous wine *S. cerevisiae* diversity within the Patagonian region was characterized using mtDNA-RFLP and killer biotype in a combined form. Six cellars with particular oenological practices located in different winegrowing areas were selected and 523 *S. cerevisiae* isolates obtained from spontaneous and guided red wine fermentations were analysed. Twenty-four commercial starters historically used in the region were also included in this study. Of the total *S. cerevisiae*, 217 showed similar patterns to commercial strains and were discarded from the study. The combination of both typing techniques demonstrated a great diversity in *S. cerevisiae* populations associated with red wine fermentations in Northwestern Patagonia (Lopes et al. 2006; Caballero et al. 2008). Similar studies were carried out with non-*Saccharomyces* indigenous isolates, as well as other yeast species such as *Kloeckera apiculata* and *Pichia guillermondii* (Rodríguez et al. 2004) (data not shown). More recently, using other methodology, this variability was

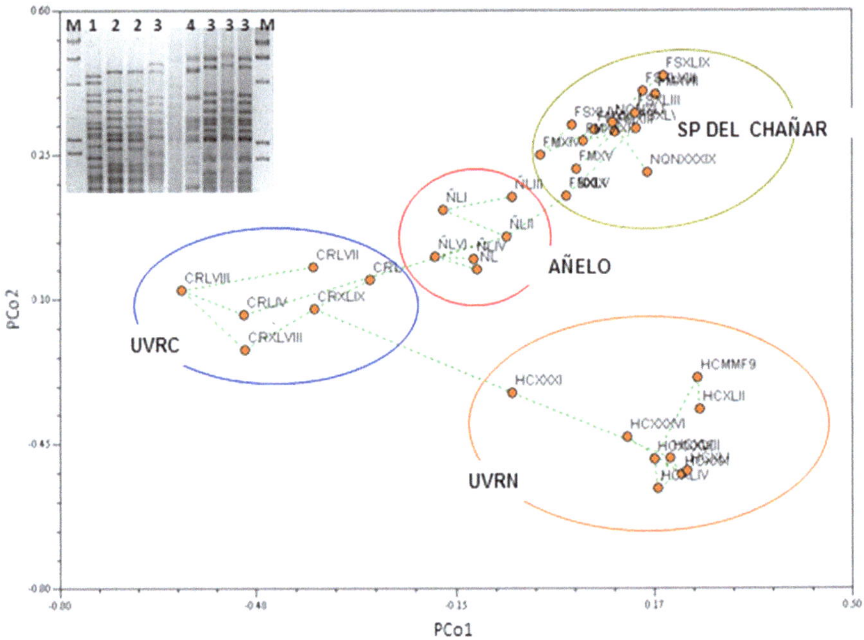

Fig. 16.3 Principal coordinates analysis of mtDNA-RFLP/ killer biotype combined patterns from Patagonian indigenous yeast populations. *Insert:* mtDNA-RFLP *Hin*fI patterns. *Lanes 1* and *2*: indigenous strains FM LVII and FM XI, respectively; *lanes 3* and *4:* commercial starters Laffort F10 and Zymaflore VLBC, respectively. Strains are identified according to the notation given by Lopes et al. 2006 and Caballero et al. 2008. *UVRN* Río Negro Upper Valley, *UVRC* Río Colorado Upper Valley, *SP del Chañar* San Patricio del Chañar

also demonstrated within indigenous *Pichia kudriavzevii* populations (del Mónaco et al. 2016). The analysis of the relatedness among indigenous *S. cerevisiae* strains using principal coordinates analysis from combined data allowed the clustering of the strains into four populations significantly related to their origin fermentations (Fig. 16.3). In spite of this conclusive result, it was difficult to determine which of the different factors involved in these fermentations, either grape must variety, cellar geographic area, or cellar oenological practice, influenced the relatedness among strains.

16.4 Yeast Selection for Patagonian Winemaking

The practical consequence of studies on microbial ecology in winemaking was the development of starter cultures of alcoholic fermentation (AF). The use of commercial starter cultures for the AF in oenology, with selected cultures of *S. cerevisiae* for the inoculation of fruit juice, has been applied since the 1960s, and it has been one of the most important technological advances in the wine industry (Molina et al. 2009; Sun et al. 2014).

The inoculation of grape musts with commercial starters was an oenological practice strongly resisted by wine producers from Europe. The most solid argument for this resistance referred to the sensory quality standard, where flattened aromatic profiles were produced in each wine fermented with those starters. In modern wineries, the use of commercial starter culture to steer fermentations is being doubted, because they often lack some advantageous oenological traits, which are present when the spontaneous fermentation is ruled by indigenous populations (Fleet and Heard 1993). For that reason, the exploitation of indigenous strains biodiversity has great importance for the characterization and selection of strains with peculiar phenotypes (Pinto et al. 2015; Capozzi et al. 2015). Although the preservation of spontaneous microflora is essential to obtain the typical flavor and aroma of wines deriving from different grape varieties (Pretorius 2000; Renouf et al. 2005), the development of starter from *S. cerevisiae* strains indigenous from each winegrowing region ended this controversy (del Mónaco et al. 2014b). The advantages to using starters of indigenous *S. cerevisiae* strains, better adapted to the ecological and technological features of each particular winegrowing area and preserving its own natural biodiversity, is now recognized by all wine producers, including the European ones.

However, the knowledge generated in recent years in the field of oenological microbiology indicates that some indigenous non-*Saccharomyces* yeasts have a significant and favourable effect on the sensory characteristics of the wine (Jolly et al. 2006; Fleet 2008) and, therefore, merit consideration as useful tools in winemaking. Because these non-*Saccharomyces* yeasts are not vigorous or competitive fermenting organisms under oenological conditions, they may be only employed as starter cultures in conjunction with strongly fermentative *S. cerevisiae* strains, ensuring in this way the completion of fermentation. The employment of non-*Saccharomyces*/*Saccharomyces* mixed starter inocula has been suggested to mimic the spontaneous alcoholic fermentations process and to avoid the risks of stuck or sluggish fermentations (Jolly et al. 2006; Fleet 2008; Capozzi et al. 2015). In this framework, yeast selection for oenological use, usually carried out within the species *S. cerevisiae*, has been today extended to non-*Saccharomyces* yeasts (Ciani and Maccarelli 1998; Soden et al. 2000; del Mónaco et al. 2010; Rodríguez et al. 2010; Benito et al. 2011; Comitini et al. 2011; Morata et al. 2012; Di Maio et al. 2012; Contreras et al. 2015).

As a consequence of the new trends in yeast selection, protocols have been also modified. Traditionally, yeasts have been selected for their fermentative power, suitable fermentative kinetics at different temperatures, low acetic acid production, and resistance to sulfur dioxide, among other qualities (Rainieri and Pretorius 2000; Lopes et al. 2007b). However, new selection criteria have emerged, and yeasts that can improve the technological properties and sensorial features of wines are now sought. These selection criteria include, among others, (1) the ability to enhance wine colour (metabolic formation of stable pigments such as vitisins and vinylphenolic pyranoanthocyanins and the scant adsorption of anthocyanins by the yeast cell wall), and the presence of anthocyanase activity; (2) glucosidase activity to enhance varietal wine aroma; (3) colloidal stabilisation in red wines by allowing over-lees aging; (4) the appropriate enhancement of aroma via the production of volatile compounds such as esters and higher alcohols, with the scant production of off-flavours; and (5) the provision of structure and body via the production of polyalcohols such as glycerol and 2,3-butanodiols, and the release of mannoproteins and yeast polysaccharides (Suárez-Lepe and Morata 2012). Yeast selection may also improve resistance to fermentative stress, the ability to ferment fructose, copper resistance, and the degradation/production of malic acid, etc., as well as aid in the production of wines with low alcohol content (a growing consumer preference). Some yeasts may have qualities that render them desirable for use in the production of sparkling wines, fortified wines, or biologically aged wines, etc. They might even be selected according to terroir specificities, particular style wines, or for the production of ecological wines (Suárez-Lepe and Morata 2012; Capozzi et al. 2015).

Microbiological studies carried out during several years in the Patagonian region as just described allowed establishing an important collection of organisms relevant for oenological application and wine starter elaboration. The oenology applied studies were divided in two basic objectives: isolation of indigenous *S. cerevisiae* strains for alcoholic fermentation (AF) and isolation of local non-*Saccharomyces* with potential use in oenology for develop pure and mixed indigenous starters with the aim to replace commercial starters, currently imported, and improve the quality of regional wines. The selection criteria for Patagonian wine yeast were chosen taking into account the type or style of wine elaborated in the region as well as certain characteristics of the grape varieties and their musts winemaking in the region.

16.4.1 Selection of **Saccharomyces cerevisiae** *Strains*

As described here, the regional wine production is mostly oriented to the elaboration of young and dry wines (total reducing sugar (RST) <2 g/l) from red vine varieties (80 %) aromatically neutral such as Malbec, Merlot, and Pinot noir.

The sensorial properties of these type of young wines are significantly affected by fermentative features of the yeasts that conduct the fermentation. In general, most studies to assess the impact of yeast strains on wine quality have been focussed solely on volatile flavour and aroma compounds (Romano et al. 2003, 2008; Clemente-Jimenez et al. 2005; Varela et al. 2009; Lopes et al. 2007a, b; Rodríguez et al. 2010). This is a very important aspect to take into account in Patagonia because of the aromatically neutral grape varieties mostly vinified. However, for red wine, nonvolatile components that influence wine colour and mouth feel are critical determinants of its style (Mercurio et al. 2010), and the role of yeasts to modulate them may be of equal importance and thus must be included in selection procedures. Among the compounds that influence these wine properties are organic nonvolatile acids and polyphenols. They have a direct impact on wine quality, and imbalances in these fractions can affect its physicochemical and sensorial properties, mainly mouth feel (Beelman and Gallander 1979; Henick-Kling 1993; Radler 1993; Gao and Fleet 1995; Gawel et al. 2007) and colour (Holt et al. 2013). Additionally, the influence of yeast on colour and total polyphenols of wine was included in the technological and qualitative properties historically addressed in selection procedures,

L(+)-tartaric and L(−)-malic acids are the most important constituents of organic nonvolatile acid fraction in grapes and grape musts, accounting for 90 % of the titratable acidity. Several factors such as grapevine variety, vineyard agricultural practices, temperature, humidity, and berry maturity degree may affect their concentrations in grape musts (Ruffner 1982; Flanzy 2000; Volschenk et al. 2001). In particular, L(−)-malic acid content, directly related to the respiratory quotient of berries, is higher in grape musts from cooler regions than those from warmer regions (Ribéreau-Gayon et al. 2003). In the Comahue region, one of the southernmost winegrowing regions of the world, malic acid concentrations account for 56 % of red grape must titratable acidity, reaching 66 % in Pinot noir (Caballero et al. 2005), the emblematic regional vine variety (Weizman 2009). Additional to its contribution to wine acidity, malic acid represents a fermentable substrate for other microorganisms that can spoil the wine before and after bottling (du Toit and Pretorius 2000). Without adjustment of acidity, the wines will be regarded as unbalanced or spoilt (Swiegers et al. 2005). For these reasons, malolactic fermentation (MLF) is a routine oenological practice in Patagonian red winemaking, and yeast–lactic acid bacteria (LAB) interactions are a great concern for winemakers and researchers. Previous studies carried out in the Patagonian region on spontaneous MLF and its relationship with the yeast biota involved in alcoholic fermentation evidenced antagonist interactions between the *S. cerevisiae* biota and indigenous LAB (Curilén et al. 2009). In this context, yeast–LAB interactions were also included within the criteria for Patagonian *S. cerevisiae* wine yeast selection. An *Oenococcus oeni* strain isolated from the Patagonian region and selected for its adequate oenological properties (see Chap. 14) was used in assays carried out to evaluate these interactions. This is the second ecological criterion included in the Patagonian yeast selection procedure program, the first being killer interactions (Lopes et al. 2007a, b). There is evidence that in guided fermentations the dominance of the starter can be subordinated to the specific winemaking conditions to settle down during the initial stages (Pretorius and Hoj 2005). Killer yeasts could be present at the beginning of the fermentation, and they could hinder the implantation of *S. cerevisiae* starter, inducing stuck fermentation (Pérez et al. 2001b). Hence, the evaluation of killer biotype became relevant in areas where the killer phenomenon is widespread, such as the Patagonian region (Sangorrín et al. 2001, 2007).

In a first step, 27 Patagonian indigenous yeasts were evaluated in their key technological and qualitative properties using small-scale (100-ml) fermentations carried out on semi-synthetic media for the performance of general assays [fermentative power (FP), volatile acidity (VA), foam production] and specific assays (SH_2 production, ethanol tolerance, killer behaviour). For comparative purposes, the commercial *S. cerevisiae F15* (Laffort), the commercial starter most widely used in Patagonian red winemaking, was also evaluated. The results obtained from small-scale fermentations showed little foam production as well as similar FP and VA

Table 16.1 Technological properties and killer biotype of commercial F15 and Patagonian F8 *Saccharomyces cerevisiae* strains assayed at laboratory scale (5 l) using Pinot noir must as substrate

		Yeast strain		
Oenological parameters	Grapes	ScF8	ScF15	*p* value*
TRS (g/l)[a]	235±0.31	0.69±0.11	2.10±0.23	<0.01
PF[b]	–	13.9±0.6	13.8±0.4	ns
Titratable acidity (g l^{-1})[c]	6.41±0.15[a]	5.7±0.25[b]	5.32±0.20[b]	ns
Volatile acidity (g l^{-1})[d]	–	0.42±0.13	0.60±0.22	ns
pH	3.57±0.01[a]	3.35±0.10[c]	3.24±0.03[b]	ns
Glicerol (gl^{-1})	–	7.12±0.58	5. 68±0.62	<0.05
Color index	–	718±2	623±3	<0,001
Tint	–	0.89±0.00	0.96±0.00	<0.001
Total phenols[e]	–	42.6±0.3	41.1±0.4	<0.01
Killer phenotype	–	Killer R	N	–

Killer R killer toxin producer and resistant to other killer factors; *N* neutral phenotype
*ANOVA and Tukey test, $n=3$
[a]Total reducing sugar
[b]Fermentative power, expressed as ml ethanol in 100 ml wine
[c]Expressed as tartaric acid
[d]Expressed as acetic acid
[e]Expressed as galic acid (g/l)

values in most of the isolates. However, a second round of assays using natural must as a substrate and a higher vinification scale (5 l) evidenced significant differences between strains, mainly in their influence on wine colour. As a result, the indigenous *S. cerevisiae* ÑIF8 strain, whose technological and ecological properties are shown in Table 16.1 and Fig. 16.4 (Caballero et al. 2008; Curilén et al. 2015), was selected to be developed as a regional Patagonian starter culture.

16.5 Selection of Patagonian Non-*Saccharomyces* Strain

According to those already described, the purposes of the study included selection of autochthonous yeasts with metabolic ability to degrade L(–)-malic acid for its potential use in equilibrated young wine elaboration. Fifty-seven Patagonian non-*Saccharomyces* yeast of oenological origin were identified by conventional molecular methods and tested in their capability to grow at the expense of L-malic acid (del Mónaco et al. 2014a). An isolate, noted as *P. kudriavzevii* ÑNI15, was able to degrade L-malic acid in microvinifications, increasing pH in 0.2–0.3 units with a minimal effect on the acid structure of wine. Additionally, this isolate was a weak producer of ethanol, an important producer of glycerol (10.41±0.48 g l^{-1}), a producer of acceptable amounts of acetic acid (0.86±0.13 g l^{-1}), as well as capable of improving the sensorial attributes of wine, increasing its fruity aroma (del Mónaco et al. 2014a). Microvinification studies were carried out using a synthetic must as a

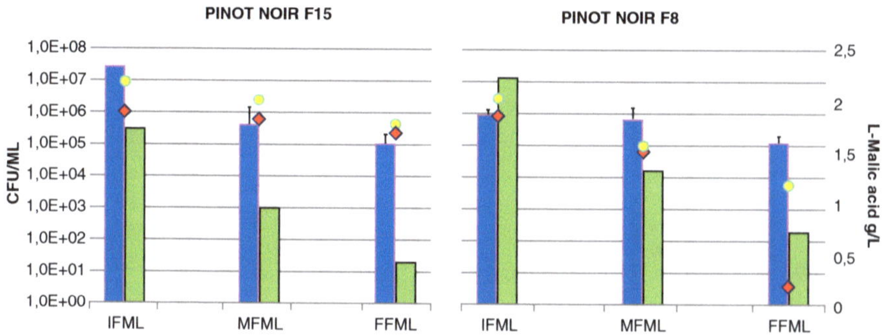

Fig. 16.4 Ecological criteria for yeast wine selection. Evaluation of yeast–lactic acid bacteria interaction during spontaneous (*blue bars* and *yellow circles*) and indigenous *Oenococcus oeni* UNQ31 guided (*green bars* and *red diamonds*) malolactic fermentation carried out at laboratory scale (1 l) using young wines obtained from alcoholic fermentations guided by commercial F15 (*left*) and indigenous Patagonian F8 (*right*) yeast strains. *Bars* represent microbial (BAL) growth, circles/diamonds represent the L(−)-malic acid concentrations in wine. *IFML* start of malolactic fermentation (MLF), *MFML* middle of MLF, *FFML* end of MLF

substrate, with similar nitrogen and acidic fraction composition to Patagonian Pinot noir juice. Cultures with indigenous *P. kudriavzevii* ÑNI15 and *S. cerevisiae* F8 were performed under anaerobic conditions, emulating wine fermentation. Figure 16.5 shows the results obtained for both yeast strains. An acceptable yield in biomass was observed in both microvinifications (Fig. 16.5a). Although both fermentations presented a similar sugar concentration at the end of the process (Fig. 16.5a), the fermentative efficiency (no data shown) as well as the sugar consumption rate were higher for *S. cerevisiae* than for *P. kudriavzevii* (Fig. 16.5a). A noteworthy fact is that *P. kudriavzevii* was capable of significantly raising medium pH with a minimal effect on the acid structure of the wine, which is evidenced by the titratable acidity value (Fig. 16.1b), whereas in the *S. cerevisiae* culture pH was maintained constant with the fermentation (Fig. 16.5b).

Sensorial analysis evidenced significant differences in aromatic perception between *P. kudriavzevii* and *S. cerevisiae* wines (del Mónaco et al. 2014a). These differences were in favour of the former, which showed a higher fruity and cooked pears aroma as well as less alcoholic flavor by mouth than the latter (Fig. 16.6).

The use of mixed starters combining the high fermentative capacity of *S. cerevisiae* and the distinctive metabolic inputs of non-*Saccharomyces* selected strains is an oenological strategy being evaluated currently to obtain wines with differential sensorial traits, more equilibrated and balanced (Ciani et al. 2010; Rodríguez et al. 2010; Suárez-Lepe and Morata 2012). Notwithstanding, inoculation protocol variants from the strains present in the mixed culture affect the evolution of the process, and as a consequence, the physicochemical and sensorial quality of the product is affected, which is why they have to be particularly studied (Ciani et al. 2010; Muñoz-González et al. 2011). Results from Pinot noir must vinification carried out at laboratory scale using different inoculation strategies for *P. kudriavzevii* I15 and

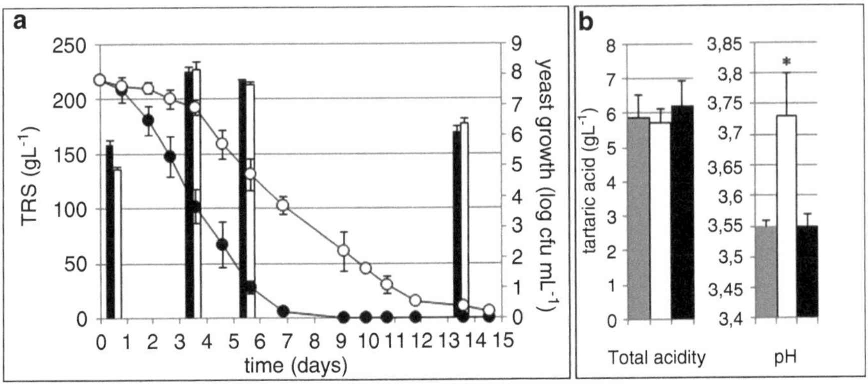

Fig. 16.5 Guided microvinifications using synthetic must. **a** Fermentation kinetic: evolution of total reducing sugars (TRS, *circles*) and yeast growth (*bars*) within 14 days of guided fermentation for *Pichia kudriavzevii* (*white symbols*) and *Saccharomyces cerevisiae* (*black symbols*), respectively. **b** Acid structure of must (*grey bars*) and wines obtained by guided *P. kudriavzevii* (*white bars*) and *S. cerevisiae* (*black bars*) microfermentations

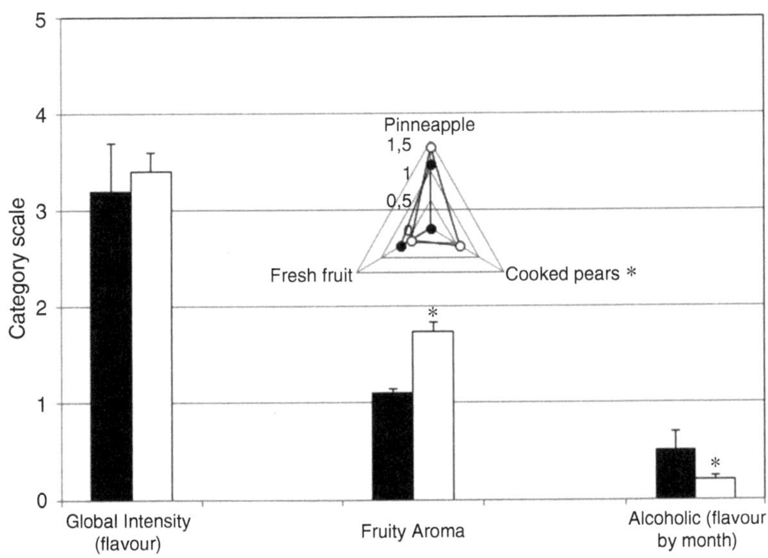

Fig. 16.6 Sensorial quality of wines obtained from synthetic must microvinifications guided by *P. kudriavzevii* (*white bars* and *circles*) or *S. cerevisiae* (*black bars* and *circles*). Category scale of five points (0 = none, 5 = extreme) anchored at different points with the corresponding references. ANOVA and HD Tukey test, $n = 12$. *Asterisks* are statistic differences ($p < 0.05$)

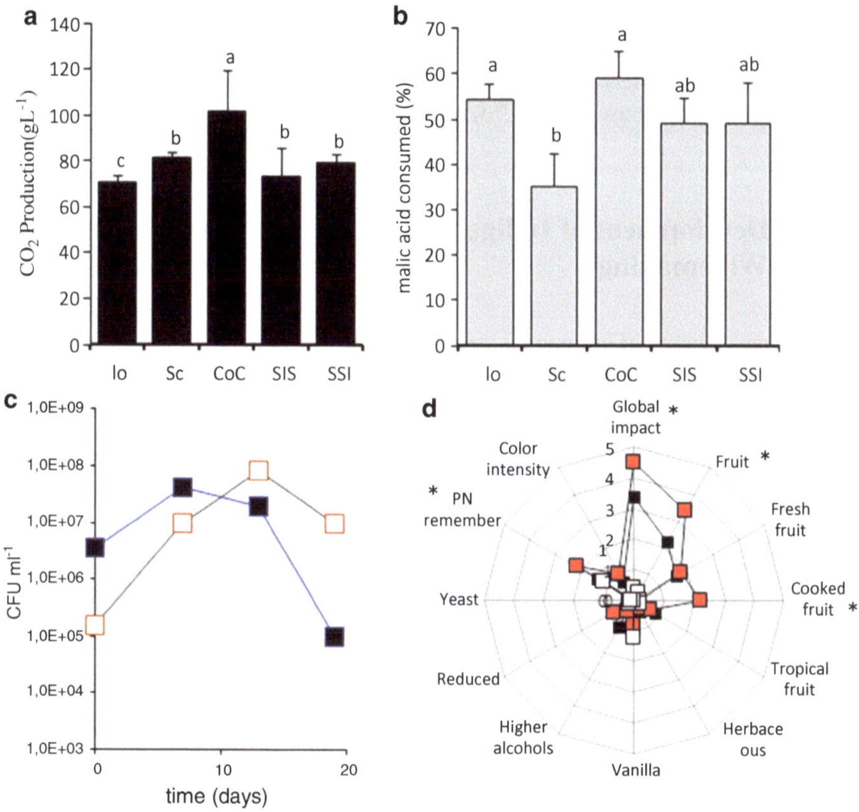

Fig. 16.7 Criteria for mixed starter selection. **a** Fermentative capacity. **b** L-malic acid consumption of *P. kudriavzavii* I15 (*Io*) and *S. cerevisiae* F8 (*Sc*) pure and mixed starters using different inoculation strategies: *SIS*, sequential Io followed by Sc (5th day); *SSI*, sequential Sc followed by Io (7th day); *CoC*, simultaneous relationship Io/Sc $10^2/1$ UFC/ml. Different letters in bars indicate significant differences (ANOVA and HD Tukey test, $p<0.01$; $n=2$). **c** Microbial growth in coculture *P. kudriavzevii* I15 (*black symbols*) and *S. cerevisiae* F8 (*white symbols*) mixed starter. **d** Colour and aroma of wine: mixed CoC starter (*red symbols*), *P. kudriavzevii* I15 pure culture (*white symbols*), and *S. cerevisiae* F8 pure culture (*black symbols*). Vinifications were carried out at laboratory scale using Pinot noir must as substrate. *Asterisks* indicate significant differences (ANOVA and HD Tukey test, $p<0.05$; $n=2$)

S. cerevisiae F8 strains (Fig. 16.7) allowed selecting the CoC protocol as the most adequate combination for mixed culture to be assayed at pilot scale. Selection criterion were (1) better must sugar fermentative efficiency (Fig. 16.7a), a property of fundamental importance for dry wine production (TRS <2 g/l); (2) better malic acid consumption (Fig. 16.7b), facilitating deacidification and microbiological stabilization for regional red wines, where malic acid is the main constituent for titratable acidity (Caballero et al. 2005); (3) appropriate implantation capacity for the non-*Saccharomyces* strain (Fig. 16.7c), being able to remain viable until the end of the fermentative process; and (4) production of physicochemically acceptable wines (data not shown), with a sensorial improvement in color and aroma (Fig. 16.7d). In particular, CoC rendered wines with the most intense color and aroma indices, significantly enhancing varietal primary aroma and improving fruity and vanilla tasting notes (Fig. 16.7d) (del Mónaco et al. 2009).

16.6 Development of Indigenous Starters for Patagonian Red Winemaking

Finally, the capacity of indigenous strains to take over the fermentation and improve the quality of wine were evaluated at pilot scale (200 l) in single form for 2012, 2013 (Malbec, Pinot noir and Merlot varieties), and 2014 (Pinot noir and Malbec varieties) vintages (validation assays) (del Mónaco et al. 2014b). *P. kudriavzevii/S. cerevisie* F8 mixed starter was only assayed for the 2014 vintage using Pinot noir and Malbec as substrates (Bravo et al. 2015).

All the winemaking trials showed kinetics of normal fermentations (data not shown), and they were mostly completed to dryness (TRS $\leq$2 g/l) (Table 16.2), but the sugar consumption rates during dryness stages were higher in F8-guided fermentations than in F15 fermentations (data not shown). As a consequence, fermentative processes guided by the indigenous starter were faster than those guided by the commercial starter. Additionally, and to evaluate the capacity of the indigenous starter to dominate the fermentations, the dynamics of the *S. cerevisiae* populations were determined by means of mtDNA-RFLP analysis. The results obtained from these studies (partially shown in Fig. 16.8) evidenced that indigenous F8 and commercial F15 *S. cerevisiae* were the strains mostly found initially (data not shown), and final stages of their respective fermentations proved their very good and similar implantation capabilities (del Mónaco et al. 2014b). Similar results were obtained in winemaking guided by the mixed starter. In this case, the non-*Saccharomyces* yeast was able to remain viable until the final stages of the process (Bravo et al. 2015).

The physicochemical parameters of the wines showed that all correspond to the standard and dry wine category (TRS $\approx$ 2 g l^{-1}) (Table 16.2). However, the indigenous *S. cerevisiae* F8 starter had a greater ability to improve the color of wine than the commercial one, showing a close relationship between the color index and the total polyphenol content. Similar results have been reported by Holt et al. 2013 in Shiraz wine. The influence of non-*Saccharomyces* yeast on these parameters is variety dependent (Table 16.2); however, its ability to intensify the color of the Pinot noir wine is relevant because it is a variety with known problems regarding this attribute.

Sensorial analysis carried out by experts and consumers using qualitative and quantitative tests, respectively, showed significant differences between F8 and F15 wines. Qualitative analysis was performed by a panel of experts using descriptive tests. As a whole, the global quality scores obtained in this analysis by F8 Pinot noir (68 = good) and Merlot (6.6 = pleasant) wines were higher than those obtained by

Table 16.2 Physicochemical and sensory evaluations of wines obtained from fermentations of Patagonian red musts conducted by mixed culture *S. cerevisiae* F8 and *Pichia kudriavzevii* I15 (CoC) and pure cultures of indigenous *S. cerevisiae* F8 (Sc F8) and commercial F15 (Sc F15) yeast strains

Evaluation	Malbec			Pinot noir			Merlot		
Physicochemical	Sc F15	Sc F8	CoC	Sc F15	Sc F8	CoC	ScF15	ScF8	*p*
TRS (g/l)	2.00±0.38	1.80±0.51	1.85±0.78	2.08±0.04	1.93±0.26	2.18±0.25	1.25±0.50	1.30±1.10	ns
Ethanol (% v/v)	15.10±0.10	15.00±0.80	14.80±0.10	14.80±0.40	14.80±0.30	14.60±0.40	15.2±0.0	15.1±0.0	ns
pH	3.93±0.04	3.90±0.02	3.87±0.18	3.88±0.17	3.89±0.09	3.84±0.08	3.78±0.01	3.78±0.03	ns
TA*	4.77±1.07	4.61±0.14	4.86±0.21	4.93±0.66	4.61±0.66	4.61±0.03	5.70±0.00	5.75±0.35	ns
VA#	0.75±0.03	0.77±0.05	0.73±0.00	0.69±0.13	0.67±0.11	0.69±0.18	0.45±0.10	0.40±0.00	ns
L-malic acid (g/l)	0.26±0.23[b]	0.23±0.11[b]	0.50±0.06[a]	1.94±0.14[b]	1.40±0.01 [b]	1.24±0.16 [a]	1.12±0.10	1.05±0.40	0.01
Colour index	1243±32.53[b]	1917±311[a]	1192±19.09[b]	644±31[c]	744±37[b]	891±71[a]	nd	nd	0.01
IPT§	43.35±1.01[a]	44.10±0.64[a]	36.80±2.83[b]	38.1±0.28[a]	39.60±0.28[b]	40.30±0.12[b]	nd	nd	0.05
Sensory preference test									
2014	48 (*n*=122)	74(*n*=122)	nd	47(*n*=119)	72(*n*=119)	–	4(*n*=17)	13 (*n*=17)	<0.05
2015	4 (*n*=27)[b]	8 (*n*=27)[b]	15 (*n*=27)[a]	2 (15)[b]	8 (15)[a]	6 (15)[a]	nd	nd	<0.05

TRS total reducing sugars; *, titratable acidity, expressed as tartaric acid (g/L); #, volatile acidity, expressed as acetic acid (g/l); §, polyphenol index, expressed as galic acid (g/l). Different letters as superscripts indicate significant differences between wines of the same variety. ANOVA and HD Tukey's test $n=2$, for physicochemical analysis, and Bonferroni's for preference test. *ns* not significant differences within each lot

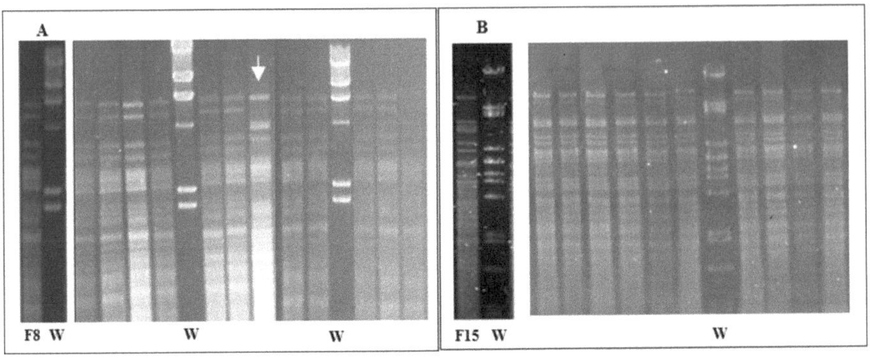

Fig. 16.8 mtDNA-RFLP patterns of indigenous F8 (**a**) and commercial F15 (**b**) starters and *S. cerevisiae* isolates obtained from F8 (**c**) and F15 (**d**) guided Pinot noir fermentations (2013 Vintage) at final stages. *Arrows* indicate isolates with mtDNA-RFLP patterns different from the inoculated starters. *W* molecular weight marker

F15 Pinot noir (52 = correct) and Merlot (5.7 = slightly pleasant) wines. Particular descriptions indicated that Pinot noir F8 wines had good color intensities, showing a red color typical for the variety but more intense than F15 wine, a result in agreement with physicochemistry parameters showed in Table 16.2, with aromas of red fruit (cherries) with notes of sherry. In mouth they were described as middling fruity, slightly rusty, sweet, and alcoholic. Meanwhile, Pinot noir F15 wines showed a limpid and bright aspect and an intense reduced aroma that did not disappear with agitation. In mouth, they were perceived as slightly fruity and bitter, astringent, and tannic. On the other hand, both F8 and F15 Merlot wines showed a limpid and bright aspect and an intense brick-red color but the F8 aroma was more intense than the F15 aroma, both aromas being of medium quality. Pepper, red fruits, butter, leather, spice, and vanilla were the aromatic descriptors highlighted in the former, and green pepper, cooked red fruits, spices, and pepper were described for the latter. In the mouth, both wines showed good acidity and body, and they were persistent. However, F8 wines were described as round and equilibrated whereas F15 wines showed a tart taste (del Mónaco et al. 2014b).

Finally, F8 wines were the favorite for the consumers ($p < 0.05$) in all preference tests (Table 16.2). A positive and significant effect of mixed starter was observed in the preference test for Malbec wine, the "brand image" of Argentine red wines.

16.7 Conclusions

The extension of the selection of yeast for oenological use among *Saccharomyces* and non-*Saccharomyces* species led to the finding of yeast strains with novel and interesting oenological characteristics that could have significant implications in the production of Patagonian young wines of improved quality. Results presented here show that *S. cerevisiae* F8 strain drives red vinifications, improving the quality of the local fermented products. On the other hand, the use of *P. kudriavzevii* ÑNI15 as wine starter would eliminate the cultural and cellar operations undertaken to adjust the musts acidity, improving wine quality and reducing production costs, mainly in Pinot noir wines, the Patagonian emblematic *Vitis vinifera* L. variety. The coinoculation of *S. cerevisiae* F8 and *P. kudriavzevii* ÑNI15 in local musts implies the oenological potential of using these strains to formulate a regional starter culture for the production of well-balanced and physicochemically stable Patagonian young red wines. The use at the industrial level of pure and mixed starter cultures constituted by indigenous yeast strains from Patagonia might become a valuable tool for the differentiation, diversification, and protection (quality certification) of regional wines.

Acknowledgments The authors gratefully acknowledge support from ANPCyT-MINCyT (PICT 2012-2804) and Universidad Nacional del Comahue (Research project PROIN L003, PIN 3 and 4)

References

Alves Z, Melo A, Figueiredo AR, Coimbra MA, Gomes AC, Rocha SM (2015) Exploring the *Saccharomyces cerevisiae* volatile metabolome: indigenous versus commercial strains. PLoS One 10(11), e0143641. doi:10.1371/journal.pone.0143641

Baleiras Couto M, Eijsma B, Hofstra H, Huis J, Vossen J (1996) Evaluation of molecular typing techniques to assign genetic diversity among *Saccharomyces cerevisiae* strains. Appl Environ Microbiol 62:41–46

Barata A, Malfeito-Ferreira M, Loureiro V (2012) The microbial ecology of wine grape berries. Int J Food Microbiol 153:243–259

Beelman RB, Gallander JF (1979) Wine deacidification. Adv Food Res 25:1–53

Benito S, Morata A, Palomero F, González MC, Suárez Lepe JA (2011) Formation of vinylphenolic pyranoanthocyanins by *Saccharomyces cerevisiae* and *Pichia guillermondii* in red wines produced following different fermentation strategies. Food Chem 124:15–23

Bokulich NA, Ohta M, Richardson PM, Mills DA (2013) Monitoring seasonal changes in winery-resident microbiota. PLoS ONE 8:e66437

Bokulich NA, Thorngate JH, Richardson PM, Mills DA (2014) Microbial biogeography of wine grapes is conditioned by cultivar, vintage, and climate. Proc Natl Acad Sci USA 111(1):E139–E148

Boundy-Mills K (2006) Methods for investigating yeast biodiversity. In: Péter G, Rosa C (eds) Biodiversity and ecophysiology of yeasts. Springer, Berlin, pp 67–100

Bravo SME, del Mónaco S; Curilén Y, Simes A, Jara J, Lucero G, Carreño V, Camacho E, Aeberhard C, Bibiloni BH, Caballero AC (2015) Biotecnologia local para la elaboracion de vinos patagónicos de calidad superior. In: AATA (eds) Actas del XV CYTAL. Buenos Aires, Argentina, pp 457, 1–6

Buzzini P, Martini A (2000) Utilization of differential killer toxin sensitivity patterns for fingerprinting and clustering yeast strains belonging to different genera. Syst Appl Microbiol 23:450–457

Caballero A, Sangorrín M, Zajonskovsky I, Lopes CA, Lavalle TL, Rodríguez ME, Giraudo MR (2004) Diversidad de levaduras nativas de la Región Vitivinícola Sur. Vino Ind 19:56–68

Caballero A, Crisóstomo B, Barbagelata R (2005) Caracterizacion fisicoquimica de mostos tintos de calidad enologica de la norpatagonia argentina". In: Crivellaro Guerra C, de Souza SS (eds) Actas X Congreso Latinoamericano de Vitivinicultura y Enología. Bento Concalves, Brasil

Caballero A, Sangorrín M, Lopes C, Rodríguez ME, Zajonskovsky I, Barbagelata R, Lavalle TL (2008). In: Caballero A (ed) Informe final PFIP 028/05. MINCYT-UNComahue Evaluación de la aptitud enológica y selección de cepas de *Saccharomyces cerevisiae* indígenas de la UNComahue. Neuquén, pp 30–42

Carrascosa AV, Bartolome B, Robredo S, Leon A, Cebollero E, Juega M, Nunez YP, Martinez MC, Martinez-Rodriguez AJ (2012) Influence of locally-selected yeast on the chemical and sensorial properties of Albariño white wines. LWT Food Sci Technol 46:319–325

Catania C, Avagnina S (2010) El Terruño Argentino. In: Caviar B (ed) Argentina. Andina Sur, Argentina, pp 113–121

Capozzi V, Garofalo C, Chiriatti M, Grieco F, Spano G (2015) Microbial terroir and food innovation: the case of yeast biodiversity in wine. Microbiol Res 181:75–83

Cavalieri D, McGovern PE, Hartl DL, Mortimer R, Polsinelli M (2003) Evidence for *S. cerevisiae* fermentation in ancient wine. J Mol Evol 57(S1):S226–S232

Chambers J, Pretorious IS (2010) Fermenting knowledge: the history of winemaking, science and yeast research. EMBO Rep 11:914–920

Ciani M, Maccarelli F (1998) Oenological properties of non-*Saccharomyces* yeasts associated with wine-making. World J Microbiol Biotechnol 14:199–203

Ciani M, Fatichenti F, Mannazzu I (2002) Yeasts in winemaking biotechnology. In: Ciani M (ed) Biodiversity and biotechnology of wine yeasts. Research Signpost, Kerala, India, pp 112–123

Ciani M, Comitini F, Mannazzu I, Domizio P (2010) Controlled mixed culture fermentation: a new perspective on the use of non-Saccharomyces yeasts in winemaking. FEMS Yeast Res 10:123–133

Clemente-Jimenez JM, Mingorance-Cazorla L, Martínez-Rodríguez S, Las Heras-Vázquez FJ, Rodríguez-Vico F (2005) Influence of sequential yeast mixtures on wine fermentation. Int J Food Microbiol 98:301–308

Comitini F, Gobbi M, Domizio P, Romani C, Lencioni L, Mannazzu I, Ciani M (2011) Selected non-*Saccharomyces* wine yeasts in controlled multistarter fermentations with *Saccharomyces cerevisiae*. Food Microbiol 28:873–882

Contreras A, Hidalgo C, Schmidt S, Henschke PA, Curtin C, Varela C (2015) The application of non-*Saccharomyces* yeast in fermentations with limited aeration as a strategy for the production of wine with reduced alcohol content. Int J Food Microbiol 205:7–15

Costantini A, García-Moruno E, Moreno-Arribas MV (2009) Biochemical transformations produced by malolactic fermentation. In: Moreno-Arribas MV, Polo MC (eds) Wine chemistry and biochemistry. Springer, Madrid, pp 27–57

Curilén YL, Barda NB, Barbagelata RJ, Bravo Ferrada B, Gallina M, Semorile LC, Caballero AC (2009) Commercial yeast inoculation effect on kinetic and microbiology of fermentative process and its relation to the sensorial quality of wine. A preliminary study in Patagonian Pinot noir winemaking. In: AATA (ed) Actas XII Congreso CYTAL, Alimentos Fermentados y Bebidas 3.8. 6 pp

Curilén YL, Gross ELA, Semorile LC, Caballero A (2015) Selección de un para levadura-bacteria láctica para el formulado de un cultivo iniciador mixto destinado a Enología. In: AATA (ed) Actas XV Congreso CYTAL, Ciudad Autónoma de Buenos Aires, Argentina 3–5 Nov 2015, ID 455

Cole VC, Noble AC (1997) Flavor chemistry and assessment. In: Lea AGH, Piggott JR (eds) Fermented beverage production. Blackie, London, pp 361–385

del Mónaco, Silvana SM, Barda NB, Caballero AC (2009) Indigenous *Issatchenkia orientalis* enhance varietal character in Pinot noir Patagonian wine. In: Proceedings 29th ISSY. Relevance of non-*Saccharomyces* yeast. Guadalajara, Méjico

del Mónaco SM, Zajonskovsky IE, Caballero AC (2010) Aislados patagónicos de *Issatchenkia orientalis* de potencial aplicación en enología. La Alimentación Latinoamericana 286:54–61 del Mónaco S, Barda N, Rubio N, Caballero A (2014a) Selection and characterization of a Patagonian *Pichia kudriavzevii* for wine deacidification. J Appl Microbiol 117(2):451–464

del Mónaco SM, Bravo SME, Curilén YL, Carreño VA, Caballero AC (2014b) A regional starter for high-quality wines: an Argentinean Patagonia experience. Bull OIV 87:217–222

del Mónaco S, Rodríguez ME, Lopes C (2016) *Pichia kudriavzevii* as a representative yeast of North Patagonian winemaking terroir. Int J Food Microbiol 230:31–39

Di Maio S, Genna G, Gandolfo V, Amore G, Ciaccio M, Oliva D (2012) Presence of *Candida zemplinina* in Sicilian musts and selection of a strain for wine mixed fermentations. S Afr J Enol Vitic 33:80–87

du Toit M, Pretorius IS (2000) Microbial spoilage and preservation of wine: using weapons from nature's own arsenal–a review. S Afr J Enol Vitic 21:74–96

Esteve-Zarzoso B, Belloch C, Uruburu F, Querol A (1999) Identification of yeasts by RFLP analysis of the 5.8S rRNA gene and two ribosomal internal transcribed spacers. Int J Syst Bacteriol 49:329–337

Flanzy C (ed) (2000) Enología: fundamentos científicos y tecnológicos. Mundi Prensa Libros, Madrid, España

Fleet GH (2003) Yeast interactions and wine flavour. Int J Food Microbiol 86:11–22

Fleet GH (2007) Yeasts in foods and beverages: impact on product quality and safety. Curr Opin Biotechnol 18:1–6

Fleet GH (2008) Wine yeasts for the future. FEMS Yeast Res 8(7):979–995

Fleet GH, Heard GM (1993) Yeasts: growth during fermentation. In: Fleet GH (ed) Wine microbiology and biotechnology. Harwood, Australia, pp 27–54

Ganter PF, Starmer WT (1992) Killer factor as a mechanism of interference competition in yeasts associated with cacti. Ecology 73:54–67

Gao C, Fleet GH (1995) Degradation of malic and tartaric acids by high density cell suspensions of wine yeasts. Food Microbiol 12:65–71

Gawel R, Francis L, Waters EJ (2007) Statistical correlations between the in-mouth textural characteristics and the chemical composition of Shiraz wines. J Agric Food Chem 55:2683–2687

Gilbert JA, van der Lelie D, Zarraonaindia I (2014) Microbial terroir for winegrapes. Proc Natl Acad Sci U S A 111(1):5–6

Henick-Kling T (1993) Malolactic fermentation. In: Fleet GH (ed) Wine microbiology and biotechnology. Harwood, Chur, pp 289–326

Holt H, Cozzolino D, McCarthy J, Abrahamse C, Holt S, Solomon M, Smith P, Chambers PJ, Curtin C (2013) Influence of yeast strain in Shiraz wine quality indicators. Int J Food Microbiol 165:302–311

Jolly NP, Augustyn OPH, Pretorius IS (2006) The role and use of non-*Saccharomyces* yeasts in wine production. S Afr J Enol Vitic 27:15–39

Jolly NP, Varela C, Pretorius IS (2014) Not your ordinary yeast: non-*Saccharomyces* yeasts in wine production uncovered. FEMS Yeast Res 14(2):215–237

Kurtzman CP, Robnett CJ (1998) Identification and phylogeny of ascomycetous yeasts from analysis of nuclear large subunit (26S) ribosomal DNA partial sequences. Antonie Van Leeuwenhoek 73:331–371

Kurtzman CP, Fell JW (eds) (1998) The yeasts: a taxonomic study. Elsevier, Amsterdam

Kurtzman CP, Fell JW, Boekhout T (eds) (2011) The yeasts: a taxonomic study, 5th edn. Elsevier, Amsterdam

Lachance MA (2003) The Phaff school of yeast ecology. Int Microbiol 6:63–167

Le Guillou B (2000) Río Negro: de la vigne dans des oasis de la Patagonie argentine. Vigne 116:112–113

Lopes CA, Lavalle TL, Querol A, Caballero A (2006) Combined use of killer biotype and mtDNA-RFLP patterns in a Patagonian wine *Saccharomyces cerevisiae* diversity study. Antonie Van Leeuwenhoek 89(1):147–156

Lopes CA, Rodríguez ME, Sangorrín M, Querol A, Caballero A (2007a) Patagonian wines: implantation of an indigenous strain of *Saccharomyces cerevisiae* in fermentations conducted in traditional and modern cellars. J Ind Microbiol Biotechnol 34:139–149
Lopes CA, Rodríguez ME, Sangorrín M, Querol A, Caballero AC (2007b) Patagonian wines: the selection of an indigenous yeast starter. J Ind Microbiol Biotechnol 34:539–546
Martini A (1993) Origin and domestication of the wine yeast *Saccharomyces cerevisiae.* J Wine Res 4(3):165–176
Martini A (2003) Biotechnology of natural and winery-associated strains of *Saccharomyces cerevisiae*. Int Microbiol 6:207–209
McGovern PE, Hartung U, Badler VR, Glusker DL, Exner LJ (1997) The beginnings of winemaking and viniculture in the ancient Near East and Egypt. Expedition 39(1):3–21
Mercado L, Dalcero A, Masuelli R, Combina M (2007) Diversity of Saccharomyces strains on grapes and winery surfaces: analysis of their contribution to fermentative flora of Malbec wine from Mendoza (Argentina) during two consecutive years. Food Microbiol 24:403–412
Mercurio MD, Dambergs RG, Cozzolino D, Herderich MJ, Smith PA (2010) Relationship between red wine grades and phenolics. 1. Tannin and total phenolics concentrations. J Agric Food Chem 58(23):12313–12319
Molina AM, Guadalupe V, Varelab C et al (2009) Differential synthesis of fermentative aroma compounds of two related commercial wine yeast strains. Food Chem 117:189–195
Morata A, Benito S, Loira I, Palomero F, Gonzalez MC, Suarez-Lepe JA (2012) Formation of pyranoanthocyanins by *Schizosaccharomyces pombe* during the fermentation of red must. Int J Food Microbiol 159:47–53
Muñoz-González C, Rodriguez-Bencomo JJ, Moreno-Arribas MV, Pozo-Bayón MA (2011) Beyond the characterization of wine aroma. Anal Bioanal Chem 40:1497–1512
OIV(2016) Coyuntura vitivinícola mundial 2015: evoluciones y tendencias. Informe de la OIV
Pérez M, Gallego F, Hidalgo P (2001a) Evaluation of molecular techniques for the genetic characterization of *Saccharomyces cerevisiae* strains. FEMS Microbiol Lett 205:375–378
Pérez F, Ramírez M, Regodón JA (2001b) Influence of killer strains of *Saccharomyces cerevisiae* on wine fermentation. Antonie Van Leeuwenhoek 79:393–399
Pinto C, Pinho D, Cardoso R et al (2015) Wine fermentation microbiome: a landscape from different Portuguese wine appellations. Front Microbiol 6:1–13
Pretorius IS (2000) Tailoring wine yeast for the new millennium: novel approaches to the ancient art of winemaking. Yeast 16:675–729
Pretorius IS, Hoj PB (2005) Grape and wine biotechnology: challenges, potential opportunities and potential benefits. Aust J Grape Wine Res 11:83–108
Querol A, Ramón D (1996) The application of molecular techniques in wine microbiology. Trends Food Sci Technol 57:183–185
Radler F (1993) Yeasts-metabolism of organic acids. In: Fleet GH (ed) Wine microbiology and biotechnology. Harwood Academic, Chur, Switzerland, pp 165–182
Rainieri S, Pretorius IS (2000) Selection and improvement of wine yeasts. Ann Microbiol 50:15–30
Renouf V, Claisse O, Lonvaud-Funel A (2005) Understanding the microbial ecosystem on the grape berry surface through numeration and identification of yeast and bacteria. Aust J Grape Wine Res 11:316–327
Ribéreau-Gayon P, Dubourdieu D, Doneche B, Lonvaud A (eds) (2003) Química del vino. Estabilización y tratamientos. Editorial Hemisferio Sur y Mundi Prensa, Buenos Aires, Argentina, pp 3–49
Rodríguez ME, Lopes CA, van Broock M et al (2004) Screening and typing of Patagonian wine yeasts for glycosidase activities. J Appl Microbiol 96:84–95
Rodríguez ME, Lopes CA, Barbagelata RJ, Barda NB, Caballero AC (2010) Influence of Candida pulcherrima Patagonian strain on alcoholic fermentation behaviour and wine aroma. Int J Food Microbiol 138:19–25

Romano P, Fiore C, Paraggio M et al (2003) Function of yeast species and strains in wine flavour. Int J Food Microbiol 86:169–180

Romano P, Capece A, Serafino V et al (2008) Biodiversity of wild strains of *Saccharomyces cerevisiae* as tool to complement and optimize wine quality. World J Microbiol Biotechnol 24:1797–1802

Rosini G (1984) Assessment of dominance of added yeast with fermentation and origin of *Saccharomyces cerevisiae* in wine-making. J Gen Appl Microbiol 30:249–256

Ruffner HP (1982) Metabolism of tartaric and malic acids. Vitis 21:247–259

Sangorrín MP, Zajonskovsky IE, Lopes CA, Rodríguez ME, Giraudo de van Broock MR, Caballero AC (2001) Killer behaviour in wild wine yeasts associated with Merlot and Malbec type musts spontaneously fermented from Northwestern Patagonia (Argentina). J Basic Microbiol 41:105–113

Sangorrín M, Zajonskovsky I, van Broock M, Caballero A (2002) The use of killer biotyping in an old Patagonian winery yeast ecological survey. World J Microbiol Biotechnol 18:115–120

Sangorrín MP, Lopes CA, Giraudo MR, Caballero AC (2007) Diversity and killer behaviour of indigenous yeasts isolated from the fermentation vat surfaces in four Patagonian wineries. Int J Food Microbiol 119:351–357

Santamaría P, Garijo P, López R, Tenorio C, Gutiérrez AR (2005) Analysis of yeast population during spontaneous fermentation. Effect of the age of the cellar and the practice of inoculation. Int J Food Microbiol 103:49–56

Schuller D, Valero E, Dequin S, Casal M (2004) Survey of molecular methods for the typing of wine yeast strains. FEMS Microbiol Lett 231:19–26

Schuller D, Alves H, Dequin S, Casal M (2005) Ecological survey of *Saccharomyces cerevisiae* strains from vineyards in the Vinho Verde Region of Portugal. FEMS Microbiol Ecol 51:167–177

Soden A, Francis IL, Oakey H, Henschke PA (2000) Effects of co-fermentation with Candida stellata and *Saccharomyces cerevisiae* on the aroma and composition of Chardonnay wine. Aust J Grape Wine Res 6:21–30

Suárez-Lepe JA, Morata A (2012) New trends in yeast selection for winemaking. Trends Food Sci Technol 23:39–50

Sun SY, Gong HS, Jiang XM, Zhao YP (2014) Selected *non-Saccharomyces* wine yeasts in controlled multistarter fermentations with *Saccharomyces cerevisiae* on alcoholic fermentation behaviour and wine aroma of cherry wines. Food Microbiol 44:15–23

Swiegers H, Bartowsky E, Henschke P, Pretorious IS (2005) Microbial modulation of wine aroma and flavour. Aust J Grape Wine Res 11:139–173

Terrell E, Cliff M, Van Vuuren HJJ (2015) Functional characterization of individual and mixed-Burgundian *Saccharomyces cerevisiae* isolates for fermentation of Pinot noir. Molecules 20:5112–5136

Tofalo R, Perpetuini G, Schirone M, Fasoli G, Aguzzi I, Corsetti A, Suzzi G (2013) Biogeographical characharacterization of *Saccharomyces cerevisiae* wine yeast bimolecular methods. Food Microbiol 4:1–13

Tufariello M, Chiriatti M, Grieco F, Perrotta C, Capone S, Rampino P (2014) Influence of autochthonous *Saccharomyces cerevisiae* strains on volatile profile of Negroamaro wines. LWT Food Sci Technol 58:35–48

van Broock M, Zajonskovsky I, Assadourian M, Lavalle L, Caballero de Castro A (1996) Wine yeasts associated to merlot type grapes from North Patagonian Region. An ecological study. Proc 10th Int. Biotechnol. Symposium and 9th ISSY Sydney, Australia, pp 13–15

van Leeuwen C, Seguin G (2006) The concept of terroir in viticulture. J Wine Res 17:1–10

Varela C, Siebert T, Cozzolino D, Rose L, Mclean H, Henschke PA (2009) Discovering a chemical basis for differentiating wines made by fermentation with 'wild' indigenous and inoculated yeasts: role of yeast volatile compounds. Aust J Grape Wine Res 15:238–248

Vaughan-Martini A, Martini A (1995) Facts, myths and legends on the prime industrial microorganism. J Ind Microbiol Biotechnol 14:514–522

Vaughan-Martini A, Cardinali G, Martini A (1996) Differential killer sensitivity as a tool for fingerprinting wine-yeast strains of *S. cerevisiae*. J Ind Microbiol 17:124–127
Vilanova M, Massneuf-Pomarede I (2005) Characterization of yeast strains from Rias Baixas (NW Spain) and their contribution to the fermentation of Albarino wine. Ann Microbiol 55:23–26
Vilanova M, Sieiro C (2006) Contribution by Saccharomyces cerevisiae yeast to fermentative flavor compounds in wines from cv. Albariño. J Ind Microbiol Biotechnol 33:929–933
Volschenk H, Viljoen-Bloom M, Subden RE, Subden RE, Van Vuuren HJJ (2001) Malo-ethanolic fermentation in grape must by recombinant strains of *Saccharomyces cerevisiae*. Yeast 18:963–970
Weizman D (2009) El mapa argentino de los sentidos. Rumbos 326:18–24

Chapter 17
Patagonian Antagonist Yeasts for Food Biopreservation

María L. Villalba, María C. Lutz, Sofía Lopez, María B. Pildain, and Marcela P. Sangorrín

Abstract Worldwide, microbial growth destroys large amounts of various products, causing yield losses in the agronomical and biotechnological industries. Traditionally, biocides have been used to manage these problems, but different disadvantages such as the establishment of resistant strains and the suppression of natural competitors have made alternatives such as biological control necessary. After harvest, many fruits are kept in cold storage to prolong their availability and shelf life. Often, this requires the application of a chemical fungicide to prevent postharvest decay from decay fungi. An alternative approach for preventing postharvest fungal decay during storage could be based on the treatment of the commodity with antagonistic yeasts. In this regard, the use of cold-adapted yeasts may offer a distinct advantage. Numerous cold-adapted yeasts species have been isolated from artificial cold environments, as well as cold-stored fruits. Recently, we isolated and identified epiphytic yeasts during the cold postharvest storage of pears and fine fruits from packinghouses in Argentinean Patagonia, and we tested their efficacy in controlling the postharvest diseases of different fruits caused by several pathogens. Additionally, killer yeasts as producers of mycocins or killer toxins that can neutralize the activities of spoilage yeasts in wines represent an interesting biocontrol strategy. Several screening studies focused to determine the occurrence of killer yeasts in winemaking environments have been carried out, and they have demonstrated the presence of killer phenotypes in yeasts from wines, cellar surfaces, and winery equipment. In previous studies carried out in our laboratory, most yeasts isolated from spontaneously fermenting grape musts evidenced killer

M.L. Villalba • M.P. Sangorrín (✉)
Instituto de Investigación y Desarrollo en Ingeniería de Procesos, Biotecnología y Energías Alternativas (PROBIEN, Consejo Nacional de Investigaciones Científicas y Técnicas de la República Argentina-Universidad Nacional del Comahue), Buenos Aires 1400, (8300), Neuquén, Argentina
e-mail: sangorrinmarcela@conicet.gov.ar

M.C. Lutz
Facultad de Ciencias Agrarias, Universidad Nacional del Comahue, Buenos Aires 1400, (8300), Neuquén, Argentina

S. Lopez • M.B. Pildain
Centro de Investigación y Extensión Forestal Andino Patagónico (CIEFAP), Chubut, Argentina

N.L. Olivera et al. (eds.), *Biology and Biotechnology of Patagonian Microorganisms*, DOI 10.1007/978-3-319-42801-7_17

character. These studies provide an exceptional source of potential antagonist yeasts to be used in biocontrol of undesired microorganisms in wine. Nonetheless, it is necessary to continue to identify new potential microorganisms and to develop a better understanding of the biology of yeast biocontrol systems to increase the potential of postharvest biocontrol as a viable alternative to synthetic postharvest fungicides and chemical preservatives against wine spoilage yeast.

17.1 Introduction

The discovery of antagonistic activities of yeasts has had a significant impact on numerous fields such as food, agriculture, medicine, and environmental protection. The use of antagonistic bacteria to inhibit pathogenic bacteria has been studied extensively over the years, but little attention has been given to yeasts with a similar role. The study and potential applications of antibacterial compounds secreted by yeasts are therefore still at an early stage of development. Antagonism of microorganisms by yeasts has been attributed primarily to (a) competition for nutrients, (b) pH changes in the medium as a result of growth-coupled ion exchange or organic acid production, (c) production of high concentrations of ethanol, (d) production of proteases degrading bacterial toxins, (e) stimulation of the immune response, (f) inhibition of attachment to intestinal cells produced by yeasts, and (g) secretion of antibacterial compounds and release of antimicrobial compounds such as killer toxins or mycocins (Hatoum et al. 2012; Lima et al. 2013).

Most fruits are marketed after harvesting without excessive handling, or after storage, and there is an increasing trend for produce with some degree of packaging and of processing in the form of ready-to-eat, peeled, fruits and fresh or fermented juices. Yeasts and mold pathogens are always associated with fruit, and they have a significant role in the spoilage of raw fruit and processed products. Yeasts form a part of the natural microbiota of most fruits and vegetables, although populations and the relative proportion of species vary from commodity to commodity, and they are influenced by environmental, harvesting, and storage conditions (Deak 2009).

This chapter provides an updated summary of some published findings regarding the antagonistic properties of yeasts and our experience with these issues.

17.2 Yeasts for Biological Control of Postharvest Diseases of Fruit

Postharvest losses of stored fruit, vegetables, and grains caused by decay by fungal pathogens can be very significant, and in fact addressing this problem is receiving increasing importance as the demand for food increases worldwide (Wilson 2013; Feliziani and Romanazzi 2013; Massart et al. 2015). Postharvest decay of fruits can be reduced by avoiding injury to the fruit during harvest and subsequent handling, stringent sanitation practices, and the use of chemical fungicides during cold or modified atmosphere storage. These beneficial practices, however, are usually not sufficient to protect harvested fruits from mold spoilage.

The production of apples and pears is a complex process involving orchard, storage, and marketing phases. Modern postharvest technologies make the long-term storage of apples and pears possible. Argentina is the largest pear-producing country and the major exporter in the Southern Hemisphere. The main pear-growing area in North Patagonia is situated in the provinces of Río Negro and Neuquén. Pears may be stored in cold for as long as 9 months in cold or modified atmosphere storage. Postharvest diseases can be a limiting factor for the long-term storage of fruit. Fruit decay and repacking of fruit because of decay cost the fruit industry millions of dollars in losses each year.

Based on the nature of infection, postharvest diseases of pear fruits can be divided into two categories: those that originate from latent fungal infection of fruit in the orchard and those which originate from infection of wounds, such as stem punctures and bruises, by fungi at harvest or during the postharvest handling and packing process. The former category is exemplified by *Phytophthora* rot (Dobra et al. 2011; Sosa et al. 2015), and calyx rot by *Botrytis cinerea* and *Alternaria* spp. In the latter category, the main diseases are exemplified by blue mold caused by *Penicillium expansum* and gray mold caused by *Botrytis cinerea* (Dobra et al. 2008); according to pear cultivar *Alternaria* spp. rot and *Cladosporium* spp. rot, and they can be important diseases. Understanding when infections occur is an essential step for developing and implementing control measures to reduce storage losses from decay.

Berry and cherry fruits are crops that rightly adapt to climate areas with fresh summers and that have a winter latency period with exposure to low temperatures as the main requirement (Sozzi 2007). This requirement is fully satisfied by the mountain range valleys of the South Patagonia: South of Río Negro, Chubut, and Santa Cruz (Scarpati et al. 2011). These fruits have a limited storage period at 0 °C post harvest, of around 45–60 days for cherries and 7 days for raspberries and blackberries because of the increase in their metabolism post harvest, which leads to an organoleptic and appearance change, accompanied with an increase in susceptibility to fungi causing rot (Crisosto et al. 1993). After cold storage, this type of fruit has a maximum of 3–4 days of lifetime for commercialization, usually at room temperature (Romanazzi 2010). Fungi causing berry and cherry fruit to rot worldwide include *Botrytis cinerea*, *Monilia fructicola*, *Penicillium expansum*, *Rhizopus stolonifer*, *Cladosporium* spp., and *Mucor* spp. (Ogawa et al. 1995; Romanazzi 2010). In Argentina, information about postharvest pathogens is meagre. A report by López et al. (2013, 2014) indicates that *Penicillium crustosum*, *Mucor piriformis*, and *B. cinerea* are associated with postharvest disease in cherries, blackberries, and raspberries in South Patagonia. *Penicillium crustosum* and *Mucor piriformis* are considered to be emerging pathogens, and in our assay they were the most abundant at storage temperature, being also less sensitive to fungicides and the most virulent. Both *Mucor* and *Penicillium* contain species considered pathogens of high risk for developing resistance to fungicides (FRAC 2010). Currently, there is no fungicide in Argentina labeled to control postharvest pathogens on cherries and raspberries.

The control of postharvest mold decays has been traditionally carried out with synthetic fungicides; however, during past decades, their effectiveness has decreased because of the appearance of resistant strains (Wisniewski et al. 2007; Schirra et al. 2011). A public demand to reduce the use of synthetic fungicides, stimulated by a growing awareness of environmental and health issues, also limits the postharvest application of these and other chemicals to agricultural products. In past decades, research efforts have focused on developing alternative control methods against postharvest diseases of fresh commodities (Baker and Cook 1974; Wilson et al. 1993; Sharma et al. 2009).

Among the alternative approaches that have been explored, the use of microbial antagonists such as yeasts, fungi, and bacteria has been demonstrated to be quite promising and has gained an increasing amount of attention (Wisniewski et al. 2007; Droby et al. 2009; Sharma et al. 2009; Teixido et al. 2011). The use of biocontrol agents (BCAs) occurring naturally on fruit surfaces has become one of the most preferred ways of controlling postharvest diseases. The majority of microbial organisms studied for their potential use as postharvest biocontrol agents belong to yeasts (Droby et al. 2009; Pimenta et al. 2009; Jijakli 2011; Schisler et al. 2011) because many of these microorganisms are considered as GRAS (generally recognized as safe).

Research on the use of yeasts as BCAs has mainly focused on their use for managing postharvest diseases, mainly of fruit; however, this application represents only a small portion of the complete spectrum of plant disease management (Fravel 2005; Nunes 2012). Only six products based on different yeast species have been registered for postharvest use: Aspire, based on the active ingredient (a.i.) *Candida oleophila* (Ecogen, Langhorne, PA, USA); Yield Plus a.i. *Cryptococcus albidus* (Lallemand, Montreal, Canada), Shemer a.i. *Metschnikowia fructicola* (AgroGreen, Asgdod, Israel); Candifruit a.i. *Candida sake* (Sipcam-Inaagri, Valencia, Spain); Nexy a.i. *C. oleophila* (Lesaffre-Bionext, France), and Boni-Protect a.i. *Aureobasidium pullulans* (Biofa, Münsingen, Germany) (Jijakli 2011; Hatoum et al. 2012).

Isolation and screening are the first crucial steps in the development of a BCA. Many different strategies, directed to select potential BCAs against postharvest pathogens in fruit, have been reported (Wilson et al. 1993; Chand Goyal and Spotts 1996; Lima, et al. 1998). The problem of that strategy used in several studies is that both isolation and bioassays are performed at room temperature (Wilson et al. 1993; Lima et al. 1998; Scherm et al. 2003; Zhang et al. 2010). The approach used in these steps will have a major impact on the species of antagonists isolated and their mode of action; to date, only BCAs from a very narrow range of yeast genera have been isolated (mainly *Aureobasidium*, *Candida*, *Cryptococcus*, *Pichia*, and *Rhodotorula*) because the methods used for isolation and screening in most studies researches are very similar. To overcome this shortcoming, we suggested the use of new screening procedures to increase the range of yeast species examined for their biocontrol efficacy under commercial conditions (Sangorrín et al. 2014).

Because most of the harvested fruit is placed in cold storage for varying amounts of time, one of the most important attributes to be considered in the selection of yeasts as BCAs of postharvest fruit diseases is the ability to be effective at low temperatures. Therefore, inclusion of this characteristic in the first step of the selection process would reduce the number of isolates that need to be tested in time-consuming and labor-intensive biocontrol assays of fruit stored at low temperatures.

Relevant to the use of yeasts as BCAs is the distinctive species composition of epiphytic yeasts isolated from fruit in cold storage, which include species of yeasts that are also members of the yeast communities found in permanently cold natural environments (e.g., glacial habitats, icy seas, cold deserts, frozen ground) (Brandao et al. 2011; Buzzini et al. 2012; Sangorrín et al. 2014). The fact that yeast species common to both artificial and natural cold environments were detected as good potential antagonists was most likely because of the isolation method utilized to obtain the yeasts.

In previous studies, we proposed a BCA selection protocol in which both yeast selection and in situ assays of each yeast isolate against the pathogens were performed at the same cold storage conditions (−1/0 °C), for pear fruit (Robiglio et al. 2011; Lutz et al. 2012) and for berry fruit (López et al. 2014, 2015). In the former study, a total of 51 psychrophilic yeast isolates (Table 17.1) were obtained from the surfaces of healthy pear fruits stored for 7 months at −1/0 °C in two different packinghouses: one with continuous use of postharvest synthesis fungicides (conventional) and another without any use of fungicides for the past 2 years (transition to certified organic management). Six different yeast species were detected in the transition packinghouse and two species in the conventional packinghouse (Table 17.1), suggesting that fungicides may decrease the microbial diversity on fruit surfaces, as has been observed by others (Janisiewicz and Korsten 2002; Telias et al. 2011; Debode et al. 2013). Taking into account that the most important postharvest pear spoilage fungi grow below −1/0 °C (storage temperature), the capacity of each yeast species to grow in the cold was evaluated and used as a pre-selection criterion for in situ biocontrol assays (Table 17.2). A foreign commercial strain of *Candida albidus* was also evaluated with comparative intent. In particular, the pear epiphytic isolates *Aureobasidium pullulans* NPCC 1281 and *Rhodotorula mucilaginosa* NPCC 1278 caused the minimum percentage of disease incidence (33 %) and the highest percentage of decay reduction (88 %) after 60 days of incubation with *Penicillium expansum*. On the other hand, no yeast was able to reduce the disease incidence caused by *Botrytis cinerea* in assayed conditions (Table 17.2). Nevertheless, the isolates *A. pullulans* NPCC 1281 and *R. mucilaginosa* NPCC 1278 showed the highest levels of decay reduction (22 % and 30 %, respectively) against this spoilage fungus. The commercial strain of *Candida albidus* was unable to control the incidence percentage and only achieved 30 % of decay reduction. The results of that study indicated that the microbiota associated with pear fruits exhibited a better biocontrol efficacy against spoilage fungi than a foreign commercial BCA based on *C. albidus* (Robiglio et al. 2011).

In subsequent work in our laboratory (Lutz et al. 2012), we proposed an improved strategy that utilizes the following steps: (1) wounded fruits are stored for 6 months at low temperatures; (2) washings from healthy wounds are obtained from the stored fruits, and these washings, along with pathogen spores, are used to inoculate fresh

Table 17.1 Identity, origin, and number of cold-growth yeast isolates from two strategies from pear

	Yeast isolates from pear surface (Robiglio et al. 2011)		Yeast isolates from selected washing waters (Lutz et al. 2012)			
	Surface		Surface		Wounds	
Species	Conventional	Transition	Transition	Organic	Transition	Organic
Aureobasidium pullulans	10	18	2	5	6	4
Candida patagonica	–	–	–	1	–	–
Cryptococcus weringae	–	–	1	–	3	1
Cryptococcus albidus	–	1	1	5	2	1
Cryptococcus diffluens	–	7	–	–	–	–
Cryptococcus tephrensis	–	–	–	4	–	–
Cryptococcus victoriae	–	–	2	–	7	1
Cystofilobasidium informominiatum	–	–	–	–	1	–
Pichia membranifaciens	–	1	–	–	–	1
Pichia philogaea	–	7	–	–	–	–
Rhodotorula laryngis	–	–	–	–	4	–
Rhodotorula glutinis	–	–	–	1	–	–
Rhodotorula mucilaginosa	1	6	1	1	–	–
Isolates total	11	40	17	7	8	23

C conventional, *T* transition, *O* organic

Table 17.2 Antagonistic efficacy of yeast isolates and a commercial yeast culture against *P. expansum* and *B. cinerea* in pear wounds

Antagonist source	Yeast isolate	NPCC	*P. expansum* DI (%)*	*P. expansum* DR (%)[#]	*B. cinerea* DI (%)*	*B. cinerea* DR (%)[#]
Pear surfaces (Robiglio et al. 2011)	*Aureobasidium pullulans*	1281	33a	88.12a	100b	22.10ab
	Cryptococcus albidus	1245	100b	8.24de	100b	0
	Cryptococcus difluens	1279	100b	66.41ab	100b	0
	Pichia membranifaciens	1280	100b	20.01d	100b	18.30ab
	Pichia philogaea	1252	100b	38.12 cd	100b	11.58bc
	Rhodotorula mucilaginosa	1278	33a	88.09[a]	100b	30.22a
Commercial	*Cryptococcus albidus*	–	100b	30c	100b	11.1a
Biocontrol selected washing waters (Lutz et al. 2012)	*Aureobasidium pullulans*	1273	0	100b	100b	19.34a
		1274	0	100b	100b	23.08a
		1262	0	100b	100b	22.72a
	Cryptococcus albidus	1248	0	100b	80ab	62.08c
	Cryptococcus victoriae	1259	0	100b	80ab	62.08c
		1263	0	100b	60a	55.94c
		1260	0	100b	100b	28.13a
	Cystofilobasidium infirmominiatum	1261	0	100b	100b	40.23b
	Pichia membranifaciens	1250	0	100b	60a	67.13c
	Rhodotorula laryngis	1264	0	100b	100b	22.72a
Control without yeasts			100	–	100b	–

*(DI) is calculated as the number of decay wounds over the total number of wounds[#](DR) represents mean percentage of decay reduction calculated as (mean lesion diameter in control – mean lesion diameter in treatment) × 100/mean lesion diameter in control Values within a column followed by the same letter are not significantly different according to Fisher's test ($p > 0.05$)

wounds in fruits incubated at low temperatures; (3) use of regional isolates of pathogens selected for their aggressiveness and resistance to fungicides during BCA evaluation; (4) washings exhibiting 50 % reduction in disease incidence after 50 days at low temperature are selected for yeast isolation; (5) yeast isolation is carried out on pear juice agar plates at low temperature; (6) subsequent bioassays of individual yeast isolates are performed under cold storage conditions for several months; and (7) yeasts exhibiting more than 80 % reduction in disease incidence are selected (Sangorrín et al. 2014).

To isolate highly effective antagonistic yeasts, a selective method as mentioned was employed (Lutz et al. 2012). The same procedure was carried out in the two packinghouses located in the Upper Valley of Río Negro and Neuquén provinces: the "transition" packinghouse has not used fungicides for the past 2 years; and "organic" was characterized by the use of organic management. A total of 116 washings were recovered from healthy wounds and pear surfaces obtained from both packinghouses. More than 27 % of the total number of washings tested reached the criteria arbitrarily established for the primary screening to be selected as a potential source of antagonists. Most washings with good antagonist activity (66.7 %) were obtained from healthy wounds (Lutz et al. 2012). The most frequently isolated yeast genera in active washings were *Aureobasidium*, *Cryptococcus*, and *Rhodotorula*. These genera have already been reported as effective BCAs against a number of postharvest fruit pathogens under diverse conditions (Chand Goyal and Spotts 1996; Yu et al. 2007; Sugar and Basile 2008; Vero et al. 2009). Lutz et al. (2012) provided the first report about the isolation of *Cryptococcus (Cr.) victoriae*, *Cr. wieringae*, *Cr. tephrensis*, *Rhodotorula (R.) laryngis*, and *Candida (C.) patagonica* from pear fruit surfaces, mainly from wound washing waters (Table 17.1). In accordance with the selective isolation method proposed by Lutz et al. (2012), 81 % of 11 yeast species have also been isolated from a variety of naturally cold environments (Sangorrín et al. 2014). In a previous study (Robiglio et al. 2011), a quite different and minor diversity microbiota was obtained from the pear surface in comparison with the last work (Table 17.1). *Cryptococcus victoriae*, *Cr. wieringae*, *Cr. tephrensis*, *Cystofilobasidium (Cy.) infirmominiatum*, *Rhodotorula laryngis*, and *Pichia membranifaciens* were obtained only from selective methods and mainly from healthy wounds; among isolates of these species the strains *P. membranifaciens* NPCC 1250 and *Cr. victoriae* NPCC 1263 were ultimately selected as best antagonist yeasts, providing excellent control of postharvest pear disease and by the multiple modes of action (Lutz et al. 2013). The fact that the most promising yeast isolates obtained in the previous study showed lower biocontrol activity than most yeast isolates obtained during the former study (Table 17.2) is further evidence for the success of the improved methodology proposed. This difference could result from the selective pressure applied by Lutz et al. (2012) during yeast isolation: healthy wounds with a greater availability of nutrients than the surface might be antagonist yeasts because pathogens in the storage room do not grow.

Additionally, indigenous mold and yeast diversity of stored sweet cherries, raspberries, and blackberries from Patagonia (Argentina) were studied to identify main rot fungi and potential antagonistic biocontrol yeasts. Growth capacity at 0 °C was evaluated for 660 native yeast isolates from berry fruit, of which 53 isolates were able to grow in the cold and were identified within 20 species (Table 17.1). All species were reported as cold-adapted species from naturally cold environments (Turchetti et al. 2008; Buzzini et al. 2012), showing that the isolation and selection method utilized to obtain the yeasts favors the appearance of psychrophilic species. *Aureobasidium pullulans*, *Cryptococcus friedmannii*, *Cryptococcus victoriae*, *Cystofilobasidium capitatum*, *Cystofilobasidium infirmominiatum*, and *Holtermanniella wattica* (i.e., *Cryptococcus wattica)* were isolated from more than one berry fruit (Table 17.3). Some of the genera identified in this study, such as *Cryptococcus*, *Cystofilobasidium*, *Mrakiella*, and *Rhodotorula*, contain species already identified as food-related psychrophilic yeasts (Buzzini and Margesin 2014).

The 12 most promising yeast strains obtained in screening bioassays (López et al. 2014) were further tested in a bioassay (Table 17.4) to determine their biocontrol activity against *Penicillium crustosum* and *Mucor piriformis*, the most virulent postharvest pathogens that have been identified (López et al. 2014). Biocontrol assays were carried out on cherries at 0 °C; the treated wounds were inoculated with conidia suspensions of each of the pathogens and the potential antagonist yeasts.

Table 17.3 Identity, origin and number of cold-growth yeast isolates from cherry and berry fruits

	Number of isolates		
Species	cherries	blackberries	Raspberries
Aureobasidium pullulans	1	1	1
Cryptococcus adeliensis	5	–	–
Cryptococcus magnus	2	–	–
Cryptococcus albidosimilis	–	1	–
Cryptococcus antarticus	–	1	–
Cryptococcus friedmannii	1	2	–
Cryptococcus stepposus	–	1	–
Cryptococcus tephensis	2	–	–
Cryptococcus victoriae	10	–	2
Cryptococcus wieringae	–	2	–
Cystofilobasidium capitatum	4	1	–
Cystofilobasidium infirmominiatum	1	2	–
Cystofilobasidium macerans	3	–	–
Filobasidium capsuligenum	1	–	–
Guehomyces pullulans	4	–	–
Holtermanniella watticus	–	1	–
Meyerozyma guillermondii	–	–	1
Mrakiella cryoconiti	1	–	–
Rhodotorula colostri	–	1	–
Rhodotorula fujisanensis	–	1	–
Total	35	14	4

Aureobasidium pullulans CIEFAP 1141 and *Cryptococcus victoriae* CIEFAP 771 reduced the decay incidence of *P. crustosum* to 53 % and the decay reduction to 61–79 % after 30 days in cherries (Table 17.4); for the pathogen *Mucor piriformis*, the yeast *A. pullulans* CIEFAP 631, *Guehomyces pullulans* CIEFAP 899, and *Cystofilobasidium capitatum* CIEFAP 1204 further reduced their decay incidence to 13–23 % and decay reduction to 70–93 % in cherries (Table 17.4). To evaluate the potential antagonist activity of yeast in situ assays in raspberries and blackberries, the effect against natural pathogen infection was assessed without conidia inoculation. The evaluation of the biocontrol capacity at 0 °C after 5 days indicated that the three strains of *Cryptococcus victoriae* evaluated and *A. pullulans* CIEFAP 631 were the most promising strains, reducing the incidence caused by natural decay by more than 90 %. On the whole, in all assays in berry fruit, *C. victoriae* CIEFAP 771 was the most promising antagonist yeast (López et al. 2015). The protection levels achieved in cherry by both *C. victoriae* NPCC 1263 and *Pichia membranifaciens* NPCC 1250 yeast strains, selected as the best biocontrol yeast in previous work on the pear (Lutz et al. 2013), were lower than those obtained by other yeasts isolated from berry fruit for the artificially inoculated control in berries. However, *P. membranifaciens* NPCC 1250 was the best biocontrol yeast in the control of natural infection in assays in raspberries (Table 17.4).

Table 17.4 Antagonistic efficacy of yeast isolates against *P. crustosum* and *M. piriformis* in berry and cherry fruits

Yeast isolate	Collection number	*Penicillium crustosum*		*Mucor piriformis*		Natural infection	
		Cherries		Cherries		Raspberries	Blackberries
		Incidence*	% DR#	Incidence	% DR	Incidence	Incidence
Aureobasidium pullulans	CIEFAP 631	63,33 ab	65,18 abc	13,33 a	93 a	3,33 ab	43,33 bcde
Aureobasidium pullulans	CIEFAP 1141	53,33 a	79,11 ab	70 f	17 de	50 d	50 cde
Cryptococcus adeliensis	CIEFAP 154	86,66 cd	37,52 de	46,66 d	53,5 cd	nd	nd
Cryptococcus adeliensis	CIEFAP 441	63,33 ab	72,14 abc	100 f	0 f	nd	nd
Cryptococcus albidosimilis	CIEFAP 636	80bcd	17 ef	40 cd	62 bc	30 cd	30 abc
Cryptococcus stepposus	CIEFAP 614	66,66 abc	61,31 bc	100 f	0 f	20 abc	20 ab
Cryptococcus victoriae	CIEFAP 30	80 bcd	56,47 cd	30 bc	70 bc	nd	nd
Cryptococcus victoriae	CIEFAP 201	86,66 cd	53,38 cde	36,66 bcd	63,35 bc	10 abc	23,33 ab
Cryptococcus victoriae	CIEFAP 771	53,33[a]	61,31 bc	40 cd	60 bc	10 abc	20 ab
Cystofilobasidium capitatum	CIEFAP 1204	60 ab	83,36 a	23,33 ab	67 bc	23,33 bc	23,33 ab
Guehomyces pullulans	CIEFAP 899	100 d	0 f	23,33 ab	77 ab	nd	nd
Meyerozyma guillermondii	CIEFAP 199	100 d	16,82 f	40 cd	60 bc	13,33 abc	13,33 a
Pichia membranifaciens	NPCC 1250	93,33 d	0 f	86,66 f	13,5 ef	0 a	36,66 abcd
Cryptococcus victoriae	NPCC 1263	96,66 d	18,18 f	96,66 f	3,5 f	10 abc	63,33 de
Control without yeast		100 d	0 f	100 f	0 f	73,33 e	83,33 e

Nd not determinate
*Incidence: calculated as the number of decay wounds over the total number of wounds[#](DR) represents mean percentage of decay reduction calculated as (mean lesion diameter in control – mean lesion diameter in treatment) × 100/mean lesion diameter in control. Values within a column followed by the same letter are not significantly different according to Fisher´s test (p>0.05)

The results obtained in our work support the hypothesis that the best strategy to isolate potential antagonists against a particular etiological agent is to look in places where a disease caused by the pathogen can be expected but it does not occur (Baker and Cook 1974). The improved methodology proposed in this work is based on obtaining microorganisms from healthy wounds of pear fruits after periods of cold storage. These organisms are probably adapted to storage conditions (low temperatures, fruit–host and postharvest treatments) and could exhibit some antagonist activity because they were isolated from healthy wounds or surfaces in berry fruits (Lutz et al. 2012).

17.3 Killer Character in the Control of Wine Spoilage Yeasts

Killer yeasts have been explored in several biotechnological applications as biocontrol agents (Meinhardt and Klassen 2009; Pfliegler et al. 2015). The use of killer strains to eliminate undesirable microbial strains has been suggested in the wine industry: killer yeasts with positive oenological characteristics can be used as starter cultures to avoid contaminations by wild yeasts present in grapes and must, while keeping the wine's proper chemical composition (Comitini et al. 2004; Santos et al. 2009, 2011; Comitini and Ciani 2011). Biocontrol strategies based on killer yeasts were also proposed for ethanol production (Ceccato-Antonini et al. 2005; Meneghin et al. 2010) and to control contaminating yeasts in the winemaking, brewing, and other fermentation industries (Goretti et al. 2009). The preservation of stored fruits from postharvest diseases (Pimenta et al. 2009), and the protection of cheese or cereals (Olstorpe et al. 2010; Rosa et al. 2010) against spoilage fungi or bacteria, can also be achieved by exploring the killer phenomenon. Strategies of preservation based on biological control solutions offer an interesting alternative to the frequently used chemical preservatives, which lead to resistance phenomena and face increased consumer resistance for environmental concerns (Antunes and Aguiar 2012).

In addition, studies on the killer phenomenon in yeasts have provided valuable insights into a number of fundamental aspects of eukaryotic cell biology and interactions of different eukaryotic cells (Schmitt and Breinig 2006). Moreover, the elucidation of molecular mechanisms of their action will be helpful to develop the strategies and design synthetic chemicals to fight harmful fungi in human, animals, and plants (Magliani et al. 2010). It should be stressed that killer yeasts among natural yeast isolates have a large biodiversity, in terms of their biochemical characteristics, genetic determinants, spectra of action, and mechanisms of their killer toxin actions.

Mycocins are extracellular proteins or glycoproteins that disrupt cell-membrane function in susceptible yeasts, which bear receptors for the compound (Golubev 2006). Their activity is directed primarily against yeasts closely related to the producer strain, which has a protective factor. Several have since been isolated, frequently where yeast populations exist in high density and in highly competitive conditions. Killer toxin production occurs among many yeast genera including *Saccharomyces*, *Candida*, *Cryptococcus*, *Debaryomyces*, *Kluyveromyces*, *Pichia*,

Torulopsis, *Williopsis*, and *Zygosaccharomyces* (Schmitt and Breinig 2002; Golubev 2006). Genetic and molecular studies have shown that the killer toxin trait may be carried on extrachromosomal elements in the form of double-stranded RNA viruses, on double-stranded linear DNA, or on a chromosome. The well-known mechanisms of the killer toxin are the interruption of cell division by blocking DNA synthesis, inhibition of synthesis of the cell-wall component β-1,3-glucan, and ion leakage caused by the formation of channels on the cytoplasmic membrane (Schmitt and Breinig 2002; Hatoum et al. 2012).

In recent years, many novel killer toxins have been purified and characterized, and their genes have been cloned. The studies on killer yeasts and their killer toxins have been intensively reviewed by many authors (Polonelli et al. 2011; Wang et al. 2012; Liu et al. 2011) and have reported on the potential applications of yeast killer toxins in medicine, whereas Schmitt and Breinig (2006) reviewed yeast viral killer toxins, their mechanism of action, and host defense systems. At the same time, in recent years, great progress in research on the molecular mechanisms of their killer toxin action has also been made. However, such progress was only considered in fragments.

Several authors have proposed the use of killer toxins produced by *Candida pyralidae*, *Kluyveromyces wickerhamii*, *Pichia membranifaciens*, and *Wickerhamomyces anomalus* (ex *Pichia anomala*), to control the development of *Dekkera/Brettanomyces*, the principal wine spoilage yeast (Comitini et al. 2004; Santos et al. 2011; Steensels et al. 2015). Killer toxins from *Tetrapisispora phaffii* (ex *Kluyveromyces phaffii*), *W. anomalus*, *Torulaspora delbrueckii*, and *Metschnikowia pulcherrima* species were also proposed as control agents against spoilage non-*Saccharomyces* yeasts present during the prefermentative stage, such as *Candida boidinii*, *Hanseniaspora uvarum*, and *Pichia guilliermondii* (Ciani and Fatichenti 2001). Nevertheless, yeast species able to control the development of other spoilage species, as some species of genera *Candida* and *Pichia*, have not been described. The non-*Saccharomyces* wild yeasts cause various types of spoilage; for example, *P. membranifaciens* is known to produce film, haze, and off-flavours such as phenolic, estery, and acidic notes (Malfeito-Ferreira 2011), and others such as *Pichia manshurica* were recently reported as producers of volatile off-flavors and biogenic amines (Barata et al. 2006; Sáez et al. 2011a; Tristezza et al. 2013).

In previous studies carried out in our laboratory, different wine spoilage yeasts able to produce high levels of volatile phenols have been detected and characterized in North Patagonia: *P. guilliermondii*, *P. manshurica*, and *P. membranifaciens* (Lopes et al. 2009; Sáez et al. 2010; 2011a, b). On the other hand, most yeasts isolated from wines, cellar surfaces, winery equipment, and spontaneously fermenting grape musts evidenced killer character (Sangorrín et al. 2002; Lopes et al. 2007; Sangorrín et al. 2007, 2008). In this context, a broad-spectrum control system effective against a wide set of the spoilage yeast species such as those just mentioned is of great importance. For this purpose, a panel of different killer strains of the wine killer yeasts belonging to NPCC was tested against four potential spoilage yeasts.

Although variability in the killer capacity of tested strains was observed (Sangorrín et al. 2008; Lopes et al. 2009), the strains *Torulaspora delbrueckii* NPCC 1033 and *Wickerhamomyces (W.) anomalus* NPCC 1027 were selected on the basis of their broadest killer spectrum (Sáez et al. 2011b; Villalba et al. 2016). Previous work carried out in our laboratory has also demonstrated that these two selected killer strains could be good candidates for spoilage wine yeast biocontrol. *T. delbrueckii* NPCC 1033, isolate 24 in Sangorrín et al. (2008), showed killer activity against *Hanseniaspora uvarum* and *Brettanomyces bruxellensis*. These same two selected killer strains had also shown good killer activity against some strains of *Pichia guilliermondii* and *P. membranifaciens* (Lopes and Sangorrín 2010). On the other hand, *W. anomalus* NPCC 1027, isolate 13 of *P. anomala* in Sangorrín et al. (2008), showed killer activity against *B. bruxellensis* strains and other wine spoilage yeasts including *Candida boidinii*, *C. patagonica*, *G. silvicola*, and *H. uvarum*. On the other hand, *T. delbrueckii* NPCC 1033 and *W. anomalus* NPCC 1027 strains were then tested by QTM against six commercial starter strains of *S. cerevisiae*; all starter strains were resistant to both killer toxin yeasts.

Different killer toxins produced by *W. anomalus* have been studied and characterized by their wide antimicrobial spectrum (Comitini et al. 2004; Wang et al. 2007; Sun et al. 2012), their variable molecular mass (e.g., from 3 to 300 kDa), their genetic codification and optimal pH, and their temperature for optimal activity (Passoth et al. 2006; Meinhardt and Klassen 2009; Walker 2011). In contrast, the killer capacity of *T. delbrueckii* was reported for the first time in a previous work from our laboratory (Sangorrín et al. 2007), and some biochemical aspects and action mechanisms of this toxin have been evaluated (Villalba et al. 2016). Because *T. delbrueckii* is an ethanol-tolerant wine yeast detected in advanced stages of wine fermentations in North Patagonia and other wine regions worldwide (Ciani and Maccarelli 1998; Sangorrín et al. 2001; Renauld et al. 2009), the killer toxins produced by this species could be interesting for wine spoilage yeasts biocontrol.

To evaluate the potential biotechnological application of the selected killer toxins, as a first stage the time course of toxin killer production by both strains was studied. *T. delbrueckii* and *W. anomalus* showed maximum killer activity in the supernatant at the beginning of the stationary growth phase. The two toxins that were active against spoilage yeasts were named TdKT (*T. delbrueckii* killer toxin) and WaKT (*W. anomalus* killer toxin). The crude extracts containing the toxins have killer ability against the four spoilage yeasts tested: (1) TdKT showed approximately 70 % inhibitory activity against the four spoilage yeasts (Villalba et al. 2016); (2) WaKT completely inhibited *P. guilliermondii* growth, showed 70 % control against *P. membranifaciens*, and demonstrated only minor inhibitory activity against *B. bruxellensis* and *P. manshurica* (Fig. 17.1).

The effect of different physicochemical stress conditions governing wine fermentation has been proposed as an important selective feature to be taken into account for killer toxin selection (Comitini et al. 2004; Santos et al. 2009; Mehlomakulu et al. 2014). To elucidate the biochemical properties of TdKT and WaKT, particularly those related to their possible use in winemaking, crude extracts containing the two toxins were subjected to a set of assays. Grape must and the wine environment are characterized by high sugar content (>250 g l^{-1}), low pH values around 3.5, and high ethanol and SO_2 concentrations. In this context, a precise characterization of the stability of killer toxins to fermentation-related factors that could

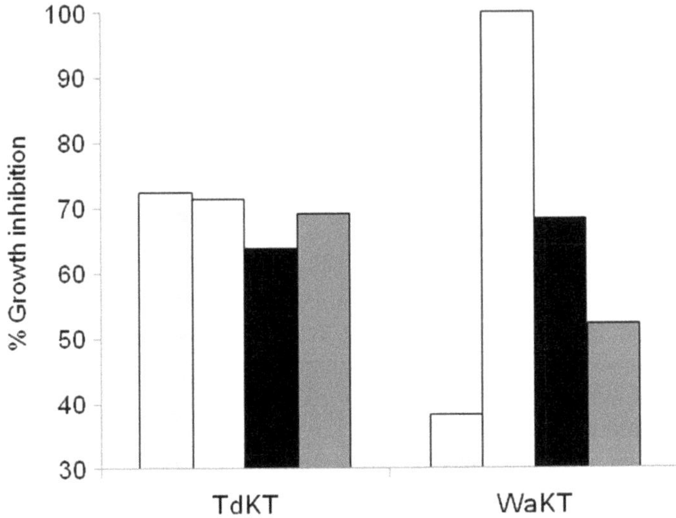

Fig. 17.1 Killer activity of the TdKT and WaKT crude extracts against different wine spoilage yeasts. Killer activity determined by the multiwell method (MWM) employed 3 AU crude extracts of *Torulaspora delbrueckii* and *W. anomala* against *Brettanomyces bruxellensis* INTA VC20 (*light gray bars*), *Pichia guilliermondii* NPCC 1055 (*white bars*), *P. membranifaciens* NPCC 1099 (*dark bars*), and *P. manshurica* NPCC 1038 (*dark gray bars*)

influence its activity would allow establishing their potential applicability in winemaking. Therefore, the stability of the killer activity under different physicochemical stress conditions typically present in wine fermentations (pH, temperature, ethanol, free sulfur dioxide and sugar concentrations) was tested. TdKT and WaKT were not affected by the higher sugar concentrations evaluated (up 250 g l^{-1}) and they were stable in the presence of SO_2, the most important antimicrobial compound used in winemaking, in concentrations ranging between 0 (control) and 150 mg l^{-1}. TdKT considerably lost activity above 37 °C but WaKT retained detectable activity even at 40 °C. Another chemical stress factor highly important during winemaking is ethanol concentration: TdKT was not affected by the ethanol concentration evaluated (0–12 % v/v), although WaKT slightly decreased in killing potential at ethanol concentrations higher than 12 % v/v. Additionally, the two killer toxins were active at acidic pH values, but although WaKT was active even at pH 6, TdKT lost its activity above pH 4.5. WaKT and TdKT were stable at pH and temperature as well as ethanol, glucose, and sulfur dioxide concentrations simulating winemaking conditions, which evidenced their potential utility.

Killer toxins generally act by binding, in a first stage, to cell-wall receptors of the sensitive cells; then, the abundance and ubiquity of this receptor could also be associated with the spectrum of action of a killer toxin (Schmitt and Breinig 2002). Receptor molecules for different killer toxins have been identified as β-glucan, manan, chitin, etc. With the aim of identifying the first binding site for TdKT and WaKT on the envelope of the sensitive yeast, competitive inhibition of killer action

Table 17.5 Competition assays carried out with pure polysaccharides from yeast cell wall to detect putative binding sites of TdKT and WaKT toxins using two different methods (WTM and MWM)

Molecules evaluated		Residual killer activity (%)			
		WTM		MWM	
Polymer	Type of molecule	TdKT	WaKT	TdKT	WaKT
Pullulan	α-1-4,1-6-Glucan	83	100	100	100
Mannan	α-1-2,1-3,1-6-Glucan	83	100	100	100
Laminarin	β-1-3,1-6-Glucan	83	100	100	100
Curdlan	β-1-3-Glucan	100	100	100	100
Chitin	β-1-4-Nac glucosamine	67	100	80	100
Pustulan	β-1-6-Glucan	0	0	0	36
Control	–	100	100	100	100

WTM well test method, *MWM* multiwell method

by different cell-wall polysaccharides was evaluated through two methods (Lopes and Sangorrín 2010; Villalba et al. 2016). The residual killer activity that was tested through the well test method (WTM) demonstrated that pustulan (β-1,6-glucan) completely inhibited both TdKT and WaKT activity, whereas chitin, as well as the mixed-linkage polysaccharides laminarin, mannan, and pullulan, showed only partial competition with TdKT (Table 17.5).

These results were partially confirmed in liquid medium using the multiwell method (MWM). Again, the pustulan binds to both toxins and the chitin only showed a low competition with TdKT (80 % residual activity); however, no effect was observed with the mixed-linkage polysaccharides (Table 17.5). The competitive inhibition of WaKT activity by cell-wall polysaccharides showed that cytocidal action was only prevented by β-1,6-glucan, behaving similarly to other reported killer toxins of *W. anomalus* (De Ingeniis et al. 2009; Farkas et al. 2012). However, TdKT was found to have affinity for both β-1,6-glucan and chitin. The ability to bind to chitin is not widespread among killer yeasts; it only appears in *Kluyveromyces lactis*, *Pichia inositovora*, *Pichia acaciae*, and *Wingea robertsiae* killer toxins (Meinhardt and Klassen 2009).

The binding to these polysaccharides could also be related to a hydrolytic activity as part of their modes of action. Colussi et al. (2005) found a *K. lactis* killer toxin having chitin-binding and chitinase activity. Additionally, strong β-1,3-glucanase activity has been associated with the primary killing mechanism of other killer toxins such as those produced by *Williopsis saturnus* var. *mrakii* MUCL 41968 (Guyard et al. 2002), *K. phaffii* (Comitini et al. 2004), and *Wickerhamomyces anomalus* (Izgü et al. 2007; Walker 2011). In this context, the β-1,3-glucanase and chitinase activities of TdKT and WaKT were investigated to address the possibility that TdKT and WaKT may exhibit some hydrolytic activity as part of their mechanism of action. β-Glucanase and chitinase activities were detected in both killer toxins when they were incubated with the corresponding polymers, and a direct relationship between β-glucanase, chitinase, and killer activities was observed for both toxins.

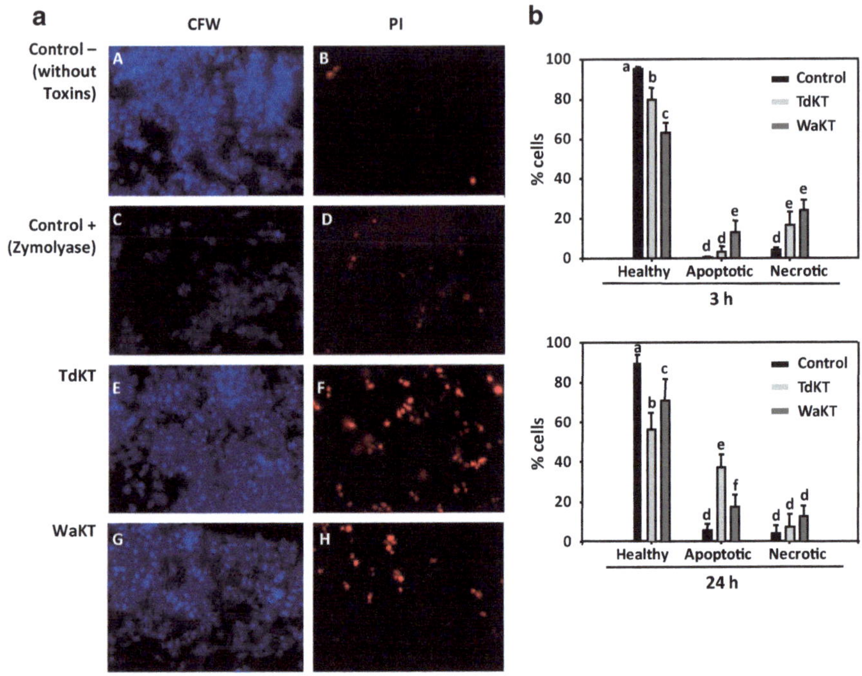

Fig. 17.2 Cell-wall damage, apoptotic and necrotic phenotype in killer toxin lethality. **a** Evaluation of possible damage of the yeast walls by double staining CFW-PI. (*A*, *B*) Untreated yeast. (*C*, *D*) Yeast after treatment for 2 h at 37 °C with zymolyase (1 mg ml^{-1}). (*E*, *F*) Yeast treated for 2 h with 6 KU of partially purified TdKT. (*G*, *H*) Yeast treated for 2 h with 6 KU of partially purified WaKT. *CPW* calcofluor white staining, *PI* propidium iodine staining. Yeast strains: *Candida glabrata*. **b** Percentages of healthy, apoptotic, or necrotic phenotypes for TdKT (6 UT, *light gray bars*), WaKT (6 UT, *dark gray bars*), and negative control (*black bars*) treatments. Bars with different letters represent statistical significance within each treatment for 3 h (*upper panel*) and 24 h (*lower panel*). (ANOVA and Tukey test, $p<0.05$, $n=7$)

To know whether the killer toxin treatment induces a cell-wall modification in the sensitive yeast (*Candida glabrata*), the cells were incubated with Calcofluor White (CW), a fluorescent dye that stains the polysaccharide chitin, and propidium iodide (PI), which stains the nucleic acids of dead or damaged cells, providing an indirect measure of cell-membrane integrity. As shown in Fig. 17.2a(A), chitin of negative control yeast (yeast without treatment) was mainly deposited at the neck of budding cells and remained on the surface of the mother cells as a bud scar after cell division, leading to a higher fluorescent signal in those sites. In contrast, yeast exposed to enzymatic digestion with zymoliase (positive control yeast) showed a reduction in the fluorescence signal and a discontinuous cell perimeter (Fig. 17.2a(C)). At the same time, abnormal patch-like depositions of materials stained with CW were observed on the surface of many yeasts treated with TdKT or WaKT, especially on premature buds, with a dimming in dye signal after exposure (Fig. 17.2a(E, G)). In yeast treated with TdKT or WaKT, the PI staining showed an increase in the dead cell number after treatment, being 11.28±3.69 and 10.99±6.65 %, respectively (Fig. 17.2a(F, H)) compared to nontreated yeast (1.28±0.70 %; Fig. 17.2a(A)) and yeast exposed to zymoliase activity (7.45±2.38 %; Fig. 17.2a(C)).

According to the results obtained by staining and fluorescence microscopy evaluation, both toxins TdKT and WaKT induce ultrastructural modifications in the cell wall of the sensitive strains (evidenced in chitin decrease in fluorescence microscopy evaluation). The decrease in fluorescence of the wall in the presence of TdKT or WaKT could be caused by the chitinase activities detected for both toxins. This effect produces a disruption in the cell-wall structure, which could result in cell death.

Finally, the effect of the two toxins on yeast viability and the mechanisms mediating cell death were evaluated with fluorescence microscopy. Sensitive yeast with the two killer toxins was treated and the arising apoptotic or necrotic phenotypes were examined using the PI-Hoescht double-staining assay. Figure 17.2b presents the percentages of apoptosis and necrosis observed after each toxin treatment, as well as the negative control treatment. Apoptotic and necrotic cell death in yeasts can be triggered by various factors such as H_2O_2, cell aging, acetic acid, or toxins (Reiter et al. 2005). In our results, two different mechanisms of action were detected for the killer toxins: TdKT provoked an increase in necrosis cell death after 3 h of treatment and then apoptotic cell death after 24 h of treatment, showing a time dependence in the mechanisms of action, whereas WaKT generated both necrosis and apoptosis induction after 3 h rot observation (Fig. 17.2b). Schmitt and Reiter (2008) and Santos et al. (2013) have defined that the killer toxins produced by *S. cerevisiae* and *P. membranifaciens* respectively may present different strategies to kill sensitive cells: at high concentrations, toxin-induced apoptosis plays a minor role, whereas at low-to-moderate toxin doses, it triggers apoptosis in susceptible target cells. According to our results, time frame was also a key factor influencing toxin effect, and both toxins represent different stimuli in sensitive yeasts. Further research is needed to elucidate the regulation on yeast apoptosis in response to TdKT and WaKT in crude extracts and also using purified toxins. Purification strategies are currently in progress in our laboratory to confirm the results obtained with crude extracts.

WaKT and TdKT exhibit the same spectrum of killer activity against the complete set of sensitive spoilage yeasts: they interact with β-1,6-glucan, they have molecular masses higher than 30 kDa, they have glucanase activity, and they prefer pH values between 3 and 5 and temperatures up to 37 °C. Our results show that WaKT is similar to several toxins of *W. anomalus*, in particular to that produced by the reference strain NCYC 434 (K5-type), but some additional features of WaKT were described for the first time: its ability to hydrolyze chitin and to provoke apoptotic or necrotic phenotypes in sensitive strains (Sáez et al. 2011b).

The results presented in Villalba et al. (2016) constitute the first steps in the characterization and elucidation of the killing mechanism of a toxin produced by the yeast species *T. delbrueckii*. The TdKT putative mode of action includes binding to β-1,6-glucan and chitin in the initial interaction of the toxin with sensitive cells, with a potential degradation of these polysaccharides (by their β-glucanase and chitinase activities), cell-wall disruption (chitin decrease in microscopy evaluation), and finally, cell death by necrosis at the initial time and by apoptosis at 24 h.

In conclusion, TdKT and WaTK have shown potential to control a broad spectrum of wine spoilage yeasts, not inhibiting the fermenting yeast *S. cerevisiae*. These killer toxins are resistant to stress conditions typically found at both initial and end stages of wine fermentations. The collective data presented in this study make these toxins potentially interesting to control spoilage yeast present during the complete winemaking process.

17.4 Conclusions

Argentinean Patagonia offers a great diversity of selective artificial and natural environments, scarcely explored, which are suitable for the bioprospection of biotechnologically relevant yeasts. The applications of yeasts in human foods and other sectors are increasing, and market demand is providing motivation to continue and to even increase research and development in this field. Only future studies will reveal the ultimate potential of these microorganisms in different fields of application. The possibility that additional unknown toxic mechanisms in other killer yeast species may occur, together with the potential of known killer toxins to be applied in the food industry as adjunct bioprotective cultures or as a component of active packaging, represents a promising strategy to reduce the use of chemical preservatives.

References

Antunes J, Aguiar C (2012) Search for killer phenotypes with potential for biological control. Ann Microbiol 62:427–433

Baker KJ, Cook RJ (1974) Biological control of plant pathogens. Freeman, San Francisco

Barata A, Nobre A, Correia P, Malfeito-Ferreira M, Loureiro V (2006) Growth, 4-ethylphenol production by 1 the yeast *Pichia guilliermondii* in grape juices. Am J Enol Viticult 57: 133–138

Brandao LR, Libkind D, Vaz AB, Espırito Santo LC, Moline M, de Garcıa V, van Broock M, Rosa CA (2011) Yeasts from an oligotrophic lake in Patagonia (Argentina): diversity, distribution and synthesis of photoprotective compounds and extracellular enzymes. FEMS Microbiol Ecol 76:1–13

Buzzini P, Margesin R (2014) Cold-adapted yeasts: biodiversity, adaptation strategies and biotechnological significance. Springer, New York

Buzzini P, Branda E, Goretti M, Turchetti B (2012) Psychrophilic yeasts from worldwide glacial habitats: diversity, adaptation strategies and biotechnological potential. FEMS Microbiol Ecol 82:217–241

Ceccato-Antonini SR, Sanino A, Araujo JC, Tosta CD (2005) The killer yeasts and the alcoholic fermentation. Braz J Food Technol 5:40–45

Chand Goyal T, Spotts RA (1996) Control of postharvest pear diseases using natural saprophytic yeast colonists and their combination with a low dosage of thiabendazole. Postharvest Biol Technol 7:51–64

Ciani M, Fatichenti F (2001) Killer toxin of *Kluyveromyces phaffii* DBVPG 6076 as a biopreservative agent to control apiculate wine yeasts. Appl Environ Microbiol 67:3058–3063

Ciani M, Maccarelli F (1998) Oenological properties of non-Saccharomyces yeasts associated with wine-making. World J Microbiol Biotechnol 14:199–203

Colussi PA, Specht CA, Taron CH (2005) Characterization of a nucleus-encoded chitinase from the yeast *Kluyveromyces lactis*. Appl Environ Microbiol 71:2862–2869

Comitini M, Ciani F (2011) Non-*Saccharomyces* wine yeasts have a promising role in biotechnological approaches to winemaking. Ann Microbiol 61:25–32

Comitini F, De Ingeniis J, Pepe L, Mannazzu L, Ciani M (2004) *Pichia anomala* and *Kluyveromyces wickerhamii* killer toxins as new tools against *Dekkera/Brettanomyces* spoilage yeasts. FEMS Microbiol Lett 238:235–240

Crisosto CH, Garner D, Doyle J, Day KR (1993) Relationship between fruit respiration, bruising susceptibility and temperature in sweet cherries. Hortic Sci 28:132–135

Deak T (2009) Ecology and biodiversity of yeasts with potential value in biotechnology. In: Satyanarayana T, Kunze G (eds) Yeast biotechnology: diversity and applications. Springer, New York

De Ingeniis J, Raffaelli N, Ciani M, Mannazzu I (2009) *Pichia anomala* DBVPG 3003 secretes a ubiquitin-like protein that has antimicrobial activity. Appl Environ Microbiol 75:1129–1134

Debode J, Hemelrijck WV, Creemers P, Maes M (2013) Effect of fungicides on epiphytic yeasts associated with strawberry. Microbiology Open 2:482–491

Dobra AC, Sosa MC, Dussi MC (2008) Low incidence of fungal and bacterial diseases in the pear production of North Patagonia, Argentina. Acta Hortic 800:907–912

Dobra AC, Sosa MC, Lutz MC, Rodriguez G, Greslebin AG, Vélez ML (2011) Fruit rot caused by phytophthora in cold storaged pears in the Valley of Rio Negro and Neuquén, Argentina. Acta Hortic 909:505–510

Droby S, Wisniewski M, Macarisin D, Wilson C (2009) Twenty years of postharvest biocontrol research: is it time for a new paradigm? Postharvest Biol Technol 52:137–145

Farkas Z, Márki-Zay J, Kucsera J, Vágvölgyi C, Golubev W, Pfeiffer I (2012) Characterization of two different toxins of *Wickerhamomyces anomalus* and *Pichia anomala,* VKM Y-159. Acta Biol Hung 63:277–287

Feliziani E, Romanazzi G (2013) Preharvest application of synthetic fungicides and alternative treatments to control postharvest decay of fruit. Stewart Postharvest Rev 3:1–6

FRAC (Fungicide Resistance Action Committee) (2010) Fungicide use guidelines 10th meeting working group. FRAC (Fungicide Resistance Action Committee), Bananas Orlando, FL

Fravel DR (2005) Commercialization and implementation of biocontrol. Annu Rev Phytopathol 43:337–359

Golubev WI (2006) Antagonistic interactions among yeasts. In: Rosa CA, Péter G (eds) The yeast handbook. Biodiversity and ecophysiology of yeasts. Springer, Heidelberg, pp 197–219

Goretti M, Turchetti B, Buratta M, Corazzi L, Vaughan-Martini A, Buzzini P (2009) In vitro antimycotic activity of a *Williopsis saturnus* killer protein against food spoilage yeasts. Int J Food Microbiol 131:178–182

Guyard C, Dehecq E, Tissier JP, Polonelli L, Dei-Cas E, Cailliez JC, Menozzi FD (2002) Involvement of β-glucans in the wide-spectrum antimicrobial activity of *Williopsis saturnus* var. *mrakii* MUCL 41968 killer toxin. Mol Med 8:686–694

Hatoum R, Labrie S, Fliss I (2012) Antimicrobial and probiotic properties of yeasts: from fundamental to novel applications. Front Microbiol 3:421

Izgü F, Altınbay D, Türeli AE (2007) *In vitro* activity of panomycocin: a novel exo-β-1,3-glucanase isolated from *Pichia anomala* NCYC 434, against dermatophytes. Mycoses 50:1–34

Janisiewicz WJ, Korsten L (2002) Biological control of postharvest diseases of fruits. Annu Rev Phytopathol 40:411–441

Jijakli HM (2011) *Pichia anomala* in biocontrol for apples: 20 years of fundamental research and practical applications. Antonie van Leeuwenhoek J Microbiol 99:93–105

Lima G, De Curtis F, Castoria R, De Cicco V (1998) Activity of the yeasts *Cryptococcus laurentii* and *Rhodotorula glutinis* against post-harvest rots on different fruits. Biocontrol Sci Technol 8:257–267

Lima JR, Barros Gonçalves LR, Rocha BL, Rosa CA, Pinto Viana FM (2013) Isolation, identification, and activity in vitro of killer yeasts against Colletotrichum gloeosporioides isolated from tropical fruits. J Basic Microbiol 53:590–599

Liu J, Michael W, Droby S, Vero S, Tian S, Hershkovitz V (2011) Glycine betaine improves oxidative stress tolerance and biocontrol efficacy of the antagonistic yeast *Cystofilobasidium infirmominiatum*. Int J Food Microbiol 146:76–83

Lopes CA, Sangorrín MP (2010) Optimization of killer assays for yeast selection protocols. Rev Argent Microbiol 42:298–306

Lopes CA, Rodríguez ME, Sangorrín MP, Querol A, Caballero A (2007) Patagonian wines: implantation of an indigenous isolate of *Saccharomyces cerevisiae* in fermentations conducted in traditional and modern cellars. J Ind Microbiol Biotechnol 34:139–149

Lopes CA, Sáez JS, Sangorrín MP (2009) Differential response *of Pichia guilliermondii* spoilage isolates to biological, physical-chemical factors prevailing in Patagonian wine fermentations. Can J Microbiol 55:801–809

López SN, Sangorrín MP, Pildain MB (2013) Patogenicidad y resistencia a fungicidas de patógenos de postcosecha en fruta fina de Patagonia. XIII Congreso AAM, II Congreso Microbiología Agrícola y Ambiental, 23 al 26 de septiembre de 2013, Buenos Aires

López SN, Sangorrín MP, Pildain MB (2014) Diversidad de levaduras en frambuesas, zarzamoras y cerezas de Patagonia. XIII Congreso Argentino de Micología-1ra Reunión de la Asociación Micológica Carlos Spegazzini Buenos Aires, 24 al 27 de agosto

López S, Sangorrín MP, Pildain MB (2015) Cold-adapted yeast as biocontrol agents of postharvest pathogens in cherries of Patagonia Sur, Argentina. V Jornadas Sudamericanas de Biología y Biotecnología de Levaduras 4-6 de Agosto, Recife, Brasil

Lutz MC, Sosa MC, Lopes CA, Sangorrín MP (2012) A new improved strategy for the selection of cold-adapted antagonist yeasts to control postharvest pear diseases. Biocontrol Sci Technol 22:1465–1483

Lutz MC, Sosa MC, Rodriguez ME, Lopes CA, Sangorrín MP (2013) Efficacy and putative mode of action of native and commercial antagonistic yeasts against postharvest pear pathogens. Int J Food Microbiol 164:166–172

Magliani W, Conti S, Giovati L, Polonelli L (2010) Yeast killer technology transfer. In: Rai M, Varma A (eds) Mycotoxins in food, feed and bioweapons. Springer, Berlin, pp 275–290

Malfeito-Ferreira M (2011) Yeasts and wine off-flavours: a technological perspective. Ann Microbiol 61:95–102

Massart S, Martinez-Medina M, Jijakli MH (2015) Biological control in the microbiome era: challenges and opportunities. Biol Control 89:98–108

Mehlomakulu NN, Sebati ME, Divol B (2014) Characterization of novel killer toxins secreted by wine-related non-*Saccharomyces* yeasts and their action on *Brettanomyces* spp. Int J Food Microbiol 188:83–91

Meinhardt M, Klassen R (2009) Yeast killer toxins, fundamentals and applications. In: Anke T, Weber D (eds) Physiology, genetics. The Mycota, XV, 1st edn. Springer, Heidelberg, pp 107–130

Meneghin MC, Vanda Renata Reis VR, Ceccato-Antonin SR (2010) Inhibition of bacteria contaminating alcoholic fermentations by killer yeasts. Braz Arch Biol Technol 53: 1043–1050

Nunes CA (2012) Biological control of postharvest diseases of fruit. Eur J Plant Pathol 133:181–196

Ogawa JM, Zehr EI, Bird GW, Ritchie DF, Uriu K, Uyemato JK (1995) Compendium of stone fruit diseases. APS Press, St. Paul

Olstorpe M, Borling J, Schnürer J, Passoth V (2010) *Pichia anomala* yeast improves feed hygiene during storage of moist crimped barley grain under Swedish farm conditions. Anim Feed Sci Technol 156:47–56

Passoth V, Fredlund E, Druvefors UA, Schnürer J (2006) Biotechnology, physiology and genetics of the yeast *Pichia anomala*. FEMS Yeast Res 6:3–13

Pfliegler WP, Pusztahelyi T, Pócsi I (2015) Mycotoxins: prevention and decontamination by yeasts. J Basic Microbiol 54:1–14

Pimenta RS, Morais B, Rosa CA, Corrêa A Jr (2009) Utilization of yeasts in biological control programs (Chapter 10). In: Satyanarayana T, Kunze G (eds) Yeast biotechnology: diversity and applications. Springer, New York

Polonelli L, Magliani W, Ciociola T, Giovato L, Conti S (2011) From *Pichia anomala* killer toxin through killer antibodies to killer peptides for a comprehensive anti-infective strategy. Antonie van Leeuwenhoek J Microbiol 99:35–41

Reiter J, Herker E, Madeo F, Schmitt MJ (2005) Viral killer toxins induce caspase-mediated apoptosis in yeast. J Cell Biol 168:353–358

Renauld P, Miot-Sertier C, Marullo P, Hernández-Orte P, Lonvaud-Funel A, Bely M (2009) Genetic characterization, phenotypic variability in *Torulaspora delbrueckii* species and potential applications in the wine industry. Int J Food Microbiol 134:201–210

Robiglio A, Sosa MC, Lutz MC, Lopes CA, Sangorrín MP (2011) Yeast biocontrol of fungal spoilage of pears stored at low temperature. Int J Food Microbiol 147:211–216

Romanazzi G (2010) Chitosan treatment for the control of postharvest decay of table grapes, strawberries and sweet cherries. New Trend Postharvest Manag Fresh Prod (Fresh Prod Spec Issue) 4:111–115

Rosa MM, Tauk-Tomisiel SM, Rampazzo PE, Ceccato-Antonini SR (2010) Evaluation of the biological control by the yeast *Torulaspora globosa* against *Colletotrichum sublineolum* in sorghum. World J Microbiol Biotechnol 6:1491–1502

Sáez JS, Lopes CA, Kirs VC, Sangorrín MP (2010) Enhanced volatile phenols in wine fermented with *Saccharomyces cerevisiae*, spoiled with *Pichia guilliermondii* and *Dekkera bruxellensis*. Lett Appl Microbiol 51:170–176

Sáez JS, Lopes CA, Kirs VC, Sangorrín MP (2011a) Production of volatile phenols by *Pichia manshurica*, *Pichia membranifaciens* isolated from spoiled wines, cellar environment in Patagonia. Food Microbiol 28:503–509

Sáez J, Villalba L, del Mónaco S, Lopes CA, Sangorrín MP (2011b) Caracterización preliminar de dos toxinas *killer* para el biocontrol de levaduras contaminantes de vinos. III Jornadas Nacionales de Biología y Biotecnología de Levaduras 30 junio-1 julio Mendoza, Argentina. Libro actas ISBN: 978-987-679-024-9

Sangorrín MP, Zajonskovsky I, Lopes CA, Rodríguez ME, van Broock MR, Caballero AC (2001) Killer behaviour in wild wine yeasts associated with Merlot and Malbec type musts spontaneously fermented from Northwestern Patagonia Argentina. J Basic Microbiol 41:105–113

Sangorrín MP, Zajonskovsky I, van Broock M, Caballero AC (2002) The use of killer biotyping in an old Patagonian winery yeast ecological survey. World J Microbiol Biotechnol 18:115–120

Sangorrín MP, Lopes CA, Giraudo MR, Caballero AC (2007) Diversity, killer behaviour of indigenous yeasts isolated from the fermentation vat surfaces in four Patagonian wineries. Int J Food Microbiol 119:351–357

Sangorrín MP, Lopes CA, Jofré V, Querol A, Caballero AC (2008) Spoilage yeasts associated with Patagonian cellars: characterization, potential biocontrol based on killer interactions. World J Microbiol Biotechnol 24:945–953

Sangorrín MP, Lopes CA, Vero S, Wisniewski M (2014) Cold-adapted yeasts as biocontrol agents: biodiversity, adaptation strategies and biocontrol potential (Chapter 20). In: Pietro B, Rosa M (eds) Cold-adapted yeasts: biodiversity, adaptation strategies and biotechnological significance. Springer, New York

Santos A, San Mauro M, Bravo E, Marquina D (2009) PMKT2, a new killer toxin from *Pichia membranifaciens*, and its promising biotechnological properties for control of the spoilage yeast *Brettanomyces bruxellensis*. Microbiology 155:624–634

Santos A, Navascués E, Bravo E, Marquina D (2011) *Ustilago maydis* killer toxin as a new tool for the biocontrol of the wine spoilage yeast *Brettanomyces bruxellensis*. Int J Food Microbiol 145:147–154

Santos A, Alonso A, Belda I, Marquina D (2013) Cell cycle arrest, apoptosis and two alternative mechanisms for PMKT2 killer activity. Fungal Genet Biol 50:44–54

Scarpati O, Maio S, Puga Y (2011) Cerezo: desarrollo de un cultivo no tradicional en Argentina. Estud Geogr 72:591–610

Scherm B, Ortu G, Muzzu A, Budroni M, Arras G, Migheli O (2003) Biocontrol activity of antagonistic yeasts against *Penicillium expansum* on apple. J Plant Pathol 85:205–213

Schirra M, D'Aquino S, Cabras P, Angioni A (2011) Control of postharvest diseases of fruit by heat and fungicides: efficacy, residue levels, and residue persistence: a review. J Agric Food Chem 59:8531–8542

Schisler DA, Janisiewicz WJ, Boekhout T, Kurtzman CP (2011) Agriculturally important yeasts: biological control of field and postharvest diseases using yeast antagonists, and yeasts as pathogens of plants. In: Kurtzman CP, Fell JW, Boekhout T (eds) The yeasts . Elsevier, Maryland Heights, pp 45–52

Schmitt MJ, Breinig F (2002) The viral killer system in yeast and from molecular biology to application. FEMS Microbiol Rev 26:257–276

Schmitt MJ, Breinig F (2006) Yeast viral killer toxins: lethality and self-protection. Nat Rev Microbiol 4:212–221

Schmitt MJ, Reiter J (2008) Viral induced yeast apoptosis. Biochim Biophys Acta 1783: 1350–1353

Sharma R, Singh D, Singh R (2009) Biological control of postharvest diseases of fruits and vegetables by microbial antagonists: a review. Biol Control 50:205–221

Sosa MC, Lutz MC, Vélez ML, Greslebin A (2015) A pre-harvest rot of pear fruits Golden Russet Bosc caused by *Phytophthora lacustris* and *P. drechsleri* in Argentina. Austr Plant Dis Notes 17:10–18

Sozzi G (2007) Dormición en árboles frutales de hojas caducas. In: Sozzi GO (ed) Árboles Frutales: ecología, cultivo y aprovechamiento. Facultad de Agronomía, Buenos Aires, pp 85–103

Steensels J, Daenen L, Malcorps P, Derdelinckx G, Verachtert H, Verstrepen KJ (2015) *Brettanomyces* yeasts from spoilage organisms to valuable contributors to industrial fermentations. Int J Food Microbiol 206:24–38

Sugar D, Basile SR (2008) Timing and sequence of postharvest fungicide and biocontrol agent applications for control of pear decay. Postharvest Biol Technol 49:107–112

Sun HY, Wang K, Chi K, Xu HM, Chi ZM (2012) Simultaneous production of single cell protein and killer toxin by *Wickerhamomyces anomalus* HN1-2 isolated from mangrove ecosystem. Proc Biochem 47:251–256

Teixido N, Torres R, Viñas I, Abadias M, Usall J (2011) Biological control of postharvest diseases in fruit and vegetables. In: Lacroix C (ed) Protective cultures, antimicrobial metabolites, and bacteriophages for food and beverage biopreservation. Elsevier, Amsterdam, pp 365–402

Telias A, White JR, Pahl DM, Ottesen AR, Walsh CS (2011) Bacterial community diversity and variation in spray water sources and the tomato fruit surface. BMC Microbiol 11:81–85

Tristezza M, Vetrano C, Bleve G, Capozzi V, Logrieco A, Mita G, Grieco F (2013) Biodiversity and safety aspects of yeast strains characterized from vineyards, spontaneous fermentations in the Apulia Region Italy. Food Microbiol 36:335–342

Turchetti B, Buzzini P, Goretti M, Branda E, Diolaiuti G, D'Agata C, Smiraglia C, Vaughan-Martini A (2008) Psychrophilic yeasts in glacial environments of Alpine glaciers. FEMS Microbiol Ecol 63:73–83

Vero S, Garmendia G, Gonzalez MB, Garat MF, Wisniewski M (2009) *Aureobasidium pullulans* as a biocontrol agent of postharvest pathogens of apples in Uruguay. Biosci Technol 19:1033–1049

Villalba ML, Sáez JS, del Monaco S, Lopes CA, Sangorrín MP (2016) TdKT, a new killer toxin produced by *Torulaspora delbrueckii* effective against wine spoilage yeasts. Int J Food Microbiol 217:94–100

Walker GM (2011) *Pichia anomala*, cell physiology and biotechnology relative to other yeasts. Antonie van Leeuwenhoek J Microbiol 99:25–34

Wang X, Chi Z, Yue L, Li J (2007) Purification and characterization of killer toxin from a marine yeast *Pichia anomala* YF07b against the pathogenic yeast in crab. Curr Microbiol 55:396–401

Wang XX, Chi Z, Peng Y, Wang XH, Ru SG, Chi ZM (2012) Purification, characterization and gene cloning of the killer toxin produced by the marine-derived yeast *Williopsis saturnus* WC91-2. Microbiol Res 167:558–563
Wilson C (2013) Establishment of a world food preservation center. Agric Food Security 2:1–4
Wilson CL, Wisniewski ME, Droby S, Chalutz E (1993) A selection strategy for microbial antagonists to control postharvest diseases of fruits and vegetables. Sci Hortic 53:183–189
Wisniewski M, Wilson C, Droby S, Chalutz E, ElGhaouth A, Stevens C (2007) Postharvest biocontrol: new concepts and applications. In: Vincent C, Goettel MS, Lazarovits G (eds) Biological control: a global perspective. CABI, Wallingford, pp 262–273
Yu T, Chen J, Chen R, Huang P, Liu D, Zheng X (2007) Biocontrol of blue and gray mold diseases of pear fruit by integration of antagonistic yeast with salicylic acid. Int J Food Microbiol 116:337–345
Zhang D, Spadaro D, Garibaldi A, Gullino ML (2010) Selection and evaluation of new antagonists for their efficacy against postharvest brown rot of peaches. Postharvest Biol Technol 55:174–181

Chapter 18
Biotechnologically Relevant Yeasts from Patagonian Natural Environments

Diego Libkind, Martin Moliné, Andrea Trochine, Nicolas Bellora, and Virginia de Garcia

Abstract The Patagonia region constitutes a vast geographic area with multiple extreme environments having one or more of these stress factors: cold, high UV incidence, desiccation, ultra-oligotrophy, acidity, and the presence of heavy metals, among others. Yeasts that constantly live under stress conditions evolve adaptive mechanisms to minimize or resist their negative effects and thus permit survival and reproduction. These specific mechanisms are promising sources of biotechnologically relevant molecules or genes. Here we summarize numerous yeast bioprospection studies performed in the conventional and extreme environments of the Argentinean Patagonia. More than 1000 yeasts and dimorphic fungi were collected and molecularly identified; when possible, relevant secondary metabolites were screened, as well as their ability to tolerate several types of stress in laboratory conditions. Screened metabolites include carotenoid pigments, mycosporines (UV sunscreens), and cold-active enzymes. In some cases, these traits could be correlated to habitat characteristics and for those (e.g., mycosporines, carotenoid pigments, heavy metal tolerance) their potential role in the adaptive mechanisms to specific stress factors was evaluated. Genome sequencing and analyses were performed for biotechnologically relevant isolates such as *Saccharomyces eubayanus*, *Saccharomyces uvarum*, and *Phaffia rhodozyma* (synonym of *Xanthophyllomyces dendrorhous*). The biotechnological potential of selected species is addressed as specific study cases. The present work represents an overview of our findings related to biotechnologically relevant yeasts from Patagonian natural environments.

D. Libkind (✉)
Andean-Patagonian Institute for Biological and Geoenvironmental Technologies (IPATEC)
CONICET, National University of Comahue, Argentina
e-mail: libkindfd@comahue-conicet.gob.ar

M. Moliné • A. Trochine • N. Bellora • V. de Garcia
Laboratorio de Microbiología Aplicada, Biotecnología y Bioinformática de Levaduras, Instituto Andino-Patagónico en Tecnologías Biológicas y Geoambientales (IPATEC), CONICET y Universidad Nacional del Comahue (CRUB), San Carlos de Bariloche, Argentina

N.L. Olivera et al. (eds.), *Biology and Biotechnology of Patagonian Microorganisms*, DOI 10.1007/978-3-319-42801-7_18

18.1 Introduction

The Patagonian Andes possess unique physical and environmental characteristics (Villarosa et al. 2008), and its natural substrates have been poorly studied at the microbiological level. Pristine natural environments are common in Patagonia, with a higher diversity of fauna and flora toward the more humid and higher areas of Andean Patagonia. Regions under very harsh environmental conditions also exist in Patagonia and can be considered as extreme environments. These environments impose difficulties for microbial colonization and growth, selecting for microorganisms adapted to such conditions that frequently have evolved special metabolic abilities, many of which may be of interest for industrial exploitation. High ultraviolet radiation (UVR), high altitude, extreme desiccation, very cold or freezing temperatures, freezing and thawing cycles during the day, ultra-oligotrophic conditions, and volcanic activity, singly or combined, are among the stress factors that influence the microbial communities in the Patagonian natural environment, in particular Andean Patagonia. Yeasts belonging to either the Ascomycota or the Basidiomycota have proven to be able to adapt to multiple types of substrates and environments, including those regarded as extreme, and display an amazing diversity of highly plastic phenotypes. Several novel species have been obtained and formally described from those environments, and their screening for industrially relevant traits has been carried out. This chapter summarizes research carried out mainly on the biodiversity and biotechnology of Patagonian native yeasts, and how their adaptive characteristics are exploited for industrial applications.

18.1.1 Yeasts in Extreme Environments

Certain yeast species have evolved metabolic adaptations that allow them to colonize extreme natural habitats such as the deep sea (Nagahama et al. 2001; Gadanho and Sampaio 2005), glacial meltwaters (Vishniac 2005; Branda et al. 2010), hypersaline lakes (Butinar et al. 2005), ultra-oligotrophic mountain lakes exposed to increased UV radiation (Libkind et al. 2009a), and natural and anthropic hyperacidic ($pH<3$) aquatic environments (Gadanho et al. 2006), among others. Yeasts living in these environments are often polyextremophiles, that is, they tolerate many different extreme conditions (low pH, high temperatures, osmotic pressures, high concentration of heavy metals, chemolithotrophic microbial activity). In this section we review the research on yeasts isolated from extreme environments in Patagonia, with special emphasis in ultra-oligotrophic mountain lakes with high UVR, highly acidic waterbodies of volcanic origin, and very cold environments (glaciers and meltwater). Table 18.1 summarizes the principal yeast species recovered from Patagonian natural environments, including those considered extreme.

Table 18.1 Yeast species in Patagonian extreme environments and their biotechnological relevance

Habitat	Substrate	Predominant species	Novel yeasts described	Biotechnologically relevant species
Freshwater bodies	Lakes, lagoons, rivers	*R. mucilaginosa*[a] *Rh. babjevae* *Rh. kratochvilovae* *Sp. salmonicolor* *Sporobolomyces*[b]	*Sp. longiusculus* *S. patagonicus* *Cy. lacus-mascardii* *Cy. macerans* *R. meli*	Pigmented yeasts, sources of carotenoids, mycosporines, polyunsaturated fatty acids, and other lipids
	High-altitude lakes (>1400 m a.s.l.)	*R. mucilaginosa* *Rh. babjevae*	*H. takashimae*	
	Oligotrophic lakes	*R. mucilaginosa*[a] *Cr. victoriae*	*Dioszegia* sp. 1[c]	
	Acidic lakes and rivers	*Cr. agrionensis* *R. mucilaginosa* *Rh. toruloides*	*Cr. agrionensis*	Acid- and metal-tolerant yeasts: bioremediation of heavy metals
Glaciers	Meltwater	*Cr. spencermartinsiae* *L. fragarium*	*Cr. spencermartinsiae* *Cr. frias* *Cr. tronadorensis*	Cold-adapted yeasts: sources of cold-active enzymes, antifreeze proteins, lipids
	Ice	*D. crocea*, new genus[c] *S. ruberrimus* *D. fristingensis*		
Nothofagus forest	*Cyttaria* spp. fungal stromata, sap exudates, bark, leaves	*Saccharomyces* spp. *Hanseniospora* spp. *Pichia* spp.[d]	*W. patagonicus* *S. eubayanus* *Cys. psychroaquaticum*	Yeasts for fermentation processes (beer, wine)
	Soil	*Cr. podzolicus* (BS, R, E) *Cr. phenolicus*, *Cr. terreus*, and *T. porosum* (BS) *Cr. aerius* and *C. maritima* (R) *Cr. aerius*, *Cr. phenolicus* and *H. wattica* (E)	*La. nothofagi*[e] *Li. rizospherae*	Yeasts for enhancing plant growth and biocontrol
	Seeds, fruits	*A. pullulans* *Cr. heveanensis*		

A., Aureobasidium; C., Candida; Cr., Cryptococcus; Cy., Cystofilobasidium; Cys., Cystobasidium; D., Dioszegia; H., Holtermanniella; L., Leucosporidium; La., Lachancea; Li., Lindnera; R., Rhodotorula; Rh., Rhodosporidium; S., Sporobolomyces; Sp., Sporidiobolus; T., Trichosporon; W., Wickerhamomyces; BS, bulk soil; E, ectomycorrhizosphere; R, rhizosphere

[a]Frequently associated with anthropically influenced lakes or sites

[b]Species of this genus were predominant in lakes with low anthropic influence

[c]Formal description in progress

[d]Selective isolation of fermentative species (Ulloa et al. 2009)

[e]Strains also found in sap and bark of *Nothofagus* sp. (Mestre et al. 2010)

18.1.1.1 Ultra-oligotrophic Aquatic Environments Under High UVR

Yeasts are common inhabitants of aquatic environments, and their density and species diversity depend on water type and purity (Hagler and Mendoça-Hagler 1981). The northern part of Andean Patagonia offers a great variety of pristine, glacially formed waterbodies, covering an ultra- to mesotrophic range of small and large lakes including small high-elevation lakes or wetlands. These habitats are normally exposed to extended daylight (latitude 41–45°) and consequently increased UVR, also the result of ozone layer depletion and a clean atmosphere. Many Patagonian aquatic environments are still further affected by these environmental factors because of their transparency and ultra-oligotrophic character (Villafañe et al. 2001). Thus, yeast survival in pristine waters in Patagonia is conditioned by low temperature, ultra-oligotrophic conditions, and, more significantly, high UVR doses (Libkind et al. 2006; de Garcia et al. 2014).

Because of their differential ability to assimilate a larger number of complex carbon sources and a better capacity to cope with harsh conditions (Sampaio 2004), basidiomycetous yeasts predominate in the ultra-oligotrophic waterbodies of Patagonia (Libkind et al. 2003; Brandão et al. 2011). The biodiversity of basidiomycetous yeasts in certain lakes in Andean Patagonia (Argentina) was investigated, focusing on species producing photoprotective compounds such as carotenoid pigments and UV sunscreens (mycosporines) (Libkind et al. 2003, 2005a, 2009a; Brandao et al. 2011), both known strategies for the minimization of UV-induced damage in organisms (Roy 2000). These secondary metabolites, which are of industrial interest for many reasons, are discussed in Sect. 2 of this chapter. These studies indicated UVR as a selective factor that favors the occurrence of more UV-resistant yeast species/strains in these kinds of lakes and thus determining their yeast community structure (Libkind et al. 2009a; Moliné et al. 2009; Moliné 2010). Also, several novel species were described, including *Rhodotorula meli*, *Sporidiobolus longiusculus*, *Sporobolomyces patagonicus*, and *Cystofilobasidium lacus-mascardii* (Libkind et al. 2005b; 2009b; 2010) (Table 18.1). These species are interesting for their ability to accumulate carotenoid pigments, and some were regarded as biotechnologically relevant (Libkind et al. 2006; Libkind and van Broock 2006; Moliné et al. 2012).

Yeast diversity and distribution, including the entire cultivable yeast community, were evaluated in the pristine water of Nahuel Huapi (NH) Lake, one of the largest lakes in Patagonia (Brandão et al. 2011). Yeast counts ranged from 22 to 141 cfu l^{-1}, typical of clean lakes (Hagler and Ahearn 1987), with the highest values corresponding to the most anthropogenically influenced sites. Isolates from NH Lake were identified as belonging to 13 genera and 34 species, with 73.8 % being basidiomycetous. *Rhodotorula mucilaginosa* and *Cryptococcus victoriae* were the most frequently found species. Some yeast species were more represented in anthropogenically influenced sites (such as *Aureobasidium pullulans* and *R. mucilaginosa*) whereas the most represented species in sites considered less affected by human activity were also components of the community of the surrounding *Nothofagus* phylloplane. The occurrence and distribution of yeasts at the studied sites showed peculiar distributional patterns that are probably influenced by inputs of allochthonous organic matter from the borders of the lake and by abiotic factors such as UVR. Photoprotective compound-producing yeasts were mainly found in pelagic points of the lake, suggesting that both carotenoids and mycosporines production capacities are important for yeast survival under high-UVR conditions.

18.1.1.2 Acidic Environments

Acidic environments can harbor numerous microorganisms, including algae, bacteria, and fungi, with distinctive capacities to survive the acidic conditions and the resulting high metal concentrations. The yeast community of a natural acidic environment located in Northwestern Patagonia was analyzed (Rio Agrio and Lake Caviahue system) (Russo et al. 2008). Yeasts were isolated from water sites with different pHs, ranging from 1.6 to 6.7. The recovery of putative autochthonous yeasts was enhanced when water from the sampling site was used in the formulation of the isolation media, compared to use of conventional yeast media. In total, 25 different species were identified, with 99 % of the isolates being Basidiomycetes. *Rhodotorula mucilaginosa*, *Rhodosporidium toruloides,* and two novel *Cryptococcus* species were the most adapted species. One of the novel species, named *Cryptococcus agrionensis* (Russo et al. 2010), is highly resistant to heavy metals and belongs to the acid rock drainage (ARD) ecoclade (Gadanho and Sampaio 2009; Libkind et al. 2014; Russo et al. 2016) (Table 18.1). The second *Cryptococcus* species (referred to as *Cryptococcus* sp. 2) was able to grow in a very narrow pH range (2.5–4.5), with an optimum at pH 3, and thus could be regarded as an acidophilic yeast. Both *R. mucilaginosa* and *Cr. agrionensis* showed wider pH growth ranges. Ongoing studies aim to analyze the molecular basis of such atypical phenotypes. Interestingly, the yeast community of the naturally originated Patagonian acidic environment resembled that of acidic aquatic environments resulting from anthropic activities such as the São Domingos mines in Portugal and the Rio Tinto in Spain (Gadanho et al. 2006). The current knowledge of yeast diversity and ecology in acidic aquatic environments is scarce and limited. Detailed studies on their metabolic features, including assessment of their ability to bioremediate heavy metals, will give further insights into their biotechnological potential. In a recent study some of the aforementioned yeasts were evaluated for their ability to capture heavy metals (such as Cu^{2+}, Ni^{2+}, and Zn^{2+}) in solution at low pH, with promising results (Russo et al. 2016).

18.1.1.3 Cold Environments

Andean Patagonia in Argentina offers a great variety of glaciers and glacially formed waterbodies that are still glacier fed. The latter include small and large oligotrophic to ultra-oligotrophic lakes, including small high-elevation lakes, sometimes surrounded by a dense native forest of trees of *Nothofagus* spp. and *Austrocedrus chilensis* trees (Quirós and Drago 1985; Díaz et al. 2000). In Mount Tronador alone, in Nahuel Huapi National Park, there are ten different glaciers

(Rabassa et al. 1978). Los Glaciares National Park (Argentina) and Patagonian Icefields (Hielos Patagónicos) are the largest temperate ice masses in the Southern Hemisphere, accounting for more than 60 % of the Southern Hemisphere glacial area outside Antarctica. Perito Moreno Glacier is located within these icefields (Stuefer et al. 2007).

Extensive studies on the occurrence of psychrophilic and psychrotolerant yeasts from the cold environments of Patagonia have been carried out since the 1990s in aquatic (freshwater, meltwaters, glacial ice, seawater) and terrestrial habitats (flowers, phylloplane, sap exudates, bark, soil, rotten wood, rhizosphere, *Cyttaria* sp. stromata) (Brizzio and van Broock 1998; Libkind et al. 2003; 2004a, b; 2006; Brizzio et al. 2007; de Garcia et al. 2007; Libkind et al. 2007, 2008a, 2009a, 2011a; Mestre et al. 2011; de Garcia et al. 2012; Fernández et al. 2012). Yeasts isolated from these cold environments belong to taxa previously described as cold adapted whereas yeast species not considered as such were also present and considered transient components of the microbial community. An almost up-to-date review on cold-adapted yeasts from Patagonia was recently published (de Garcia et al. 2014). Since then a few novel investigations have been published, including the taxonomic reorganization of the psychrotolerant yeasts of the genus *Leucosporidium* (de Garcia et al. 2015). Members of this genus are particularly important as potential sources of extracellular enzymes that are active at low temperatures (cold enzymes), antifreeze proteins, and have the ability to biodegrade phenol and phenol-related compounds (Bergauer et al. 2005; Sampaio 2011a, b; de Garcia et al. 2012). In Patagonia the predominant species is *Leucosporidium creatinivorum*, a member of the *L. scotti* species complex (de Garcia et al. 2015). A considerable percentage (25–40 %) of the yeast species recovered from different cold substrates belonged to undescribed taxa. The most recently described is *Cystobasidium psychroaquaticum*, isolated in Patagonia from glacier meltwater and from the phylloplane of high-altitude trees (Yurkov et al. 2015). Including those previously mentioned, a total of 11 cold-adapted new yeast species have been formally described from Patagonian environments (Libkind et al. 2005b, 2009b; de Garcia et al. 2010a, b; Libkind et al. 2010, 2011a; de Garcia et al. 2012; Yurkov et al. 2015).

Regarding the number of cultivable yeast cells detected in aquatic cold environments of Patagonia, average yeast counts of 10^2 to 10^3 cfu l^{-1} were found in freshwater mountain lakes (Libkind et al. 2003, 2009a; Brandão et al. 2011), 1×10^2 to 3×10^2 cfu l^{-1} in meltwater rivers (de Garcia et al. 2007), and 1×10^3 to 5×10^3 cfu l^{-1} in continental glacial ice (de Garcia et al. 2012). Again, basidiomycetous yeasts are the predominant group in these environments. Similar results from different cold environments worldwide have been reported (Frisvad 2008; Buzzini et al. 2012). Notably, a relatively higher richness index of taxa among ice and meltwater samples was observed (de Garcia et al. 2012), compared to the values reported for soil samples in Patagonian forest (Mestre et al. 2011). Brandão et al. (2011) mentioned similar richness index values for water samples from Nahuel Huapi Lake (coast sites, $H = 2.2$, and pelagic sites, $H = 2.8$).

The occurrence of cold-adapted yeasts was evaluated in the dominant tree genus in Andean Patagonia, *Nothofagus* spp., in different terrestrial substrates including leaves, seeds, bark, rotting wood, sap, soil, rhizosphere, flowers, and the stromata of *Cyttaria* spp. In contrast to the studies of oligotrophic environments, the major component of the yeast community in terrestrial substrates were ascomycetous fungi. *Aureobasidium* was the most frequently isolated genus in the phylloplane and other substrates (Muñoz et al. 2013). Other interesting yeasts of the genera *Saccharomyces* and *Phaffia* (synonym *Xanthophyllomyces*) were also detected in such forests and are reviewed in the next section.

The diversity, distribution, and physiological properties of yeasts inhabiting different substrates related to *Nothofagus* forests (seeds, bulk soil, rhizosphere, ectomycorrhizosphere) were recently published (Mestre et al. 2011; Fernández et al. 2012; Mestre et al. 2014). For example, *Cryptococcus* species such as *Cr. podzolicus*, *Cr. phenolicus*, and *Cr. aerius* were the species most frequently occurring in *N. pumilio* (Mestre et al. 2011). Recently, some of these psychrotolerant yeasts (i.e., *Aureobasidium pullulans*, *Holtermaniella takashimae*, *Candida maritima*) were shown to possess plant growth-enhancing features, such as production of auxin-like compounds and siderophores, and the ability to solubilize inorganic phosphate and to reduce the growth of common plant pathogens (Mestre et al. 2016).

18.2 Biotechnologically Relevant Traits in Patagonian Native Yeasts

18.2.1 Carotenoid Pigments: Biological Function and Biotechnological Applications

Carotenoids are yellow to red natural pigments formed by a C-40 chain, which is considered the backbone of the molecule. This chain has several conjugated double bonds (7–15) where the p electrons are highly delocalized, conferring a low-energy excited state giving their characteristic yellow to dark red color (Britton 1995). Furthermore, several modifications can occur in this basic skeleton, giving rise to more than 700 types of naturally occurring carotenes and xanthophylls (Britton 2004). Different functions have been attributed to these pigments in fungi, including protecting against reactive oxygen species (ROS) and ultraviolet radiation (UVR), being precursors of hormones (in Mucorales), being associated with membrane permeability modifications, and providing resistance to heat, radiation, and oxidation (Lampila et al. 1985; Britton 1995; Schroeder and Johnson 1995; Johnson and Schroeder 1996; Britton 2008). Regardless of their biological function, carotenoids are important for their benefits to human health, and are known to act as provitamin A (Olson 1989; Johnson and Schroeder 1996), antioxidants (Krinsky 1979; Sies and Stahl 1995), and antimutagens and anticarcinogens (Rao and Agarwal 2000; Donaldson 2004; Rao and Rao 2007). For these reasons they are of interest to the pharmaceutical, chemical, food, and feed industries (Ausich 1997; Rodríguez-Sáiz et al. 2010).

Yeast as well as other fungi can synthesize and accumulate carotenoid compounds. Most of these yeasts are called the "red yeasts," given the red to salmon-pink color of their colonies when grown in the laboratory. Despite the large number of possible pigments, red yeasts only synthesize a few carotenoid molecules, and three pigments are common for most yeast species; torulene, γ-carotene, and β-carotene. These carotenoids can be produced by species of the subphyla Agaricomycotina, Pucciniomycotina, Ustilaginomycotina, Taphrinomycotina, and Pezizomycotina (Kurtzman et al. 2011). The carotenoid pigment torularhodin (a xanthophyll product of the oxidation of torulene) is less common and is found associated with Pucciniomycotina species (Davoli et al. 2004; Buzzini et al. 2007; Sperstad et al. 2006). Torularhodin has been reported to have provitamin A activity in vitro, better antioxidant activity against singlet oxygen, and a more potent effect on the scavenging of peroxyl radicals than β-carotene (Simpson 1983; Sakaki et al. 2001; Ungureanu and Ferdes 2012). Other rare yeast carotenoids, produced by different species, are xanthophylls such as 16-hydroxytorulene, torularhodinaldehyde, plectaniaxanthin, and 2-hydroxyplectaniaxanthin, with hitherto unknown applications. Finally, the biotechnologically relevant astaxanthin can only be synthesized by one yeast species: *Phaffia rhodozyma.*

Among Patagonian yeast mycobiota, species producing carotenoids were found in all the environments tested so far (water, soil, phylloplane, glaciers), and *Rhodotorula mucilaginosa* was the most common species. In our studies, the freshwater environment had the higher proportion (>50 %) of red yeasts. In a survey of five high-altitude water bodies located in the Nahuel Huapi National Park, Libkind et al. (2009b) revealed that carotenogenic yeasts prevail in lakes with higher transparency. More than 24 yeast species were recovered in this study, and 12 corresponded to red yeasts (classified into seven genera). *Rhodotorula mucilaginosa* was the most frequently isolated species (representing more than 50 % of the total isolates), followed by *Rhodosporidium babjevae*. Other less frequent pigmented species such as *Sporobolomyces ruberrimus*, *Cystobasidium laryngis*, *Cystibasidium minutum* (ex *Rhodotorula minuta*), *Sporobolomyces marcillae*, *Rhodosporidium diobovatum*, and three *Dioszegia* species have also been isolated from Patagonian lakes (Libkind, et al. 2003, 2009a). In another study Brandão et al. (2011) revealed the occurrence of 47–74 % of pigmented yeasts in pelagic sites (more transparent) of the Nahuel Huapi Lake. In Patagonian glacier meltwater and ice, the number of species bearing carotenoids was less representative; however, the isolation of species such as *Dioszegia crocea*, *Dioszegia fristingensis*, and *Sporobolomyces ruberrimus* was common, reaching more than 30 % of the isolates (de Garcia et al. 2012). In contrast, pigmented strains are unusual in soil, and only three species, namely, *Cystofilobasidium infirmominiatum*, *Cystofilobasidium capitatum*, and *Rhodotorula colostri*, were reported from this environment (Mestre et al. 2014). Given the photoprotective function of carotenoids, our results suggest that the higher abundance of carotenogenic yeasts in substrates with a higher exposition to UVR is the result of their higher tolerance to this damaging stress factor (Libkind et al. 2006, 2009a; Moliné et al. 2011a).

The proportion and type of carotenoids produced by Patagonian yeasts was highly variable and depended on the phylogenetic group to which they belonged. For example, in *Rhodotorula mucilaginosa* strains, carotenoid content ranged from 60 to 301 mg g^{-1} of dry biomass (Libkind and van Broock 2006), and torularhodin was the most important carotenoid, representing 61–98 % of the total carotenoid content (Moliné et al. 2011a, 2012). *Phaffia rhodozyma* carotenoids ranged from 98 to 415 mg g^{-1} of dry biomass, and astaxanthin was the principal carotenoid (Libkind et al. 2008b), indicating that quantitative pigment production is a strain-related feature. Moreover, the origin of the strains seemed to be a relevant aspect when considering carotenoid accumulation, because the total carotenoid content of Patagonian strains (mainly isolated from extreme environments) was in almost all cases higher than values observed for type strains of the same or related species (Libkind et al. 2004b; 2009b; Libkind and van Broock 2006; Moliné 2010; Moliné et al. 2011a).

Pigment analysis of the Patagonian yeasts revealed that torulene, torularhodin, and β-carotene were the most important carotenoids present in the genera *Rhodotorula*, *Rhodosporidium*, *Sporobolomyces*, and *Sporidiobolus* (Libkind and van Broock 2006; Buzzini et al. 2007; Moliné et al. 2011a) (Fig. 18.1). Another carotenoid observed only in a few isolates was γ-carotene. Other yeast genera such as *Cystofilobasidium* and *Dioszegia* were also found to synthesize the pigments torulene and β-carotene; however, in *Cystofilobasidium* there are other unknown carotenoids representing the major compounds (probably 16′-hydroxytorulene, torularhodinaldehyde, and β-apo-2′-carotenal, based on Herz et al. (2007), and in *Dioszegia* the principal carotenoid identified was plectaniaxanthin (Moliné 2010). Finally, the most

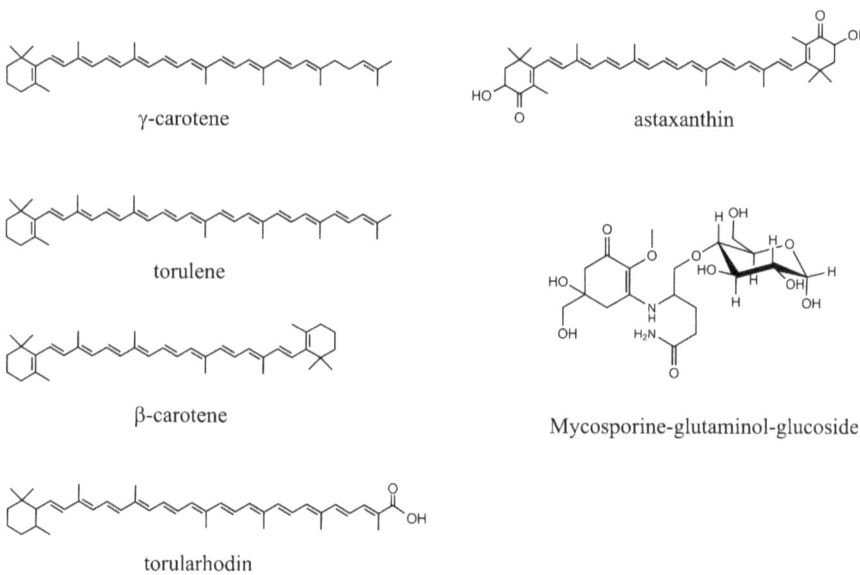

Fig. 18.1 Chemical structure of the principal carotenoids and mycosporines found in yeasts

relevant pigment identified in Patagonian yeasts was the xhantophyll astaxanthin produced by *Phaffia rhodozyma* (Fig. 18.1). Evidence has been gathered that demonstrates astaxanthin protects cells from oxidative stress including that caused by photogenerated reactive oxygen species (Schroeder and Johnson 1995). Astaxanthin is probably the most important yeast carotenoid for its high commercial value in pharmaceutics, nutraceutics, and cosmetics (Bhosale and Bernstein 2005). In addition to its antioxidant properties, astaxanthin has antiinflammatory and neuroprotective properties (Naguib 2000; Lee et al. 2011). Furthermore, astaxanthin is also used in aquaculture for pigmentation of fish and crustaceans, being the most expensive feed ingredient (Johnson and Schroeder 1995). Therefore, we address further *Phaffia rhodozyma* and its importance in aquaculture in Sect. 3.3 of this chapter and in Chapter 13, Microorganisms from Patagonian Aquatic Environments for Use in Aquaculture.

Carotenoid accumulation, the concentration and relative proportion of each pigment, is also affected by the culture media and by physical and chemical factors. For example, in *Phaffia rhodozyma*, a high carbon:nitrogen (C/N) ratio favors carotenoid production (Yamane et al. 1997; Flores-Cotera et al. 2001), and the same occurs for strains isolated from Patagonia (Moliné 2010), whereas in *Rhodotorula mucilaginosa* strains, the specific carotenoid production is not affected by the C/N ratio (Libkind et al. 2004a; Libkind and van Broock 2006). Light is one of the most important environmental forces triggering the synthesis of carotenoids in yeasts (Tada and Shiroishi 1982; An and Johnson 1990; Sakaki et al. 2001; Bhosale 2004). Exposure to photosynthetically active radiation (PAR) and UV-A was found to produce different responses in the synthesis of carotenoid pigments in yeast, with increases from 6 % up to 800 % depending on the species. The increasing effect of photostimulation in carotenoid synthesis after exposure to PAR + UV-A was negatively correlated to the basal concentration of carotenoids, suggesting that yeasts with high constitutive levels of intracellular carotenoids were less responsive (Libkind et al. 2004b). The reason is probably that they already possess sufficient carotenoids functioning as photoprotective agents to cope with the UVR-damaging effects imposed in our experiments. Evidence that one of the roles of carotenoid pigments in yeast cells could be photoprotection has been accumulated for decades (Maxwell et al. 1966; Tsimako et al. 2002). Using Patagonian carotenogenic isolates and naturally occurring albino strains, we experimentally compared pigmented and albino strains of *Cystofilobasidium capitatum* and *Sporobolomyces ruberrimus*. Albino strains invariably were less tolerant to UV-B exposure than pigmented strains, and a direct relationship between carotenoid content and survivorship in *Cy. capitatum* was observed (Libkind et al. 2006; Moliné et al. 2009). Afterward, using a set of *R. mucilaginosa* and *P. rhodozyma* strains, we established a significant positive relationship between intracellular carotenoid concentration and UV-B survival (Moliné 2010; Moliné et al. 2011a). Analysis of carotenoid content pointed out that torularhodin in *R. mucilaginosa* and astaxanthin in *P. rhodozyma* provided effective photoprotection, whereas other carotenoids such as β-carotene showed a lack of correlation with survival to UV-B.

The studies described here show that the environmental niches for the isolation of red yeasts were identified, as well as the carotenoid pigments synthesized, and the biological function of such pigments was experimentally confirmed. In summary, red yeasts from Patagonia represent an interesting source of carotenoid pigments with biotechnological value, and further studies are needed to determine their potential for industrial applications.

18.2.2 Natural Sunscreens: Mycosporines

Mycosporines are water-soluble compounds composed by a cyclohexenone attached to an amino acid (or amino alcohol). There are more than a dozen different mycosporines, but in fungi only mycosporine serine, alanine, α-amino alcohol serinol, pyroglutamic acid, and the related pairs glutamine–glutaminol and glutamic acid–glutamicol were described (Young and Patterson 1982; Bernillon et al. 1984; Leite and Nicholson 1992; Volkmann et al. 2003; Sommaruga et al. 2004). Because these compounds absorb light in the UV spectrum with a maximum at 310 nm, the primary function of mycosporines was to act as photoprotective UV filters (Shick and Dunlap 2002; Torres et al. 2004). However, other functions were also attributed, including antioxidant activity, osmoregulation, resistance to thermal stress, and serving as intracellular nitrogen storage (Oren and Gunde-Cimerman 2007). Only recently was mycosporine synthesis reported for yeasts by our laboratory. Different basidiomycetous yeasts, most isolated from Patagonia lakes, were found to synthesize a UV-absorbing compound when grown under photosynthetically active radiation (PAR) (Libkind et al. 2004b). In basidiomycetous yeasts, mycosporines were reported in different species of the subphyla Pucciniomycotina (Libkind et al. 2011b) and Agaricomycotina (Libkind et al. 2005a, 2011c). In both groups there are taxa with and without the ability to produce mycosporines, suggesting that this trait might be plesiomorphic. Yeasts and dimorphic fungi of the Ascomycota able to produce mycosporines are classified in the orders Dothideales, Capnodiales, and Taphrinales (Gunde-Cimerman and Plemenitaš 2006; Kogej et al. 2006). In most yeasts so far tested a main mycosporine was detected, mycosporine-glutaminol-glucoside (MGG) (Sommaruga et al. 2004), which consists of a cyclohexanone attached to a glutaminol and glucose molecule, with a molecular weight of 464.5 g mol^{-1} and a characteristic absorbance at 310 nm (Fig. 18.1).

The biodiversity of basidiomycetous MGG-producing yeasts was investigated by Libkind et al. (2003, 2005a, 2009a) in certain lakes in Andean Patagonia (Argentina). The occurrence of MGG-positive yeast in waterbodies goes through a wide range, from 14% to near 90% of total cultivable yeast community. MGG synthesis was more frequent in yeasts that were not able to accumulate carotenoid pigments, and only red yeasts such as *Rhodotorula minuta*, *R. laryngis*, and *Dioszegia* spp. were positive for MGG. As for red yeasts, the abundance of mycosporine-positive species was higher in highly transparent lakes or in pelagic zones (Brandão et al. 2011). In glacier meltwater and ice, yeasts able to synthesize MGG are less frequent, including such species as *Dioszegia crocea* and *D. fristingensis* as the most important ones (de Garcia et al. 2012).

As for carotenoids, the biological role of mycosporines in yeasts was evaluated. Libkind et al. (2004b) suggested a photo-protective role, based in the strong response of mycosporine production to radiation. Using a set of *Cryptococcus stepposus* and *Phaffia rhodozyma* strains isolated from different Patagonian environments, we found a high positive correlation between survival to UV-B and MGG concentration. The fact that MGG accumulation protects yeasts against the effects of UVR avoiding the direct damage of DNA was also experimentally demonstrated (Moliné 2010; Moliné et al. 2011b). In addition, biochemical characterization of MGG from yeast revealed that it possesses high photostability and antioxidant properties (Moliné et al. 2011b). Thus, MGG appears as an interesting compound for multipurpose UV sunscreens, and yeasts become a valuable biotechnological source of these natural UV protectants. Several patents related to the production and usage of mycosporines sensu lato from different types of microorganisms have been published (for review, see Colabella et al. 2014) among which so far only one covers the use of yeasts (van Broock et al. 2009). In this patent, MGG from yeasts was purified and incorporated into base creams that were tested for UVR sunscreen. The product, with MGG concentration between 0.1 % and 5 %, showed a reduction in the UV-B flux (315 nm) for all the concentrations tested. UVB reduction values for 5 % were comparable with those obtained for commercial sunscreens with SPF 15 and 30.

These results are promising and lead to the investigation of the ability of different yeast species and strains from Patagonia to produce MGG in laboratory conditions. The concentration of MGG produced by Patagonian yeasts was highly variable; quantitative studies on MGG accumulation showed that differences in the production of this compound occur between different yeast species but also between strains. MGG accumulation varied among species, ranging from 2.5 to more than 50 mg g^{-1} dry weight. For example, different strains of *Cryptococcus stepposus* produce from 2.5 to a maximum of 5 mg g^{-1} whereas other species of the genera *Dioszegia* and *Aureobasidium* produce between 35 and more than 50 mg g^{-1} (Libkind et al. 2005a; Moliné 2010; Moliné et al. 2011b; Muñoz et al. 2013). Such production yields are larger than those observed for similar compounds (other mycosporines) in Cyanobacteria (2–9 mg g^{-1}) (Scherer et al. 1988; Portwich and Garcia-Pichel 1999). Using a set of 20 strains of *Phaffia rhodozyma* obtained either from culture collections or from Patagonian natural environments, we observed that the MGG production from the latter source ranged from 16 to 39 mg g^{-1} (average, 25 mg g^{-1}), whereas in the former MGG production ranged from 8 to 26 mg g^{-1} (average, 18 mg g^{-1}) (Moliné 2010). Thus, certain Patagonian yeasts are interesting sources of natural sunscreens for use in human photoprotection products.

MGG production in yeasts depends on illumination conditions and culture media composition. Yeast mycosporinogenesis is a phenomenon triggered by photostimulation (Libkind et al. 2004b; 2011c; Moliné 2010); however, significant constitutive synthesis has been found for certain yeast species. In *Cryptococcus stepposus*, increase in MGG accumulation after photostimulation (PAR) is 2- to 3 fold (Moliné et al. 2011b), whereas in *P. rhodozyma* increases are 10 fold and in *Cystobasidium minutum* (ex *Rhodotorula minuta*) the increases are 20 fold higher. When UVR rather than PAR is used, the latter increase goes up to 34 fold, indicating that the type and intensity of light modulates the response in MGG production in yeasts (Libkind et al. 2006).

Active compounds used in commercial sunscreens are, in general, lipophilic organic molecules produced by chemistry synthesis. These compounds absorb light in the UV-A to UV-C spectrum and are commonly added in concentrations up to 15 % to sunscreen products for skin protection (Wang et al. 2010; Loden et al. 2011). Because of growing public concern about skin damage by UV light, the demand for UV screens is increasing; however, several studies have brought to light the negative effects of these compounds to the environment and human health. Chemically synthesized UV screens are relevant environmental contaminants (Balmer et al. 2005), and for some active compounds androgenic and estrogenic effects have been detected (Krause et al. 2012). Further, common active ingredients such as 3-benzophenone of zinc oxide have been shown to produce reactive oxygen species and a photoallergenic effect in humans (Brezová et al. 2005; Hanson et al. 2006; Scheuer and Warshaw 2006). Consequently, the industry is seeking for more innocuous and natural compounds to replace current chemically synthesized and controversial ingredients. Today, a few examples of sunscreen products containing mycosporine-like amino acids are known, such as 'Helioguard' and 'Helionori' (Colabella et al. 2014). Although none of these contains MGG or any yeast-derived compound, we anticipate that natural sunscreen compounds such as yeast MGG represent a new and natural alternative to be used in commercial photoprotection products.

18.2.3 Cool Applications from Cold-Adapted Yeasts: Extracellular Enzymes

Cold-adapted microorganisms, including bacteria, archaea, filamentous fungi, algae, and yeasts, are being studied as sources for cold-active enzymes. The biotechnological value of cold-adapted enzymes stems from their high catalytic activity at low to moderate temperatures, their high thermolability at elevated temperatures, and their ability to function in organic solvents (Gerday et al. 1997). Applications include their use in cheese, wine, and juice production (pectinases), dough fermentation (xylanases), animal feed (cellulases), pulp bleaching (alpha amylases), detergents (lipases, peptidases), molecular biology (DNA, RNA polymerases), and nutrition (phytases) (Cavicchioli et al. 2011). Yeasts are heterotrophic organisms with the ability to degrade organic macromolecules by means of extracellular hydrolytic enzymes. Low molecular weight compounds are subsequently transported to serve in both catabolic and anabolic reactions. Cold-adapted yeasts secrete extracellular enzymes that can catalyze these reactions at low temperatures, mainly because of their highly flexible structures (Gerday et al. 1997). Other extracellular proteins function in cell-wall remodeling and anti-freezing, among other reactions (Crevel et al. 2002). The presence of some of these proteins was evaluated in a number of surveys with yeasts isolated from cold environments in Patagonia. In a first survey, 78 yeast strains were analyzed for extracellular enzymatic activities (EEA) (de Garcia et al. 2007). The ability of the strains to degrade starch, proteins, lipids, pectin, cellulose, and chitin was evaluated. Ninety-five percent of the tested strains showed at least one extracellular enzyme activity at either 4 °C or 20 °C. Lipolysis was the most frequent extracellular enzyme activity whereas none of the strains showed the ability to hydrolyze chitin or cellulose. In a subsequent screening, yeasts from glacial and subglacial water (belonging to the genera *Cryptococcus*, *Leucosporidiella*, *Dioszegia*, *Mrakia*, *Rhodotorula*, *Rhodosporidium*, *Sporobolomyces*, *Sporidiobolus*, *Cystofilobasidium*, and *Udeniomyces*) were studied (Brizzio et al. 2007). Most of the 91 studied isolates exhibited amilolytic, protease, and lipase activities that were higher at 4 °C than at 20 °C. In a more recent survey, five enzymatic activities were analyzed in 212 yeast strains isolated from ice and meltwater (de Garcia et al. 2012). At least one enzymatic activity was present in 85 % of the strains, whereas 18 % showed positive tests for all the five activities (degradation of starch, caseine, pectin, carboxymethyl cellulose, and Tween-80). As a consequence of these studies, a number of biotechnologically relevant strains have emerged. Ongoing studies, including genomic and proteomic analyses of the extracellular proteins, will allow finding interesting new cold enzymes. Detailed studies of their enzymatic capabilities are also necessary to reveal possible applications.

18.3 Biotechnologically Relevant Yeasts from Patagonia: Three Hot Cases from the Cold

In this section we review the three most interesting cases of yeasts isolated from Patagonian natural environments that clearly have potential for biotechnological exploitation: these include the recently described new species of *Saccharomyces*, *S. eubayanus*, the ancestor of the lager brewing hybrid yeast. Also, the discovery of novel populations of *S. uvarum* in Patagonia is addressed together with their contribution to the origin of domesticated lineages used for wine and cider production. Finally, the finding and characterization of the astaxanthinogenic and mycosporinogenic yeast *P. rhodozyma* from Patagonia are reviewed.

18.3.1 The Discovery of S. eubayanus*, the Ancestor of the Lager Brewing Yeast*

Beer is the most common fermented beverage in the world and can be classified as ale or lager, depending on the fermentation conditions and yeasts used. Lager beer is the most common commercially produced beer worldwide (94 % of total beer market) and yet, the genetic origin of the yeast strains that brew them has been full of mystery and controversy. Compared with conventional ale-style beers, which are generally brewed with *Saccharomyces cerevisiae* (Hornsey 2003), lagers are brewed at colder temperatures with allopolyploid hybrid yeasts of *Saccharomyces cerevisiae* × *S. eubayanus* (known as *S. pastorianus*). *S. eubayanus* was only recently discovered and formally described as a result of yeast biodiversity surveys in Patagonian *Nothofagus* forests, from which two cold-adapted *Saccharomyces* species were recovered mainly from bark and soil samples but particularly from the stromata of *Cyttaria* fungus (Libkind et al. 2011a). *S. eubayanus* was the first yeast species to be formally described based on a complete genome sequence (Libkind et al. 2011a). The first draft genome of the type strain (PYCC 6148) was obtained for comparison with the non-*S. cerevisiae* subgenome of the lager yeast (see Sect. 4.1), revealing the 99.5 % similarity that lead to the conclusion that *S. eubayanus* was the ancestor species. Today, additional isolates have been obtained from other parts of the world, generating discussion about the actual origin of the population that gave rise to the hybdrid. However, so far none of the new sources of *S. eubayanus* showed the high frequency of recovery and the large genetic diversity found in Patagonia (addressed in Sect. 4.1). The second species of *Saccharomyces* found in Andean Patagonia *Nothofagus* forests was *S. uvarum*, the sister species of *S. eubayanus*, which is discussed next.

In Patagonia, *S. eubayanus* seems to be partially restricted to certain species of *Nothofagus* such as *N. pumilio* and *N. antarctica*; however Rodríguez et al. (2014) found several isolates from a different tree in the north of Patagonia: *Araucaria araucana*. These tree species are endemic from Patagonia; thus in other parts of the world *S. eubayanus* was collected mainly from oak trees, or other types such as *Cedrus* sp., *Pinus taeda*, and *Fagus* sp. (Peris et al. 2014; 2016).

18.3.2 Saccharomyces uvarum: *Wine and Cider*

Thus, in Patagonian habitats *S. eubayanus* and *S. uvarum* (two sister species), exist in apparent sympatry in *Nothofagus* (Southern beech) forests, but are isolated genetically through intrinsic postzygotic barriers (Libkind et al. 2011a). Previously, it was shown that sympatric *Saccharomyces* species tend to have different growth temperature preferences, as is true for *S. cerevisiae* (thermotolerant) and *Saccharomyces kudriavzevii* (psychrotolerant) co-occurring in Mediterranean regions, as well as *Saccharomyces paradoxus* (thermotolerant) and *S. uvarum* (psychrotolerant) coinhabiting temperate Europe and North America (Sampaio and Gonçalves 2008). Another particular characteristic of Patagonian environments in contrast to North Hemisphere counterparts is the almost complete occupancy of the *Nothofagus* niche by psychrotolerant *Saccharomyces* species. Although less than 50 % of the isolates from bark and soil samples from the North Hemisphere belong to *Saccharomyces*, in Patagonia these values range from 64 % to 95 % for both *S. uvarum* and *S. eubayanus*. The substrate of greatest occupancy is *Cyttaria* stromata (~95 %), which is in agreement with its high content of simple sugars.

Furthermore, a putative ecological isolation through host preference was detected, given that *S. eubayanus* was found in association with *N. antarctica* and *N. pumilio*, whereas *S. uvarum* was associated with *N. dombeyi*. This finding might explain the co-existence of these two hitherto phenotypically undistinguishable species.

18.3.3 *The Colorful Case:* **Phaffia rhodozyma**

Phaffia rhodozyma (synonym of *Xanthophyllomyces dendrorhous*) represents an exceptional fungal case of the basidiomycota, given that it combines the production of orange-colored colonies with the ability to ferment simple sugars. The main carotenoid pigment synthesized by *Phaffia* is astaxanthin, another exclusive characteristic of this yeast species (Andrewes et al. 1976) and the main reason for which it is currently being exploited biotechnologically as a natural source of astaxanthin in aquaculture feed (Rodríguez-Sáiz et al. 2010). The first isolates of *P. rhodozyma* were found in association with spring sap flows of various broad-leaved trees in Japan, Alaska, and Russia (Phaff et al. 1972; Golubev et al. 1977). Later, more strains were recovered from beech trees in central Europe (Weber et al. 2006) and the United States (US) (Fell et al. 2007). The range of *P. rhodozyma* was significantly expanded when a South American population associated with *Nothofagus* trees (southern beech), particularly the stromata of its biotrophic fungal parasite *Cyttaria* spp., was discovered (Libkind et al. 2007, 2008b, 2011d). The Patagonian isolates were found to be genetically different from the Northern Hemisphere strains based on DNA–DNA reassociation experiments, micro/mini-satellite-primed (MSP)–polymerase chain reaction (PCR) fingerprinting, as well as internal transcribed spacer (ITS) and intergenic spacer (IGS) rRNA gene sequencing (Libkind et al. 2007, 2011b). Finally, thes differences were confirmed using multi-locus sequence typing (David-Palma et al. 2014) and later using complete genome analyzes (Bellora et al., in press). However, Patagonian strains appeared to be genetically uniform (minor differences were found using the L41 gene as marker) and could be included into a distinct population, supporting the hypothesis that geographic isolation and association with different host species has determined genetically different *P. rhodozyma* populations worldwide (David-Palma et al. 2014). Weber et al. (2008) described a novel isolate from Chile with marked ITS and LSU sequence differences from the other known populations. This single isolate was obtained from a leaf of the Tasmanian blue gum tree (*Eucalyptus globulus*) in the Mediterranean climate at Concepci6n, a tree species originally from Australasia. David-Palma et al. (2014) reported the isolation of highly divergent lineages of *P. rhodozyma* from *Nothofagus* forests in Australia and New Zealand, expanding the known geographic distribution of this yeast and its genetic diversity. Two of these lineages deserve the assignation to distinct species and will be described in the near future as novel taxa in the genus *Phaffia* (unpublished results).

It is evident that *P. rhodozyma* possesses a greater genetic variability and geographic distribution than previously thought, generating the necessity to uncover it and assess its potential for the astaxanthin and UV sunscreen industry. However, difficulties for the isolation of this yeast hinder extensive environmental surveys. Our group developed a new and innovative strategy for improving *P. rhodozyma* recovery rate and identification from environmental samples (Tognetti et al. 2013), as well as a new PCR-based method for the rapid identification of *P. rhodozyma* isolates (Colabella and Libkind 2016).

18.4 Genomic Approaches to the Study of Patagonian Yeasts

As already demonstrated, yeasts from Patagonian natural environments include biotechnologically useful species and strains relevant to the production of beer, wines, antioxidants, photoprotective and cryoprotective compounds, among others. Genome assemblies using next-generation sequencing (NGS) and comparative genomic analyses were performed for the most interesting species, namely, *Saccharomyces eubayanus* (brewing), *S. uvarum* (wine and cider), and *Phaffia rhodozyma* (astaxanthin and mycosporines). The phylogenetic analyses in combination with geographic information enlightened the relationships between strains from the Northern and Southern Hemispheres and contributed to a better understanding of the origin and complexity of domesticated genomes. On the other hand, genome mining allowed detecting and characterizing specific genes and variants implicated in pathways of processes of biotechnological relevance. A remarkable outcome of such studies was the discovery of the wild genetic stock of domesticated yeasts currently used in major fermented beverages industries such as beer, wine, and cider (Table 18.2).

18.4.1 Phylogenomic and Phylogeographic Studies of Industrially Relevant Yeasts from Patagonia

In recent years, several genomics studies on the phylogeography of biotechnological relevant yeasts from Patagonia were conducted, including the two psychrotolerant biologically recognized species of *Saccharomyces*: *S. eubayanus* and *S. uvarum*.

The draft genome sequence of *S. eubayanus* was reported in 2011 by Libkind and colleagues (Libkind et al. 2011a) from wild Patagonian isolates that showed some genetic resemblance to the currently controversial species *S. bayanus*. Surprisingly, the authors found a 99.5 % of identity of the wild *S. eubayanus* to the non-*S. cerevisiae* subgenome of the allopolyploid lager brewing yeast: *S. pastorianus*. No evidence of introgression, hybridization, or horizontal gene transfer indicated that *S. eubayanus* represented a pure lineage. On the other hand, *S. pastorianus* as well

Table 18.2 Available genomes of Patagonian strains that belong to biotechnologically relevant yeast species

Species	Genome size (Mb)	Genomes available for Patagonian strains (other strains)	Biotechnology application	Remarks
S. eubayanus	11.6	18 (5)	Brewing	The largest diversity is in Patagonia
S. uvarum	11.5	16 (38)	Wine, cider	The largest diversity is in Patagonia
P. rhodozyma	18.9	1 (2)	Asthaxantin, mycosporines	Unique population in Patagonia

S., Saccharomyces; P., Phaffia

as *S. bayanus* are not biologically recognized "species" but hybrids, products of the artificial brewing environment with no occurrence in nature (Libkind et al. 2011a; Baker et al. 2015). The genomic complexity of *S. bayanus*, a hybrid yeast frequently associated with contaminated beer, could be explained by the contribution of mixtures of regions from *S. uvarum*, *S. eubayanus*, and *S. cerevisiae*. *S. eubayanus* and *S. uvarum* are sister species and form a basal clade distant to *S. cerevisiae* within the genus. Their genome-wide divergence is ~7 % and is the lowest between *Saccharomyces* species to result in genetic isolation (Libkind et al. 2011a). The two species coexist in Patagonian *Nothofagus* forests. Later, a higher-quality genome of *S. eubayanus* was released together with the first mitochondrial complete sequence (Baker et al. 2015). In this study, *S. eubayanus* subgenomes of lager brewing yeasts were shown to have experienced increased rates of evolution since hybridization, and that certain genes involved in metabolism may have been particularly affected (see following section). Interestingly, it was demonstrated that the *S. eubayanus* subgenome underwent an especially strong shift in selection regimes, consistent with more extensive domestication of the *S. cerevisiae* parent before hybridization. In contrast to recent proposals that lager brewing yeasts were domesticated following a single hybridization event (Walther et al. 2014; Wendland 2014), the radically different neutral site divergences between the subgenomes of the two major lager yeast lineages strongly favor at least two independent origins for the *S. cerevisiae* × *S. eubayanus* hybrids that brew lager beers (Baker et al. 2015). The recent isolation of newly *S. eubayanus* strains and their genomic characterization fueled the controversy of which population gave rise to the lager hybrid. Peris et al. (2016), using genome sequence data, examined the relationships of a larger set of wild *S. eubayanus* strains to each other and to domesticated lager strains. Results supported the existence of a relatively low diversity lineage of *S. eubayanus* whose distribution stretches across the Holarctic region and includes wild isolates from Tibet (Bing et al. 2014), new wild isolates from North America (Peris et al. 2014; 2016), and the *S. eubayanus* parents of lager yeasts. This clade is closely related to a high-diversity population that is found primarily in South America but includes some widely distributed isolates in the US (Peris et al. 2016) and New Zealand (Gayevskiy and Goddard 2016). It was further shown that no single Holarctic isolate was the sole closest relative of lager yeasts and that the wild Holarctic population of *S. eubayanus* is responsible for genetic variation still segregating among modern lager brewing hybrids. These observations suggest that lager yeast origins were more complex than we thought and stress the need for further investigations in the Northern Hemisphere.

Almeida and colleagues studied 54 *S. uvarum* domesticated (from wine and cider) and wild strains from different geographic areas of the globe based on high-quality polymorphic sites and resolved the strains into three main groups (Almeida et al. 2014): a clade that contained all Holarctic isolates (including the domesticated ones) and a few from South America, a second with only South American isolates, and a third with a few recently found Australasian strains. Holarctic populations when compared to Patagonian ones were extremely low in genomic diversity. Then, the highest diversity of strains of *S. uvarum* was thus detected in Patagonian environments. The Australasian lineage instead diverged significantly from the other two groups (by

4.4%), forming a sister group to the two other clades and being closer to *S. uvarum* than any other *Saccharomyces* species. It yielded similar divergences to those found between North American and European populations of *S. paradoxus* (Liti et al. 2006), so it was considered conspecific, although possibly going through a process of allopatric speciation. Phylogenomic analyses support the view that a restricted subset of one of the two Patagonian *S. uvarum* populations gave rise to the Holarctic population, although vectors and mechanisms of this migration into the Northern Hemisphere remain to be elucidated (Almeida et al. 2014).

In a recent study, the genomes of the Patagonian and the type strain of the astaxanthin- and mycosporine-producing yeast *P. rhodozyma* were assembled and compared (Bellora et al. 2016). The Patagonian strain showed 4.4% of genomic divergence toward two sequenced Holarctic strains (0.073% between them) (CBS 7918 T and CBS 6938), indicating that an allopatric speciation process might be occurring and that the former deserves to be assigned at least to a distinct variety. A considerable number of exclusive genes were present in the Patagonian strain but not in the European strains. Other interesting observations include a high occurrence of introns in *P. rhodozyma* and other Cystofilobasidiales and new insights into fungal homothallism.

18.4.2 Genes Related to Biotechnologically Relevant Pathways and Genomic Footprints of Domestication

The study of the effect of domestication in the lager brewing yeast was only possible once the genome of the parental *S. eubayanus* become available. Genetic changes detected in the *S. pastorianus* genome in comparison to the parental strain that seem to have been favored by the brewing environment include extra copies of *S. cerevisiae IMA1* (isomaltase, cleavage of disaccharide isomaltose), inactivation of the *SUL1* gene (encode high-affinity transporters of sulfate, the metabolic precursor of sulfite, a known antioxidant and flavor stabilizer), favoring *SUL2* which improves sulfite production in brewing conditions, and several gene expression regulators related to genes that allow alcohol utilization (i.e., *ADR1*), glucose-repressed genes such as *MAL* genes (i.e., *REG2*), among others (Libkind et al. 2011a; Baker et al. 2015). Moreover, a similar mtDNA gene arrangement and sequence established *S. eubayanus* as the main mitochondrial donor of lager yeast of the Frohberg lineage, harboring CDSs with ~98.6% identity.

Almeida et al. (2014) detected that European *S. uvarum* domesticated strains (wine and cider) consistently contained several *S. eubayanus* introgressions. These introgressions were absent in the large majority of wild strains, and gene ontology analyses indicated that several genes included in the introgressed regions were relevant for wine fermentation. Many of the introgressions were subtelomeric and contained genes such as ASP1, a gene encoding the cytosolic L-asparaginase used to degrade asparagine to be used as nitrogen source, and FZF1, a transcription factor that regulates several genes, including Ssu1, that encode an efflux pump involved in sulfite resistance (Almeida et al. 2014). These results represented the first clear indication of domestication in the yeast *S. uvarum*, used for wine and cider production worldwide. Genome mining of *P. rhodozyma* allowed detection and annotation of all genes of the astaxanthin synthesis as well as previously unknown putative regulatory enzymes of the metabolic pathway (Bellora et al., in press). Additionally, genes homologous to those reported to be implicated in the synthesis of mycosporines in Cyanobacteria (Balskus and Walsh 2010) were detected in a similar cluster disposition in the Patagonian strain of *P. rhodozyma*. One of the Holarctic strains (the type strain of the species) possessed the same cluster arrangement but a second European strain (Sharma et al. 2015) lacked the complete set of genes. In agreement with this finding, it was later demonstrated that the former *P. rhodozyma* strain (CBS 6938) lacked the ability to synthesize MGG (Bellora et al., in press).

Acknowledgments All the authors are staff members of the Argentinean National Research Council (CONICET). The work at the Applied Microbiology, Biotechnology and Bioinformatics of yeasts (MABBlev) is supported by grants from CONICET (PIP 11220130100392CO), the National Agency for Science and Technology Promotion of Argentina (PICT 2542, 2483, 3081 and 3677), and The Universidad Nacional del Comahue (project B171). We thank the authorities of National Parks for sampling authorization. We also thank our collaborators: Jose Paula Sampaio (Portugal), Chris Hittinger (USA), and Carlos Rosa (Brazil).

References

Almeida P, Gonçalves C, Teixeira S, Libkind D, Bontrager M, Masneuf-Pomarède I, Albertin W, Durrens P, Sherman DJ, Marullo P, Hittinger CT, Gonçalves P, Sampaio P (2014) A Gondwanan imprint on global diversity and domestication of wine and cider yeast *Saccharomyces uvarum*. Nat Commun 5:4044

An GH, Johnson EA (1990) Influence of light on growth and pigmentation of the yeast *Phaffia rhodozyma*. Antonie van Leeuwenhoek 57:191–203

Andrewes AG, Phaff HJ, Starr MP (1976) Carotenoids of *Phaffia rhodozyma*, a red-pigmented fermenting yeast. Phytochemistry 15:1003–1007

Ausich RL (1997) Commercial opportunities for carotenoid production by biotechnology. Pure Appl Chem 69:2169–2174

Baker EC, Wang B, Bellora N, Peris D, Hulfachor AB, Koshalek JA, Adams M, Libkind D, Hittinger CT (2015) The genome sequence of *Saccharomyces eubayanus* and the domestication of lager-brewing yeasts. Mol Biol Evol 32:2818–2831

Balmer ME, Buser HR, Müller MD, Poiger T (2005) Occurrence of some organic UV filters in wastewater, in surface waters, and in fish from Swiss lakes. Environ Sci Technol 39:953–962

Bellora N, Moliné M, David-Palma M, Coelho MA, Hittinger CT, Sampaio JP, Libkind D (in press). Comparative genomics provides new insights into the diversity, physiology, and sexuality of the only industrially exploited tremellomycete: *Phaffia* rhodozyma. BMC Genomics

Bergauer P, Fonteyne PA, Nolard N, Schinner F, Margesin R (2005) Biodegradation of phenol and phenol-related compounds by psychrophilic and cold-tolerant alpine yeasts. Chemosphere 59:909–918

Bernillon J, Bouillant M-L, Pittet J-L, Favre-Bonvin J, Arpin N (1984) Mycosporine glutamine and related mycosporines in the fungus *Pyronema omphalodes*. Phytochemistry 23:1083–1087

Bhosale P (2004) Environmental and cultural stimulants in the production of carotenoids from microorganisms. Appl Microbiol Biotechnol 63:351–361

Bhosale P, Bernstein PS (2005) Microbial xanthophylls. Appl Micriobiol Biotechnol 68:445–455
Bing J, Han PJ, Liu WQ, Wang QM, Bai FY (2014) Evidence for a Far East Asian origin of lager beer yeast. Curr Biol 24:380–381
Balskus EP, Walsh CT (2010) The genetic and molecular basis for sunscreen biosynthesis in Cyanobacteria. Science 329:1653–1656. doi:10.1126/science.1193637
Branda E, Turchetti B, Diolaiuti G, Pecci M, Smiraglia C, Buzzini P (2010) Yeast and yeast-like diversity in the southernmost glacier of Europe (Calderone Glacier, Apennines, Italy). FEMS Microbiol Ecol 72:354–369
Brandão LR, Libkind D, Vaz ABM, Espírito Santo L, Moliné M, de Garcia V, van Broock M, Rosa CA (2011) Yeasts from an oligotrophic lake in Patagonia (Argentina): diversity, distribution and synthesis of photo-protective compounds and extracellular enzymes. FEMS Microbiol Ecol 76:1–13
Brezová V, Gabcová S, Dvoranová D, Stasko A (2005) Reactive oxygen species produced upon photoexcitation of sunscreens containing titanium dioxide (an EPR study). J Photochem Photobiol B Biol 79:121–134
Britton G (1995) Structure and properties of carotenoids in relation to function. FASEB J 9:1551–1558
Britton G (ed) (2004) Carotenoids handbook. Birkhäuser, Basel
Britton G (2008) Functions of intact carotenoids. In: Britton G, Liaaen-Jensen S, Pfander H (eds) Carotenoids, vol 4. Birkhäuser, Basel, pp 189–212
Brizzio S, van Broock M (1998) Characteristics of wild yeast killer from Nahuel Huapi Park (Patagonia, Argentina). Food Technol Biotechnol 36:273–278
Brizzio S, Turchetti B, de Garcia V, Libkind D, Buzzini P, van Broock M (2007) Extracellular enzymatic activities of basidiomycetous yeasts isolated from glacial and subglacial waters of northwest Patagonia (Argentina). Can J Microbiol 53:519–525
Butinar L, Santos S, Spencer-Martins I, Oren A, Gunde-Cimerman N (2005) Yeast diversity in hypersaline habitats. FEMS Microbiol Lett 244:229–234
Buzzini P, Innocenti M, Turchetti B, Libkind D, van Broock M, Mulinacci N (2007) Carotenoid profiles of yeasts belonging to the genera *Rhodotorula*, *Rhodosporidium*, *Sporobolomyces*, and *Sporidiobolus*. Can J Microbiol 53:1024–1031
Buzzini P, Branda E, Goretti M, Turchetti B (2012) Psychrophilic yeasts from worldwide glacial habitats: diversity, adaptation strategies and biotechnological potential. FEMS Microbiol Ecol 82:217–241
Cavicchioli R, Charlton T, Ertan H, Mohd Omar S, Siddiqui KS, Williams TJ (2011) Biotechnological uses of enzymes from psychrophiles. Microb Biotechnol 4:449–460
Colabella F, Libkind D (2016) Un método basado en la PCR para la identificación rápida de levaduras acumuladoras de astaxantina (*Phaffia* spp.). Rev Arg Microbiol 48:15–20
Colabella F, Moliné M, Libkind D (2014) UV sunscreens of microbial origin: mycosporines and mycosporine-like amino acids. Recent Pat Biotechnol 8:189–193
Crevel RW, Fedyk JK, Spurgeon MJ (2002) Antifreeze proteins: characteristics, occurrence and human exposure. Food Chem Toxicol 40:899–903
David-Palma M, Libkind D, Sampaio JP (2014) Global distribution, diversity hotspots and niche transitions of an astaxanthin-producing eukaryotic microbe. Mol Ecol 23:921–932
Davoli P, Mierau V, Weber RWS (2004) Carotenoids and fatty acids in red yeasts *Sporobolomyces roseus* and *Rhodotorula glutinis*. App Biochem Microbiol 40:392–397
de Garcia V, Brizzio S, Libkind D, Buzzini P, van Broock M (2007) Biodiversity of cold-adapted yeasts from glacial meltwater rivers in Patagonia, Argentina. FEMS Microbiol Ecol 59:331–341
de Garcia V, Brizzio S, Libkind D, Rosa CA, van Broock M (2010a) *Wickerhamomyces patagonicus* sp. nov., an ascomycetous yeast species from Patagonia, Argentina. Int J Syst Evol Microbiol 60:1693–1696
de Garcia V, Brizzio S, Russo G, Rosa CA, Boekhout T, Theelen B, Libkind D, van Broock M (2010b) *Cryptococcus spencermartinsiae* sp. nov., a basidiomycetous yeast isolated from glacial waters and apple fruits. Int J Syst Evol Microbiol 60:707–711

de Garcia V, Brizzio S, van Broock MR (2012) Yeasts from glacial ice of Patagonian Andes, Argentina. FEMS Microbiol Ecol 82:540–550
de Garcia V, Libkind D, Moline M, Rosa CA, Giraudo MR (2014) Cold-adapted yeasts in Patagonian habitats. In: Buzzini P, Margesin R (eds) Cold-adapted yeasts. Springer, Berlin, pp 123–148
De Garcia V, Coelho MA, Maia TM, Rosa LH, Vaz AM, Rosa CA, Sampaio JP, Goncalves P, van Broock M, Libkind D (2015) Sex in the cold: taxonomic reorganization of psychrotolerant yeasts in the order leucosporidiales. FEMS Yeast Res 15(4):fov019
Díaz M, Pedrozo F, Baccala N (2000) Summer classification of Southern Hemisphere temperate lakes (Patagonia, Argentina). Lakes Reserv 5:213–229
Donaldson MS (2004) Nutrition and cancer: a review of the evidence for an anti-cancer diet. Nutr J 3:19
Fell JW, Scorzetti G, Statzell-Tallman A, Boundy-Mills K (2007) Molecular diversity and intragenomic variability in the yeast genus *Xanthophyllomyces*: the origin of *Phaffia rhodozyma*? FEMS Yeast Res 7:1399–1408
Fernández NV, Mestre MC, Marchelli P, Fontenla SB (2012) Yeast and yeast-like fungi associated with dry indehiscent fruits of *Nothofagus nervosa* in Patagonia, Argentina. FEMS Microbiol Ecol 80:179–192
Flores-Cotera L, Martín R, Sánchez S (2001) Citrate, a possible precursor of astaxanthin in *Phaffia rhodozyma*: influence of varying levels of ammonium, phosphate and citrate in a chemically defined medium. Appl Microbiol Biotechnol 55:341–347
Frisvad JC (2008) Fungi in cold ecosystems. In: Margesin R, Schinner F, Marx JC, Gerday C (eds) Psychrophiles: from biodiversity to biotechnology. Springer, Berlin, pp 137–156
Gadanho M, Sampaio JP (2005) Occurrence and diversity of yeasts in the mid-Atlantic ridge hydrothermal fields near the Azores Archipelago. Microb Ecol 50:408–417
Gadanho M, Sampaio JP (2009) *Cryptococcus ibericus* sp. nov., *Cryptococcus aciditolerans* sp. nov. and *Cryptococcus metallitolerans* sp. nov., a new ecoclade of anamorphic basidiomycetous yeast species from an extreme environment associated with acid rock drainage in Sao Domingos pyrite mine, Portugal. Int J Syst Evol Microbiol 59:2375–2379
Gadanho M, Libkind D, Sampaio JP (2006) Yeast diversity in the extreme acidic environments of the Iberian Pyrite Belt. Microb Ecol 52:552–563
Gayevskiy V, Goddard MR (2016) *Saccharomyces eubayanus* and *Saccharomyces arboricola* reside in North Island native New Zealand forests. Environ Microbiol 18:1137–1147
Gerday C, Aittaleb M, Arpigny JL, Baise E, Chessa JP, Garsoux G, Petrescu I, Feller G (1997) Psychrophilic enzymes: a thermodynamic challenge. Biochim Biophys Acta 1342:119–131
Golubev VI, Babéva IP, Blagodatskaya VM, Reshetova IS (1977) Taxonomic study of yeasts isolated from spring sap-flows. Microbiology 46:461–466
Gunde-Cimerman N, Plemenitaš A (2006) Ecology and molecular adaptations of the halophilic black yeast *Hortaea werneckii*. Rev Environ Sci Biotechnol 5:323–331
Hagler AN, Ahearn DG (1987) Ecology of aquatic yeasts. In: Rose AH, Harrison JS (eds) The yeasts, vol 2, Yeasts and the environment. Academic Press, London, pp 181–205
Hagler AN, Mendonca-Hagler LC (1981) Yeasts from marine and estuarine waters with different levels of pollution in the state of Rio de Janeiro, Brazil. Appl Environ Microbiol 41: 173–178
Hanson KM, Gratton E, Bardeen CJ (2006) Sunscreen enhancement of UV-induced reactive oxygen species in the skin. Free Radic Biol Med 41:1205–1212
Herz S, Weber RWS, Anke H, Mucci A, Davoli P (2007) Intermediates in the oxidative pathway from torulene to torularhodin in the red yeasts *Cystofilobasidium infirmominiatum* and *C. capitatum* (Heterobasidiomycetes, Fungi). Phytochemistry 68:2503–2511
Hornsey IS (2003) A history of beer and brewing. Royal Society of Chemistry, London
Johnson EA, Schroeder WA (1995) Astaxanthin from the yeast *Phaffia rhodozyma*. Stud Mycol 38:81–90
Johnson EA, Schroeder WA (1996) Microbial carotenoids. In: Downstream processing biosurfactants carotenoids. Advances in Biochemical Engineering/Biotechnology, vol 53. Springer, Berlin, pp 119–178

Kogej T, Gostinčar C, Volkmann M, Gorbushina AA, Gunde-Cimerman N (2006) Mycosporines in extremophilic fungi: novel complementary osmolytes? Environ Chem 3:105–110
Krause M, Klit A, Blomberg Jensen M, Søeborg T, Frederiksen H, Schlumpf M, Lichtensteiger W, Skakkebaek NE, Drzewiecki KT (2012) Sunscreens: are they beneficial for health? An overview of endocrine disrupting properties of UV filters. Int J Androl 35:424–436
Krinsky NI (1979) Carotenoid protection against oxidation. Pure Appl Chem 51:649–660
Kurtzman C, Fell JW, Boekhout T (eds) (2011) The yeasts: a taxonomic study. Elsevier, Amsterdam
Lampila LE, Wallen SE, Bullerman LB (1985) A review of factors affecting biosynthesis of carotenoids by the order Mucorales. Mycopathology 90:65–80
Lee DH, Kim CS, Lee YJ (2011) Astaxanthin protects against MPTP/MPP^+-induced mitochondrial dysfunction and ROS production *in vivo* and *in vitro*. Food Chem Toxicol 49:271–280
Leite B, Nicholson RL (1992) Mycosporine-alanine: a self-inhibitor of germination from the conidial mucilage of *Colletotrichum graminicola*. Exp Mycol 16:76–86
Libkind D, van Broock MR (2006) Biomass and carotenoid pigments production by Patagonian native yeasts. World J Microbiol Biotechnol 22:687–692
Libkind D, Brizzio S, Ruffini A, Gadanho M, van Broock MR, Sampaio JP (2003) Molecular characterization of carotenogenic yeasts from aquatic environments in Patagonia, Argentina. Antonie Van Leeuwenhoek 84:313–322
Libkind D, Brizzio S, van Broock M (2004a) *Rhodotorula mucilaginosa*, a carotenoid-producing yeast strain from a Patagonian high-altitude lake. Folia Microbiol 49:19–25
Libkind D, Perez P, Sommaruga R, Dieguez Mdel C, Ferraro M, Brizzio S, Zagarese H, van Broock M (2004b) Constitutive and UV-inducible synthesis of photoprotective compounds (carotenoids and mycosporines) by freshwater yeasts. Photochem Photobiol Sci 3:281–286
Libkind D, Sommaruga R, Zagarese H, van Broock M (2005a) Mycosporines in carotenogenic yeasts. Syst Appl Microbiol 28:749–754
Libkind D, Gadanho M, van Broock M, Sampaio JP (2005b) *Sporidiobolus longiusculus* sp. nov. and *Sporobolomyces patagonicus* sp. nov., novel yeasts of the Sporidiobolales isolated from aquatic environments in Patagonia, Argentina. Int J Syst Evol Microbiol 55:503–509
Libkind D, Dieguez MC, Moliné M, Perez P, Zagarese HE, van Broock M (2006) Occurrence of photoprotective compounds in yeasts from freshwater ecosystems of northwestern Patagonia (Argentina). Photochem Photobiol 82:972–980
Libkind D, Ruffini A, van Broock M, Alves L, Sampaio JP (2007) Biogeography, host-specificity, and molecular phylogeny of *Phaffia rhodozyma* and its sexual form, *Xanthophyllomyces dendrorhous*. Appl Environ Microbiol 73:1120–1125
Libkind D, Gadanho M, van Broock M, Sampaio JP (2008a) Studies on the heterogeneity of the carotenogenic yeast *Rhodotorula mucilaginosa* from Patagonia, Argentina. J Basic Microbiol 48:93–98
Libkind D, Moliné M, de Garcia V, Fontenla S, van Broock M (2008b) Characterization of a novel South American population of the astaxanthin producing yeast *Xanthophyllomyces dendrorhous* (*Phaffia rhodozyma*). J Ind Microbiol Biotechnol 35:151–158
Libkind D, Moliné M, Sampaio JP, van Broock M (2009a) Yeasts from high-altitude lakes: influence of UV radiation. FEMS Microbiol Ecol 69:353–362
Libkind D, Gadanho M, van Broock M, Sampaio JP (2009b) *Cystofilobasidium lacus-mascardii* sp. nov., a basidiomycetous yeast species isolated from aquatic environments of the Patagonian Andes, and *Cystofilobasidium macerans* sp. nov., the sexual stage of *Cryptococcus macerans*. Int J Syst Evol Microbiol 59:622–630
Libkind D, Sampaio JP, van Broock M (2010) Cystobasidiomycetes yeasts from Patagonia (Argentina): description of *Rhodotorula meli* sp. nov. from glacial meltwater. Int J Syst Evol Microbiol 60:2251–2256
Libkind D, Hittinger CT, Valerio E, Goncalves C, Dover J, Johnston M, Goncalves P, Sampaio JP (2011a) Microbe domestication and the identification of the wild genetic stock of lager-brewing yeast. Proc Natl Acad Sci USA 108:14539–14544
Libkind D, Moliné M, Sommaruga R, Sampaio JS, van Broock M (2011b) Phylogenetic distribution of fungal mycosporines within Pucciniomycotina (Basidiomycota). Yeast 28:619–627

Libkind D, Moliné M, van Broock M (2011c) Production of the UVB absorbing compound mycosporine-glutaminol-glucoside by *Xanthophyllomyces dendrorhous* (*Phaffia rhodozyma*). FEMS Yeast Res 11(1):52–59

Libkind D, Tognetti C, Ruffini A, Sampaio JP, van Broock M (2011d) *Xanthophyllomyces dendrorhous* (*Phaffia rhodozyma*) on stromata of *Cyttaria hariotii* in Patagonian *Nothofagus* forests. Rev Argent Microbiol 43:226–232

Libkind D, Russo G, van Broock MR (2014) Yeasts from extreme aquatic environments: hyperacidic freshwaters. In: Jones EBG, Hyde KD, Pang K-L (eds) Freshwater fungi and fungal-like organisms. De Gruyter, Göttingen, pp 443–464

Liti G, Barton DBH, Louis EJ (2006) Sequence diversity, reproductive isolation and species concepts in *Saccharomyces*. Genetics 174:839–850

Loden M, Beitner H, Gonzalez H, Edström DW, Åkerström U, Austad J, Buraczewska-Norin I, Matsson M, Wulf HC (2011) Sunscreen use: controversies, challenges and regulatory aspects. Br J Dermatol 165:255–262

Maxwell WA, Macmillan JD, Chichester C (1966) Function of carotenoids in protection of *Rhodotorula glutinis* against irradiation from a gas laser. Photochem Photobiol 5: 567–577

Mestre MC, Ulloa JR, Rosa CA, Lachance MA, Fontenla S (2010) *Lachancea nothofagi* sp. nov., a yeast associated with *Nothofagus* species in Patagonia, Argentina. Int J Syst Evol Microbiol 60:2247–2250

Mestre MC, Rosa CA, Safar SV, Libkind D, Fontenla SB (2011) Yeast communities associated with the bulk-soil, rhizosphere and ectomycorrhizosphere of a *Nothofagus pumilio* forest in northwestern Patagonia, Argentina. FEMS Microbiol Ecol 78:531–541

Mestre MC, Fontenla S, Rosa CA (2014) Ecology of cultivable yeasts in pristine forests in northern Patagonia (Argentina) influenced by different environmental factors. Can J Microbiol 60:371–382

Mestre MC, Fontenla S, Bruzone MC, Fernandez NV, Dames J (2016) Detection of plant growth enhancing features in psychrotolerant yeasts from Patagonia (Argentina). J Basic Microbiol. doi: 10.1002/jobm.201500728.

Moliné M (2010) Producción de compuestos fotoprotectores (carotenoides y micosporinas) por levaduras. PhD thesis, Universidad Nacional de Tucumán, Tucumán, Argentina.

Moliné M, Libkind D, Dieguez M del C, van Broock M (2009) Photoprotective role of carotenoids in yeasts: response to UV-B of pigmented and naturally-occurring albino strains. J Photochem Photobiol B 95:156–161

Moliné M, Flores MR, Libkind D, Dieguez M del C, Farias ME, van Broock M (2011a) Photoprotection by carotenoid pigments in the yeast *Rhodotorula mucilaginosa*: the role of torularhodin. Photochem Photobiol Sci 9:1145–1151

Moliné M, Arbeloa EM, Flores MR, Libkind D, Farías ME, Bertolotti SG, Churio MS, van Broock MR (2011b) UVB photoprotective role of mycosporines in yeast: photostability and antioxidant activity of mycosporine-glutaminol-glucoside. Radiat Res 175:44–50

Moliné M, Libkind D, van Broock M (2012) Production of torularhodin, torulene, and β-carotene by *Rhodotorula* yeasts. In: Barredo J-L (ed) Microbial carotenoids from fungi, vol 898, Methods Mol Biol. Humana Press, New York, pp 275–283

Muñoz M, Moliné M, Libkind D (2013) Comparación de técnicas para el aislamiento y recuento de levaduras y hongos dimórficos del filoplano de *Nothofagus pumilio*. Bol Soc Argent Bot 48:183–191

Nagahama T, Hamamoto M, Nakase T, Horikoshi K (2001) *Rhodotorula lamellibrachii* sp. nov., a new yeast species from a tubeworm collected at the deep-sea floor in Sagami Bay and its phylogenetic analysis. Antonie van Leeuwenhoek 80:317–323

Naguib YM (2000) Antioxidant activities of astaxanthin and related carotenoids. J Agric Food Chem 48:1150–1154

Olson JH (1989) Provitamin A function of carotenoids: the conversion of β-carotene into vitamin A. J Nutr 119:105–108

Oren A, Gunde-Cimerman N (2007) Mycosporines and mycosporine-like amino acids: UV protectants or multipurpose secondary metabolites? FEMS Microbiol Lett 269:1–10

Peris D, Sylvester K, Libkind D, Gonçalves P, Sampaio JP, Alexander WG, Hittinger CT (2014) Population structure and reticulate evolution of *Saccharomyces eubayanus* and its lager-brewing hybrids. Mol Ecol 23:2031–2045

Peris D, Langdon Q, Moriarty RV, Sylvester K, Bontrager M, Charron G, Leducq J-B, Landry CR, Libkind D, Hittinger CT (2016) Complex origins of lager-brewing hybrids were shaped by standing variation in the wild yeast *Saccharomyces eubayanus*. PLoS Genet 12(7):e1006155

Phaff HJ, Miller MW, Yoneyama M, Soneda MA (1972) Comparative study of the yeast florae associated with trees on the Japanese Islands and on the West Coasts of North America. In: Terui G (ed) Proceedings of the 4th IFS: fermentation technology today meeting. Society of Fermentation Technology, Osaka, pp 759–774

Portwich A, Garcia-Pichel F (1999) Ultraviolet and osmotic stresses induce and regulate the synthesis of mycosporines in the cyanobacterium *Chlorogloeopsis* PCC 6912. Arch Microbiol 172:187–192

Quirós R, Drago E (1985) Relaciones entre variables físicas, morfométricas y climáticas en lagos patagónicos. Rev Asoc Cienc Nat Litoral 16:181–199

Rabassa J, Rubilis S, Suárez J (1978) Los Glaciares del Monte Tronador Parque Nacional Nahuel Huapi (Río Negro, Argentina). An Parques Nacionales 14:261–316

Rao AV, Agarwal S (2000) Role of antioxidant lycopene in cancer and heart disease. J Am Coll Nutr 19:563–569

Rao AV, Rao LG (2007) Carotenoids and human health. Pharmacol Res 55:207–216

Rodríguez ME, Pérez-Través L, Sangorrín MP, Barrio E, Lopes CA (2014) *Saccharomyces eubayanus* and *Saccharomyces uvarum* associated with the fermentation of *Araucaria araucana* seeds in Patagonia. FEMS Yeast Res 14:948–965

Rodríguez-Sáiz M, de la Fuente JL, Barredo JL (2010) *Xanthophyllomyces dendrorhous* for the industrial production of astaxanthin. Appl Microbiol Biotechnol 88:645–658

Roy S (2000) Strategies for the minimisation of UV-induced damage. In: de Mora S, Demers S, Vernet M (eds) The effects of UV radiation in the marine environment. Cambridge University Press, Cambridge, pp 177–205

Russo G, Libkind D, Sampaio JP, van Broock MR (2008) Yeast diversity in the acidic Rio Agrio–Lake Caviahue volcanic environment (Patagonia, Argentina). FEMS Microbiol Ecol 65:415–424

Russo G, Libkind D, Ulloa RJ, de Garcia V, Sampaio JP, van Broock MR (2010) *Cryptococcus agrionensis* sp. nov., a basidiomycetous yeast of the acidic rock drainage ecoclade, isolated from an acidic aquatic environment of volcanic origin. Int J Syst Evol Microbiol 60:996–1000

Russo G, Libkind D, Giraudo MR, Delgado O (2016) Metal capture by autochthonous yeasts from a volcanic influenced environment of Patagonia. J Basic Microbiol 56:1–9

Sakaki H, Nakanishi T, Tada A, Miki W, Komemushi S (2001) Activation of torularhodin production by *Rhodotorula glutinis* using weak white light irradiation. J Biosci Bioeng 92:294–297

Sampaio JP (2004) Diversity, phylogeny and classification of basidiomycetous yeasts. In: Agerer R, Piepenbring M, Blanz P (eds) Frontiers in basidiomycote mycology. IHW Verlag, Eching, Germany, pp 49–80

Sampaio JP (2011a) *Leucosporidiella* Sampaio (2003). In: Kurtzman CP, Fell JW, Boekhout T (eds) The yeasts: a taxonomic study, 5th edn. Elsevier, Amsterdam, pp 1802–1806

Sampaio JP (2011b) Leucosporidium Fell, Statzell, I.L. Hunter, Phaff (1969). In: Kurtzman CP, Fell JW, Boekhout T (eds) The Yeasts, a Taxonomic Study. 5th edn. Elsevier, Amsterdam, pp 1485–1494

Sampaio JP, Gonçalves P (2008) Natural populations of *Saccharomyces kudriavzevii* in Portugal are associated with oak bark and are sympatric with *S. cerevisiae* and *S. paradoxus*. Appl Environ Microbiol 74:2144–2152

Scherer S, Chen TW, Böger P (1988) A new UV-A/B protecting pigment in the terrestrial cyanobacterium *Nostoc commune*. Plant Physiol 88:1055–1057

Scheuer E, Warshaw E (2006) Sunscreen allergy: a review of epidemiology, clinical characteristics, and responsible allergens. Dermatitis 17:3–11

Schroeder WA, Johnson EA (1995) Carotenoids protect *Phaffia rhodozyma* against singlet oxygen damage. J Ind Microbiol 14:502–507
Sharma R, Gassel S, Steiger S, Xia X, Bauer R, Sandmann G, Thines M (2015) The genome of the basal agaricomycete *Xanthophyllomyces dendrorhous* provides insights into the organization of its acetyl-CoA derived pathways and the evolution of Agaricomycotina. BMC Genomics 16:233
Shick JM, Dunlap WC (2002) Mycosporine-like amino acids and related gadusols: biosynthesis, accumulation, and UV-protective functions in aquatic organisms. Annu Rev Physiol 64:223–262
Sies H, Stahl W (1995) Vitamins E and C, beta-carotene, and other carotenoids as antioxidants. Am J Clin Nutr 62:1315–1321
Simpson KL (1983) Relative value of carotenoids as precursors of vitamin A. Proc Nutr Soc 42:7–17
Sommaruga R, Libkind D, van Broock M, Whitehead K (2004) Mycosporine-glutaminolglucoside, a UV-absorbing compound of two *Rhodotorula* yeast species. Yeast 21:1077–1081
Sperstad S, Lutnæs BF, Stormo SK, Liaaen-Jensen S, Landfald B (2006) Torularhodin and torulene are the major contributors to the carotenoid pool of marine *Rhodosporidium babjevae* (Golubev). J Ind Microbiol Biotechnol 33:269–273
Stuefer M, Rott H, Skvarca P (2007) Glaciar Perito Moreno, Patagonia: climate sensitivities and characteristics preceding 2003/04 and 2005/06 damming events. J Glaciol 53:13
Tada M, Shiroishi M (1982) Mechanism of photoregulated carotenogenesis in *Rhodotorula minuta*. I. Photocontrol of carotenoid production. Plant Cell Physiol 23:541–547
Tognetti C, Moliné M, van Broock M, Libkind D (2013) Favored isolation and rapid identification of the astaxanthin-producing yeast *Xanthophyllomyces dendrorhous* (*Phaffia rhodozyma*) from environmental samples. J Basic Microbiol 53:766–772
Torres A, Hochberg M, Pergament I, Smoum R, Niddam V, Dembitsky VM, Temina M, Dor I, Lev O, Srebnik M (2004) A new UV-B absorbing mycosporine with photo protective activity from the lichenized ascomycete *Collema cristatum*. Eur J Biochem 271:780–784
Tsimako M, Guffogg S, Thomas-Hall S, Watson K (2002) Resistance to UVB radiation in Antarctic yeasts. Redox Rep 7:312–314
Ulloa J, Libkind D, Fontenla S, van Broock M (2009) Levaduras fermentadoras aisladas de *Cyttaria hariotii* (Fungi) en bosques Andino-Patagónicos (Argentina). Bol Soc Argent Bot 44:239–248
Ungureanu C, Ferdes M (2012) Evaluation of antioxidant and antimicrobial activities of torularhodin. Adv Sci Lett 18:50–53
van Broock M, Libkind D, Moliné M (2009) Composiciones que absorben radiaciones UVB y antioxidantes. Procedimientos y Usos. Argentinean Patent P090103845
Villafañe VE, Helbling EW, Zagarese HE (2001) Solar ultraviolet radiation and its impact on aquatic systems of Patagonia, South America. Ambio 30:112–117
Villarosa G, Outes V, Masiokas M, Villalba R, Rivas S (2008) El Monte Tronador: historias de hielo y fuego. In: Rastelli D (ed) Sitios de Interés Geológico de la República Argentina. SEGEMAR (Servicio de Geología y Minería de la República Argentina), Buenos Aires, pp 627–641
Vishniac H (ed) (2005) Yeast biodiversity in the Antarctic. Yeast handbook. Biodiversity and ecophysiology of yeasts. Springer, Berlin, pp 419–440
Volkmann M, Whitehead K, Rütters H, Rullkötter J, Gorbushina AA (2003) Mycosporineglutamicolglucoside: a natural UV-absorbing secondary metabolite of rock-inhabiting microcolonial fungi. Rapid Commun Mass Spectrom 17:897–902
Walther A, Hesselbart A, Wendland J (2014) Genome sequence of *Saccharomyces carlsbergensis*, the world's first pure culture lager yeast. G3 (Bethesda) 4:783–793
Wang SQ, Balagula Y, Osterwalder U (2010) Photoprotection: a review of the current and future technologies. Dermatol Ther 23:31–47
Weber RWS, Davoli P, Anke H (2006) A microbial consortium involving the astaxanthin producer *Xanthophyllomyces dendrorhous* on freshly cut birch stumps in Germany. myCOL 20:57–61
Weber RWS, Becerra J, Silva MJ, Davoli P (2008) An unusual *Xanthophyllomyces* strain from leaves of *Eucalyptus globulus* in Chile. Mycol Res 112:861–867
Wendland J (2014) Lager yeast comes of age. Eukaryot Cell 13:1256–1265

Yamane Y, Higashida K, Nakashimada Y, Kakizono T, Nishio N (1997) Astaxanthin production by *Phaffia rhodozyma* enhanced in fed-batch culture with glucose and ethanol feeding. Biotechnol Lett 19:1109–1111

Young H, Patterson VJ (1982) A UV protective compound from *Glomerella cingulata*, a mycosporine. Phytochemistry 21:1075–1077

Yurkov AM, Kachalkin AV, Daniel HM, Groenewald M, Libkind D, de Garcia V, Zalar P, Gouliamova DE, Boekhout T, Begerow D (2015) Two yeast species, *Cystobasidium psychroaquaticum* f.a. sp. nov. and *Cystobasidium rietchieii* f.a. sp. nov. isolated from natural environments, and the transfer of *Rhodotorula minuta* clade members to the genus *Cystobasidium*. Antonie van Leeuwenhoek 107:173–185

Index

N.L. Olivera et al. (eds.), *Biology and Biotechnology of Patagonian Microorganisms*, DOI 10.1007/978-3-319-42801-7

MIX
Papier aus verantwortungsvollen Quellen
Paper from responsible sources
FSC® C105338

If you have any concerns about our products,
you can contact us on
ProductSafety@springernature.com

In case Publisher is established outside the EU,
the EU authorized representative is:
Springer Nature Customer Service Center GmbH
Europaplatz 3, 69115 Heidelberg, Germany

Printed by Libri Plureos GmbH
in Hamburg, Germany